Elements of X-RAY DIFFRACTION

SECOND EDITION

Elements of X-RAY DIFFRACTION

SECOND EDITION

B. D. CULLITY
Department of Metallurgical Engineering and Materials Science
University of Notre Dame

ADDISON-WESLEY PUBLISHING COMPANY, INC.
Reading, Massachusetts · Menlo Park, California
London · Amsterdam · Don Mills, Ontario · Sydney

This book is in the
Addison-Wesley Series in Metallurgy and Materials

Morris Cohen
Consulting Editor

 Printed in the United States of America. Published simultaneously in Canada. Library of Congress Catalog Card No. 77-73950.

ISBN 0-201-01174-3

34 35 36 37 MA 04 03 02 01 00

Preface

X-ray diffraction is a tool for the investigation of the fine structure of matter. This technique had its beginnings in von Laue's discovery in 1912 that crystals diffract x-rays, the manner of the diffraction revealing the structure of the crystal. At first, x-ray diffraction was used only for the determination of crystal structure. Later on, however, other uses were developed, and today the method is applied not only to structure determination, but to such diverse problems as chemical analysis and stress measurement, to the study of phase equilibria and the measurement of particle size, to the determination of the orientation of one crystal or the ensemble of orientations in a polycrystalline aggregate.

The purpose of this book is to acquaint the reader who has no previous knowledge of the subject with the theory of x-ray diffraction, the experimental methods involved, and the main applications. Because the author is a metallurgist, the majority of these applications are described in terms of metals and alloys. However, little or no modification of experimental method is required for the examination of nonmetallic materials, inasmuch as the physical principles involved do not depend on the material investigated. This book should therefore be useful to metallurgists, chemists, physicists, ceramists, mineralogists, etc., namely, to all who use x-ray diffraction purely as a laboratory tool for the sort of problems already mentioned.

Members of this group, unlike x-ray crystallographers, are not normally concerned with the determination of complex crystal structures. For this reason the rotating-crystal method and space-group theory, the two chief tools in the solution of such structures, are described only briefly.

This is a book of principles and methods intended for the student, and not a reference book for the advanced research worker. Thus no metallurgical data are given beyond those necessary to illustrate the diffraction methods involved. For example, the theory and practice of determining preferred orientation are treated in detail, but the reasons for preferred orientation, the conditions affecting its development, and actual orientations found in specific metals and alloys are not described, because these topics are adequately covered in existing books. In short, x-ray diffraction is stressed rather than metallurgy.

The book is divided into three main parts: fundamentals, experimental methods, and applications. The subject of crystal structure is approached through, and based on, the concept of the point lattice (Bravais lattice), because the point lattice of a substance is so closely related to its diffraction pattern. X-ray diffraction phenomena are rather sharply divisible into those effects that are understandable in terms of the Bragg law and those that require a more advanced treatment, based on the reciprocal lattice. This book is written entirely in terms of the Bragg law and can be read without any knowledge of the reciprocal lattice. My experience with teaching x-ray diffraction to senior students in metallurgy, for many of whom this book represents a terminal course in the subject, is that there is insufficient time to attain both a real facility for "reciprocal thinking" and a good knowledge of the many applications of diffraction. I therefore prefer the Bragg-law approach for a first course. Those instructors who wish to introduce the reciprocal lattice at the beginning can interpose Appendix 1, which contains the rudiments of the subject, between Chapters 2 and 3.

Chapters on chemical analysis by x-ray diffraction and x-ray spectroscopy are included because of the industrial importance of these analytical methods. Electron and neutron diffraction are treated in appendices.

This second edition includes an account of new developments made possible by the semiconductor detector and pulse-height analysis, namely, energy-dispersive spectrometry and diffractometry. Applications of position-sensitive detectors are also described.

A new section is devoted to x-ray topography and other x-ray methods of assessing the quality of single crystals. Other additions include a quantitative treatment of the temperature factor and descriptions of the Auger effect, microcameras and Guinier cameras, and microanalysis in the electron microscope. References to original papers are now given, and the tables of wavelengths and absorption coefficients have been expanded.

This edition contains more material on the measurement of preferred orientation and residual stress than the first edition, but the former chapter on chemical analysis by x-ray absorption has been dropped, as being of minor interest to most readers.

The first edition carried the following acknowledgements:

Like any author of a technical book, I am greatly indebted to previous writers on this and allied subjects. I must also acknowledge my gratitude to two of my former teachers at the Massachusetts Institute of Technology, Professor B. E. Warren and Professor John T. Norton: they will find many an echo of their own lectures in these pages. Professor Warren has kindly allowed me to use many problems of his devising, and the advice and encouragement of Professor Norton has been invaluable. My colleague at Notre Dame, Professor G. C. Kuczynski, has read the entire book as it was written, and his constructive criticisms have been most helpful. I would also like to thank the following, each of whom has read one or more chapters and offered valuable suggestions: Paul A. Beck, Herbert Friedman, S. S. Hsu, Lawrence Lee, Walter C. Miller, William Parrish, Howard Pickett, and Bernard Waldman. I am also indebted to C. G. Dunn for the loan of illustrative material and to many graduate students, August Freda in particular, who have helped with the preparation of diffraction patterns. Finally, but not perfunctorily, I wish to thank Miss Rose Kunkle for her patience and diligence in preparing the typed manuscript.

In the preparation of the second edition I have been helped in many ways by Charles W. Allen, A. W. Danko, Ron Jenkins, Paul D. Johnson, A. R. Lang, John W. Mihelich, J. B. Newkirk, Paul S. Prevey, B. E. Warren, Carl Cm. Wu, and Leo Zwell. To all these, my best thanks.

Notre Dame, Indiana
November 1977

B. D. Cullity

Contents

Chapter 4 Diffraction II: Intensities of Diffracted Beams

EXPERIMENTAL METHODS

Chapter 5 Laue Photographs

Chapter 6 Powder Photographs

Chapter 7 Diffractometer and Spectrometer Measurements

Chapter 10 Determination of Crystal Structure

Chapter 11 Precise Parameter Measurements

Chapter 12 Phase-Diagram Determination

Chapter 13 Order-Disorder Transformations

Chapter 14 Chemical Analysis by X-ray Diffraction

Fundamentals

1

Properties of X-rays

1–1 INTRODUCTION

X-rays were discovered in 1895 by the German physicist Roentgen and were so named because their nature was unknown at the time. Unlike ordinary light, these rays were invisible, but they traveled in straight lines and affected photographic film in the same way as light. On the other hand, they were much more penetrating than light and could easily pass through the human body, wood, quite thick pieces of metal, and other "opaque" objects.

It is not always necessary to understand a thing in order to use it, and x-rays were almost immediately put to use by physicians and, somewhat later, by engineers, who wished to study the internal structure of opaque objects. By placing a source of x-rays on one side of the object and photographic film on the other, a shadow picture, or *radiograph*, could be made, the less dense portions of the object allowing a greater proportion of the x-radiation to pass through than the more dense. In this way the point of fracture in a broken bone or the position of a crack in a metal casting could be located.

Radiography was thus initiated without any precise understanding of the radiation used, because it was not until 1912 that the exact nature of x-rays was established. In that year the phenomenon of x-ray *diffraction* by crystals was discovered, and this discovery simultaneously proved the wave nature of x-rays and provided a new method for investigating the fine structure of matter. Although radiography is a very important tool in itself and has a wide field of applicability, it is ordinarily limited in the internal detail it can resolve, or disclose, to sizes of the order of 10^{-1} cm. Diffraction, on the other hand, can indirectly reveal details of internal structure of the order of 10^{-8} cm in size, and it is with this phenomenon, and its applications to metallurgical problems, that this book is concerned. The properties of x-rays and the internal structure of crystals are here described in the first two chapters as necessary preliminaries to the discussion of the diffraction of x-rays by crystals which follows.

1–2 ELECTROMAGNETIC RADIATION

We know today that x-rays are electromagnetic radiation of exactly the same nature as light but of very much shorter wavelength. The unit of measurement in the x-ray region is the angstrom (Å), equal to 10^{-8} cm, and x-rays used in diffraction have wavelengths lying approximately in the range 0.5–2.5 Å, whereas the wavelength of visible light is of the order of 6000 Å. X-rays therefore occupy the region

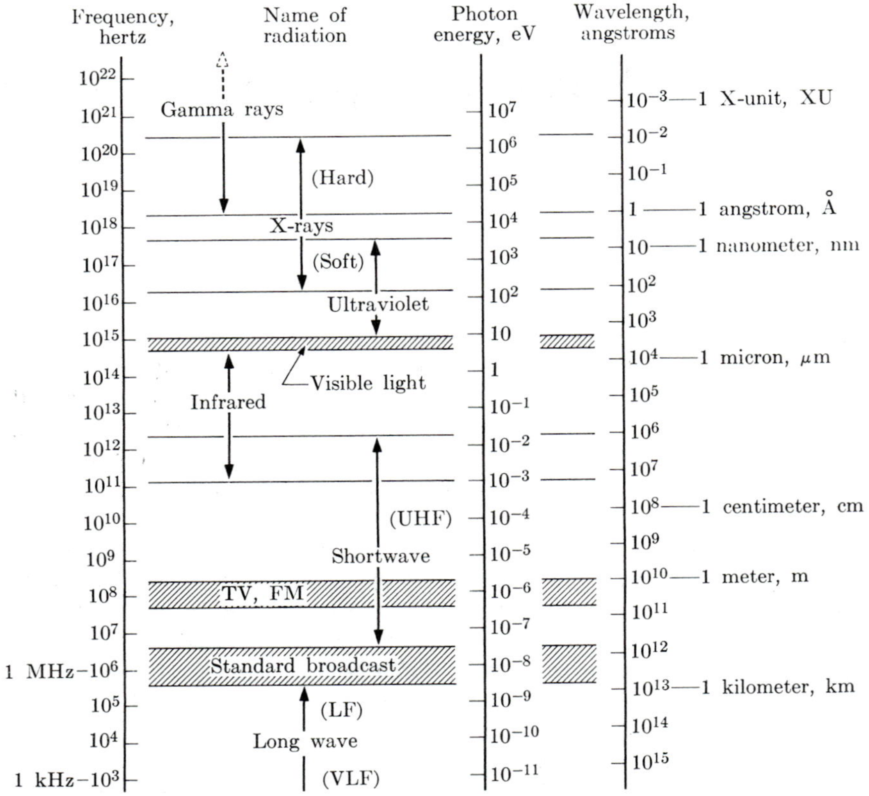

Fig. 1–1 The electromagnetic spectrum. The boundaries between regions are arbitrary, since no sharp upper or lower limits can be assigned. (H. A. Enge, M. R. Wehr, J. A. Richards, *Introduction to Atomic Physics*, Addison-Wesley Publishing Company, Inc., Reading, Mass., 1972).

between gamma and ultraviolet rays in the complete electromagnetic spectrum (Fig. 1–1). Other units sometimes used to measure x-ray wavelength are the X unit (XU) and the kilo X unit (kX = 1000 XU). The kX unit, whose origin will be described in Sec. 3–4, is only slightly larger than the angstrom. The approved SI unit for wavelengths in the x-ray region is the nanometer:

$$1 \text{ nanometer} = 10^{-9} \text{ m} = 10 \text{ Å}.$$

This unit has not become popular.

It is worth while to review briefly some properties of electromagnetic waves. Suppose a monochromatic beam of x-rays, i.e., x-rays of a single wavelength, is traveling in the x direction (Fig. 1–2). Then it has associated with it an electric field **E** in, say, the y direction and, at right angles to this, a magnetic field **H** in the z direction. If the electric field is confined to the xy-plane as the wave travels along, the wave is said to be plane-polarized. (In a completely unpolarized wave, the

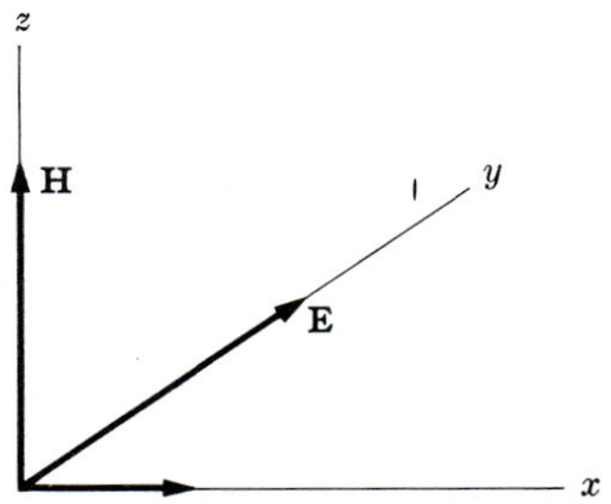

Fig. 1–2 Electric and magnetic fields associated with a wave moving in the x-direction.

electric field vector **E** and hence the magnetic field vector **H** can assume all directions in the yz-plane.) The magnetic field is of no concern to us here and we need not consider it further.

In the plane-polarized wave considered, **E** is not constant with time but varies from a maximum in the $+y$ direction through zero to a maximum in the $-y$ direction and back again, at any particular point in space, say $x = 0$. At any instant of time, say $t = 0$, **E** varies in the same fashion with distance along the x-axis. If both variations are assumed to be sinusoidal, they may be expressed in the one equation

$$\mathbf{E} = A \sin 2\pi \left(\frac{x}{\lambda} - \nu t \right), \tag{1–1}$$

where A = amplitude of the wave, λ = wavelength, and ν = frequency. The variation of **E** is not necessarily sinusoidal, but the exact form of the wave matters little; the important feature is its periodicity. Figure 1–3 shows the variation of **E** graphically. The wavelength and frequency are connected by the relation

$$\lambda = \frac{c}{\nu}, \tag{1–2}$$

where c = velocity of light = 3.00×10^8 m/sec.

Electromagnetic radiation, such as a beam of x-rays, carries energy, and the rate of flow of this energy through unit area perpendicular to the direction of motion of the wave is called the *intensity I*. The average value of the intensity is proportional to the square of the amplitude of the wave, i.e., proportional to A^2. In absolute units, intensity is measured in joules/m^2/sec, but this measurement is a difficult one and is seldom carried out; most x-ray intensity measurements are made on a relative basis in arbitrary units, such as the degree of blackening of a photographic film exposed to the x-ray beam.

An accelerated electric charge radiates energy. The acceleration may, of course, be either positive or negative, and thus a charge continuously oscillating about some mean position acts as an excellent source of electromagnetic radiation. Radio waves, for example, are produced by the oscillation of charge back and forth in the broadcasting antenna, and visible light by oscillating electrons in the atoms

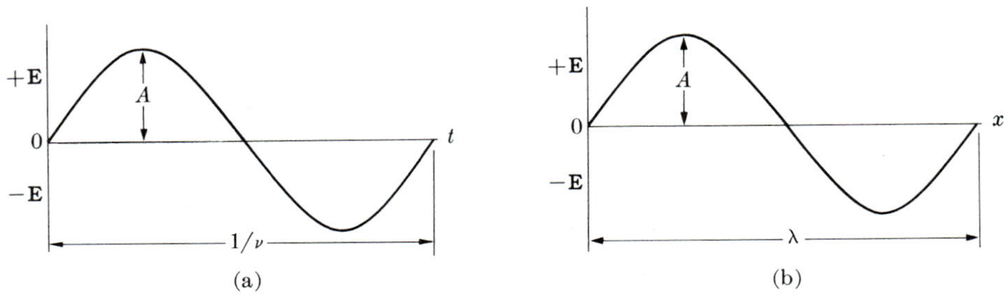

Fig. 1–3 The variation of **E**, (a) with t at a fixed value of x and (b) with x at a fixed value of t.

of the substance emitting the light. In each case, the frequency of the radiation is the same as the frequency of the oscillator which produces it.

Up to now we have been considering electromagnetic radiation as *wave* motion in accordance with classical theory. According to the quantum theory, however, electromagnetic radiation can also be considered as a stream of *particles* called quanta or photons. Each photon has associated with it an amount of energy $h\nu$, where h is Planck's constant (6.63×10^{-34} joule·sec). A link is thus provided between the two viewpoints, because we can use the frequency of the wave motion to calculate the energy of the photon. Radiation thus has a dual wave-particle character, and we will use sometimes one concept, sometimes the other, to explain various phenomena, giving preference in general to the classical wave theory whenever it is applicable.

1–3 THE CONTINUOUS SPECTRUM

X-rays are produced when any electrically charged particle of sufficient kinetic energy is rapidly decelerated. Electrons are usually used for this purpose, the radiation being produced in an *x-ray tube* which contains a source of electrons and two metal electrodes. The high voltage maintained across these electrodes, some tens of thousands of volts, rapidly draws the electrons to the anode, or *target*, which they strike with very high velocity. X-rays are produced at the point of impact and radiate in all directions. If e is the charge on the electron (1.60×10^{-19} coulomb) and V the voltage across the electrodes, then the kinetic energy (in joules) of the electrons on impact is given by the equation

$$\text{KE} = eV = \tfrac{1}{2}mv^2, \tag{1-3}$$

where m is the mass of the electron (9.11×10^{-31} kg) and v its velocity in m/sec just before impact. At a tube voltage of 30,000 volts, this velocity is about one-third that of light. Most of the kinetic energy of the electrons striking the target is converted into heat, less than 1 percent being transformed into x-rays.

When the rays coming from the target are analyzed, they are found to consist

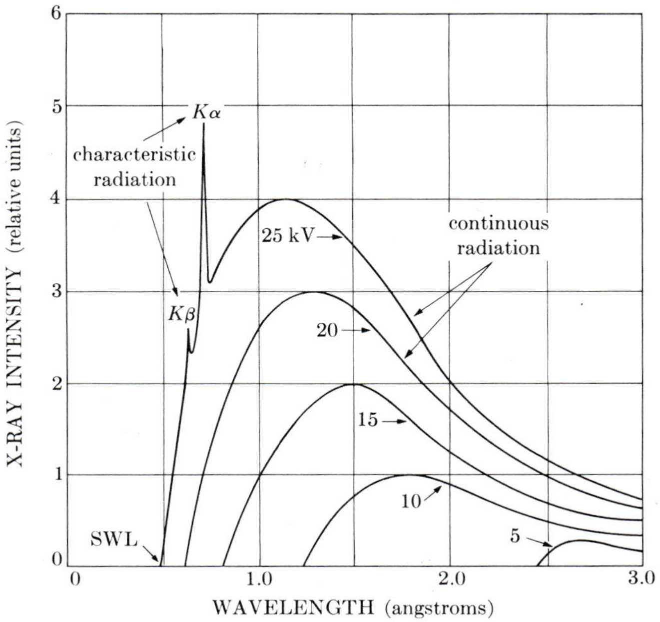

Fig. 1-4 X-ray spectrum of molybdenum as a function of applied voltage (schematic). Line widths not to scale.

of a mixture of different wavelengths, and the variation of intensity with wavelength is found to depend on the tube voltage. Figure 1-4 shows the kind of curves obtained. The intensity is zero up to a certain wavelength, called the *short-wavelength limit* (λ_{SWL}), increases rapidly to a maximum and then decreases, with no sharp limit on the long wavelength side. When the tube voltage is raised, the intensity of all wavelengths increases, and both the short-wavelength limit and the position of the maximum shift to shorter wavelengths. We are concerned now with the smooth curves in Fig. 1-4, those corresponding to applied voltages of 20 kV or less in the case of a molybdenum target. The radiation represented by such curves is called *heterochromatic*, *continuous*, or *white* radiation, since it is made up, like white light, of rays of many wavelengths. White radiation is also called *bremsstrahlung*, German for "braking radiation," because it is caused by electron deceleration.

The continuous spectrum is due to the rapid deceleration of the electrons hitting the target since, as mentioned above, any decelerated charge emits energy. Not every electron is decelerated in the same way, however; some are stopped in one impact and give up all their energy at once, while others are deviated this way and that by the atoms of the target, successively losing fractions of their total kinetic energy until it is all spent. Those electrons which are stopped in one impact will give rise to photons of maximum energy, i.e., to x-rays of minimum wavelength. Such electrons transfer all their energy eV into photon energy and we may write

$$eV = h\nu_{max},$$

$$\lambda_{SWL} = \lambda_{min} = \frac{c}{\nu_{max}} = \frac{hc}{eV},$$

$$\lambda_{SWL} = \frac{(6.626 \times 10^{-34})(2.998 \times 10^{8})}{(1.602 \times 10^{-19})\,V} \text{ meter},$$

$$\boxed{\lambda_{SWL} = \frac{12.40 \times 10^{3}}{V}}\,. \tag{1–4}$$

This equation gives the short-wavelength limit (in angstroms) as a function of the applied voltage V. If an electron is not completely stopped in one encounter but undergoes a glancing impact which only partially decreases its velocity, then only a fraction of its energy eV is emitted as radiation and the photon produced has energy less than $h\nu_{max}$. In terms of wave motion, the corresponding x-ray has a frequency lower than ν_{max} and a wavelength longer than λ_{SWL}. The totality of these wavelengths, ranging upward from λ_{SWL}, constitutes the continuous spectrum.

We now see why the curves of Fig. 1–4 become higher and shift to the left as the applied voltage is increased, since the number of photons produced per second and the average energy per photon are both increasing. The total x-ray energy emitted per second, which is proportional to the area under one of the curves of Fig. 1–4, also depends on the atomic number Z of the target and on the tube current i, the latter being a measure of the number of electrons per second striking the target. This total x-ray intensity is given by

$$I_{\text{cont. spectrum}} = AiZV^{m}, \tag{1–5}$$

where A is a proportionality constant and m is a constant with a value of about 2. Where large amounts of white radiation are desired, it is therefore necessary to use a heavy metal like tungsten ($Z = 74$) as a target and as high a voltage as possible. Note that the material of the target affects the intensity but not the wavelength distribution of the continuous spectrum.

1–4 THE CHARACTERISTIC SPECTRUM

When the voltage on an x-ray tube is raised above a certain critical value, characteristic of the target metal, sharp intensity maxima appear at certain wavelengths, superimposed on the continuous spectrum. Since they are so narrow and since their wavelengths are characteristic of the target metal used, they are called *characteristic lines*. These lines fall into several sets, referred to as K, L, M, etc., in the order of increasing wavelength, all the lines together forming the *characteristic spectrum* of the metal used as the target. For a molybdenum target the K lines have wavelengths of about 0.7 Å, the L lines about 5 Å, and the M lines still longer wavelengths. Ordinarily only the K lines are useful in x-ray diffraction, the longer-wavelength lines being too easily absorbed. There are several lines in the K set, but only the three strongest are observed in normal diffraction work. These are the $K\alpha_1$, $K\alpha_2$, and $K\beta_1$, and for molybdenum their wavelengths are approximately:

$$\begin{aligned} K\alpha_1&: \quad 0.709 \text{ Å}, \\ K\alpha_2&: \quad 0.714, \\ K\beta_1&: \quad 0.632. \end{aligned}$$

The α_1 and α_2 components have wavelengths so close together that they are not always resolved as separate lines; if resolved, they are called the *$K\alpha$ doublet* and, if not resolved, simply the *$K\alpha$ line*.* Similarly, $K\beta_1$ is usually referred to as the *$K\beta$ line*, with the subscript dropped. $K\alpha_1$ is always about twice as strong as $K\alpha_2$, while the intensity ratio of $K\alpha_1$ to $K\beta_1$ depends on atomic number but averages about 5/1.

These characteristic lines may be seen in the uppermost curve of Fig. 1–4. Since the critical *K excitation voltage*, i.e., the voltage necessary to excite K characteristic radiation, is 20.01 kV for molybdenum, the K lines do not appear in the lower curves of Fig. 1–4. An increase in voltage above the critical voltage increases the intensities of the characteristic lines relative to the continuous spectrum but *does not change their wavelengths.* Figure 1–5 shows the spectrum of molybdenum at 35 kV on a compressed vertical scale relative to that of Fig. 1–4; the increased voltage has shifted the continuous spectrum to still shorter wavelengths and increased the intensities of the K lines relative to the continuous spectrum but has not changed their wavelengths.

The intensity of any characteristic line, measured above the continuous spectrum, depends both on the tube current i and the amount by which the applied voltage V exceeds the critical excitation voltage for that line. For a K line, the intensity is given approximately by

$$I_{K \text{ line}} = Bi(V - V_K)^n, \tag{1-6}$$

where B is a proportionality constant, V_K the K excitation voltage, and n a constant with a value of about 1.5. (Actually, n is not a true constant but depends on V and varies from 1 to 2.) The intensity of a characteristic line can be quite large: for example, in the radiation from a copper target operated at 30 kV, the $K\alpha$ line has an intensity about 90 times that of the wavelengths immediately adjacent to it in the continuous spectrum. Besides being very intense, characteristic lines are also very narrow, most of them less than 0.001 Å wide measured at half their maximum intensity, as indicated in Fig. 1–5. The existence of this strong sharp $K\alpha$ line is what makes a great deal of x-ray diffraction possible, because many diffraction

* The wavelength of an unresolved $K\alpha$ doublet is usually taken as the weighted average of the wavelengths of its components, $K\alpha_1$ being given twice the weight of $K\alpha_2$, since it is twice as strong. Thus the wavelength of the unresolved Mo $K\alpha$ line is

$$\tfrac{1}{3}(2 \times 0.709 + 0.714) = 0.711 \text{ Å}.$$

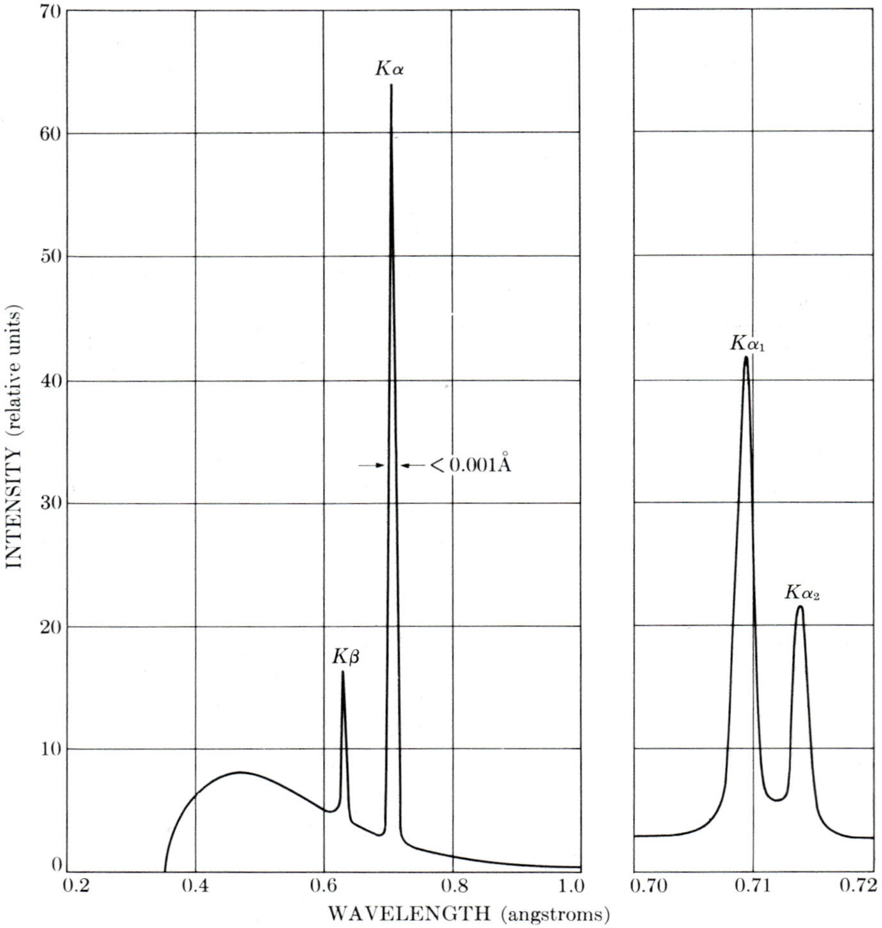

Fig. 1–5 Spectrum of Mo at 35 kV (schematic). Line widths not to scale. Resolved $K\alpha$ doublet is shown on an expanded wavelength scale at right.

experiments require the use of monochromatic or approximately monochromatic radiation.

The characteristic x-ray lines were discovered by W. H. Bragg and systematized by H. G. Moseley. The latter found that the wavelength of any particular line decreased as the atomic number of the emitter increased. In particular, he found a linear relation (Moseley's law) between the square root of the line frequency ν and the atomic number Z:

$$\sqrt{\nu} = C(Z - \sigma), \tag{1–7}$$

where C and σ are constants. This relation is plotted in Fig. 1–6 for the $K\alpha_1$ and $L\alpha_1$ lines, the latter being the strongest line in the L series. These curves show,

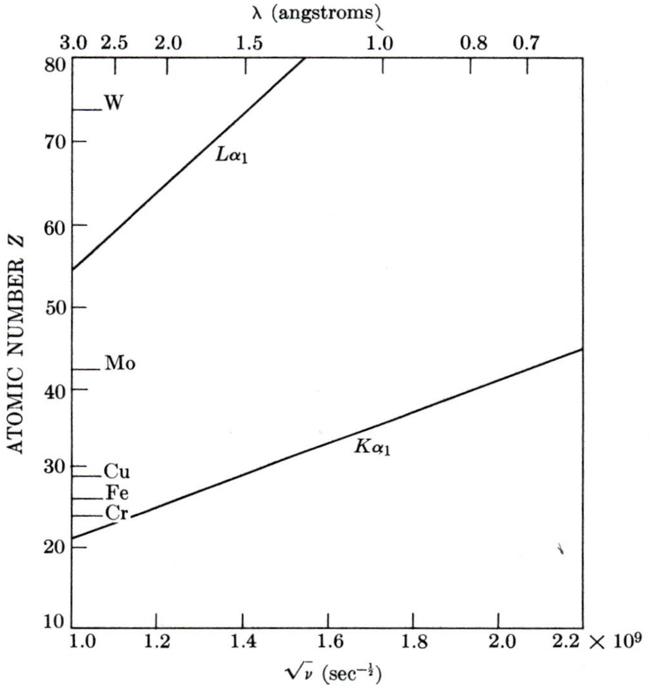

Fig. 1–6 Moseley's relation between $\sqrt{\nu}$ and Z for two characteristic lines.

incidentally, that L lines are not always of long wavelength: the $L\alpha_1$ line of a heavy metal like tungsten, for example, has about the same wavelength as the $K\alpha_1$ line of copper, namely about 1.5 Å. The wavelengths of the characteristic x-ray lines of almost all the known elements have been precisely measured, mainly by M. Siegbahn and his associates, and a tabulation of these wavelengths for the strongest lines of the K and L series will be found in Appendix 7. Data on weaker lines can be found in Vol. 4 of the *International Tables for X-Ray Crystallography* [G.11].*

While the continuous spectrum is caused by the rapid deceleration of electrons by the target, the origin of the characteristic spectrum lies in the atoms of the target material itself. To understand this phenomenon, it is enough to consider an atom as consisting of a central nucleus surrounded by electrons lying in various shells (Fig. 1–7), where the designation $K, L, M, \ldots$ corresponds to the principal quantum number $n = 1, 2, 3, \ldots$. If one of the electrons bombarding the target has sufficient kinetic energy, it can knock an electron out of the K shell, leaving the atom in an excited, high-energy state. One of the outer electrons immediately falls into the vacancy in the K shell, emitting energy in the process, and the atom is

* Numbers in square brackets relate to the references at the end of the book. "G" numbers are keyed to the General References.

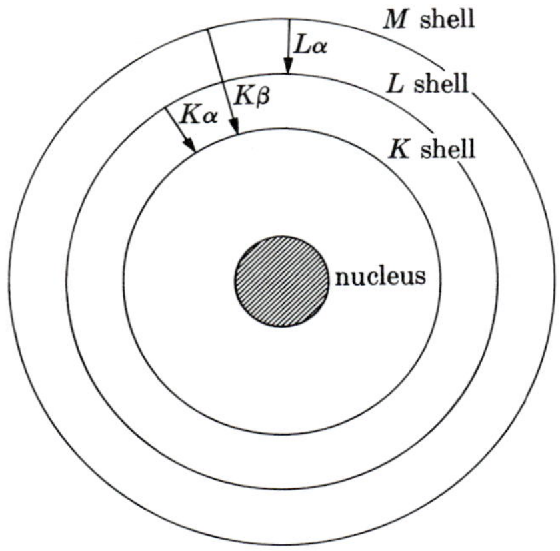

Fig. 1–7 Electronic transitions in an atom (schematic). Emission processes indicated by arrows.

once again in its normal energy state. The energy emitted is in the form of radiation of a definite wavelength and is, in fact, characteristic *K* radiation.

The *K*-shell vacancy may be filled by an electron from any one of the outer shells, thus giving rise to a series of *K* lines; $K\alpha$ and $K\beta$ lines, for example, result from the filling of a *K*-shell vacancy by an electron from the *L* or *M* shells, respectively. It is *possible* to fill a *K*-shell vacancy from either the *L* or *M* shell, so that one atom of the target may be emitting $K\alpha$ radiation while its neighbor is emitting $K\beta$; however, it is more *probable* that a *K*-shell vacancy will be filled by an *L* electron than by an *M* electron, and the result is that the $K\alpha$ line is stronger than the $K\beta$ line. It also follows that it is impossible to excite one *K* line without exciting all the others. *L* characteristic lines originate in a similar way: an electron is knocked out of the *L* shell and the vacancy is filled by an electron from some outer shell.

We now see why there should be a critical excitation voltage for characteristic radiation. *K* radiation, for example, cannot be excited unless the tube voltage is such that the bombarding electrons have enough energy to knock an electron out of the *K* shell of a target atom. If W_K is the work required to remove a *K* electron, then the necessary kinetic energy of the electrons is given by

$$\tfrac{1}{2}mv^2 = W_K. \tag{1–8}$$

It requires less energy to remove an *L* electron than a *K* electron, since the former is farther from the nucleus; it therefore follows that the *L* excitation voltage is less than the *K* and that *K* characteristic radiation cannot be produced without *L*, *M*, etc., radiation accompanying it.

1–5 ABSORPTION

Further understanding of the electronic transitions which can occur in atoms can be gained by considering not only the interaction of electrons and atoms, but also the interaction of x-rays and atoms. When x-rays encounter any form of matter, they are partly transmitted and partly absorbed. Experiment shows that the fractional decrease in the intensity I of an x-ray beam as it passes through any homogeneous substance is proportional to the distance traversed x. In differential form,

$$-\frac{dI}{I} = \mu\, dx, \tag{1–9}$$

where the proportionality constant μ is called the *linear absorption coefficient* and is dependent on the substance considered, its density, and the wavelength of the x-rays. Integration of Eq. (1–9) gives

$$I_x = I_0 e^{-\mu x}, \tag{1–10}$$

where I_0 = intensity of incident x-ray beam and I_x = intensity of transmitted beam after passing through a thickness x.

The linear absorption coefficient μ is proportional to the density ρ, which means that the quantity μ/ρ is a constant of the material and independent of its physical state (solid, liquid, or gas). This latter quantity, called the *mass absorption coefficient*, is the one usually tabulated. Equation (1–10) may then be rewritten in a more usable form:

$$\boxed{I_x = I_0 e^{-(\mu/\rho)\rho x}}. \tag{1–11}$$

Values of the mass absorption coefficient μ/ρ are given in Appendix 8 for various characteristic wavelengths used in diffraction.

It is occasionally necessary to know the mass absorption coefficient of a substance containing more than one element. Whether the substance is a mechanical mixture, a solution, or a chemical compound, and whether it is in the solid, liquid, or gaseous state, its mass absorption coefficient is simply the weighted average of the mass absorption coefficients of its constituent elements. If w_1, w_2, etc., are the weight fractions of elements 1, 2, etc., in the substance and $(\mu/\rho)_1$, $(\mu/\rho)_2$, etc., their mass absorption coefficients, then the mass absorption coefficient of the substance is given by

$$\frac{\mu}{\rho} = w_1\left(\frac{\mu}{\rho}\right)_1 + w_2\left(\frac{\mu}{\rho}\right)_2 + \cdots. \tag{1–12}$$

The way in which the absorption coefficient varies with wavelength gives the clue to the interaction of x-rays and atoms. The lower curve of Fig. 1–8 shows this variation for a nickel absorber; it is typical of all materials. The curve consists of two similar branches separated by a sharp discontinuity called an *absorption edge*.

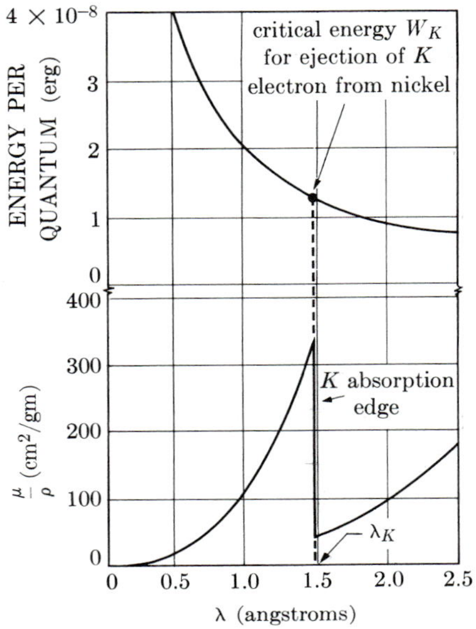

Fig. 1–8 Variation with wavelength of the energy per x-ray quantum and of the mass absorption coefficient of nickel.

Along each branch the absorption coefficient varies with wavelength approximately according to a relation of the form

$$\frac{\mu}{\rho} = k\lambda^3 Z^3, \tag{1-13}$$

where k = a constant, with a different value for each branch of the curve, and Z = atomic number of absorber. Short-wavelength x-rays are therefore highly penetrating and are termed *hard*, while long-wavelength x-rays are easily absorbed and are said to be *soft*.

Matter absorbs x-rays in two distinct ways, by scattering and by true absorption, and these two processes together make up the total absorption measured by the quantity μ/ρ. The *scattering* of x-rays by atoms is similar in many ways to the scattering of visible light by dust particles in the air. It takes place in all directions, and since the energy in the scattered beams does not appear in the transmitted beam, it is, so far as the transmitted beam is concerned, said to be absorbed (Fig. 1–9). The phenomenon of scattering will be discussed in greater detail in Chap. 4; it is enough to note here that, except for the very light elements, it is responsible for only a small fraction of the total absorption. *True absorption* is caused by electronic transitions within the atom and is best considered from the viewpoint of the quantum theory of radiation. Just as an electron of sufficient energy can knock a K electron, for example, out of an atom and thus cause the emission of K characteristic radiation, so also can an incident quantum of x-rays,

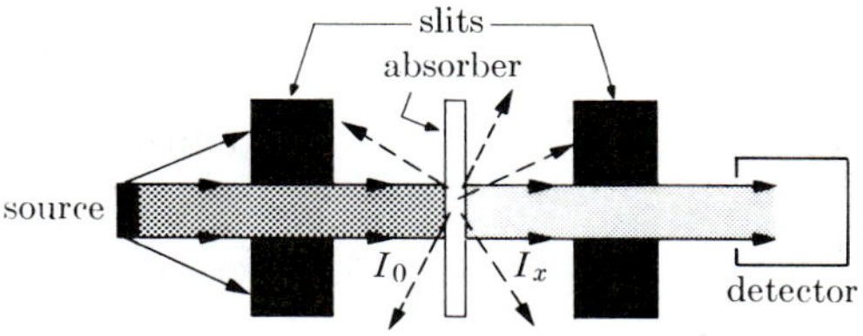

Fig. 1-9 Experimental arrangement for measuring absorption. Narrow slits or pinholes define the beam. The detector measures the intensity I_0 of the incident beam when the absorber is removed and the intensity I_x of the transmitted beam when the absorber is in place. Although the scattered radiation (dashed lines) does not represent energy absorbed *in* the specimen, it does constitute energy removed from the beam and accordingly forms part of the total absorption represented by the coefficient μ/ρ.

provided it has the same minimum amount of energy W_K. In the latter case, the ejected electron is called a *photoelectron* and the emitted characteristic radiation is called *fluorescent radiation.* It radiates in all directions and has exactly the same wavelength as the characteristic radiation caused by electron bombardment of a metal target. (In effect, an atom emits the same K radiation no matter how the K-shell vacancy was originally created.) This phenomenon is the x-ray counterpart of the photoelectric effect in the ultraviolet region of the spectrum; there, photoelectrons can be ejected from the outer shells of a metal atom by the action of ultraviolet radiation, provided the latter has a wavelength less than a certain critical value.

To say that the energy of the incoming quanta must exceed a certain value W_K is equivalent to saying that the wavelength must be less than a certain value λ_K, since the energy per quantum is $h\nu$ and wavelength is inversely proportional to frequency. These relations may be written

$$W_K = h\nu_K = \frac{hc}{\lambda_K}, \tag{1-14}$$

where ν_K and λ_K are the frequency and wavelength, respectively, of the K absorption edge. Now consider the absorption curve of Fig. 1-8 in light of the above. Suppose that x-rays of wavelength 2.5 Å are incident on a sheet of nickel and that this wavelength is continuously decreased. At first the absorption coefficient is about 180 cm^2/gm, but, as the wavelength decreases, the frequency increases and so does the energy per quantum, as shown by the upper curve, thus causing the absorption coefficient to decrease, since the greater the energy of a quantum the more easily it passes through an absorber. When the wavelength is reduced just below the critical value λ_K, which is 1.488 Å for nickel, the absorption coefficient suddenly increases about eightfold in value. True K absorption is now occurring and a large fraction of the incident quanta simply disappear, their energy being converted into K fluorescent radiation and the kinetic energy of ejected photoelectrons. Since energy must be conserved in the process, it follows that the energy per quantum of the fluorescent radiation must be less than that of the incident radiation, or that the wavelength λ_K of the K absorption edge must be shorter than that of any K characteristic line. (The eight-fold increase in μ/ρ mentioned above means a tremendous decrease in transmitted intensity, because of the exponential nature of Eq. (1–11). If the transmission factor I_x/I_0 of a particular nickel sheet is 0.1 for a wavelength just longer than λ_K, then it is only 10^{-8} for a wavelength just shorter.)

As the wavelength of the incident beam is decreased below λ_K, the absorption coefficient begins to decrease again, even though the production of K fluorescent radiation and photoelectrons is still occurring. At a wavelength of 1.0 Å, for example, the incident quanta have more than enough energy to remove an electron from the K shell of nickel. But the more energetic the quanta become, the greater is their probability of passing right through the absorber, with the result that less and less of them take part in the ejection of photoelectrons.

If the absorption curve of nickel is plotted for longer wavelengths than 2.5 Å, i.e., beyond the limit of Fig. 1–8, other sharp discontinuities will be found. These are the L, M, N, etc., absorption edges; in fact, there are three closely spaced L edges (L_I, L_{II}, and L_{III}), five M edges, etc. (Fig. 1–10). Each of these discontinuities marks the wavelength of the incident beam whose quanta have just sufficient energy to eject an L, M, N, etc., electron from the atom. The right-hand branch of the curve of Fig. 1–8, for example, lies between the K and L absorption edges; in this wavelength region incident x-rays have enough energy to remove L, M, etc., electrons from nickel but not enough to remove K electrons. Absorption-edge wavelengths vary with the atomic number of the absorber in the same way, but not quite as exactly, as characteristic emission wavelengths, that is, according to Moseley's law. Values of the K and L absorption-edge wavelengths are given in Appendix 7.

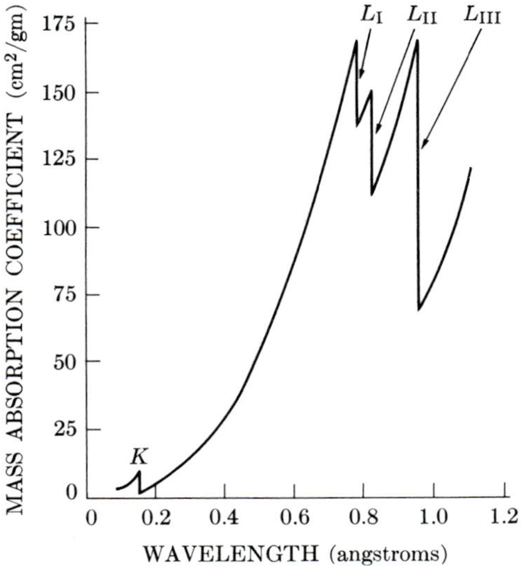

Fig. 1–10 Absorption coefficients of lead, showing K and L absorption edges [1.6].

The measured values of the absorption edges can be used to construct an energy-level diagram for the atom, which in turn can be used in the calculation of characteristic-line wavelengths. For example, if we take the energy of the neutral atom as zero, then the energy of an ionized atom (an atom in an excited state) will be some positive quantity, since work must be done to pull an electron away from the positively charged nucleus. If a *K* electron is removed, work equal to W_K must be done and the atom is said to be in the *K* energy state. The energy W_K may be calculated from the wavelength of the *K* absorption edge by the use of Eq. (1–14). Similarly, the energies of the *L*, *M*, etc., states can be calculated from the wavelengths of the *L*, *M*, etc., absorption edges and the results plotted in the form of an energy-level diagram for the atom (Fig. 1–11).

Although this diagram is simplified, in that the substructure of all the levels is not shown, it illustrates the main principles. The arrows show the transitions of the *atom*, and their directions are therefore just the opposite of the arrows in Fig. 1–7,

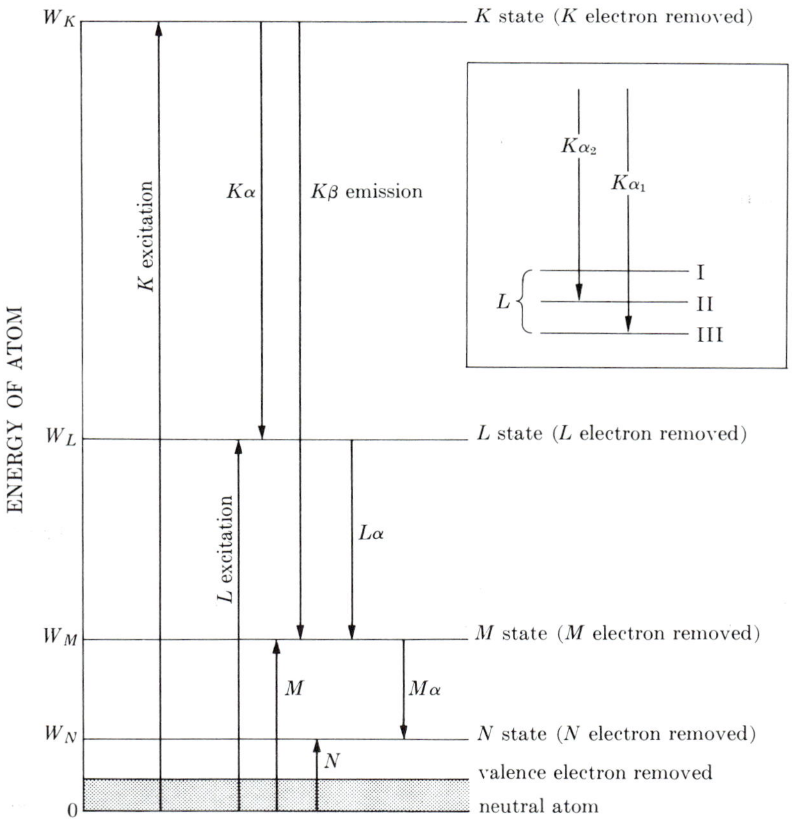

Fig. 1–11 Atomic energy levels (schematic). Excitation and emission processes indicated by arrows. The insert at top right shows the fine structure of the *L* state. After Barrett [1.7].

which shows the transitions of the *electron*. Thus, if a K electron is removed from an atom (whether by an incident electron or x-ray), the atom is raised to the K state. If an electron then moves from the L to the K level to fill the vacancy, the atom undergoes a transition from the K to the L state. This transition is accompanied by the emission of $K\alpha$ characteristic radiation and the arrow indicating $K\alpha$ emission is accordingly drawn *from* the K state *to* the L state.

Figure 1–11 shows clearly how the wavelengths of characteristic emission lines can be calculated, since the difference in energy between two states will equal $h\nu$, where ν is the frequency of the radiation emitted when the atom goes from one state to the other. Consider the $K\alpha_1$ characteristic line, for example. The "L level" of an atom is actually a group of three closely spaced levels (L_{I}, L_{II}, and L_{III}), and the emission of the $K\alpha_1$ line is due to a $K \rightarrow L_{\text{III}}$ transition. The frequency $\nu_{K\alpha_1}$ of this line is therefore given by the equations

$$h\nu_{K\alpha_1} = W_K - W_{L_{\text{III}}},$$

$$h\nu_{K\alpha_1} = h\nu_K - h\nu_{L_{\text{III}}},$$

$$\frac{1}{\lambda_{K\alpha_1}} = \frac{1}{\lambda_K} - \frac{1}{\lambda_{L_{\text{III}}}}, \tag{1–15}$$

where the subscripts K and L_{III} refer to absorption edges and the subscript $K\alpha_1$ to the emission line.

Excitation voltages can be calculated by a relation similar to Eq. (1–4). To excite K radiation, for example, in the target of an x-ray tube, the bombarding electrons must have energy equal to W_K. Therefore

$$eV_K = W_K = h\nu_K = \frac{hc}{\lambda_K},$$

$$V_K = \frac{hc}{e\lambda_K},$$

$$V_K = \frac{12.40 \times 10^3}{\lambda_K}, \tag{1–16}$$

where V_K is the K excitation voltage and λ_K is the K absorption edge wavelength (in angstroms).

Figure 1–12 summarizes some of the relations developed above. This curve gives the short-wavelength limit of the continuous spectrum as a function of applied voltage. Because of the similarity between Eqs. (1–4) and (1–16), the same curve also enables us to determine the critical excitation voltage from the wavelength of an absorption edge.

Auger effect [1.1, 1.2]. It might be inferred, from the last two sections, that every atom that has a vacancy in, for example, the K shell will always emit K radiation. That is not so. An atom with a K-shell vacancy is in an ionized, high-energy state. It can lose this excess energy and return to its normal state in two ways: (1) by emitting K radiation ("normal" production of characteristic radiation), or (2) by emitting an electron (*Auger effect*). In the Auger process a K-shell vacancy is filled from, say, the L_{II} level; the

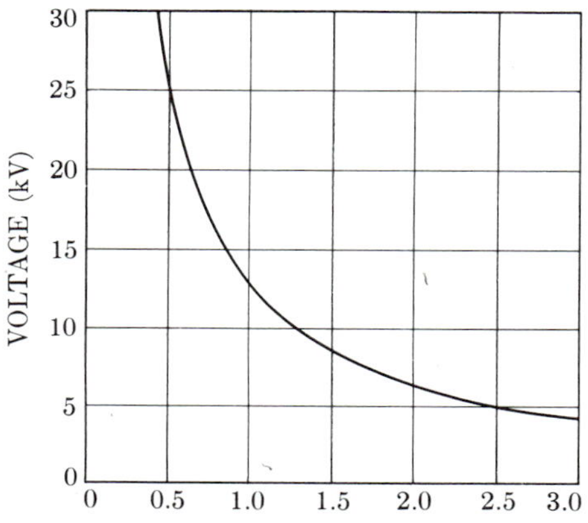

Fig. 1–12 Relation between the voltage applied to an x-ray tube and the short-wavelength limit of the continuous spectrum, and between the critical excitation voltage of any metal and the wavelength of its absorption edge.

resulting K radiation does not escape from the atom but ejects an electron from, say, the L_{III} level. The ejected electron, called an *Auger electron*, has a kinetic energy related to the energy difference between the K and L_{II} states.

The Auger effect is by no means a minor one. In fact, atoms with an atomic number Z less than 31 (gallium) are more likely to eject Auger electrons than to emit x-rays. The likelihood of the Auger process can be found from the fluorescence yield ω, which is defined, for the K shell, by

$$\omega_K = \frac{\text{number of atoms that emit } K \text{ radiation}}{\text{number of atoms with a } K\text{-shell vacancy}}. \tag{1–17}$$

(This quantity is called the *fluorescence* yield, whether the vacancy is caused by incident x-rays or by electrons.) Some values of ω_K are 0.03 for Mg ($Z = 12$), 0.41 for Cu ($Z = 29$), and 0.77 for Mo ($Z = 42$) [G.31, p. 131]. The probability of the Auger process occurring is $(1 - \omega_K)$, which amounts to some 97 percent for Mg and 23 percent for Mo.

Electrons of moderate energy like Auger electrons cannot travel very far in a solid, and an Auger electron emitted by one atom in a solid specimen cannot escape from the specimen unless the atom is situated within about 10 Å of the surface. The electrons that do escape have kinetic energies related to the differences between energy levels of the parent atom, i.e., their energies are characteristic of that atom. Means are available for measuring these energies, and we therefore have a method for chemical analysis of very thin surface layers, called *Auger electron spectroscopy*, used in studies of catalysts, corrosion, impurity segregation at surfaces, etc.

1–6 FILTERS

Many x-ray diffraction experiments require radiation which is as closely monochromatic as possible. However, the beam from an x-ray tube operated at a voltage above V_K contains not only the strong $K\alpha$ line but also the weaker $K\beta$ line and the continuous spectrum. The intensity of these undesirable components can be decreased relative to the intensity of the $K\alpha$ line by passing the beam through a *filter* made of a material whose K absorption edge lies between the $K\alpha$ and $K\beta$ wavelengths of the target metal. Such a material will have an atomic number one less than that of the target metal, for metals with Z near 30.

A filter so chosen will absorb the $K\beta$ component much more strongly than the $K\alpha$ component, because of the abrupt change in its absorption coefficient between these two wavelengths. The effect of filtration is shown in Fig. 1–13, in which the partial spectra of the unfiltered and filtered beams from a copper target ($Z = 29$) are shown superimposed on a plot of the mass absorption coefficient of the nickel filter ($Z = 28$).

The thicker the filter the lower the ratio of intensity of $K\beta$ to $K\alpha$ in the transmitted beam. But filtration is never perfect, of course, no matter how thick the filter, and one must compromise between reasonable suppression of the $K\beta$ component and the inevitable weakening of the $K\alpha$ component which accompanies it. In practice it is found that a reduction in the intensity of the $K\alpha$ line to about half its original value will decrease the ratio of intensity of $K\beta$ to $K\alpha$ from about $\frac{1}{9}$ in the incident beam to about $\frac{1}{500}$ in the transmitted beam; this level is sufficiently low for most purposes. Table 1–1 shows the filters used in conjunction with the common target metals, the thicknesses required, and the transmission factors for the $K\alpha$ line. Filter materials are usually used in the form of thin foils. If it is not

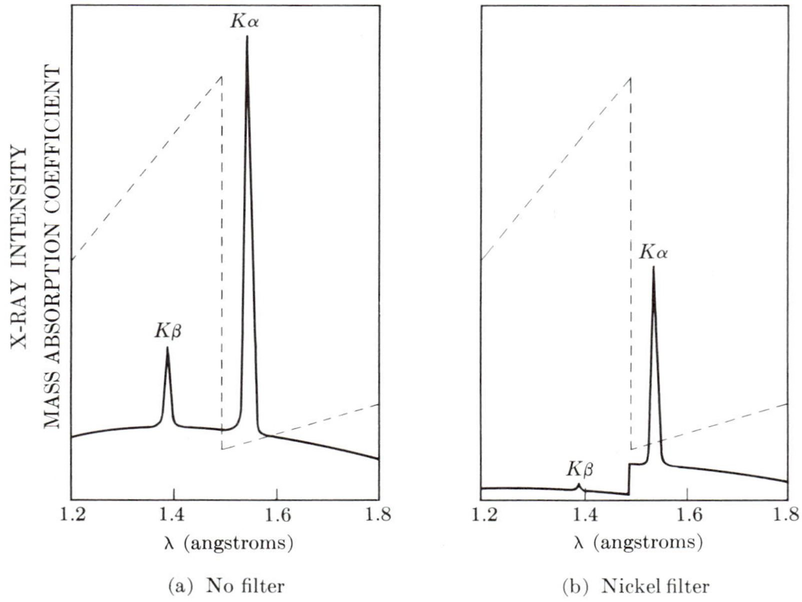

Fig. 1–13 Comparison of the spectra of copper radiation (a) before and (b) after passage through a nickel filter (schematic). The dashed line is the mass absorption coefficient of nickel.

Table 1–1
Filters for Suppression of $K\beta$ Radiation

Target	Filter	Incident beam* $\frac{I(K\alpha)}{I(K\beta)}$	Filter thickness for $\frac{I(K\alpha)}{I(K\beta)} = \frac{500}{1}$ in trans. beam		$\frac{I(K\alpha) \text{ trans.}}{I(K\alpha) \text{ incident}}$
			mg/cm^2	in.	
Mo	Zr	5.4	77	0.0046	0.29
Cu	Ni	7.5	18	0.0008	0.42
Co	Fe	9.4	14	0.0007	0.46
Fe	Mn	9.0	12	0.0007	0.48
Cr	V	8.5	10	0.0006	0.49

* This is the intensity ratio *at the target* [G.11, Vol. 3, p. 71]. This ratio outside the x-ray tube will be changed somewhat by the differential absorption of $K\alpha$ and $K\beta$ by the tube window, typically beryllium, 0.01 inch (0.25 mm) thick.

possible to obtain a given metal in the form of a stable foil, the oxide of the metal may be used. The powdered oxide is mixed with a suitable binder and spread on a paper backing, the required mass of metal per unit area being given in Table 1–1.

1–7 PRODUCTION OF X-RAYS

We have seen that x-rays are produced whenever high-speed electrons collide with a metal target. Any x-ray tube must therefore contain (*a*) a source of electrons, (*b*) a high accelerating voltage, and (*c*) a metal target. Furthermore, since most of the kinetic energy of the electrons is converted into heat in the target, the latter is almost always water-cooled to prevent its melting.

All x-ray tubes contain two electrodes, an anode (the metal target) maintained, with few exceptions, at ground potential, and a cathode, maintained at a high negative potential, normally of the order of 30,000 to 50,000 volts for diffraction work. X-ray tubes may be divided into two basic types, according to the way in which electrons are provided: gas tubes, in which electrons are produced by the ionization of a small quantity of gas (residual air in a partly evacuated tube), and filament tubes, in which the source of electrons is a hot filament.

Gas Tubes

These resemble the original x-ray tube used by Roentgen. They are now obsolete.

Filament Tubes

These were invented by Coolidge in 1913. They consist of an evacuated glass envelope which insulates the anode at one end from the cathode at the other, the cathode being a tungsten filament and the anode a water-cooled block of copper containing the desired target metal as a small insert at one end. Figure 1–14 is a photograph of such a tube, and Fig. 1–15 shows its internal construction. One lead of the high-voltage transformer is connected to the filament and the other to

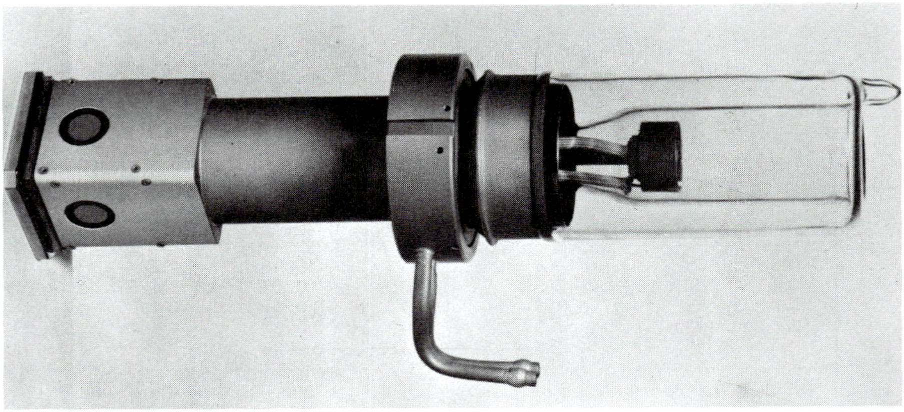

Fig. 1–14 Sealed-off filament x-ray tube. Cooling-water tubes at center connect with internal ducts leading to anode at left end. Three windows: two for projecting square focal spots and one for projecting a line focal spot. Focal spots of three sizes are available with this tube (Type A-5): 1.2 × 12.5 mm, 0.75 × 12.5 mm, and 0.45 × 12.5 mm. (Courtesy of Machlett Laboratories, Inc.)

ground, the target being grounded by its own cooling-water connection. The filament is heated by a *filament current* of about 3 amp and emits electrons which are rapidly drawn to the target by the high voltage across the tube. Surrounding the filament is a small metal cup maintained at the same high (negative) voltage as the filament: it therefore repels the electrons and tends to focus them into a narrow region of the target, called the *focal spot*. X-rays are emitted from the focal spot in all directions and escape from the tube through two or more windows in the tube housing. Since these windows must be vacuum tight and yet highly transparent to x-rays, they are usually made of beryllium.

Although one might think that an x-ray tube would operate only from a dc source, since the electron flow must occur only in one direction, it is actually possible to operate a tube from an ac source such as a transformer because of the rectifying properties of the tube itself. Current exists during the half-cycle in which the filament is negative with respect to the target; during the reverse half-cycle the filament is positive, but no electrons can flow since only the filament is hot enough to emit electrons. Thus a simple circuit such as shown in Fig. 1–16 suffices for many installations, although more elaborate circuits, containing rectifying tubes, smoothing capacitors, and voltage stabilizers, are often used, particularly when the x-ray intensity must be kept constant within narrow limits. In Fig. 1–16, the voltage applied to the tube is controlled by the autotransformer which controls the voltage applied to the primary of the high-voltage transformer. The voltmeter shown measures the input voltage but may be calibrated, if desired, to read the output voltage applied to the tube. The milliammeter measures the *tube current*, i.e., the flow of electrons from filament to target. This current is normally of the order of 10 to 25 mA and is controlled by the filament rheostat. The rheostat controls

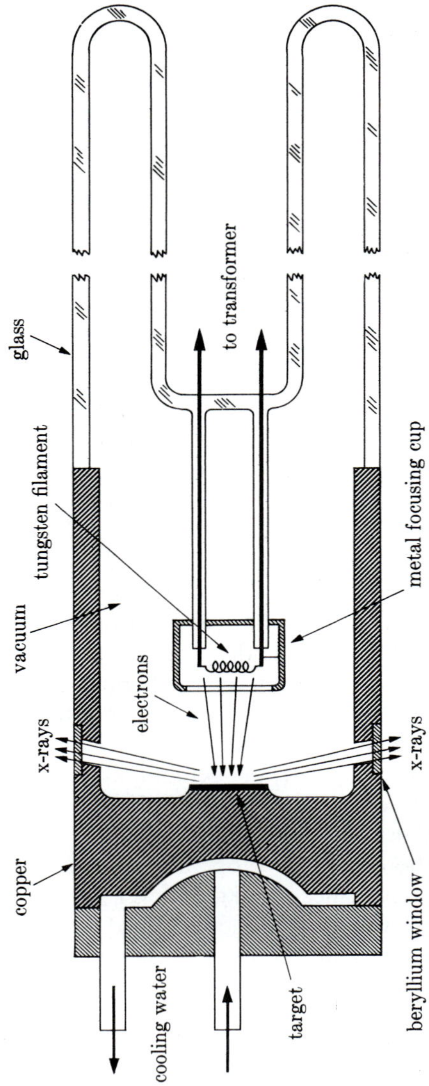

Fig. 1–15 Cross section of sealed-off filament x-ray tube (schematic).

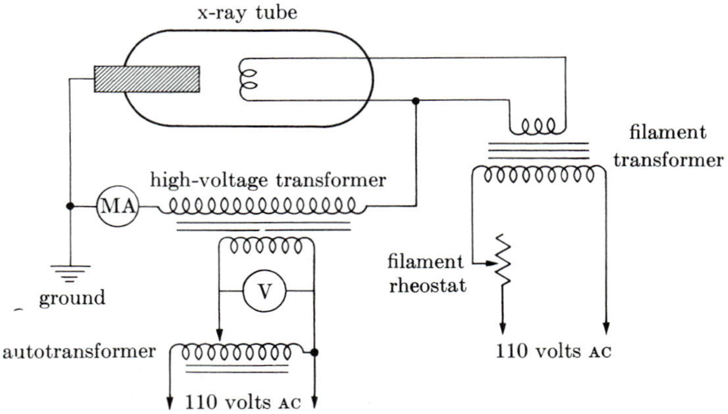

Fig. 1–16 Wiring diagram for self-rectifying filament tube.

the output voltage of the filament transformer; this voltage determines the filament current and, in turn, the temperature of the filament and the number of electrons it can emit per second. Although the filament transformer is a low-voltage step-down transformer, since it need apply only about 5 volts to the filament, it is itself at a high negative voltage relative to ground and must be well insulated.

Two kinds of filament tube exist: sealed-off and demountable. A sealed-off tube is evacuated and sealed off at the factory. It is by far the easier kind to operate, since no high-vacuum pumping equipment is needed; however, it is expensive (one needs as many tubes as there are target metals required), and the life of the tube is determined by the life of the filament. In demountable tubes, which are used nowadays only for special purposes, both the filament and the target are accessible for replacement; burned-out filaments can be replaced and targets can be interchanged at will. However, the demountable tube must be pumped out continuously during operation, and both a diffusion and a mechanical pump are necessary to obtain the high vacuum required.

The old gas tube, although tricky to operate, had the advantage of producing the purest radiation available, since the target never became contaminated with a foreign metal. In filament tubes, on the other hand, some tungsten occasionally evaporates from the filament and deposits on the target, and the tungsten then emits characteristic L radiation (the L excitation voltage of tungsten is only 10,200 volts), as well as the radiation characteristic of the target metal itself.

Focal Spot

The size and shape of the focal spot of an x-ray tube is one of its most important characteristics. Within limits, it should be as small as possible in order to concentrate the electron energy into a small area of the target and so produce an x-ray source of high intensity.

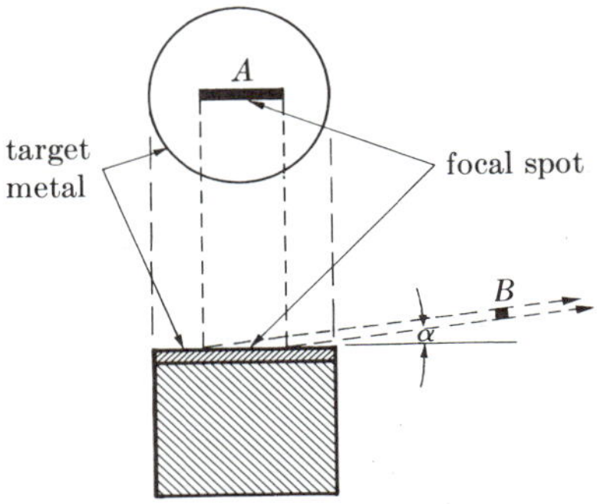

Fig. 1–17 Reduction in apparent size of focal spot.

Filament tubes usually have the filament wound in a helix in order to produce a so-called "line focus" which is actually a narrow rectangle (Fig. 1–17). The total electron energy is thus spread over a rather large focal spot A, which helps to dissipate the heat formed; yet the cross section B of the beam issuing at a small target-to-beam angle α is that of a small square, and this beam is of greater intensity than one leaving the focal spot at some larger angle α. The best value of α is about 6°, and a good tube will have a projected focal-spot size at this angle of less than 1 mm square. If the tube has a window so arranged that a beam can issue from the focal spot A almost normal to the plane of the drawing and at a small angle α, then the cross section of the beam will be an extremely narrow line; such a beam is quite useful in some diffraction experiments.

Power Rating

All x-ray tubes have a maximum power rating which cannot be exceeded without injury to the tube. This limit is fixed by the amount of heat that can be dissipated by the target and is usually stated by the manufacturer in terms of the maximum allowable tube current (in mA) for a given tube voltage (in kV).

Rotating-anode Tubes

Since an x-ray tube is less than 1 percent efficient in producing x-rays and since the diffraction of x-rays by crystals is far less efficient than this, it follows that the intensities of diffracted x-ray beams are extremely low. In fact, it may require as much as several hours exposure to a photographic film in order to detect them at all. Constant efforts are therefore being made to increase the intensity of the x-ray source. One solution to this problem is the rotating-anode tube, in which rotation of the anode continuously brings fresh target metal into the focal-spot area and so allows a greater power input without excessive heating of the anode. Figure 1–18 shows two designs that have been used successfully; the shafts rotate through vacuum-tight seals in the tube housing. Such tubes can operate at a power level 5 to 10 times higher than that of a fixed-focus tube, with corresponding reductions in exposure time. They are common in the area of radiography but not often used for diffraction.

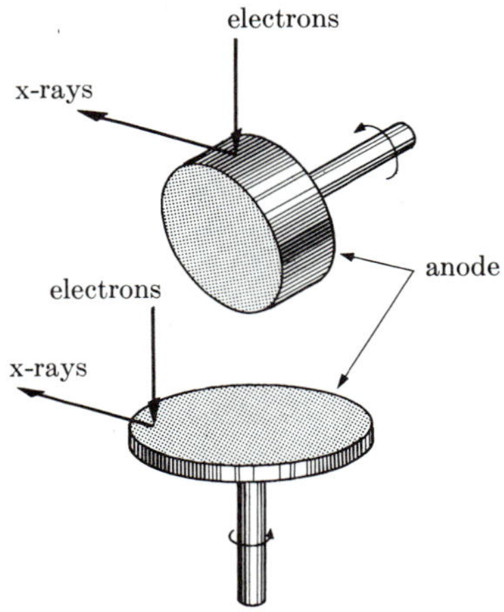

Fig. 1–18 Schematic drawings of two types of rotating anode for high-power x-ray tubes.

Microfocus Tubes

Some diffraction methods require extremely fine x-ray beams. Such beams are most efficiently produced by special demountable x-ray tubes, called microfocus tubes, in which special attention is paid to achieving a very small focal spot. The design problem—fine focusing of the electron beam—is similar to that of the electron microscope or the x-ray microprobe. One focusing method is electrostatic and consists simply in maintaining the focusing cup around the filament at a potential of a few hundred volts more negative than the filament, thus concentrating the electrons into a narrower beam.

The focal spots of these tubes have areas of less than 1 percent of those of conventional tubes. Typical sizes are 0.1×1 mm for a line focus and 0.05 mm ($=50$ μm) diameter for a circular focus.

Pulsed (or Flash) Tubes

The maximum power at which an x-ray tube can operate continuously is limited by the rate at which the target can be cooled. But if the tube is operated for only a small fraction of a second, a pulse of x-rays can be obtained at a very high power level without any cooling. This can be done by slowly charging a bank of capacitors and then abruptly discharging them across a special x-ray tube. In this way an x-ray pulse lasting about 30 nanoseconds at a peak voltage of 300 kV and a peak current of 5000 amperes has been produced [1.8]. (Such a brief flash of x-rays is useful only if its results, in radiography or diffraction, can be recorded. One example of high-speed recording is described in Sec. 8–5.)

Miniature Tubes

If increased attention is given, during the design of an x-ray tube, to focusing of the electron beam and to the shape and placement of the target, the intensity of the beam issuing from the tube can be made about as large as that from a conventional tube, but with a power input of one-tenth or less. As a result, water cooling is not needed; air cooling is sufficient. This feature is important for portable apparatus.

Such tubes have been made experimentally [1.9, 1.10] and commercially [1.11]. They are small, only about 4 to 8 in. (10 to 20 cm) in length, and operate typically at a voltage of about 50 kV and a tube current of the order of 1 mA, as compared to 10 mA or more in conventional tubes.

1-8 DETECTION OF X-RAYS

The principal means used to detect x-ray beams are fluorescent screens, photographic film, and counters.

Fluorescent Screens

Fluorescent screens are made of a thin layer of zinc sulfide, containing a trace of nickel, mounted on a cardboard backing. Under the action of x-rays, this compound fluoresces in the visible region, i.e., emits visible light, in this case yellow light. Although most diffracted beams are too weak to be detected by this method, fluorescent screens are widely used in diffraction work to locate the position of the primary beam when adjusting apparatus.

Photographic Film

Photographic film is affected by x-rays in much the same way as by visible light. However, the emulsion on ordinary film is too thin to absorb much of the incident x-radiation, and only absorbed x-rays can be effective in blackening the film. For this reason, x-ray films are made with rather thick layers of emulsion on both sides in order to increase the total absorption. (Division of the total emulsion thickness into two layers permits easier penetration of the film-processing solutions.) The grain size is also made large for the same purpose: this has the unfortunate consequence that x-ray films are grainy, do not resolve fine detail, and cannot stand much enlargement.

Because the mass absorption coefficient of any substance varies with wavelength, it follows that film sensitivity, i.e., the amount of blackening caused by x-ray beams of the same intensity, depends on their wavelength. This should be borne in mind whenever white radiation is recorded photographically; for one thing, this sensitivity variation alters the effective shape of the continuous spectrum. Figure 1–19(a) shows the intensity of the continuous spectrum as a function of wavelength and (b) the variation of film sensitivity. This latter curve is merely a plot of the mass absorption coefficient of silver bromide, the active ingredient of the emulsion, and is marked by discontinuities at the K absorption edges of silver and bromine. (Note, incidentally, how much more sensitive the film is to the K radiation from copper than to the K radiation from molybdenum, other things being equal.) Curve (c) of Fig. 1–19 shows the net result, namely the amount of film blackening caused by the various wavelength components of the continuous spectrum, or what might be called the "effective photographic intensity" of the continuous spectrum. These curves are only approximate, however, and in practice it is almost impossible to measure photographically the relative intensities

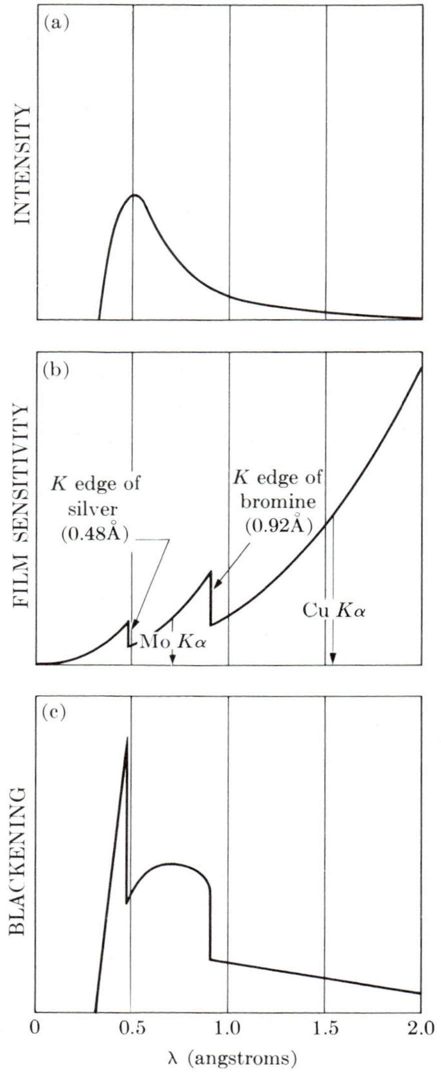

Fig. 1–19 Relation between film sensitivity and effective shape of continuous spectrum (schematic): (a) continuous spectrum from a tungsten target at 40 kV; (b) film sensitivity; (c) blackening curve for spectrum shown in (a).

of two beams of different wavelength. On the other hand, the relative intensities of beams of the *same* wavelength can be accurately measured by photographic means, and such measurements are described in Chap. 6.

The Polaroid Land rapid-process system of photography has been adapted to some kinds of diffraction equipment. The Polaroid film is backed by an intensifying screen (Sec. 5–2) which converts x-rays to visible light that can darken the film. X-ray exposures are about one tenth of those required by x-ray film, and finished prints are available about ten seconds after the x-ray exposure.

Counters

X-ray counters are devices that convert x-rays into a pulsating electric current, and the number of current pulses per unit of time is proportional to the intensity of the x-rays entering the counter. Three types are currently in use: proportional, scintillation, and semiconductor. They will be described in Chap. 7.

General

Fluorescent screens are used only for the detection of x-ray beams, while photographic film and the various kinds of counters permit both detection and measurement of intensity. Photographic film has the advantage of being able to record a number of diffracted beams at one time and their relative positions in space, and the film can be used as a basis for intensity measurements if desired. Intensities can be measured much more rapidly with counters, and these instruments are more popular for quantitative work. However, most counters record only one diffracted beam at a time.

1–9 SAFETY PRECAUTIONS

The operator of x-ray apparatus is exposed to two obvious dangers, electric shock and radiation injury, but both of these hazards can be reduced to negligible proportions by proper design of equipment and reasonable care on the part of the user. Nevertheless, it is only prudent for the x-ray worker to be continually aware of these hazards.

Electric Shock

The danger of electric shock is always present around high-voltage apparatus. The anode end of most x-ray tubes is usually grounded and therefore safe, but the cathode end is a source of danger. X-ray tubes of the nonshockproof variety (such as the one shown in Fig. 1–14) must be so mounted that their cathode end is absolutely inaccessible to the user during operation; this may be accomplished by placing the cathode end below a table top, in a box, behind a screen, etc. The installation should be so contrived that it is impossible for the operator to touch the high-voltage parts without automatically disconnecting the high voltage. Shockproof sealed-off tubes are also available: these are encased in a grounded metal covering, and an insulated, shockproof cable connects the cathode end to the transformer. Being shockproof, such a tube has the advantage that it need not be permanently fixed in position but may be set up in various positions as required for particular experiments.

Radiation Hazard

The radiation hazard is due to the fact that x-rays can kill human tissue; in fact, it is precisely this property which is utilized in x-ray therapy for the killing of cancer cells. The biological effects of x-rays include burns (due to localized high-intensity beams), radiation sickness (due to radiation received generally by the whole body), and, at a lower level of radiation intensity, genetic mutations. The burns are painful and may be difficult, if not impossible, to heal. Slight exposures to x-rays are not cumulative, but above a certain level called the "tolerance dose," they do have a cumulative effect and can produce permanent injury. The x-rays used in diffraction are particularly harmful because they have relatively long wavelengths and are therefore easily absorbed by the body.

There is no excuse today for receiving serious injuries as early x-ray workers did through ignorance. There would probably be no accidents if x-rays were visible and produced an immediate burning sensation, but they are invisible and burns may not be immediately felt. If the body has received general radiation above the tolerance dose, the first noticeable effect will be a lowering of the white-blood-cell count, so periodic blood counts are advisable if there is any doubt about the general level of intensity in the laboratory.

Portable counters, called *radiation survey meters*, are available for surveying various areas around x-ray equipment for possible radiation leaks. Film badges should be worn on the torso or wrist of persons who spend a large fraction of their working day near x-ray equipment. Government regulations regarding radiation safety are becoming increasingly severe; Jenkins and Haas [1.3] describe some of these regulations and give useful information about radiation units, survey meters, tolerance levels, and reported accidents. Certain government booklets should also be consulted [1.4, 1.5].

The safest procedure for the experimenter to follow is: first, to locate the *primary* beam from the tube with a small fluorescent screen fixed to the end of a rod and thereafter avoid it; and second, to make sure that he is well shielded by lead or lead-glass screens from the radiation *scattered* by the camera or other apparatus which may be in the path of the primary beam. Strict and constant attention to these precautions will ensure safety.

PROBLEMS

* **1–1** What is the frequency (per second) and energy per quantum (in joules) of x-ray beams of wavelength 0.71 Å (Mo $K\alpha$) and 1.54 Å (Cu $K\alpha$)?

1–2 Calculate the velocity and kinetic energy with which the electrons strike the target

---

* Answers to starred problems are given at the back of the book.

of an x-ray tube operated at 50,000 volts. What is the short-wavelength limit of the continuous spectrum emitted and the maximum energy per quantum of radiation?

1–3 Show that the velocity with which electrons strike the target of an x-ray tube depends only on the voltage between anode (target) and cathode and not on the distance between them. [The force on a charge e (coulombs) by a field E (volts/m) is eE newtons.]

1–4 Graphically verify Moseley's law for the $K\beta_1$ lines of Cu, Mo, and W.

1–5 Plot the ratio of transmitted to incident intensity vs. thickness of lead sheet for Mo $K\alpha$ radiation and a thickness range of 0.00 and 0.02 mm.

* **1–6** Graphically verify Eq. (1–13) for a lead absorber and Mo $K\alpha$, Rh $K\alpha$, and Ag $K\alpha$ radiation. (The mass absorption coefficients of lead for these radiations are 122.8, 84.13, and 66.14 cm^2/gm, respectively.) From the curve, determine the mass absorption coefficient of lead for the shortest wavelength radiation from a tube operated at 30,000 volts.

1–7 Lead screens for the protection of personnel in x-ray diffraction laboratories are usually at least 1 mm thick. Calculate the "transmission factor" ($I_{trans.}/I_{incident}$) of such a screen for Cu $K\alpha$, Mo $K\alpha$, and the shortest wavelength radiation from a tube operated at 30,000 volts.

* **1–8** (a) Calculate the mass and linear absorption coefficients of air for Cr $K\alpha$ radiation. Assume that air contains 80 percent nitrogen and 20 percent oxygen by weight and has a density of 1.29×10^{-3} g/cm^3. (b) Plot the transmission factor of air for Cr $K\alpha$ radiation and a path length of 0 to 20 cm.

* **1–9** Calculate the K excitation voltage of copper.

1–10 Calculate the wavelength of the L_{III} absorption edge of molybdenum.

***1–11** Calculate the wavelength of the Cu $K\alpha_1$ line.

1–12 Plot the curve shown in Fig. 1–12 and save it for future reference.

***1–13** What voltage must be applied to a molybdenum-target tube in order that the emitted x-rays excite K fluorescent radiation from a piece of copper placed in the x-ray beam? What is the wavelength of the fluorescent radiation?

In Problems 14 and 15 take the intensity ratios of $K\alpha$ to $K\beta$ in unfiltered radiation from Table 1–1.

1–14 Suppose that a nickel filter is required to produce an intensity ratio of Cu $K\alpha$ to Cu $K\beta$ of 100/1 in the filtered beam. Calculate the thickness of the filter and the transmission factor for the Cu $K\alpha$ line.

***1–15** Filters for Co K radiation are usually made of iron oxide (Fe_2O_3) powder rather than iron foil. If a filter contains 5 mg Fe_2O_3/cm^2, what is the transmission factor for the Co $K\alpha$ line? What is the intensity ratio of Co $K\alpha$ to Co $K\beta$ in the filtered beam?

1–16 A copper-target x-ray tube is operated at 40,000 volts and 25 mA. The efficiency of an x-ray tube is so low that, for all practical purposes, one may assume that all the input energy goes into heating the target. If there were no dissipation of heat by water-cooling, conduction, radiation, etc., how long would it take a 100-gm copper target to melt? (Melting point of copper = 1083°C, mean specific heat = 6.65 cal/mole/°C, latent heat of fusion = 3220 cal/mole.)

***1–17** Assume that the sensitivity of x-ray film is proportional to the mass absorption coefficient of the silver bromide in the emulsion for the particular wavelength involved. What, then, is the ratio of film sensitivities to Cu $K\alpha$ and Mo $K\alpha$ radiation?

2

Geometry of Crystals

2-1 INTRODUCTION

Turning from the properties of x-rays, we must now consider the geometry and structure of crystals in order to discover what there is about crystals in general that enables them to diffract x-rays. We must also consider particular crystals of various kinds and how the very large number of crystals found in nature are classified into a relatively small number of groups. Finally, we will examine the ways in which the orientation of lines and planes in crystals can be represented in terms of symbols or in graphical form.

Crystallography is a very broad subject. In this book we are concerned only with its simpler aspects: how atoms are arranged in some common crystals and how this arrangement determines the way in which a particular crystal diffracts x-rays. Readers who need a deeper knowledge of crystallography should consult such books as those by Phillips [G.38], Buerger [G.35], and Kelly and Groves [G.33].

2-2 LATTICES

A crystal may be defined as *a solid composed of atoms arranged in a pattern periodic in three dimensions.* As such, crystals differ in a fundamental way from gases and liquids because the atomic arrangements in the latter do not possess the essential requirement of periodicity. Not all solids are crystalline, however; some are *amorphous*, like glass, and do not have any regular interior arrangement of atoms. There is, in fact, no essential difference between an amorphous solid and a liquid, and the former is often referred to as an "undercooled liquid."

In thinking about crystals, it is often convenient to ignore the actual atoms composing the crystal and their periodic arrangement in space, and to think instead of a set of imaginary points which has a fixed relation in space to the atoms of the crystal and which may be regarded as a sort of framework or skeleton on which the actual crystal is built.

This set of points can be formed as follows. Imagine space to be divided by three sets of planes, the planes in each set being parallel and equally spaced. This division of space will produce a set of cells each identical in size, shape, and orientation to its neighbors. Each cell is a parallelepiped, since its opposite faces are parallel and each face is a parallelogram. The space-dividing planes will intersect each other in a set of lines (Fig. 2-1), and these lines in turn intersect in the

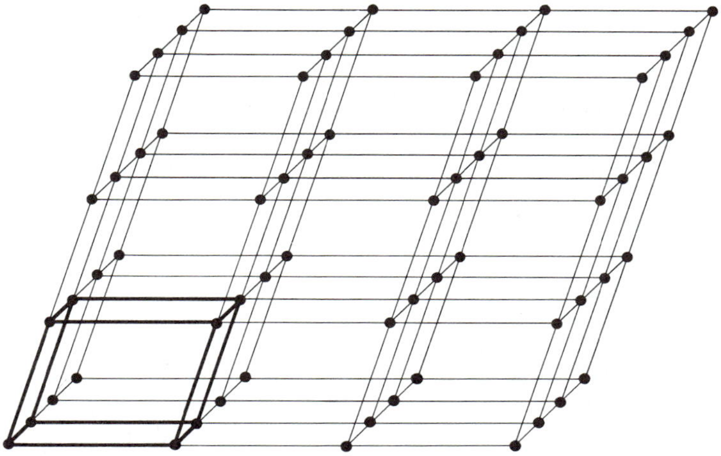

Fig. 2–1 A point lattice.

set of points referred to above. A set of points so formed has an important property: it constitutes a *point lattice*, which is defined as *an array of points in space so arranged that each point has identical surroundings*. By "identical surroundings" we mean that the lattice of points, when viewed in a particular direction from one lattice point, would have exactly the same appearance when viewed in the same direction from any other lattice point.

Since all the cells of the lattice shown in Fig. 2–1 are identical, we may choose any one, for example the heavily outlined one, as a *unit cell*. The size and shape of the unit cell can in turn be described by the three vectors* **a**, **b**, and **c** drawn from one corner of the cell taken as origin (Fig. 2–2). These vectors define the cell and are called the *crystallographic axes* of the cell. They may also be described in terms of their lengths (a, b, c) and the angles between them (α, β, γ). These lengths and angles are the *lattice constants* or *lattice parameters* of the unit cell.

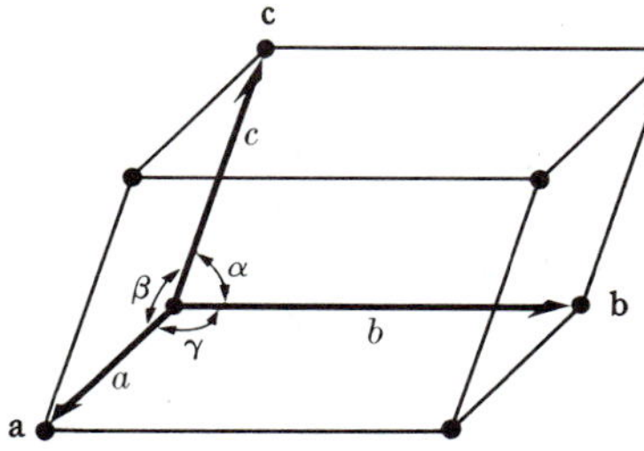

Fig. 2–2 A unit cell.

* Vectors are here represented by boldface symbols. The same symbol in italics stands for the absolute value of the vector.

Note that the vectors **a**, **b**, **c** define, not only the unit cell, but also the whole point lattice through the translations provided by these vectors. In other words, the whole set of points in the lattice can be produced by repeated action of the vectors **a**, **b**, **c** on one lattice point located at the origin, or, stated alternatively, the vector coordinates of any point in the lattice are $P\mathbf{a}$, $Q\mathbf{b}$, and $R\mathbf{c}$, where P, Q, and R are whole numbers. It follows that the arrangement of points in a point lattice is absolutely periodic in three dimensions, points being repeated at regular intervals along any line one chooses to draw through the lattice.

2-3 CRYSTAL SYSTEMS

In dividing space by three sets of planes, we can of course produce unit cells of various shapes, depending on how we arrange the planes. For example, if the planes in the three sets are all equally spaced and mutually perpendicular, the unit cell is cubic. In this case the vectors **a**, **b**, **c** are all equal and at right angles to one another, or $a = b = c$ and $\alpha = \beta = \gamma = 90°$. By thus giving special values to the axial lengths and angles, we can produce unit cells of various shapes and therefore various kinds of point lattices, since the points of the lattice are located at the cell corners. It turns out that only seven different kinds of cells are necessary to include all the possible point lattices. These correspond to the seven *crystal systems* into which all crystals can be classified. These systems are listed in Table 2-1. (Some writers consider the rhombohedral system as a subdivision of the hexagonal, thus reducing the number of crystal systems to six.)

Seven different point lattices can be obtained simply by putting points at the corners of the unit cells of the seven crystal systems. However, there are other arrangements of points which fulfill the requirements of a point lattice, namely, that each point have identical surroundings. The French crystallographer Bravais worked on this problem and in 1848 demonstrated that there are fourteen possible point lattices and no more; this important result is commemorated by our use of the terms *Bravais lattice* and *point lattice* as synonymous. For example, if a point is placed at the center of each cell of a cubic point lattice, the new array of points also forms a point lattice. Similarly, another point lattice can be based on a cubic unit cell having lattice points at each corner and in the center of each face.

The fourteen Bravais lattices are described in Table 2-1 and illustrated in Fig. 2-3, where the symbols P, F, I, etc., have the following meanings. We must first distinguish between *simple*, or *primitive*, cells (symbol P or R) and *nonprimitive* cells (any other symbol): primitive cells have only one lattice point per cell while nonprimitive have more than one. A lattice point in the interior of a cell "belongs" to that cell, while one in a cell face is shared by two cells and one at a corner is shared by eight. The number of lattice points per cell is therefore given by

$$N = N_i + \frac{N_f}{2} + \frac{N_c}{8}, \tag{2-1}$$

where N_i = number of interior points, N_f = number of points on faces, and N_c = number of points on corners. Any cell containing lattice points on the

Table 2–1
Crystal Systems and Bravais Lattices
(The symbol ≠ means that equality is not required by symmetry. Accidental equality may occur, as shown by an example in Sec. 2–4.)

System	Axial lengths and angles	Bravais lattice	Lattice symbol
Cubic	Three equal axes at right angles $a = b = c,\ \alpha = \beta = \gamma = 90^\circ$	Simple Body-centered Face-centered	P I F
Tetragonal	Three axes at right angles, two equal $a = b \neq c,\ \alpha = \beta = \gamma = 90^\circ$	Simple Body-centered	P I
Orthorhombic	Three unequal axes at right angles $a \neq b \neq c,\ \alpha = \beta = \gamma = 90^\circ$	Simple Body-centered Base-centered Face-centered	P I C F
Rhombohedral*	Three equal axes, equally inclined $a = b = c,\ \alpha = \beta = \gamma \neq 90^\circ$	Simple	R
Hexagonal	Two equal coplanar axes at 120°, third axis at right angles $a = b \neq c,\ \alpha = \beta = 90^\circ,\ \gamma = 120^\circ$	Simple	P
Monoclinic	Three unequal axes, one pair not at right angles $a \neq b \neq c,\ \alpha = \gamma = 90^\circ \neq \beta$	Simple Base-centered	P C
Triclinic	Three unequal axes, unequally inclined and none at right angles $a \neq b \neq c,\ \alpha \neq \beta \neq \gamma \neq 90^\circ$	Simple	P

* Also called trigonal.

corners only is therefore primitive, while one containing additional points in the interior or on faces is nonprimitive. The symbols F and I refer to face-centered and body-centered cells, respectively, while A, B, and C refer to base-centered cells, centered on one pair of opposite faces A, B, or C. (The A face is the face defined by the b and c axes, etc.) The symbol R is used especially for the rhombohedral system. In Fig. 2–3, axes of equal length in a particular system are given the same symbol to indicate their equality, e.g., the cubic axes are all marked a, the two equal tetragonal axes are marked a and the third one c, etc.

At first glance, the list of Bravais lattices in Table 2–1 appears incomplete. Why not, for example, a base-centered tetragonal lattice? The full lines in Fig. 2–4 delineate such a cell, centered on the C face, but we see that the same array of lattice points can be referred to the simple tetragonal cell shown by dashed lines, so that the base-centered arrangement of points is not a new lattice. However, the base-centered cell is a perfectly good unit cell and, if we wish, we may choose to use it rather than the simple cell. Choice of one or the other has certain consequences, which are described later (Problem 4–3).

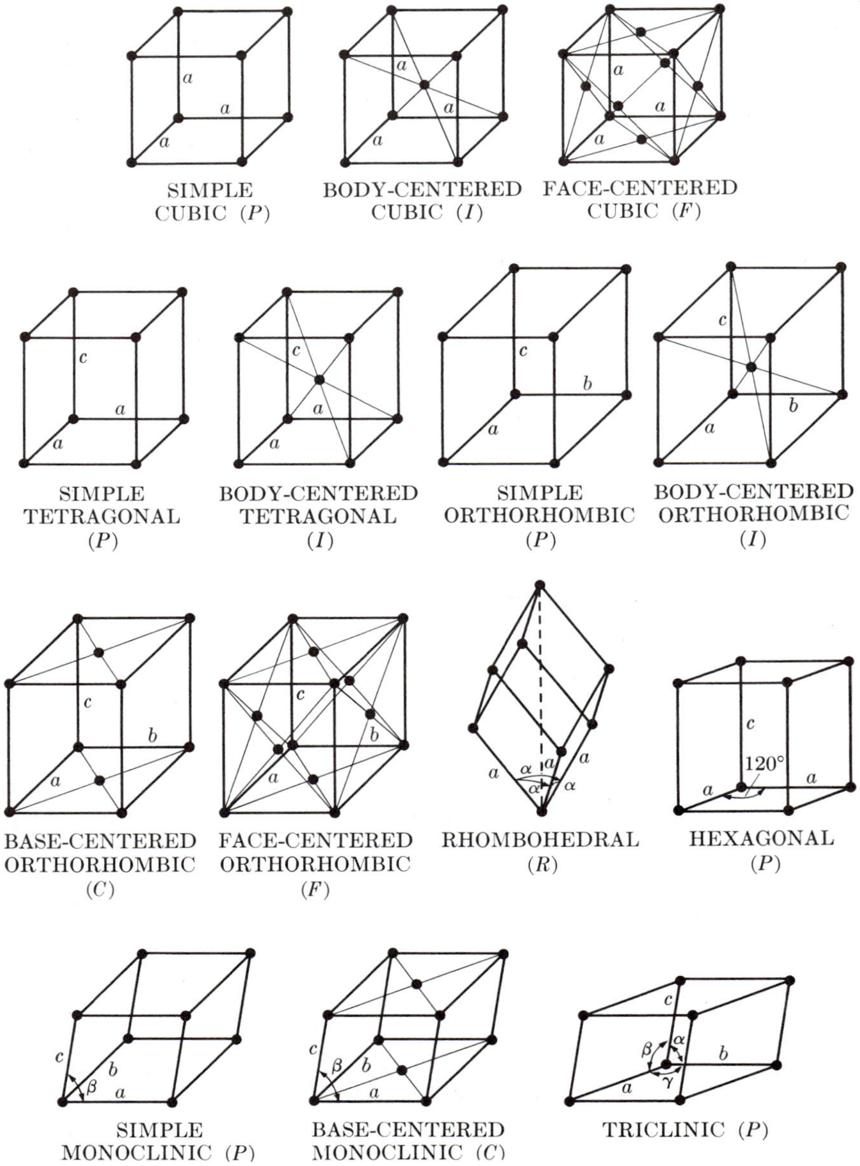

Fig. 2–3 The fourteen Bravais lattices.

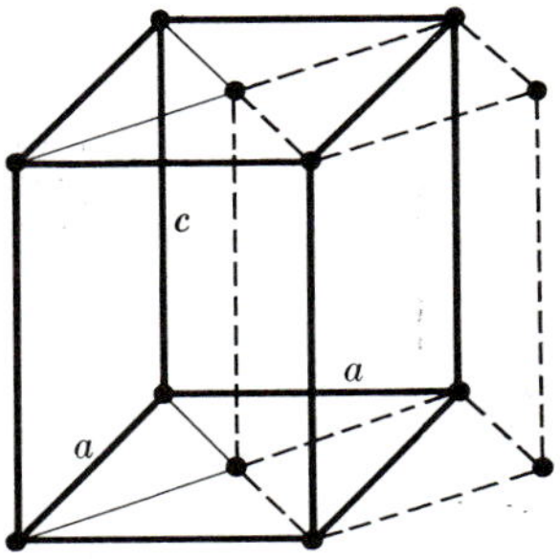

Fig. 2-4 Relation of tetragonal C lattice (full lines) to tetragonal P lattice (dashed lines).

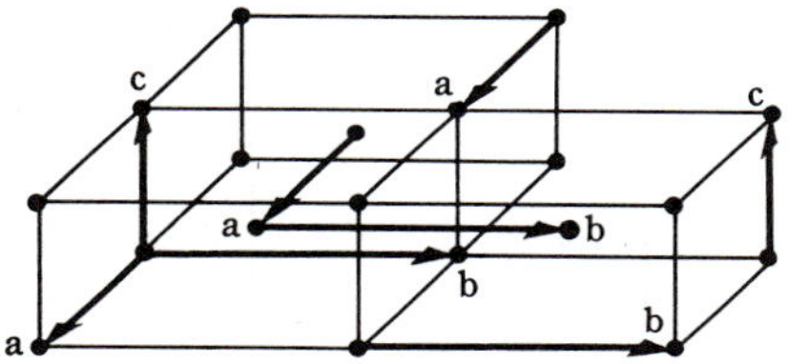

Fig. 2-5 Extension of lattice points through space by the unit cell vectors **a, b, c**.

The lattice points in a nonprimitive unit cell can be extended through space by repeated applications of the unit-cell vectors **a**, **b**, **c** just like those of a primitive cell. We may regard the lattice points associated with a unit cell as being translated one by one or as a group. In either case, equivalent lattice points in adjacent unit cells are separated by one of the vectors **a**, **b**, **c**, wherever these points happen to be located in the cell (Fig. 2-5).

2-4 SYMMETRY

Both Bravais lattices and the real crystals which are built up on them exhibit various kinds of symmetry. A body or structure is said to be symmetrical when its component parts are arranged in such balance, so to speak, that certain operations can be performed on the body which will bring it into coincidence with itself. These are termed *symmetry operations*. For example, if a body is symmetrical with respect to a plane passing through it, then reflection of either half of the body in the plane as in a mirror will produce a body coinciding with the other half. Thus a cube has several planes of symmetry, one of which is shown in Fig. 2-6(a).

There are in all four macroscopic* symmetry operations or elements: *reflection,*

---

* So called to distinguish them from certain microscopic symmetry operations with which we are not concerned here. The macroscopic elements can be deduced from the angles between the faces of a well-developed crystal, without any knowledge of the atom arrangement inside the crystal. The microscopic symmetry elements, on the other hand, depend entirely on atom arrangement, and their presence cannot be inferred from the external development of the crystal.

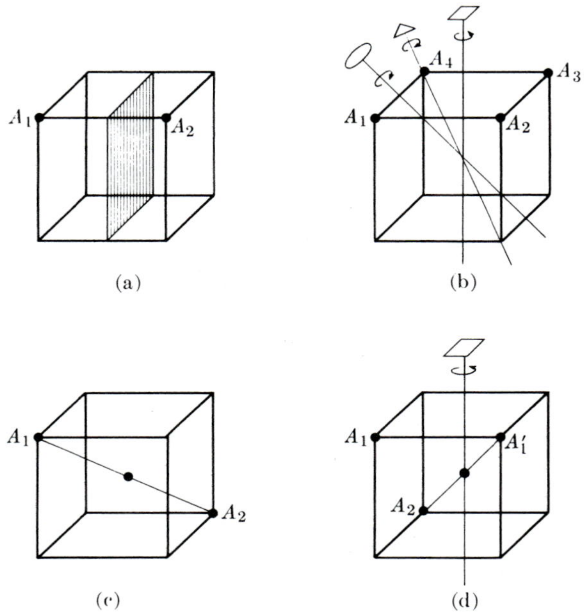

Fig. 2–6 Some symmetry elements of a cube. (a) Reflection plane. A_1 becomes A_2. (b) Rotation axes. 4-fold axis: A_1 becomes A_2; 3-fold axis: A_1 becomes A_3; 2-fold axis: A_1 becomes A_4. (c) Inversion center. A_1 becomes A_2. (d) Rotation-inversion axis. 4-fold axis: A_1 becomes A_1'; inversion center: A_1' becomes A_2.

rotation, *inversion*, and *rotation-inversion*. A body has n-fold rotational symmetry about an axis if a rotation of $360°/n$ brings it into self-coincidence. Thus a cube has a 4-fold rotation axis normal to each face, a 3-fold axis along each body diagonal, and 2-fold axes joining the centers of opposite edges. Some of these are shown in Fig. 2–6(b) where the small plane figures (square, triangle, and ellipse) designate the various kinds of axes. In general, rotation axes may be 1-, 2-, 3-, 4-, or 6-fold. A 1-fold axis indicates no symmetry at all, while a 5-fold axis or one of higher degree than 6 is impossible, in the sense that unit cells having such symmetry cannot be made to fill up space without leaving gaps.

A body has an inversion center if corresponding points of the body are located at equal distances from the center on a line drawn through the center. A body having an inversion center will come into coincidence with itself if every point in the body is inverted, or "reflected," in the inversion center. A cube has such a center at the intersection of its body diagonals [Fig. 2–6(c)]. Finally, a body may have a rotation-inversion axis, either 1-, 2-, 3-, 4-, or 6-fold. If it has an n-fold rotation-inversion axis, it can be brought into coincidence with itself by a rotation of $360°/n$ about the axis followed by inversion in a center lying on the axis. Figure 2–6(d) illustrates the operation of a 4-fold rotation-inversion axis on a cube.

Now, the possession of a certain minimum set of symmetry elements is a

Table 2–2
Symmetry Elements

System	Minimum symmetry elements
Cubic	Four 3-fold rotation axes
Tetragonal	One 4-fold rotation (or rotation-inversion) axis
Orthorhombic	Three perpendicular 2-fold rotation (or rotation-inversion) axes
Rhombohedral	One 3-fold rotation (or rotation-inversion) axis
Hexagonal	One 6-fold rotation (or rotation-inversion) axis
Monoclinic	One 2-fold rotation (or rotation-inversion) axis
Triclinic	None

fundamental property of each crystal system, and one system is distinguished from another just as much by its symmetry elements as by the values of its axial lengths and angles. In fact, *these are interdependent*. For example, the existence of 4-fold rotation axes normal to the faces of a cubic cell *requires* that the cell edges be equal in length and at 90° to one another. On the other hand, a tetragonal cell has only one 4-fold axis, and this symmetry requires that only two cell edges be equal, namely, the two that are at right angles to the rotation axis.

The minimum number of symmetry elements possessed by each crystal system is listed in Table 2–2. Some crystals may possess more than the minimum symmetry elements required by the system to which they belong, but none may have less. The existence of certain symmetry elements often implies the existence of others. For example, a crystal with three 4-fold rotation axes necessarily has, in addition, four 3-fold axes and falls in the cubic system.

Symmetry operations apply not only to the unit cells shown in Fig. 2–3, considered merely as geometric shapes, but also to the point lattices associated with them. The latter condition rules out the possibility that the cubic system, for example, could include a base-centered point lattice, since such an array of points would not have the minimum set of symmetry elements required by the cubic system, namely four 3-fold rotation axes. Such a lattice would be classified in the tetragonal system, which has no 3-fold axes and in which accidental equality of the a and c axes is allowed.

Crystals in the rhombohedral (trigonal) system can be referred to either a rhombohedral or a hexagonal lattice. Appendix 4 gives the relation between these two lattices and the transformation equations which allow the Miller indices of a plane (see Sec. 2–6) to be expressed in terms of either set of axes.

2–5 PRIMITIVE AND NONPRIMITIVE CELLS

In any point lattice a unit cell may be chosen in an infinite number of ways and may contain one or more lattice points per cell. It is important to note that unit cells do not "exist" as such in a lattice: they are a mental construct and can accordingly be chosen at our convenience. The conventional cells shown in Fig. 2–3 are chosen simply for convenience and to conform to the symmetry elements of the lattice.

Any of the fourteen Bravais lattices may be referred to a primitive unit cell. For example, the face-centered cubic lattice shown in Fig. 2–7 may be referred to the primitive cell indicated by dashed lines. The latter cell is rhombohedral, its axial angle α is 60°, and each of its axes is $1/\sqrt{2}$ times the length of the axes of the cubic cell. Each cubic cell has four lattice points associated with it, each rhombohedral cell has one, and the former has, correspondingly, four times the volume of the latter. Nevertheless, it is usually more convenient to use the cubic cell rather than the rhombohedral one because the former immediately suggests the cubic symmetry which the lattice actually possesses. Similarly, the other centered nonprimitive cells listed in Table 2–1 are preferred to the primitive cells possible in their respective lattices.

Why then do the centered lattices appear in the list of the fourteen Bravais lattices? If the two cells in Fig. 2–7 describe the same set of lattice points, as they do, why not eliminate the cubic cell and let the rhombohedral cell serve instead? The answer is that this cell is a *particular* rhombohedral cell with an axial angle α of 60°. In the general rhombohedral lattice no restriction is placed on the angle α; the result is a lattice of points with a single 3-fold symmetry axis. When α becomes equal to 60°, the lattice has four 3-fold axes, and this symmetry places it in the cubic system. The general rhombohedral cell is still needed.

If nonprimitive lattice cells are used, the vector from the origin to any point in the lattice will now have components which are nonintegral multiples of the unit-cell vectors **a**, **b**, **c**. The position of any lattice point in a cell may be given in terms of its *coordinates*; if the vector from the origin of the unit cell to the given point has components $x\mathbf{a}$, $y\mathbf{b}$, $z\mathbf{c}$, where x, y, and z are fractions, then the coordinates of the point are $x\ y\ z$. Thus, point A in Fig. 2–7, taken as the origin, has coordinates 0 0 0 while points B, C, and D, when referred to cubic axes, have coordinates $0\ \frac{1}{2}\ \frac{1}{2}$, $\frac{1}{2}\ 0\ \frac{1}{2}$, and $\frac{1}{2}\ \frac{1}{2}\ 0$, respectively. Point E has coordinates $\frac{1}{2}\ \frac{1}{2}\ 1$ and is equivalent to point D, being separated from it by the vector **c**. The coordinates

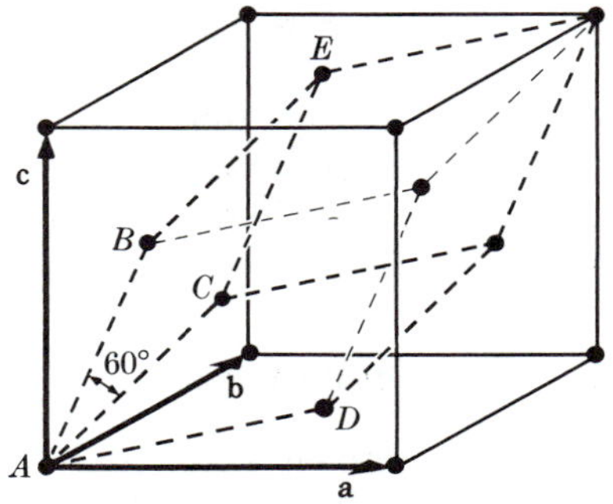

Fig. 2–7 Face-centered cubic point lattice referred to cubic and rhombohedral cells.

of equivalent points in different unit cells can always be made identical by the addition or subtraction of a set of integral coordinates; in this case, subtraction of 0 0 1 from $\frac{1}{2}\ \frac{1}{2}$ 1 (the coordinates of E) gives $\frac{1}{2}\ \frac{1}{2}$ 0 (the coordinates of D).

Note that the coordinates of a body-centered point, for example, are always $\frac{1}{2}\ \frac{1}{2}\ \frac{1}{2}$ no matter whether the unit cell is cubic, tetragonal, or orthorhombic, and whatever its size. The coordinates of a point position, such as $\frac{1}{2}\ \frac{1}{2}\ \frac{1}{2}$, may also be regarded as an operator which, when "applied" to a point at the origin, will move or translate it to the position $\frac{1}{2}\ \frac{1}{2}\ \frac{1}{2}$, the final position being obtained by simple addition of the operator $\frac{1}{2}\ \frac{1}{2}\ \frac{1}{2}$ and the original position 0 0 0. In this sense, the positions 0 0 0, $\frac{1}{2}\ \frac{1}{2}\ \frac{1}{2}$ are called the "body-centering translations," since they will produce the two point positions characteristic of a body-centered cell when applied to a point at the origin. Similarly, the four point positions characteristic of a face-centered cell, namely, 0 0 0, 0 $\frac{1}{2}\ \frac{1}{2}$, $\frac{1}{2}$ 0 $\frac{1}{2}$, and $\frac{1}{2}\ \frac{1}{2}$ 0, are called the face-centering translations. The base-centering translations depend on which pair of opposite faces are centered; if centered on the C face, for example, they are 0 0 0, $\frac{1}{2}\ \frac{1}{2}$ 0. These centering translations, summarized below, should be memorized:

$$\begin{aligned} \text{body-centering} &= 0\ 0\ 0,\ \tfrac{1}{2}\ \tfrac{1}{2}\ \tfrac{1}{2} \\ \text{face-centering} &= 0\ 0\ 0,\ 0\ \tfrac{1}{2}\ \tfrac{1}{2},\ \tfrac{1}{2}\ 0\ \tfrac{1}{2},\ \tfrac{1}{2}\ \tfrac{1}{2}\ 0 \\ \text{base-centering} &= 0\ 0\ 0,\ \tfrac{1}{2}\ \tfrac{1}{2}\ 0. \end{aligned}$$

The inclusion of 0 0 0 may appear trivial, in that it does not move the point at the origin on which it acts, but its inclusion does remind us that cells so centered contain 2, 4, and 2 lattice points, respectively.

2-6 LATTICE DIRECTIONS AND PLANES

The direction of any line in a lattice may be described by first drawing a line through the origin parallel to the given line and then giving the coordinates of any point on the line through the origin. Let the line pass through the origin of the unit cell and any point having coordinates $u\ v\ w$, where these numbers are not necessarily integral. (This line will also pass through the points $2u\ 2v\ 2w$, $3u\ 3v\ 3w$, etc.) Then $[uvw]$, written in square brackets, are the *indices* of the direction of the line. They are also the indices of any line parallel to the given line, since the lattice is infinite and the origin may be taken at any point. Whatever the values of u, v, w, they are always converted to a set of smallest integers by multiplication or division throughout: thus, $[\frac{1}{2}\ \frac{1}{2}\ 1]$, [112], and [224] all represent the same direction, but [112] is the preferred form. Negative indices are written with a bar over the number, e.g., $[\bar{u}vw]$. Direction indices are illustrated in Fig. 2-8. Note how one can mentally shift the origin, to avoid using the adjacent unit cell, in finding a direction like $[1\bar{2}0]$.

Directions related by symmetry are called *directions of a form*, and a set of these are represented by the indices of one of them enclosed in angular brackets; for example, the four body diagonals of a cube, [111], $[1\bar{1}1]$, $[\bar{1}\bar{1}1]$, and $[\bar{1}11]$, may all be represented by the symbol $\langle 111 \rangle$.

The orientation of planes in a lattice may also be represented symbolically, according to a system popularized by the English crystallographer Miller. In the

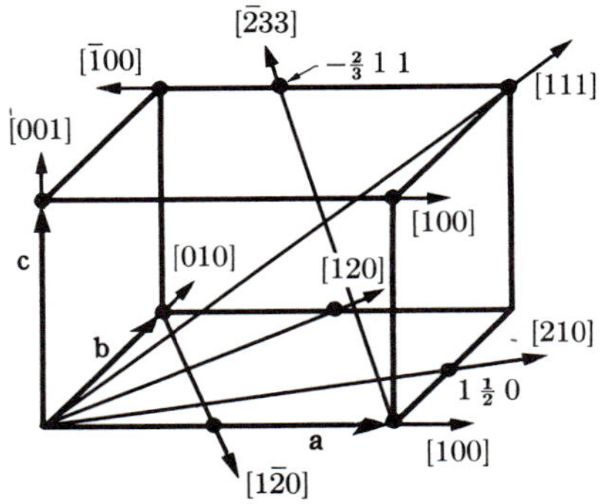

Fig. 2–8 Indices of directions.

general case, the given plane will be tilted with respect to the crystallographic axes, and, since these axes form a convenient frame of reference, we might describe the orientation of the plane by giving the actual distances, measured from the origin, at which it intercepts the three axes. Better still, by expressing these distances as fractions of the axial lengths, we can obtain numbers which are independent of the particular axial lengths involved in the given lattice. But a difficulty then arises when the given plane is parallel to a certain crystallographic axis, because such a plane does not intercept that axis, i.e., its "intercept" can only be described as "infinity." To avoid the introduction of infinity into the description of plane orientation, we can use the reciprocal of the fractional intercept, this reciprocal being zero when the plane and axis are parallel. We thus arrive at a workable symbolism for the orientation of a plane in a lattice, the *Miller indices*, which are defined as *the reciprocals of the fractional intercepts which the plane makes with the crystallographic axes.* For example, if the Miller indices of a plane are (hkl), written in parentheses, then the plane makes fractional intercepts of $1/h$, $1/k$, $1/l$ with the axes, and, if the axial lengths are a, b, c, the plane makes actual intercepts of a/h, b/k, c/l, as shown in Fig. 2–9(a). Parallel to any plane in any lattice, there is a whole set of parallel equidistant planes, one of which passes through the origin; the Miller indices (hkl) usually refer to that plane in the set which is nearest the origin, although they may be taken as referring to any other plane in the set or to the whole set taken together.

We may determine the Miller indices of the plane shown in Fig. 2–9(b) as follows:

Axial lengths	4 Å	8 Å	3 Å
Intercept lengths	1 Å	4 Å	3 Å
Fractional intercepts	$\frac{1}{4}$	$\frac{1}{2}$	1
Miller indices	4	2	1

As stated earlier, if a plane is parallel to a given axis, its fractional intercept on that axis is taken as infinity and the corresponding Miller index is zero. If a plane cuts

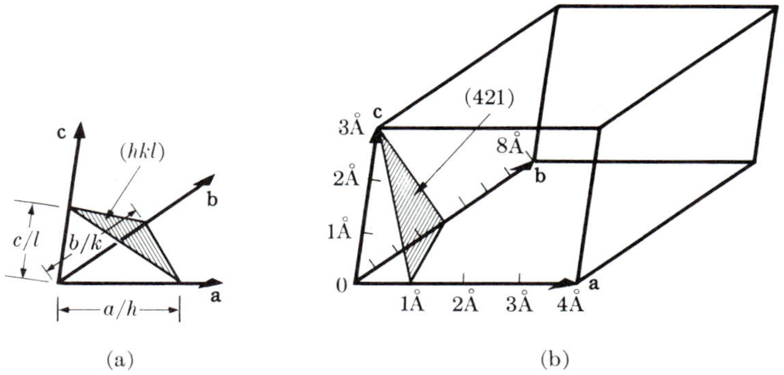

Fig. 2-9 Plane designation by Miller indices.

a negative axis, the corresponding index is negative and is written with a bar over it. Planes whose indices are the negatives of one another are parallel and lie on opposite sides of the origin, e.g., $(\bar{2}10)$ and $(2\bar{1}0)$. The planes $(nh\ nk\ nl)$ are parallel to the planes (hkl) and have $1/n$th the spacing. The same plane may belong to two different sets, the Miller indices of one set being multiples of those of the other; thus the same plane belongs to the (210) set and the (420) set, and, in fact, the planes of the (210) set form every second plane in the (420) set. In the cubic system, it is convenient to remember that a direction $[hkl]$ is always perpendicular to a plane (hkl) of the same indices, but this is not generally true in other systems. Further familiarity with Miller indices can be gained from a study of Fig. 2-10.

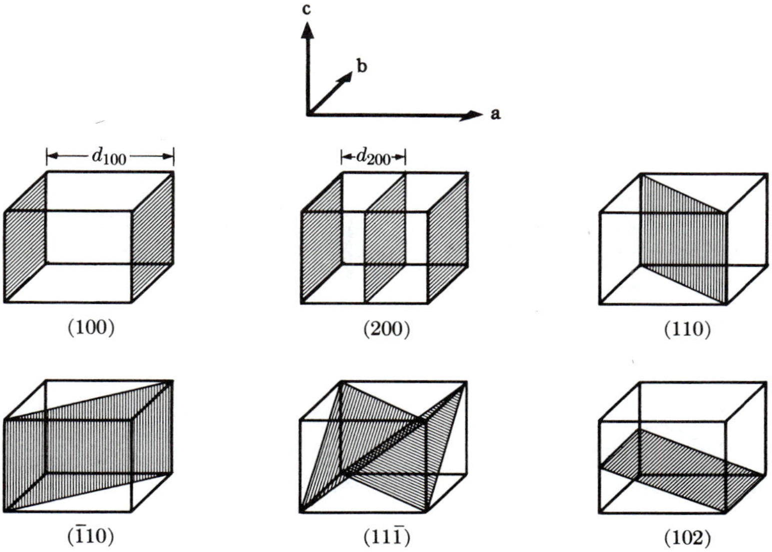

Fig. 2-10 Miller indices of lattice planes. The distance d is the plane spacing.

A slightly different system of plane indexing is used in the hexagonal system. The unit cell of a hexagonal lattice is defined by two equal and coplanar vectors $\mathbf{a}_1$ and $\mathbf{a}_2$, at 120° to one another, and a third axis $\mathbf{c}$ at right angles [Fig. 2–11(a)]. The complete lattice is built up, as usual, by repeated translations of the points at the unit cell corners by the vectors $\mathbf{a}_1$, $\mathbf{a}_2$, $\mathbf{c}$. Some of the points so generated are shown in the figure, at the ends of dashed lines, in order to exhibit the hexagonal symmetry of the lattice, which has a 6-fold rotation axis parallel to $\mathbf{c}$. The third axis $\mathbf{a}_3$, lying in the basal plane of the hexagonal prism, is so symmetrically related to $\mathbf{a}_1$ and $\mathbf{a}_2$ that it is often used in conjunction with the other two. Thus the indices of a plane in the hexagonal system, called Miller–Bravais indices, refer to *four* axes and are written $(hkil)$. The index i is the reciprocal of the fractional intercept on the $\mathbf{a}_3$ axis. Since the intercepts of a plane on $\mathbf{a}_1$ and $\mathbf{a}_2$ determine its intercept on $\mathbf{a}_3$, the value of i depends on the values of h and k. The relation is

$$h + k = -i. \tag{2-2}$$

Since i is determined by h and k, it is sometimes replaced by a dot and the plane symbol written $(hk \cdot l)$. Sometimes even the dot is omitted. However, this usage defeats the purpose for which Miller–Bravais indices were devised, namely, to give similar indices to similar planes. For example, the side planes of the hexagonal prism in Fig. 2–11(b) are all similar and symmetrically located, and their relationship is clearly shown in their full Miller–Bravais symbols: $(10\bar{1}0)$, $(01\bar{1}0)$, $(\bar{1}100)$, $(\bar{1}010)$, $(0\bar{1}10)$, $(1\bar{1}00)$. On the other hand, the abbreviated symbols of these planes, $(10 \cdot 0)$, $(01 \cdot 0)$, $(\bar{1}1 \cdot 0)$, $(\bar{1}0 \cdot 0)$, $(0\bar{1} \cdot 0)$, $(1\bar{1} \cdot 0)$ do not immediately suggest this relationship.

Directions in a hexagonal lattice are best expressed in terms of the *three* basic vectors $\mathbf{a}_1$, $\mathbf{a}_2$, and $\mathbf{c}$. Figure 2–11(b) shows several examples of both plane and direction indices. Another system, involving four indices, is sometimes used to designate directions. The required direction is broken up into four component vectors, parallel to $\mathbf{a}_1$, $\mathbf{a}_2$, $\mathbf{a}_3$, and $\mathbf{c}$ and so chosen that the third index is the

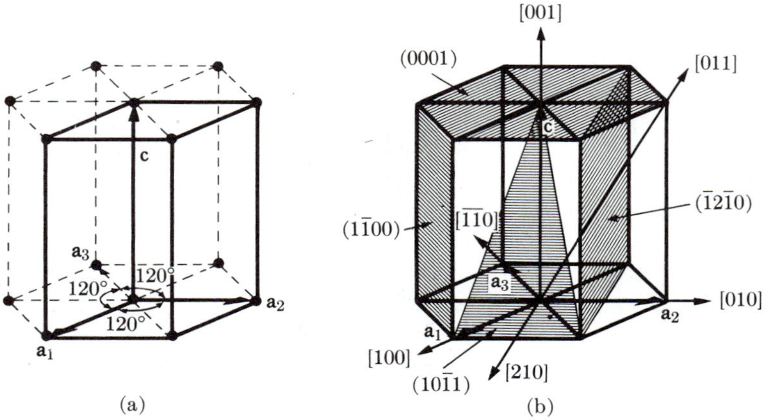

Fig. 2–11 (a) The hexagonal unit cell (heavy lines) and (b) indices of planes and directions.

negative of the sum of the first two. Then, if $[UVW]$ are the indices of a direction referred to three axes and $[uvtw]$ the four-axis indices, the two are related as follows:

$$U = u - t \qquad u = (2U - V)/3$$
$$V = v - t \qquad v = (2V - U)/3$$
$$W = w \qquad t = -(u + v) = -(U + V)/3$$
$$w = W.$$

Thus, [100] becomes $[2\bar{1}\bar{1}0]$, [210] becomes $[10\bar{1}0]$, etc.

Note that the indices of a plane or direction are meaningless unless the orientation of the unit-cell axes is given. This means that the indices of a particular lattice plane depend on the unit cell chosen. For example, consider the right-hand vertical plane of the cell shown by full lines in Fig. 2–4; the indices of this plane are of the form {100} for the base-centered cell and {110} for the simple cell.

In any crystal system there are sets of equivalent lattice planes related by symmetry. These are called *planes of a form*, and the indices of any one plane, enclosed in braces $\{hkl\}$, stand for the whole set. In general, planes of a form have the same spacing but different Miller indices. For example, the faces of a cube, (100), (010), $(\bar{1}00)$, $(0\bar{1}0)$, (001), and $(00\bar{1})$, are planes of the form {100}, since all of them may be generated from any one by operation of the 4-fold rotation axes perpendicular to the cube faces. In the tetragonal system, however, only the planes (100), (010), $(\bar{1}00)$, and $(0\bar{1}0)$ belong to the form {100}; the other two planes, (001) and $(00\bar{1})$, belong to the different form {001}; the first four planes mentioned are related by a 4-fold axis and the last two by a 2-fold axis.*

Planes of a zone are planes which are all parallel to one line, called the *zone axis*, and the zone, i.e., the set of planes, is specified by giving the indices of the zone axis. Such planes may have quite different indices and spacings, the only requirement being their parallelism to a line. Figure 2–12 shows some examples. If the axis of a zone has indices $[uvw]$, then any plane belongs to that zone whose indices (hkl) satisfy the relation

$$hu + kv + lw = 0. \tag{2–3}$$

(A proof of this relation is given in Sec. 3 of Appendix 1.) Any two nonparallel planes are planes of a zone since they are both parallel to their line of intersection. If their indices are $(h_1k_1l_1)$ and $(h_2k_2l_2)$, then the indices of their zone axis $[uvw]$ are given by the relations

$$\begin{aligned} u &= k_1l_2 - k_2l_1, \\ v &= l_1h_2 - l_2h_1, \\ w &= h_1k_2 - h_2k_1. \end{aligned} \tag{2–4}$$

* Certain important crystal planes are often referred to by name without any mention of their Miller indices. Thus, planes of the form {111} in the cubic system are often called octahedral planes, since these are the bounding planes of an octahedron. In the hexagonal system, the (0001) plane is called the basal plane, planes of the form $\{10\bar{1}0\}$ are called prismatic planes, and planes of the form $\{10\bar{1}1\}$ are called pyramidal planes.

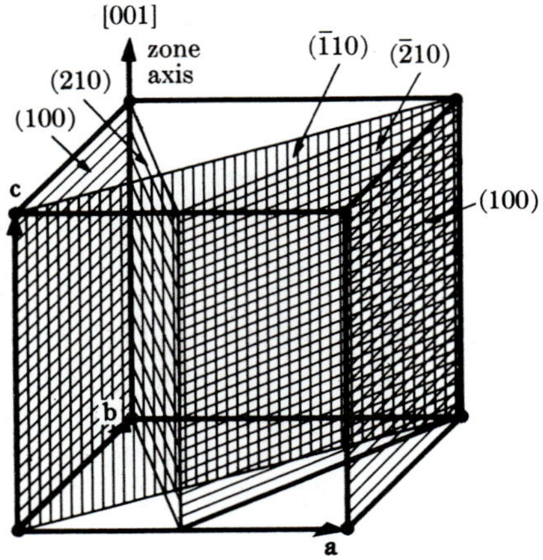

Fig. 2–12 All shaded planes in the cubic lattice shown are planes of the zone [001].

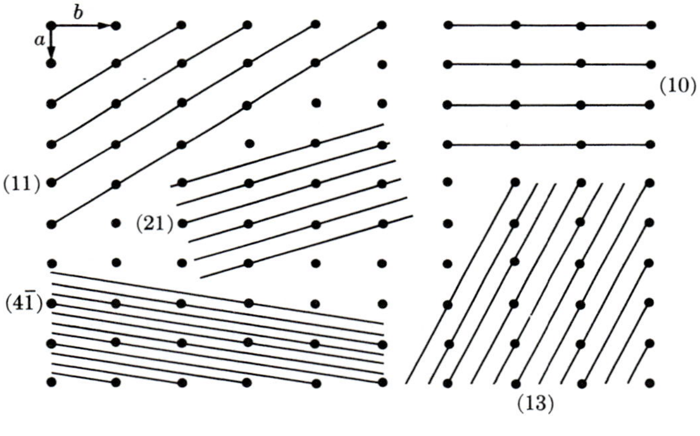

Fig. 2–13 Two-dimensional lattice, showing that lines of lowest indices have the greatest spacing and the greatest density of lattice points.

The various sets of planes in a lattice have various values of interplanar spacing. The planes of large spacing have low indices and pass through a high density of lattice points, whereas the reverse is true of planes of small spacing. Figure 2–13 illustrates this for a two-dimensional lattice, and it is equally true in three dimensions. The interplanar spacing d_{hkl}, measured at right angles to the planes, is a function both of the plane indices (hkl) and the lattice constants $(a, b, c, \alpha, \beta, \gamma)$. The exact relation depends on the crystal system involved and for the cubic system takes on the relatively simple form

$$\text{(Cubic)} \quad d_{hkl} = \frac{a}{\sqrt{h^2 + k^2 + l^2}}. \tag{2-5}$$

In the tetragonal system the spacing equation naturally involves both a and c since these are not generally equal:

$$\text{(Tetragonal)} \quad d_{hkl} = \frac{a}{\sqrt{h^2 + k^2 + l^2 (a^2/c^2)}}. \tag{2–6}$$

Interplanar spacing equations for all systems are given in Appendix 3.

2–7 CRYSTAL STRUCTURE

So far we have discussed topics from the field of *mathematical (geometrical) crystallography* and have said practically nothing about actual crystals and the atoms of which they are composed. In fact, all of the above was well known long before the discovery of x-ray diffraction, i.e., long before there was any certain knowledge of the interior arrangements of atoms in crystals.

It is now time to describe the structure of some actual crystals and to relate this structure to the point lattices, crystal systems, and symmetry elements discussed above. The cardinal principle of crystal structure is that *the atoms of a crystal are set in space either on the points of a Bravais lattice or in some fixed relation to those points.* It follows from this that the atoms of a crystal will be arranged periodically in three dimensions and that this arrangement of atoms will exhibit many of the properties of a Bravais lattice, in particular many of its symmetry elements.

The simplest crystals one can imagine are those formed by placing atoms of the same kind *on* the points of a Bravais lattice. Not all such crystals exist but, fortunately for metallurgists, many metals crystallize in this simple fashion, and Fig. 2–14 shows two common structures based on the body-centered cubic (BCC) and face-centered cubic (FCC) lattices. The former has two atoms per unit cell and the latter four, as we can find by rewriting Eq. (2–1) in terms of the number of atoms, rather than lattice points, per cell and applying it to the unit cells shown.

The next degree of complexity is encountered when two or more atoms of the same kind are "associated with" each point of a Bravais lattice, as exemplified by the hexagonal close-packed (HCP) structure common to many metals. This structure is simple hexagonal and is illustrated in Fig. 2–15. There are two atoms per unit cell, as shown in (a), one at 0 0 0 and the other at $\frac{2}{3}\,\frac{1}{3}\,\frac{1}{2}$ (or at $\frac{1}{3}\,\frac{2}{3}\,\frac{1}{2}$, which is an equivalent position). Figure 2–15(b) shows the same structure with the origin of the unit cell shifted so that the point 1 0 0 in the new cell is midway between

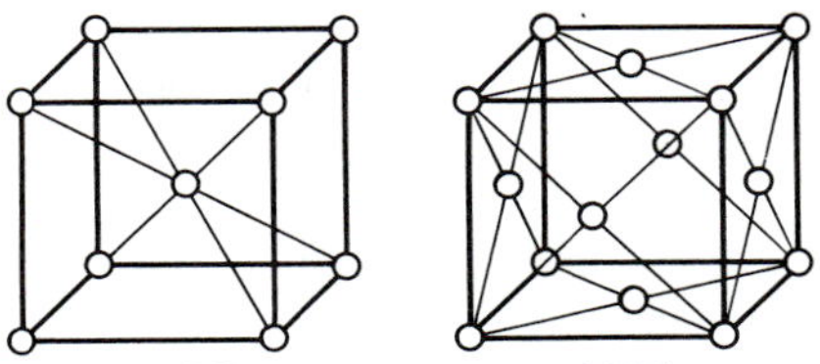

Fig. 2–14 Structures of some common metals. Body-centered cubic: α-Fe, Cr, Mo, V, etc.; face-centered cubic: γ-Fe, Cu, Pb, Ni, etc.

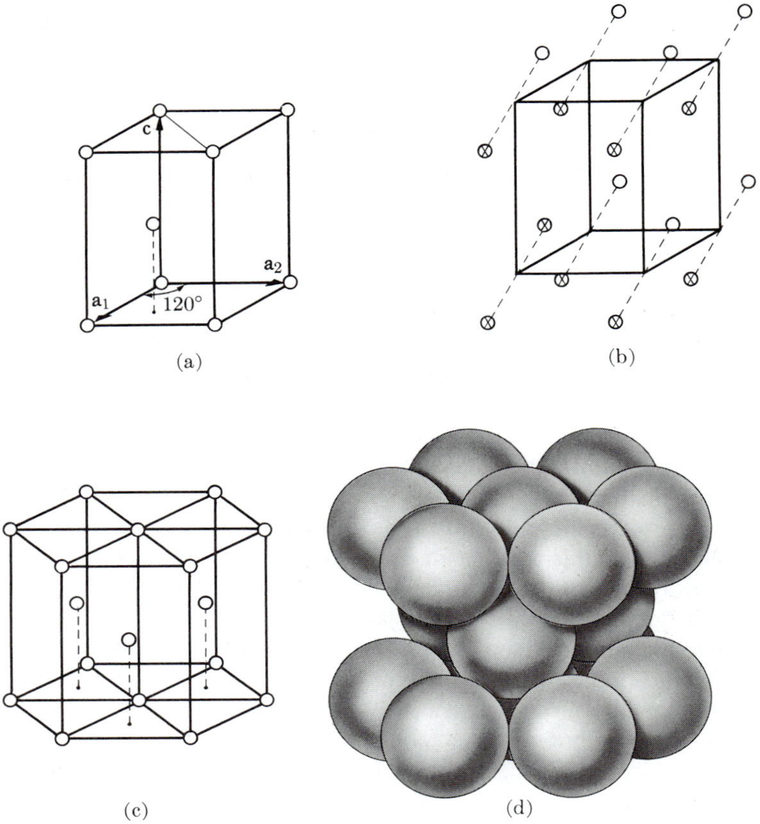

Fig. 2–15 The hexagonal close-packed structure, shared by Zn, Mg, Be, α-Ti, etc.

the atoms at 1 0 0 and $\frac{2}{3}\ \frac{1}{3}\ \frac{1}{2}$ in (a), the nine atoms shown in (a) corresponding to the nine atoms marked with an X in (b). The "association" of pairs of atoms with the points of a simple hexagonal Bravais lattice is suggested by the dashed lines in (b). Note, however, that the atoms of a close-packed hexagonal structure do not themselves form a point lattice, the surroundings of an atom at 0 0 0 being different from those of an atom at $\frac{2}{3}\ \frac{1}{3}\ \frac{1}{2}$. Figure 2–15(c) shows still another representation of the HCP structure: the three atoms in the interior of the hexagonal prism are directly above the centers of alternate triangles in the base and, if repeated through space by the vectors $\mathbf{a}_1$ and $\mathbf{a}_2$, would also form a hexagonal array just like the atoms in the layers above and below.

The HCP structure is so called because it is one of the two ways in which spheres can be packed together in space with the greatest possible density and still have a periodic arrangement. Such an arrangement of spheres in contact is shown in Fig. 2–15(d). If these spheres are regarded as atoms, then the resulting picture of an HCP metal is much closer to physical reality than is the relatively open

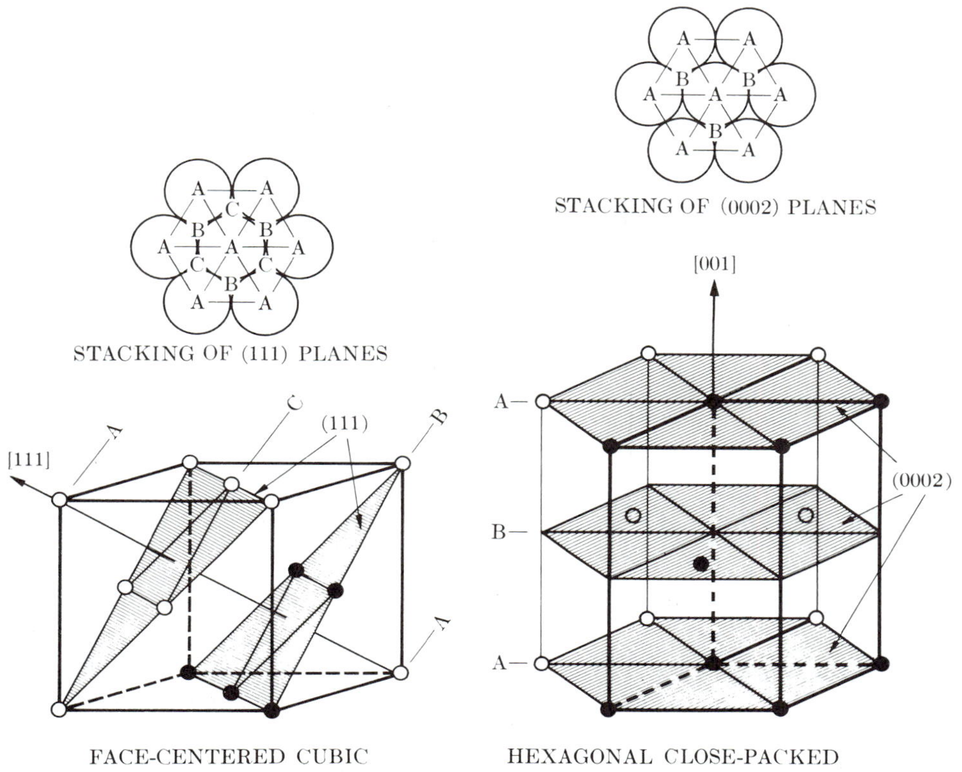

Fig. 2–16 Comparison of FCC and HCP structures. The black atoms in the FCC drawing delineate half a hexagon, which is completed on the same plane extended into the next unit cell below (not shown).

structure suggested by the drawing of Fig. 2–15(c), and this is true, generally, of all crystals. On the other hand, it may be shown that the ratio of c to a in an HCP structure formed of spheres in contact is 1.633 whereas the c/a ratio of metals having this structure varies from about 1.58 (Be) to 1.89 (Cd). As there is no reason to suppose that the atoms in these crystals are not in contact, it follows that they must be ellipsoidal in shape rather than spherical.

The FCC structure is an equally close-packed arrangement. Its relation to the HCP structure is not immediately obvious, but Fig. 2–16 shows that the atoms on the (111) planes of the FCC structure are arranged in a hexagonal pattern just like the atoms on the (0002) planes of the HCP structure. The only difference between the two structures is the way in which these hexagonal sheets of atoms are arranged above one another. In an HCP metal, the atoms in the second layer are above the hollows in the first layer and the atoms in the third layer are above the atoms in the first layer, so that the layer stacking sequence can be summarized as *A B A B A B*.... The first two atom layers of an FCC metal are put down in the same way, but the atoms of the third layer are so placed in the hollows of the second

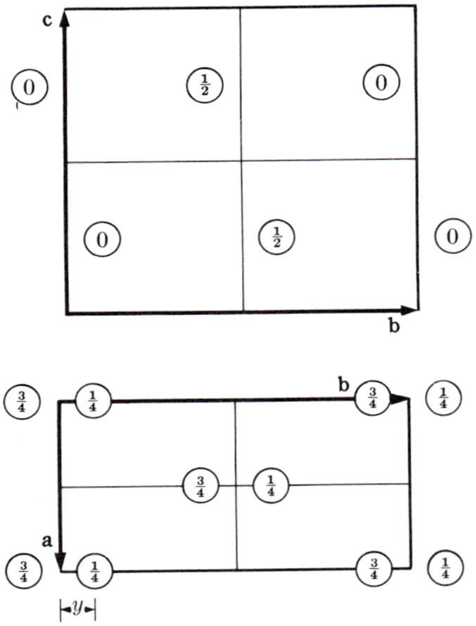

Fig. 2–17 The structure of α-uranium, after Jacob and Warren [2.1].

layer that not until the fourth layer does a position repeat. FCC stacking therefore has the sequence *A B C A B C*.... These stacking schemes are indicated in the plan views shown in Fig. 2–16.

Another example of the "association" of more than one atom with each point of a Bravais lattice is given by uranium. The structure of the form stable at room temperature, α-uranium, is illustrated in Fig. 2–17 by plan and elevation drawings. In such drawings, the height of an atom (expressed as a fraction of the axial length) above the plane of the drawing (which includes the origin of the unit cell and two of the cell axes) is given by the numbers marked on each atom. The Bravais lattice is base-centered orthorhombic, centered on the *C* face, and Fig. 2–17 shows how the atoms occur in pairs through the structure, each pair associated with a lattice point. There are four atoms per unit cell, located at $0\ y\ \frac{1}{4}$, $0\ \bar{y}\ \frac{3}{4}$, $\frac{1}{2}\ (\frac{1}{2} + y)\ \frac{1}{4}$, and $\frac{1}{2}\ (\frac{1}{2} - y)\ \frac{3}{4}$. Here we have an example of a variable parameter y in the atomic coordinates. Crystals often contain such variable parameters, which may have any fractional value without destroying any of the symmetry elements of the structure. A quite different substance might have exactly the same structure as uranium except for slightly different values of a, b, c, and y. For uranium y is 0.105 ± 0.005.

Turning to the crystal structure of *compounds* of unlike atoms, we find that the structure is built up on the skeleton of a Bravais lattice but that certain other rules must be obeyed, precisely because there are unlike atoms present. Consider, for example, a crystal of A_xB_y which might be an ordinary chemical compound, an intermediate phase of relatively fixed composition in some alloy system, or an ordered solid solution. Then the arrangement of atoms in A_xB_y must satisfy the

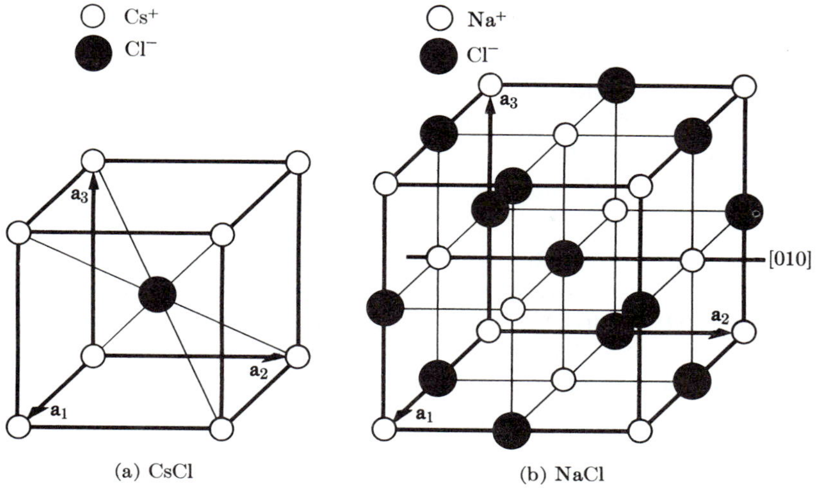

Fig. 2–18 The structures of (a) CsCl (common to CsBr, NiAl, ordered β-brass, ordered CuPd, etc.) and (b) NaCl (common to KCl, CaSe, PbTe, etc.).

following conditions:

1. Body-, face-, or base-centering translations, if present, must begin and end on atoms of the same kind. For example, if the structure is based on a body-centered Bravais lattice, then it must be possible to go from an A atom, say, to *another* A atom by the translation $\frac{1}{2}\,\frac{1}{2}\,\frac{1}{2}$.

2. The set of A atoms in the crystal and the set of B atoms must separately possess the same symmetry elements as the crystal as a whole, since in fact they make up the crystal. In particular, the operation of any symmetry element present must bring a given atom, A for example, into coincidence with another atom of the same kind, namely A.

Suppose we consider the structures of a few common crystals in light of the above requirements. Figure 2–18 illustrates the unit cells of two ionic compounds, CsCl and NaCl. These structures, both cubic, are common to many other crystals and, wherever they occur, are referred to as the "CsCl structure" and the "NaCl structure." In considering a crystal structure, one of the most important things to determine is its Bravais lattice, since that is the basic framework on which the crystal is built and because, as we shall see later, it has a profound effect on the way in which that crystal diffracts x-rays.

What is the Bravais lattice of CsCl? Figure 2–18(a) shows that the unit cell contains two atoms, ions really, since this compound is completely ionized even in the solid state: a caesium ion at 0 0 0 and a chlorine ion at $\frac{1}{2}\,\frac{1}{2}\,\frac{1}{2}$. The Bravais lattice is obviously not face-centered, but we note that the body-centering translation $\frac{1}{2}\,\frac{1}{2}\,\frac{1}{2}$ connects two atoms. However, these are unlike atoms and the lattice is therefore *not* body-centered. It is, by elimination, simple cubic. If one wishes, one may think of both ions, the caesium at 0 0 0 and the chlorine at $\frac{1}{2}\,\frac{1}{2}\,\frac{1}{2}$, as being associated with the lattice point at 0 0 0. It is not possible, however, to associate any one caesium ion with any particular chlorine ion and refer to them as a CsCl molecule; the term "molecule" therefore has no real physical significance in such a crystal, and the same is true of most inorganic compounds and alloys.

Close inspection of Fig. 2–18(b) will show that the unit cell of NaCl contains 8 ions, located as follows:

$$4\ Na^+ \text{ at } 0\,0\,0,\ \tfrac{1}{2}\,\tfrac{1}{2}\,0,\ \tfrac{1}{2}\,0\,\tfrac{1}{2},\ \text{and } 0\,\tfrac{1}{2}\,\tfrac{1}{2}$$

$$4\ Cl^- \text{ at } \tfrac{1}{2}\,\tfrac{1}{2}\,\tfrac{1}{2},\ 0\,0\,\tfrac{1}{2},\ 0\,\tfrac{1}{2}\,0,\ \text{and } \tfrac{1}{2}\,0\,0.$$

The sodium ions are clearly face-centered, and we note that the face-centering translations ($0\,0\,0,\ \tfrac{1}{2}\,\tfrac{1}{2}\,0,\ \tfrac{1}{2}\,0\,\tfrac{1}{2},\ 0\,\tfrac{1}{2}\,\tfrac{1}{2}$), when applied to the chlorine ion at $\tfrac{1}{2}\,\tfrac{1}{2}\,\tfrac{1}{2}$, will reproduce all the chlorine-ion positions. The Bravais lattice of NaCl is therefore face-centered cubic. The ion positions, incidentally, may be written in summary form as:

$$4\ Na^+ \text{ at } 0\,0\,0 + \text{face-centering translations.}$$

$$4\ Cl^- \text{ at } \tfrac{1}{2}\,\tfrac{1}{2}\,\tfrac{1}{2} + \text{face-centering translations.}$$

Note also that in these, as in all other structures, the operation of any symmetry element possessed by the lattice must bring similar atoms or ions into coincidence. For example, in Fig. 2–18(b), 90° rotation about the 4-fold [010] rotation axis shown brings the chlorine ion at $0\ 1\ \tfrac{1}{2}$ into coincidence with the chlorine ion at $\tfrac{1}{2}\ 1\ 1$, the sodium ion at 0 1 1 with the sodium ion at 1 1 1, etc.

Elements and compounds often have closely similar structures. Figure 2–19 shows the unit cells of diamond and the zinc-blende form of ZnS. Both are face-centered cubic. Diamond has 8 atoms per unit cell, located at

$$0\,0\,0 + \text{face-centering translations}$$

$$\tfrac{1}{4}\,\tfrac{1}{4}\,\tfrac{1}{4} + \text{face-centering translations.}$$

The atom positions in zinc blende are identical with these, but the first set of positions is now occupied by one kind of atom (S) and the other by a different kind (Zn).

Note that diamond and a metal like copper have quite dissimilar structures, although both are based on a face-centered cubic Bravais lattice. To distinguish between these two, the terms "diamond cubic" and "face-centered cubic" are usually used. The industrially important semiconductors, silicon and germanium, have the diamond cubic structure.

Instead of referring to a structure by name, such as the "NaCl structure," one can use the designations introduced years ago in *Strukturbericht* [G.1]. These consist of a letter and a number: the letter A indicates an element, B an AB compound, C an AB_2 compound, etc. The structure of copper, for example, is called the A1 structure, α-Fe is A2, zinc is A3, diamond is A4, NaCl is B1, etc. A full list is given by Pearson [G.16, Vol. 1, p. 85].

Some rather complex crystals can be built on a cubic lattice. For example, the ferrites, which are magnetic and are used as memory cores in digital computers, have the formula $MO \cdot Fe_2O_3$, where M is a divalent metal ion like Mn, Ni, Fe, Co, etc. Their structure is related to that of the mineral spinel. The Bravais lattice of the ferrites is face-centered cubic, and the unit cell contains 8 "molecules" or a

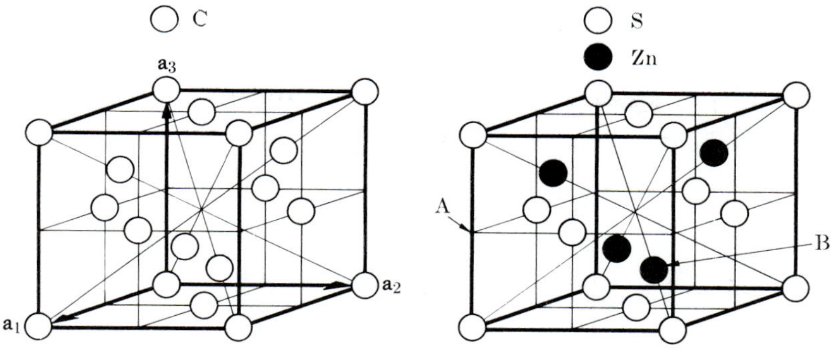

(a) Diamond (b) Zinc blende

Fig. 2–19 The structures of (a) diamond (common to Si, Ge, and gray Sn) and (b) the zinc-blende form of ZnS (common to HgS, CuI, AlSb, BeSe, etc.).

total of $8 \times 7 = 56$ ions. There are therefore 56/4 or 14 ions associated with each lattice point.

The number of atoms per unit cell in any crystal is partially dependent on its Bravais lattice. For example, the number of atoms per unit cell in a crystal based on a body-centered lattice must be a multiple of 2, since there must be, for any atom in the cell, a corresponding atom of the same kind at a translation of $\frac{1}{2}\,\frac{1}{2}\,\frac{1}{2}$ from the first. The number of atoms per cell in a base-centered lattice must also be a multiple of 2, as a result of the base-centering translations. Similarly, the number of atoms per cell in a face-centered lattice must be a multiple of 4.

The reverse of these propositions is not true. It would be a mistake to assume, for example, that if the number of atoms per cell is a multiple of 4, then the lattice is necessarily face-centered. The unit cell of the intermediate phase AuBe, for example (Fig. 2–20), contains 8 atoms and yet it is based on a simple cubic Bravais lattice. The atoms are located as follows:

4 Au at

$$u\,u\,u,\ (\tfrac{1}{2}+u)\,(\tfrac{1}{2}-u)\,\bar{u},\ \bar{u}\,(\tfrac{1}{2}+u)\,(\tfrac{1}{2}-u),\ (\tfrac{1}{2}-u)\,\bar{u}\,(\tfrac{1}{2}+u),$$

4 Be at

$$w\,w\,w,\ (\tfrac{1}{2}+w)\,(\tfrac{1}{2}-w)\,\bar{w},\ \bar{w}\,(\tfrac{1}{2}+w)\,(\tfrac{1}{2}-w),\ (\tfrac{1}{2}-w)\,\bar{w}\,(\tfrac{1}{2}+w),$$

where $u = 0.100$ and $w = 0.406$, each ± 0.005. If the parameter u is put equal to zero, the atomic coordinates of the gold atoms become those of a face-centered cubic cell. The structure of AuBe may therefore be regarded as distorted face-centered cubic, in which the presence of the beryllium atoms has forced the gold atoms out of their original positions by a distance $\pm u$, $\pm u$, $\pm u$. These translations are all in directions of the form $\langle 111 \rangle$, i.e., parallel to body diagonals of the cube, and are shown as dotted lines in Fig. 2–20.

It should now be apparent that the term "simple," when applied to a Bravais lattice, is used in a very special, technical sense and that some very complex structures can be built up on a "simple" lattice. In fact, they may contain more than a hundred atoms per unit cell. The only workable definition of a simple

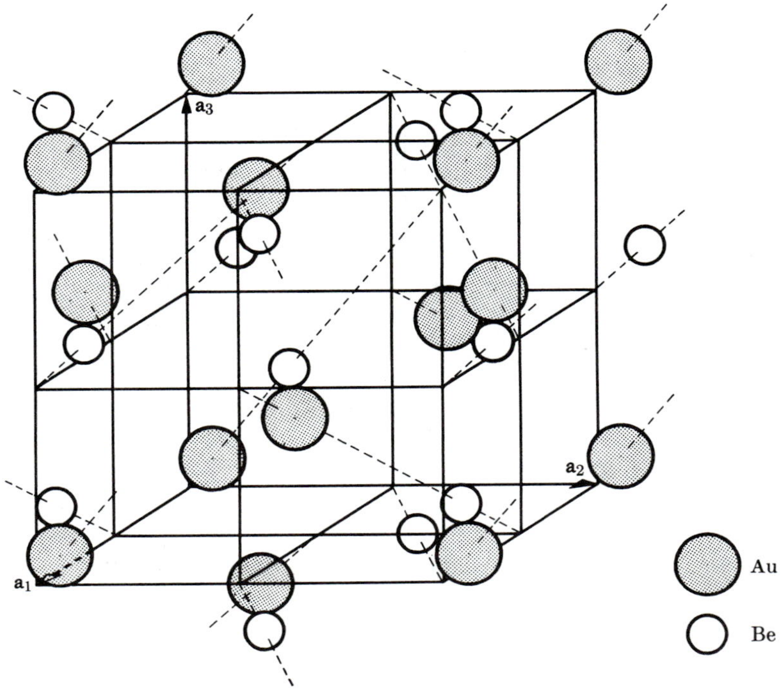

Fig. 2–20 The structure of AuBe, shared by FeSi, NiSi, CoSi, MnSi, etc. It is known as the FeSi structure [2.2].

lattice is a negative one: a given lattice is simple if it is neither body-, base-, nor face-centered; these latter possibilities can be ruled out by showing that the set of atomic positions does not contain the body-, base-, or face-centering translations. There is no rule governing the allowable number of atoms per cell in a simple lattice: this number may take on any one of the values 1, 2, 3, 4, 5, etc., although not in every crystal system and not every higher integer is permitted. Incidentally, not every theoretical possibility known to mathematical crystallography is realized in nature; for example, no known element crystallizes with a simple hexagonal lattice containing one atom per unit cell.

There is one other way of arranging unlike atoms on a point lattice besides those considered so far and that is exemplified by the structure of *solid solutions*. These solutions are of two types, substitutional and interstitial; in the former, solute atoms substitute for, or replace, solvent atoms on the lattice of the solvent, while in the latter, solute atoms fit into the interstices of the solvent lattice. The interesting feature of these structures is that the solute atoms are distributed more or less at random. For example, consider a 10 atomic percent solution of molybdenum in chromium, which has a BCC structure. The molybdenum atoms can occupy either the corner or body-centered positions of the cube in a random, irregular manner, and a small portion of the crystal might have the appearance of Fig. 2–21(a). Five adjoining unit cells are shown there, with a total of 29 atoms,

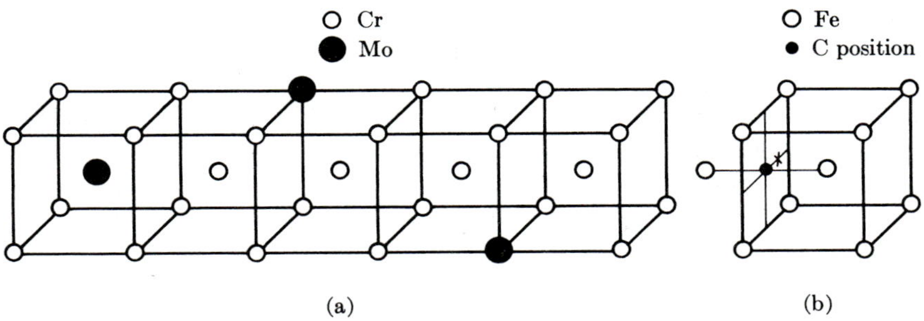

Fig. 2–21 Structure of solid solutions: (a) Mo in Cr (substitutional); (b) C in α-Fe (interstitial).

3 of which are molybdenum. This section of the crystal therefore contains somewhat more than 10 atomic percent molybdenum, but the next five cells would probably contain somewhat less. Such a structure does not obey the ordinary rules of crystallography: for example, the right-hand cell of the group shown does not have cubic symmetry, and one finds throughout the structure that the translation given by one of the unit cell vectors may begin on an atom of one kind and end on an atom of another kind. All that can be said of this structure is that it is BCC *on the average*, and experimentally we find that it displays the x-ray diffraction effects proper to a BCC lattice. This is not surprising since the x-ray beam used to examine the crystal is so large compared to the size of a unit cell that it observes, so to speak, millions of unit cells at the same time and so obtains only an average "picture" of the structure.

The above remarks apply equally well to interstitial solid solutions. These form whenever the solute atom is small enough to fit into the solvent lattice without causing too much distortion. Ferrite, the solid solution of carbon in α-iron, is a good example.* In the unit cell shown in Fig. 2–21(b), there are two kinds of "holes" in the lattice: one at $\frac{1}{2}\ 0\ \frac{1}{2}$ (marked •) and equivalent positions in the centers of the cube faces and edges, and one at $\frac{1}{4}\ 0\ \frac{1}{2}$ (marked ×) and equivalent positions. All the evidence at hand points to the fact that the carbon atoms in ferrite are located in the holes at $\frac{1}{2}\ 0\ \frac{1}{2}$ and equivalent positions. On the average, however, no more than about 1 of these positions in 500 unit cells is occupied, since the maximum solubility of carbon in ferrite is only about 0.1 atomic percent.

Still another type of structure worth noting is that of *ordered solid solutions*. As described above, a typical substitutional solid solution has solute atoms distributed more or less at random on the lattice points of the solvent.† On the other hand, there are solutions in which this is true only at elevated temperatures; when cooled to lower temperatures, the solute atoms take up an orderly, periodic

* Note the double meaning of the word *ferrite*: (1) metallurgical, for the metallic solid solution mentioned above, and (2) mineralogical, for the oxide $MO \cdot Fe_2O_3$ previously described.

† Of course, when the solution becomes concentrated, there is no real distinction between "solvent" and "solute." There is only one lattice, with two or more kinds of atoms distributed on it.

arrangement while still remaining on the lattice points of the solvent. The solid solution is then said to be *ordered* and to possess a *superlattice*. The alloy $AuCu_3$ is a classic example: at high temperatures the copper and gold atoms are located more or less at random on face-centered cubic lattice sites, while at low temperature the gold atoms occupy only the cube corner positions and the copper atoms only the face-centered positions. In its temperature range of stability then, an ordered solid solution resembles a chemical compound, with atoms of one kind on one set of lattice sites and atoms of a different kind on another set. But an ordered solid solution is a "half-hearted compound" because, when heated, it disorders before it melts; a real compound, like NaCl, remains ordered right up to the melting point. Crystallographically, the structures of the disordered and ordered solid solutions are quite different; disordered $AuCu_3$ is, on the average, face-centered cubic while the ordered form is simple cubic. Such structures will be discussed more fully in Chap. 13.

2-8 ATOM SIZES AND COORDINATION

When two or more unlike atoms unite to form a chemical compound, intermediate phase, or solid solution, the kind of structure formed is dependent, in part, on the relative sizes of the atoms involved. But what is meant by the size of an atom? To regard an atom as something like a billiard ball with a sharply defined bounding surface is surely an oversimplification, since we know that the electron density decreases gradually at the "surface" of the atom and that there is a small but finite probability of finding an electron at quite large distances from the nucleus. And yet the only practical way we have of defining atomic size lies in considering a crystal as a collection of rigid spheres in contact. The size of an atom, then, is given by the distance of closest approach of atom centers in a crystal of the element, and this distance can be calculated from the lattice parameters.

For example, the lattice parameter a of α-iron is 2.87 Å, and in a BCC lattice the atoms are in contact only along the diagonals of the unit cube. The diameter of an iron atom is therefore equal to one half the length of the cube diagonal, or $(\sqrt{3}/2)a = 2.48$ Å. The following formulas give the distance of closest approach in the three common metal structures:

$$
\begin{aligned}
\text{BCC} &= \frac{\sqrt{3}}{2}a, \\
\text{FCC} &= \frac{\sqrt{2}}{2}a, \\
\text{HCP} &= a \quad \text{(between atoms in basal plane)}, \\
&= \sqrt{\frac{a^2}{3} + \frac{c^2}{4}} \quad \text{(between atom in basal plane and neighbors above or below)}.
\end{aligned}
\tag{2–7}
$$

Values of the distance of closest approach, together with the crystal structures and lattice parameters of the elements, are tabulated in Appendix 5.

To a first approximation, the size of an atom is a constant. In other words, an iron atom has about the same size whether it occurs in pure iron, an intermediate phase, or a solid solution. This is a very useful fact to remember when investigating unknown crystal structures, for it enables us to predict roughly how large a hole is necessary in a proposed structure to accommodate a given atom. More precisely, it is known that the size of an atom has a slight dependence on its *coordination number*, which is the number of nearest neighbors of the given atom and which depends on crystal structure. The coordination number of an atom in the FCC or HCP structures is 12, in BCC 8, and diamond cubic 4. The smaller the coordination number, the smaller the volume occupied by a given atom, and the approximate amount of contraction to be expected with decrease in coordination number is found to be:

Change in coordination	Size contraction, percent
$12 \rightarrow 8$	3
$12 \rightarrow 6$	4
$12 \rightarrow 4$	12

This means, for example, that the diameter of an iron atom is greater if the iron is dissolved in FCC copper than if it exists in a crystal of BCC α-iron or is dissolved in BCC vanadium. If it were dissolved in copper, its diameter would be approximately 2.48/0.97, or 2.56 Å.

The size of an atom in a crystal also depends on whether its binding is ionic, covalent, metallic, or van der Waals, and on its state of ionization. The more electrons are removed from a neutral atom the smaller it becomes, as shown strikingly for iron, whose atoms and ions Fe, Fe^{++}, Fe^{+++} have diameters of 2.48, 1.66, and 1.34 Å, respectively.

The spatial arrangement of atoms about a given point is often described by words such as *octahedral* and *tetrahedral*. For example, in the NaCl structure of Fig. 2-18(b) the central Cl^- ion at $\frac{1}{2}\,\frac{1}{2}\,\frac{1}{2}$ is said to be octahedrally surrounded by Na^+ ions, because the six Na^+ ions in the face-centered positions lie on the corners of an octahedron, a solid bounded by eight triangular sides. In the zinc blende structure of Fig. 2-19(b) the empty position marked A is octahedrally surrounded by sulphur atoms, of which only four are in the cell shown, and would be referred to as an octahedral hole in the structure. This group of atoms is shown separately in Fig. 2-22. In the same structure the Zn atom at $\frac{3}{4}\,\frac{3}{4}\,\frac{1}{4}$, marked B in Fig. 2-19(b), is surrounded by four S atoms at the corners of a tetrahedron, a solid bounded by four triangular sides (Fig. 2-22). In fact, all four of the Zn atoms in the unit cell have tetrahedral S surroundings. Also in the ZnS structure the reader can demonstrate, by sketching three cells adjacent to the one shown, that the hole at A is tetrahedrally surrounded by Zn atoms. Thus, the hole at A has both octahedral (S) and tetrahedral (Zn) surroundings, an unusual circumstance.

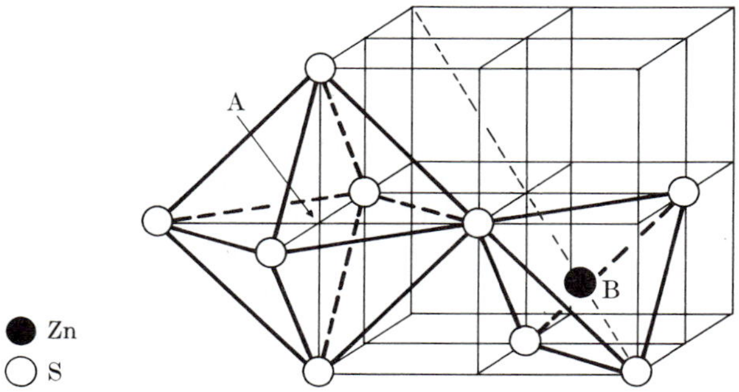

Fig. 2–22 Portion of the zinc blende structure. Compare Fig. 2–19(b). The hole at *A* has octahedral surroundings. The Zn atom at *B* has tetrahedral surroundings.

2–9 CRYSTAL SHAPE

We have said nothing so far about the shape of crystals, preferring to concentrate instead on their interior structure. However, the shape of crystals is, to the layman, perhaps their most characteristic property, and nearly everyone is familiar with the beautifully developed flat faces exhibited by natural minerals or crystals artificially grown from a supersaturated salt solution. In fact, it was with a study of these faces and the angles between them that the science of crystallography began.

Nevertheless, the shape of crystals is really a secondary characteristic, since it depends on, and is a consequence of, the interior arrangement of atoms. Sometimes the external shape of a crystal is rather obviously related to its smallest building block, the unit cell, as in the little cubical grains of ordinary table salt (NaCl has a cubic lattice) or the six-sided prisms of natural quartz crystals (hexagonal lattice). In many other cases, however, the crystal and its unit cell have quite different shapes; gold, for example, has a cubic lattice, but natural gold crystals are octahedral in form, i.e., bounded by eight planes of the form $\{111\}$.

An important fact about crystal faces was known long before there was any knowledge of crystal interiors. It is expressed as the *law of rational indices*, which states that the indices of naturally developed crystal faces are always composed of small whole numbers, rarely exceeding 3 or 4. Thus, faces of the form $\{100\}$, $\{111\}$, $\{1\bar{1}00\}$, $\{210\}$, etc., are observed but not such faces as $\{510\}$, $\{719\}$, etc. We know today that planes of low indices have the largest density of lattice points, and it is a law of crystal growth that such planes develop at the expense of planes with high indices and few lattice points.

To a metallurgist, however, crystals with well-developed faces are in the category of things heard of but rarely seen. They occur occasionally on the free surface of castings, in some electrodeposits, or under other conditions of no external constraint. To a metallurgist, a crystal is most usually a "grain," seen through a microscope in the company of many other grains on a polished section. If he has an isolated single crystal, it will have been artificially grown either from the melt, and thus have the shape of the crucible in which it solidified, or by recrystallization, and thus have the shape of the starting material, whether sheet, rod, or wire.

The shapes of the grains in a polycrystalline mass of metal are the result of several kinds of forces, all of which are strong enough to counteract the natural tendency of each grain to grow with well-developed flat faces. The result is a grain roughly polygonal in shape with no obvious aspect of crystallinity. Nevertheless, that grain is a crystal and just as "crystalline" as, for example, a well-developed prism of natural quartz, since the essence of crystallinity is a periodicity of inner atomic arrangement and not any regularity of outward form.

2–10 TWINNED CRYSTALS

Some crystals have two parts symmetrically related to one another. These, called twinned crystals, are fairly common both in minerals and in metals and alloys. For a detailed discussion of twinning, see Kelly and Groves [G.33] and Barrett and Massalski [G.25].

The relationship between the two parts of a twinned crystal is described by the symmetry operation which will bring one part into coincidence with the other or with an extension of the other. Two main kinds of twinning are distinguished, depending on whether the symmetry operation is 180° rotation about an axis, called the twin axis, or reflection across a plane, called the twin plane. The plane on which the two parts of a twinned crystal are united is called the composition plane. In the case of a reflection twin, the composition plane may or may not coincide with the twin plane.

Of most interest to metallurgists, who deal mainly with FCC, BCC, and HCP structures, are the following kinds of twins:

1. Annealing twins, such as occur in FCC metals and alloys (Cu, Ni, α-brass, Al, etc.), which have been cold-worked and then annealed to cause recrystallization.
2. Deformation twins, such as occur in deformed HCP metals (Zn, Mg, Be, etc.) and BCC metals (α-Fe, W, etc.).

Annealing Twins

Annealing twins in FCC metals are rotation twins, in which the two parts are related by a 180° rotation about a twin axis of the form ⟨111⟩. Because of the high symmetry of the cubic lattice, this orientation relationship is also given by a 60° rotation about the twin axis or by reflection across the {111} plane normal to the twin axis. In other words, FCC annealing twins may also be classified as reflection twins. The twin plane is also the composition plane.

Occasionally, annealing twins appear under the microscope as in Fig. 2–23(a), with one part of a grain (B) twinned with respect to the other part (A). The two parts are in contact on the composition plane (111) which makes a straight-line trace on the plane of polish. More common, however, is the kind shown in Fig. 2–23(b). The grain shown consists of three parts: two parts (A_1 and A_2) of identical orientation separated by a third part (B) which is twinned with respect to A_1 and A_2. B is known as a twin band.

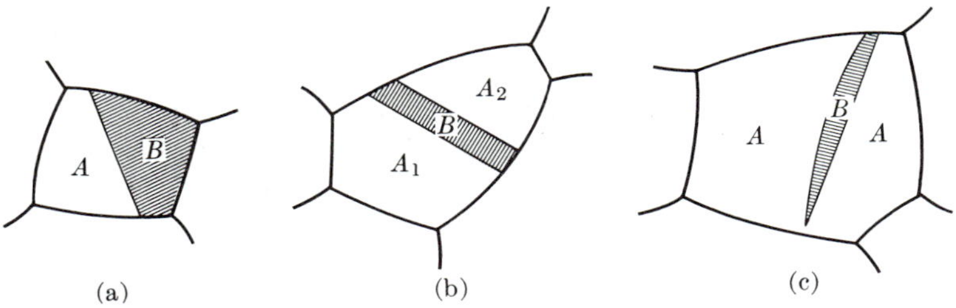

Fig. 2–23 Twinned grains: (a) and (b) FCC annealing twins; (c) HCP deformation twin.

Figure 2–24 illustrates the structure of an FCC twin band. The plane of the main drawing is $(1\bar{1}0)$, the (111) twin plane is perpendicular to this plane, and the [111] twin axis lies in it. Open circles represent atoms in the plane of the drawing and filled circles those in the layers immediately above or below. The reflection symmetry across the twin plane is suggested by the dashed lines connecting several pairs of atoms.

The statement that a rotation twin of this kind is related to the parent crystal by a 180° rotation about the twin axis is merely an expression of the orientation relationship between the two and is not meant to suggest that a twin is formed by a physical rotation of one part of the crystal with respect to another. Actually, FCC annealing twins are formed by a change in the normal growth mechanism. Suppose that, during normal grain growth following recrystallization, a grain boundary is roughly parallel to (111) and is advancing in a direction approximately normal to this boundary, namely [111]. To say that the boundary is advancing is to say that atoms are leaving the lattice of the consumed grain and joining that of the growing grain. The grain is therefore growing by the addition of layers of atoms parallel to (111), and we already know that these layers are piled up in the sequence *A B C A B C* . . . in an FCC crystal. If, however, a mistake should occur and this sequence become altered to *C B A C B A* . . . , the crystal so formed would still be FCC but it would be a twin of the former. If a similar mistake occurred later, a crystal of the original orientation would start growing and a twin band would be formed. With this symbolism, we may indicate a twin band as follows:

A B C A B C | *B A C B A C* | *A B C A B C*

parent crystal →	← twin band →	← parent crystal

In this terminology, the symbols themselves are imaged in the mirror *C*, the twin plane. At the left of Fig. 2–24 the positional symbols *A*, *B*, *C* are attached to various (111) planes to show the change in stacking which occurs at the boundaries of the twin band. Parenthetically, it should be remarked that twin bands visible under the light microscope are thousands of times thicker than the one shown in this drawing.

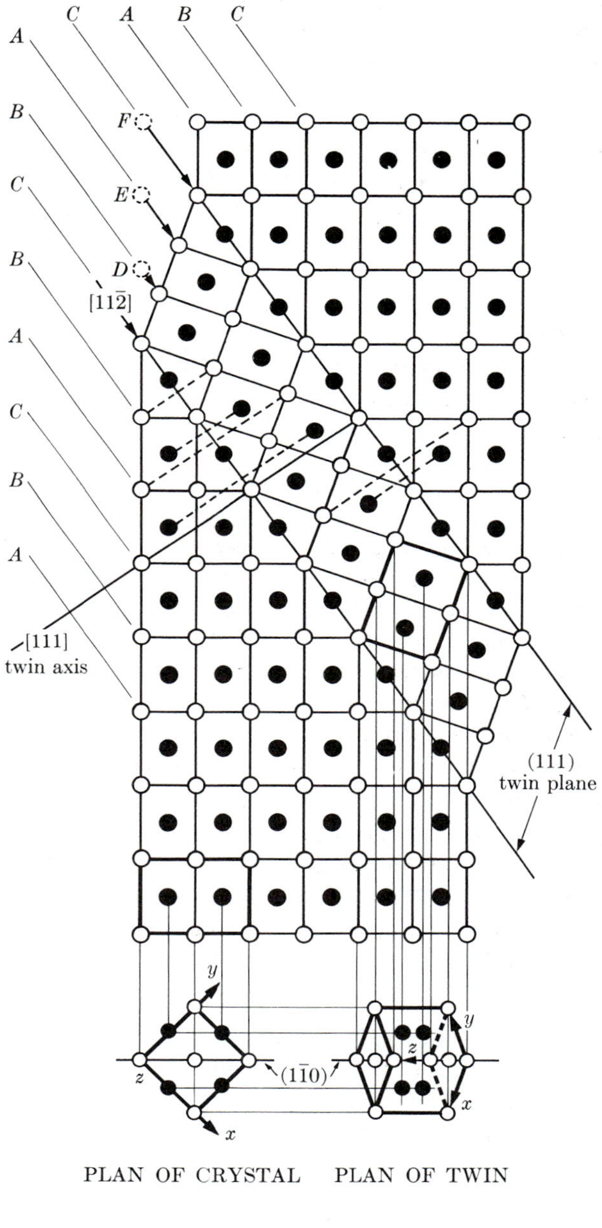

Fig. 2–24 Twin band in FCC lattice. Plane of main drawing is (1$\bar{1}$0).

There is still another way of *describing* the orientation relationship between an FCC crystal and its twin: the (111) layers of the twin are in positions which would result from homogeneous shear in a $[11\bar{2}]$ direction, each layer moving by an amount proportional to its distance from the twin plane. In Fig. 2–24, this shear is indicated by the arrows going from initial positions *D*, *E*, *F* to final positions in the twin. Although it has been frequently suggested that such twins are *formed* by deformation, no convincing evidence for this view has been advanced and it is generally held that annealing twins are the result of the growth process described above. Nevertheless, this hypothetical shear is sometimes a useful way of describing the orientation relationship between a crystal and its twin.

Deformation Twins

Deformation twins are found in both BCC and HCP lattices and are all that their name implies, since, in both cases, the cause of twinning is deformation. In each case, the orientation relationship between parent crystal and twin is that of reflection across a plane.

In BCC structures, the twin plane is (112) and the twinning shear is in the direction $[11\bar{1}]$. The only common example of such twins is in α-iron (ferrite) deformed by impact, where they occur as extremely narrow twin bands called Neumann bands. It should be noted that, in cubic lattices, both $\{112\}$ and $\{111\}$ reflection twinning produce the same orientation relationship; however, they differ in the interatomic distances produced, and an FCC lattice can twin by reflection on $\{111\}$ with less distortion than on $\{112\}$, while for the same reason $\{112\}$ is the preferred plane for BCC lattices.

In HCP metals, the twin plane is normally $(10\bar{1}2)$. The twinning shear is not well understood; in a gross sense, it takes place in the direction $[\bar{2}\bar{1}1]$ for metals with c/a ratios less than $\sqrt{3}$ (Be, Ti, Mg) and in the reverse direction $[21\bar{1}]$ for metals with c/a larger than $\sqrt{3}$ (Zn, Cd), but the direction of motion of individual atoms during shear is not definitely known. Figure 2–23(c) illustrates the usual form of a twin band in HCP metals, and it will be noted that the composition "plane," although probably parallel or nearly parallel to the twin plane, is not quite flat but often exhibits appreciable curvature.

General

Twins, in general, can form on different planes in the same crystal. For example, there are four $\{111\}$ planes of different orientation on which twinning can take place in an FCC crystal. Accordingly, in the microstructure of recrystallized copper, for example, one often sees twin bands running in more than one direction in the same grain.

A crystal may also twin repeatedly, producing several new orientations. If crystal *A* twins to form *B*, which twins to form *C*, etc., then *B*, *C*, etc., are said to be first-order, second-order, etc., twins of the parent crystal *A*. Not all these orientations are new. In Fig. 2–23(b), for example, *B* may be regarded as the first-order twin of A_1, and A_2 as the first-order twin of *B*. A_2 is therefore the second-order twin of A_1 but has the same orientation as A_1.

2–11 THE STEREOGRAPHIC PROJECTION

Crystal drawings made in perspective or in the form of plan and elevation, while they have their uses, are not suitable for displaying the angular relationship between lattice planes and directions. But frequently we are more interested in these angular relationships than in any other aspect of the crystal, and we then need a kind of drawing on which the angles between planes can be accurately measured and which will permit graphical solution of problems involving such angles. The stereographic projection fills this need. For details not given below, see Barrett and Massalski [G.25], Kelly and Groves [G.33], and Johari and Thomas [G.28].

The orientation of any plane in a crystal can be just as well represented by the inclination of the normal to that plane relative to some reference plane as by the inclination of the plane itself. All the planes in a crystal can thus be represented by a set of plane normals radiating from some one point within the crystal. If a reference sphere is now described about this point, the plane normals will intersect the surface of the sphere in a set of points called *poles*. This procedure is illustrated in Fig. 2–25, which is restricted to the {100} planes of a cubic crystal. The pole of a plane represents, by its position on the sphere, the orientation of that plane.

A plane may also be represented by the trace the extended plane makes in the surface of the sphere, as illustrated in Fig. 2–26, where the trace *ABCDA* represents the plane whose pole is P_1. This trace is a *great circle*, i.e., a circle of maximum diameter, if the plane passes through the center of the sphere. A plane not passing through the center will intersect the sphere in a *small circle*. On a ruled globe, for example, the longitude lines (meridians) are great circles, while the latitude lines, except the equator, are small circles.

The angle α between two planes is evidently equal to the angle between their great circles or to the angle between their normals (Fig. 2–26). But this angle, in degrees, can also be measured on the surface of the sphere along the great circle *KLMNK* connecting the poles P_1 and P_2 of the two planes, if this circle has been divided into 360 equal parts. The measurement of an angle has thus been transferred from the planes themselves to the surface of the reference sphere.

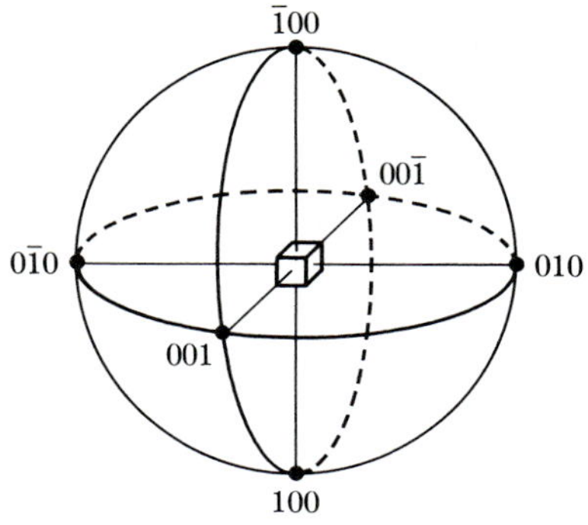

Fig. 2–25 {100} poles of a cubic crystal.

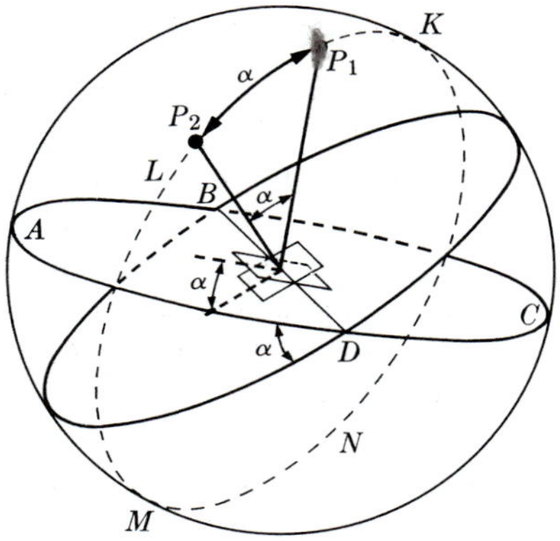

Fig. 2–26 Angle between two planes.

Preferring, however, to measure angles on a flat sheet of paper rather than on the surface of a sphere, we find ourselves in the position of the geographer who wants to transfer a map of the world from a globe to a page of an atlas. Of the many known kinds of projections, he usually chooses a more or less equal-area projection so that countries of equal area will be represented by equal areas on the map. In crystallography, however, we prefer the equiangular stereographic projection since it preserves angular relationships faithfully although distorting areas. It is made by placing a plane of projection normal to the end of any chosen diameter of the sphere and using the other end of that diameter as the point of projection. In Fig. 2–27 the projection plane is normal to the diameter AB, and the projection is made from the point B. If a plane has its pole at P, then the stereographic projection of P is at P', obtained by drawing the line BP and extending it until it meets the projection plane. Alternately stated, the stereographic projection of the pole P is the shadow cast by P on the projection plane when a light source is placed at B. The observer, incidentally, views the projection from the side opposite the light source.

The plane $NESW$ is normal to AB and passes through the center C. It therefore cuts the sphere in half and its trace in the sphere is a great circle. This great circle projects to form the *basic circle* $N'E'S'W'$ on the projection, and all poles on the left-hand hemisphere will project within this basic circle. Poles on the right-hand hemisphere will project outside this basic circle, and those near B will have projections lying at very large distances from the center. If we wish to plot such poles, we move the point of projection to A and the projection plane to B and distinguish the new set of points so formed by minus signs, the previous set (projected from B) being marked with plus signs. Note that movement of the projection plane along AB or its extension merely alters the magnification; we usually make it tangent to the sphere, as illustrated, but we can also make it pass through the center of the sphere, for example, in which case the basic circle becomes identical with the great circle $NESW$.

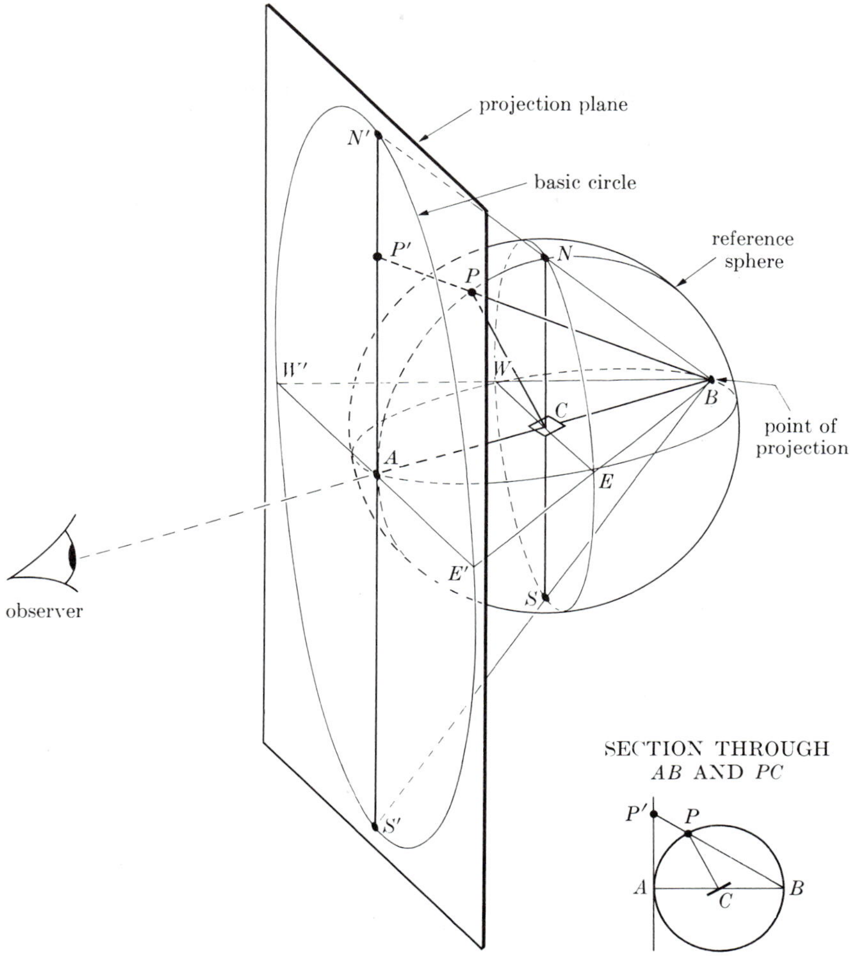

Fig. 2-27 The stereographic projection.

A lattice plane in a crystal is several steps removed from its stereographic projection, and it may be worth-while at this stage to summarize these steps:

1. The plane C is represented by its normal CP.
2. The normal CP is represented by its pole P, which is its intersection with the reference sphere.
3. The pole P is represented by its stereographic projection P'.

After gaining some familiarity with the stereographic projection, the student will be able mentally to omit these intermediate steps and he will then refer to the projected point P' as the pole of the plane C or, even more directly, as the plane C itself.

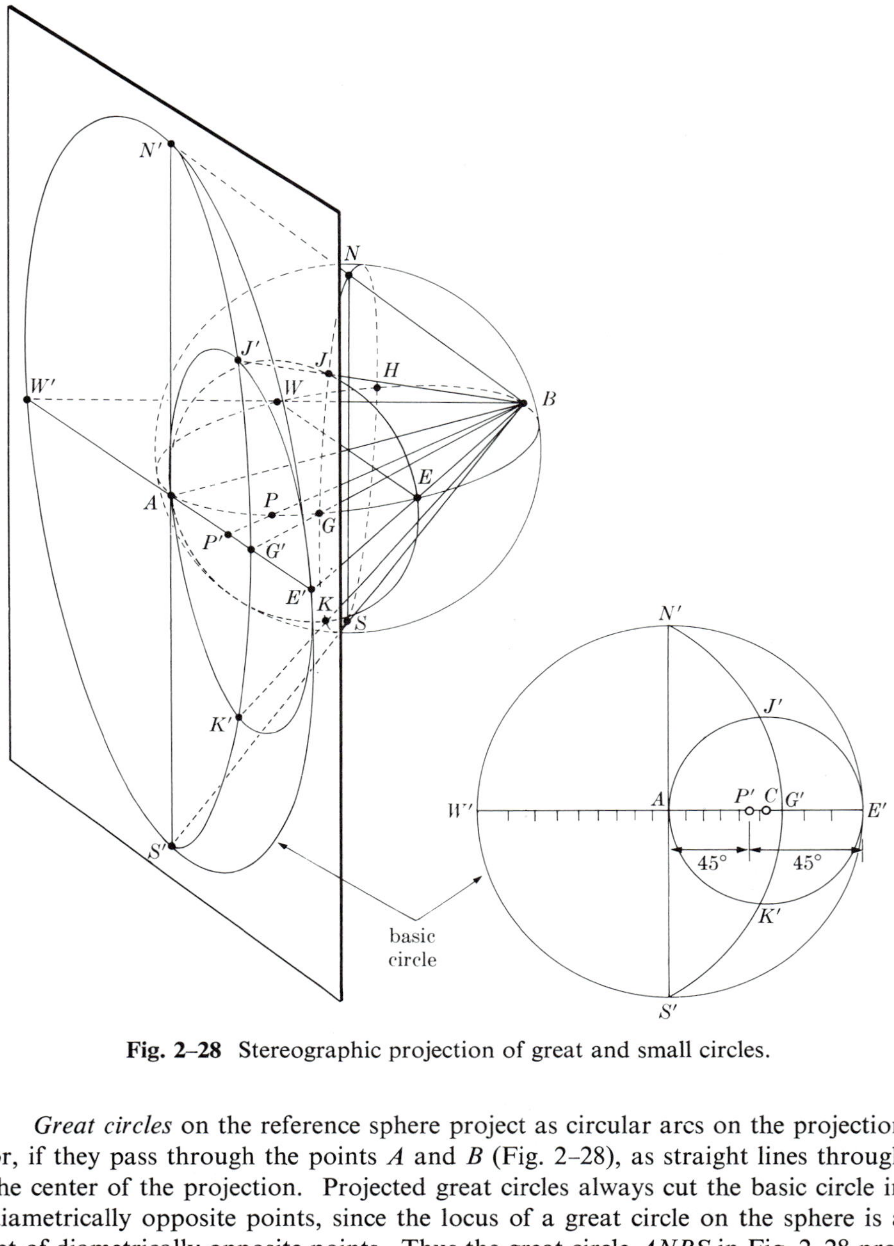

Fig. 2–28 Stereographic projection of great and small circles.

Great circles on the reference sphere project as circular arcs on the projection or, if they pass through the points A and B (Fig. 2–28), as straight lines through the center of the projection. Projected great circles always cut the basic circle in diametrically opposite points, since the locus of a great circle on the sphere is a set of diametrically opposite points. Thus the great circle $ANBS$ in Fig. 2–28 projects as the straight line $N'S'$ and $AWBE$ as $W'E'$; the great circle $NGSH$, which is inclined to the plane of projection, projects as the circle arc $N'G'S'$. If the half great circle WAE is divided into 18 equal parts and these points of division projected on $W'AE'$, we obtain a graduated scale, at 10° intervals, on the equator of the basic circle.

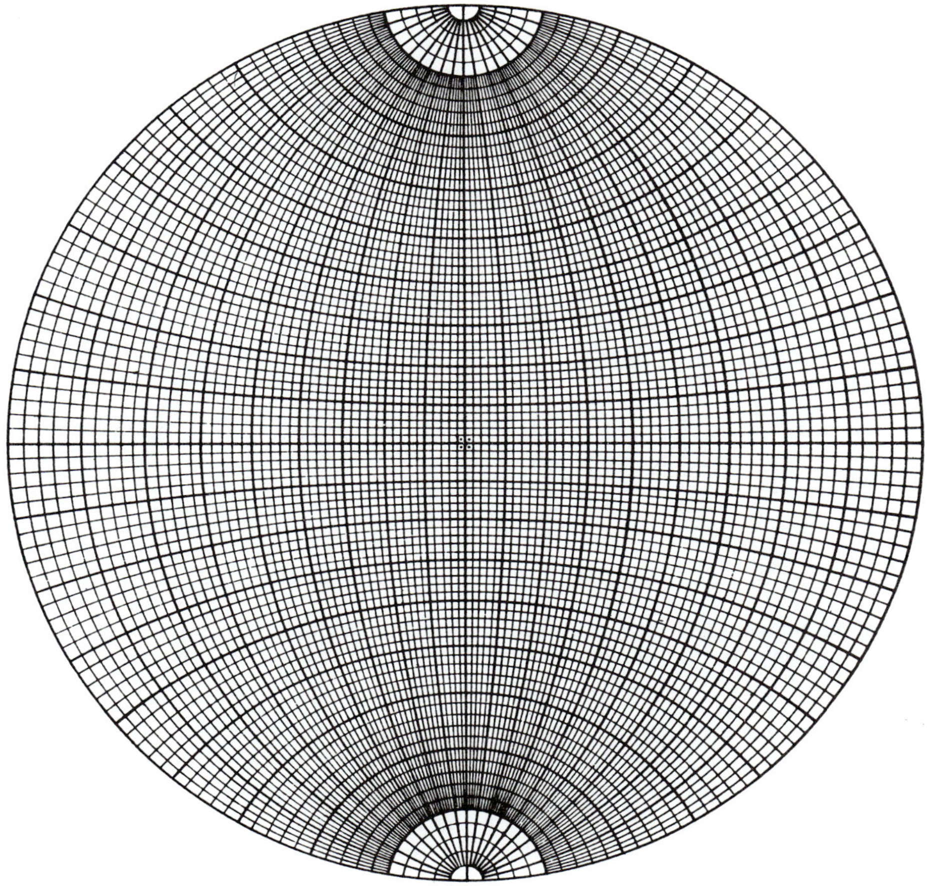

Fig. 2–29 Wulff net drawn to 2° intervals.

Small circles on the sphere also project as circles, but their projected center does not coincide with their center on the projection. For example, the circle *AJEK* whose center *P* lies on *AEBW* projects as *AJ′E′K′*. Its center *on the projection* is at *C*, located at equal distances from *A* and *E′*, but its *projected center* is at *P′*, located an equal number of degrees (45° in this case) from *A* and *E′*.

The device most useful in solving problems involving the stereographic projection is the *Wulff net* shown in Fig. 2–29. It is the projection of a sphere ruled with parallels of latitude and longitude on a plane parallel to the north-south axis of the sphere. The latitude lines on a Wulff net are small circles extending from side to side and the longitude lines (meridians) are great circles connecting the north and south poles of the net. These nets are available in various sizes [2.3], one of 18-cm diameter giving an accuracy of about one degree, which is satisfactory for most problems; to obtain greater precision, either a larger net or mathematical calculation must be used. Wulff nets are used by making the stereographic pro-

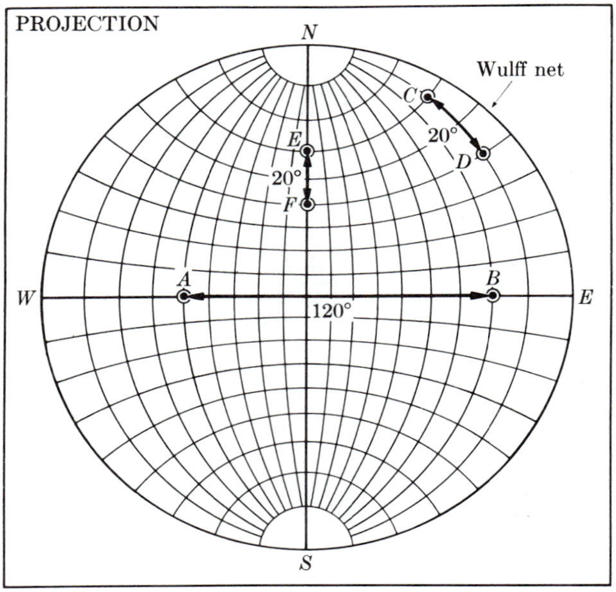

Fig. 2–30 Stereographic projection superimposed on Wulff net for measurement of angle between poles. For illustrative purposes this net is graduated at 10° intervals.

jection on tracing paper and with the basic circle of the same diameter as that of the Wulff net; the projection is then superimposed on the Wulff net, with the centers always coinciding.

To return to our problem of the measurement of the angle between two crystal planes, we saw in Fig. 2–26 that this angle could be measured on the surface of the sphere along the great circle connecting the poles of the two planes. This measurement can also be carried out on the stereographic projection *if, and only if, the projected poles lie on a great circle.* In Fig. 2–30, for example, the angle between the planes* *A* and *B* or *C* and *D* can be measured directly, simply by counting the number of degrees separating them along the great circle on which they lie. Note that the angle *C–D* equals the angle *E–F*, there being the same difference in latitude between *C* and *D* as between *E* and *F*.

If the two poles do not lie on a great circle, then the projection is rotated relative to the Wulff net until they do lie on a great circle, where the desired angle measurement can then be made. Figure 2–31(a) is a projection of the two poles P_1 and P_2 shown in perspective in Fig. 2–26, and the angle between them is found by the rotation illustrated in Fig. 2–31(b). This rotation of the projection is equivalent to rotation of the poles on latitude circles of a sphere whose north-south axis is perpendicular to the projection plane.

As shown in Fig. 2–26, a plane may be represented by its trace in the reference

* We are here using the abbreviated terminology referred to above.

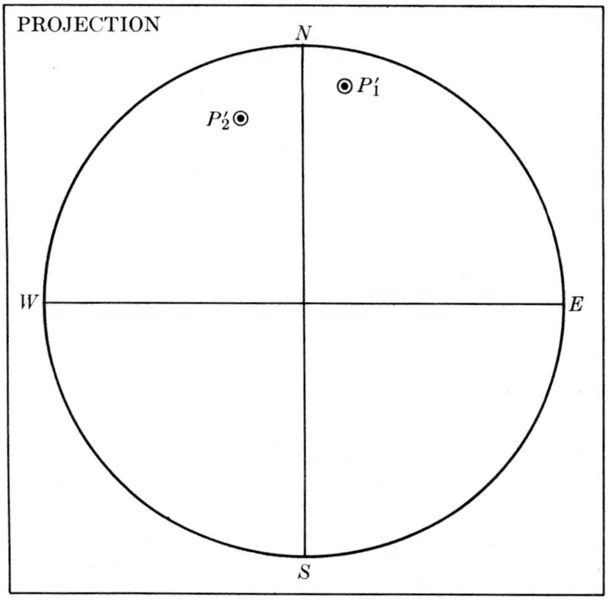

(a)

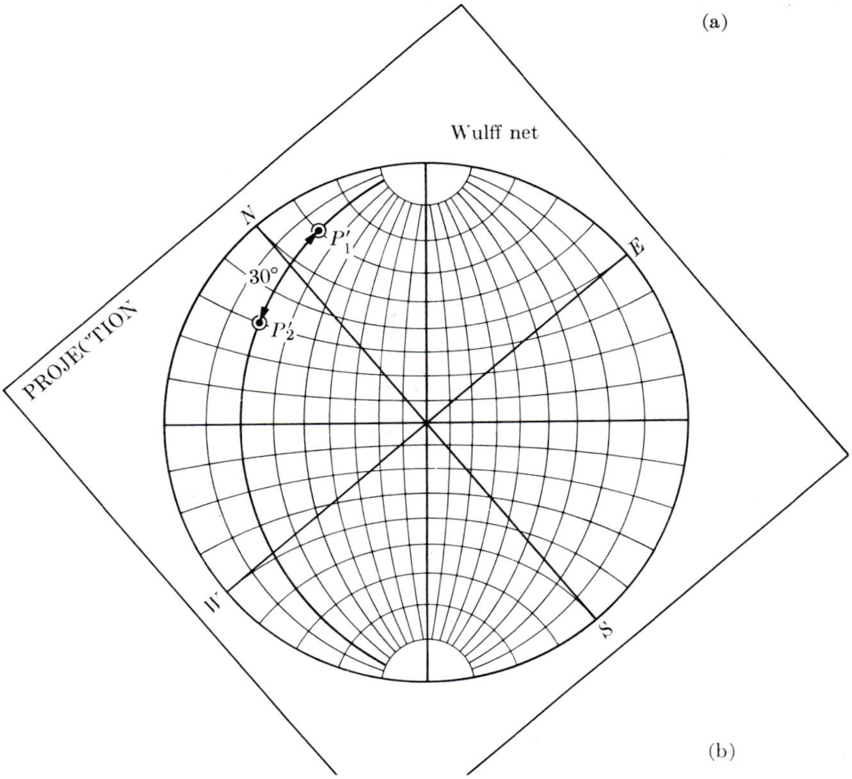

(b)

Fig. 2–31 (a) Stereographic projection of poles P_1 and P_2 of Fig. 2–26. (b) Rotation of projection to put poles on same great circle of Wulff net. Angle between poles = 30°.

sphere. This trace becomes a great circle in the stereographic projection. Since every point on this great circle is 90° from the pole of the plane, the great circle may be found by rotating the projection until the pole falls on the equator of the underlying Wulff net and tracing that meridian which cuts the equator 90° from the pole, as illustrated in Fig. 2–32. If this is done for two poles, as in Fig. 2–33, the angle between the corresponding planes may also be found from the angle of intersection of the two great circles corresponding to these poles; it is in this sense that the stereographic projection is said to be angle-true. This method of angle measurement is not as accurate, however, as that shown in Fig. 2–31(b).

We often wish to rotate poles around various axes. We have already seen that rotation about an axis normal to the projection is accomplished simply by rotation of the projection around the center of the Wulff net. Rotation about an axis lying in the plane of the projection is performed by, first, rotating the *axis* about the center of the Wulff net until it coincides with the north-south axis if it does not already do so, and, second, moving the poles involved along their respective latitude circles the required number of degrees. Suppose it is required to rotate the poles A_1 and B_1 shown in Fig. 2–34 by 60° about the NS axis, the direction of motion being from W to E on the projection. Then A_1 moves to A_2 along its

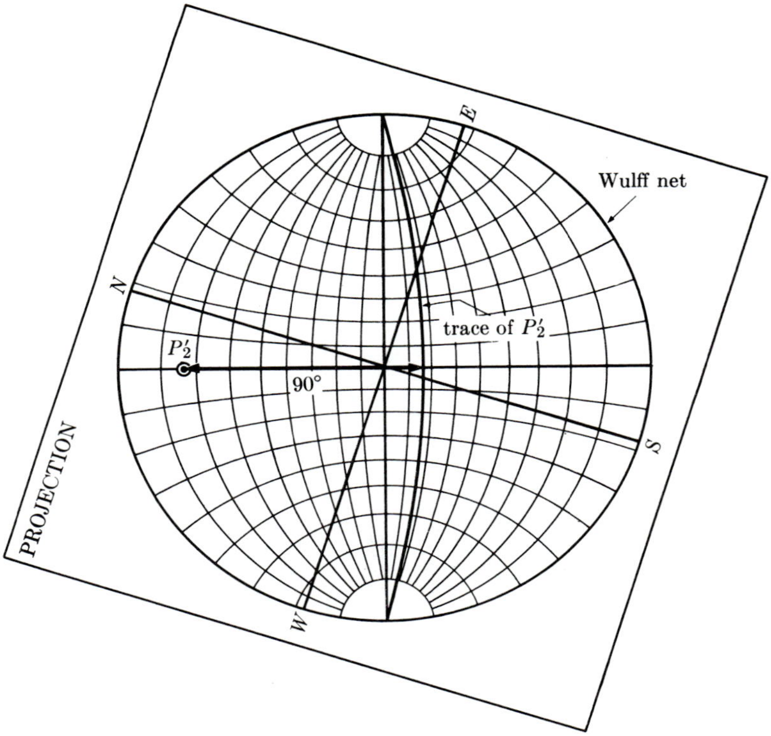

Fig. 2–32 Method of finding the trace of a pole (the pole P_2' in Fig. 2–31).

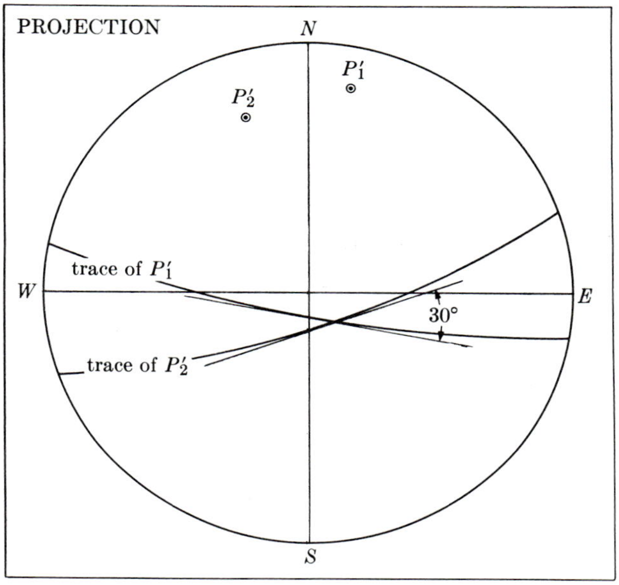

Fig. 2-33 Measurement of an angle between two poles (P_1 and P_2 of Fig. 2-26) by measurement of the angle of intersection of the corresponding traces.

latitude circle as shown. B_1, however, can rotate only 40° before finding itself at the edge of the projection; we must then imagine it to move 20° in from the edge to the point B_1' on the other side of the projection, staying always on its own latitude circle. The final position of this pole on the positive side of the projection is at B_2 diametrically opposite B_1'.

(The student should carefully note that the angle between A_1 and A_2, for example, in Fig. 2-34 is *not* 60°. The pole A_2 is the position of A_1 after a 60° rotation about *NS*, which is not the same thing. Consider the two great circles NA_1S and NA_2S; these are the traces of two planes between which there is a true dihedral angle of 60°. Any pole initially on NA_1S will be on NA_2S after a 60° rotation about *NS*, but the angle between the initial and final positions of the poles will be less than 60°, unless they lie on the equator, and will approach zero as the poles approach *N*.)

Rotation about an axis inclined to the plane of projection is accomplished by compounding rotations about axes lying in and perpendicular to the projection plane. In this case, the given axis must first be rotated into coincidence with one or the other of the two latter axes, the given rotation performed, and the axis then rotated back to its original position. Any movement of the given axis must be accompanied by a similar movement of all the poles on the projection.

For example, we may be required to rotate A_1 about B_1 by 40° in a clockwise direction (Fig. 2-35). In (a) the pole to be rotated A_1 and the rotation axis B_1 are shown in their initial position. In (b) the projection has been rotated to bring B_1

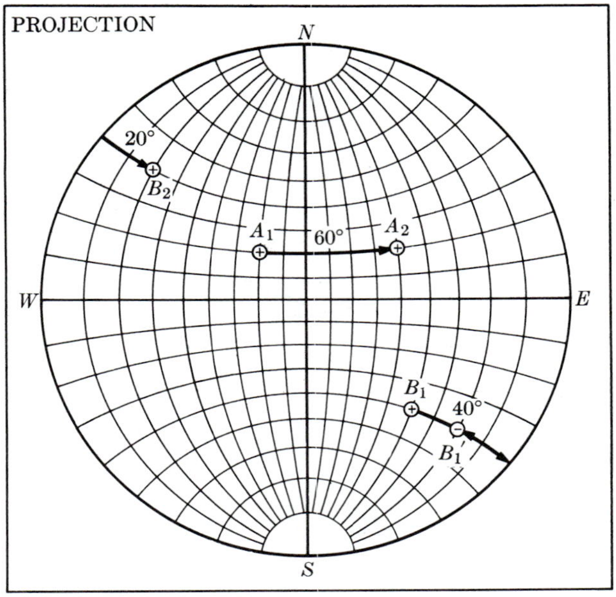

Fig. 2–34 Rotation of poles about *NS* axis of projection.

to the equator of a Wulff net. A rotation of 48° about the *NS* axis of the net brings B_1 to the point B_2 at the center of the net; at the same time A_1 must go to A_2 along a parallel of latitude. The rotation axis is now perpendicular to the projection plane, and the required rotation of 40° brings A_2 to A_3 along a circular path centered on B_2. The operations which brought B_1 to B_2 must now be reversed in order to return B_2 to its original position. Accordingly, B_2 is brought to B_3 and A_3 to A_4, by a 48° reverse rotation about the *NS* axis of the net. In (c) the projection has been rotated back to its initial position, construction lines have been omitted, and only the initial and final positions of the rotated pole are shown. During its rotation about B_1, A_1 moves along the small circle shown. This circle is centered at C on the projection and not at its projected center B_1. To find C we use the fact that all points on the circle must lie at equal *angular* distances from B_1; in this case, measurement on a Wulff net shows that both A_1 and A_4 are 76° from B_1. Accordingly, we locate any other point, such as D, which is 76° from B_1, and, knowing three points on the required circle, we can locate its center C by the methods of plane geometry.

In dealing with problems of crystal orientation a *standard projection* is of very great value, since it shows at a glance the relative orientation of all the important planes in the crystal. Such a projection is made by selecting some important crystal plane of low indices as the plane of projection [e.g., (100), (110), (111), or (0001)] and projecting the poles of various crystal planes onto the selected plane. The construction of a standard projection of a crystal requires a knowledge of the

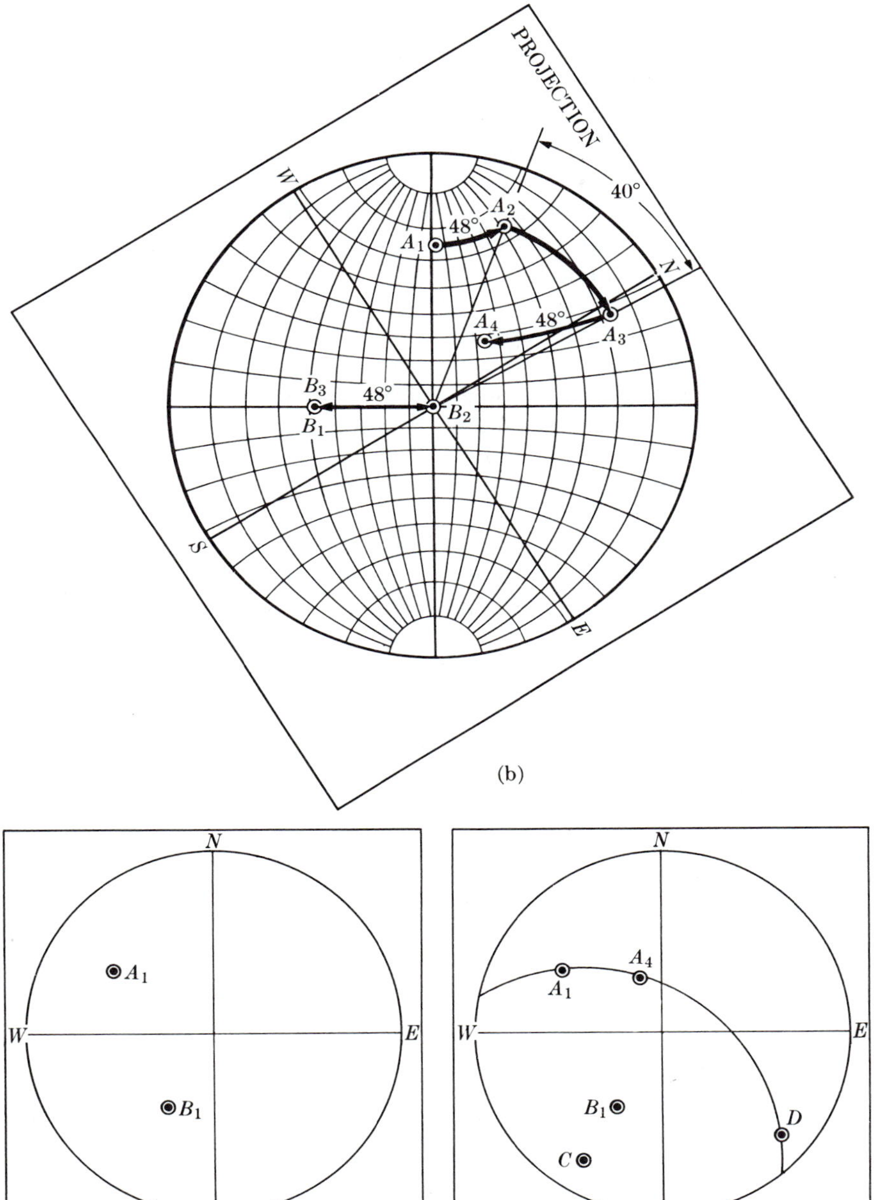

Fig. 2–35 Rotation of a pole about an inclined axis.

interplanar angles for all the principal planes of the crystal. A set of values applicable to all crystals in the cubic system is given in Table 2–3, but those for crystals of other systems depend on the particular axial ratios involved and must be calculated for each case by the equations given in Appendix 3. Much time can be saved in making standard projections by making use of the zonal relation: the normals to all planes belonging to one zone are coplanar and at right angles to the zone axis. Consequently, the poles of planes of a zone will all lie on the same great circle on the projection, and the axis of the zone will be at 90° from this great circle. Furthermore, important planes usually belong to more than one zone and their poles are therefore located at the intersection of zone circles. It is also helpful to remember that important directions, which in the cubic system are normal to planes of the same indices, are usually the axes of important zones.

Figure 2–36(a) shows the principal poles of a cubic crystal projected on the (001) plane of the crystal or, in other words, a standard (001) projection. The location of the {100} cube poles follows immediately from Fig. 2–25. To locate the {110} poles we first note from Table 2–3 that they must lie at 45° from {100} poles, which are themselves 90° apart. In this way we locate (011), for example, on the great circle joining (001) and (010) and at 45° from each. After all the {110} poles are plotted, we can find the {111} poles at the intersection of zone circles. Inspection of a crystal model or drawing or use of the zone relation given by Eq. (2–3) will show that (111), for example, belongs to both the zone [$\bar{1}$01] and the zone [0$\bar{1}$1]. The pole of (111) is thus located at the intersection of the zone circle through (0$\bar{1}$0), (101), and (010) and the zone circle through ($\bar{1}$00), (011), and (100). This location may be checked by measurement of its angular distance from (010) or (100), which should be 54.7°. The (011) standard projection shown in Fig. 2–36(b) is plotted in the same manner. Alternatively, it may be constructed by rotating all the poles in the (001) projection 45° to the left about the *NS* axis of the projection, since this operation will bring the (011) pole to the center. In both of these projections symmetry symbols have been given each pole in conformity with Fig. 2–6(b), and it will be noted that the projection itself has the symmetry of the axis perpendicular to its plane, Figs. 2–36(a) and (b) having 4-fold and 2-fold symmetry, respectively.

Figure 2–37 is a standard (001) projection of a cubic crystal with considerably more detail and a few important zones indicated. A standard (0001) projection of a hexagonal crystal (zinc) is given in Fig. 2–38.

It is sometimes necessary to determine the *Miller indices of a given pole* on a crystal projection, for example the pole *A* in Fig. 2–39(a), which applies to a cubic crystal. If a detailed standard projection is available, the projection with the unknown pole can be superimposed on it and its indices will be disclosed by its coincidence with one of the known poles on the standard. Alternatively, the method illustrated in Fig. 2–39 may be used. The pole *A* defines a direction in space, normal to the plane (*hkl*) whose indices are required, and this direction makes angles ρ, σ, τ with the coordinate axes **a**, **b**, **c**. These angles are measured on the projection as shown in (a). Let the perpendicular distance between the origin and the (*hkl*) plane nearest the origin be *d* [Fig. 2–39(b)], and let the

Table 2–3
Interplanar Angles (in degrees) in Cubic Crystals between Planes of the Form $\{h_1k_1l_1\}$ and $\{h_2k_2l_2\}$

$\{h_2k_2l_2\}$	$\{h_1k_1l_1\}$						
	100	110	111	210	211	221	310
100	0 90						
110	45 90	0 60 90					
111	54.7	35.3 90	0 70.5 109.5				
210	26.6 63.4 90	18.4 50.8 71.6	39.2 75.0	0 36.9 53.1			
211	35.3 65.9	30 54.7 73.2 90	19.5 61.9 90	24.1 43.1 56.8	0 33.6 48.2		
221	48.2 70.5	19.5 45 76.4 90	15.8 54.7 78.9	26.6 41.8 53.4	17.7 35.3 47.1	0 27.3 39.0	
310	18.4 71.6 90	26.6 47.9 63.4 77.1	43.1 68.6	8.1 58.1 45	25.4 49.8 58.9	32.5 42.5 58.2	0 25.9 36.9
311	25.2 72.5	31.5 64.8 90	29.5 58.5 80.0	19.3 47.6 66.1	10.0 42.4 60.5	25.2 45.3 59.8	17.6 40.3 55.1
320	33.7 56.3 90	11.3 54.0 66.9	36.9 80.8	7.1 29.8 41.9	25.2 37.6 55.6	22.4 42.3 49.7	15.3 37.9 52.1
321	36.7 57.7 74.5	19.1 40.9 55.5	22.2 51.9 72.0 90	17.0 33.2 53.3	10.9 29.2 40.2	11.5 27.0 36.7	21.6 32.3 40.5
331	46.5	13.1	22.0				
510	11.4						
511	15.6						
711	11.3						

Largely from R. M. Bozorth, *Phys. Rev.* **26**, 390 (1925); rounded off to the nearest 0.1°. A much longer list is given on p. 120–122 of Vol. 2 of [G.11].

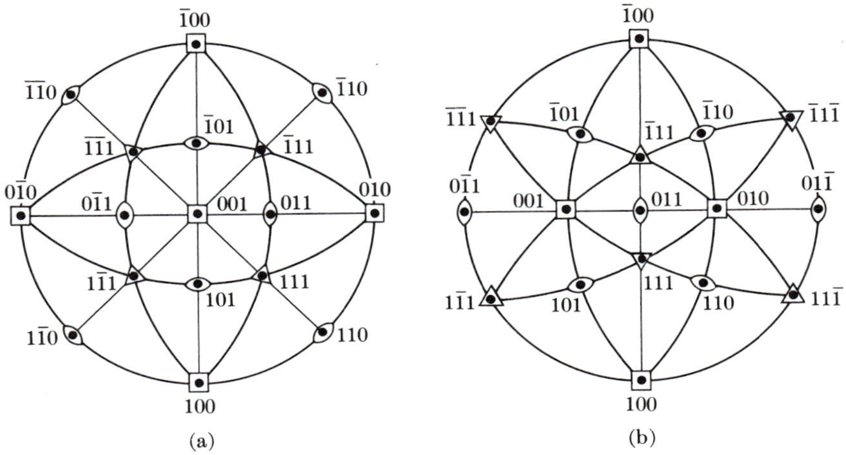

Fig. 2–36 Standard projections of cubic crystals, (a) on (001) and (b) on (011).

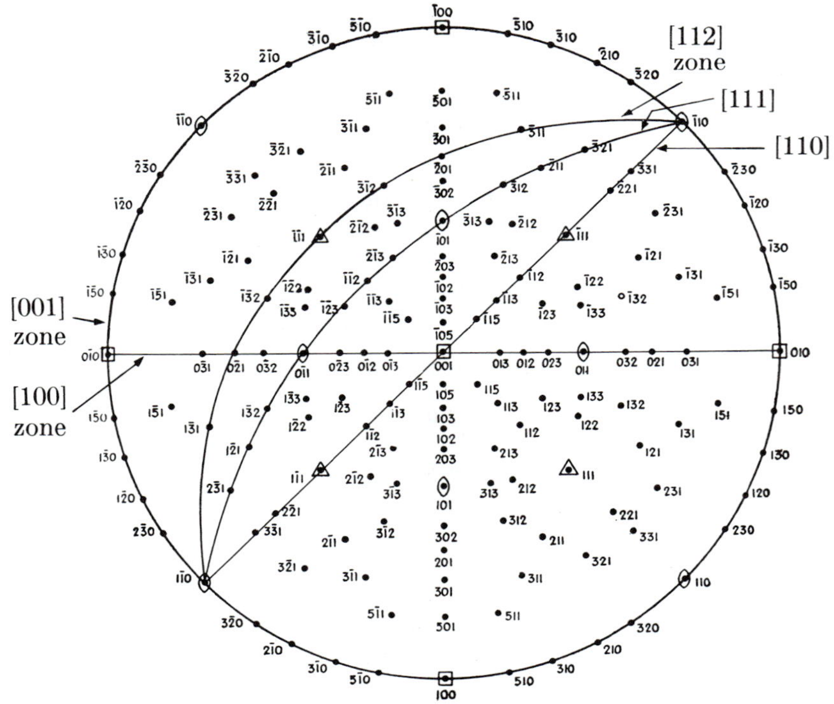

Fig. 2–37 Standard (001) projection of a cubic crystal, after Barrett [1.7].

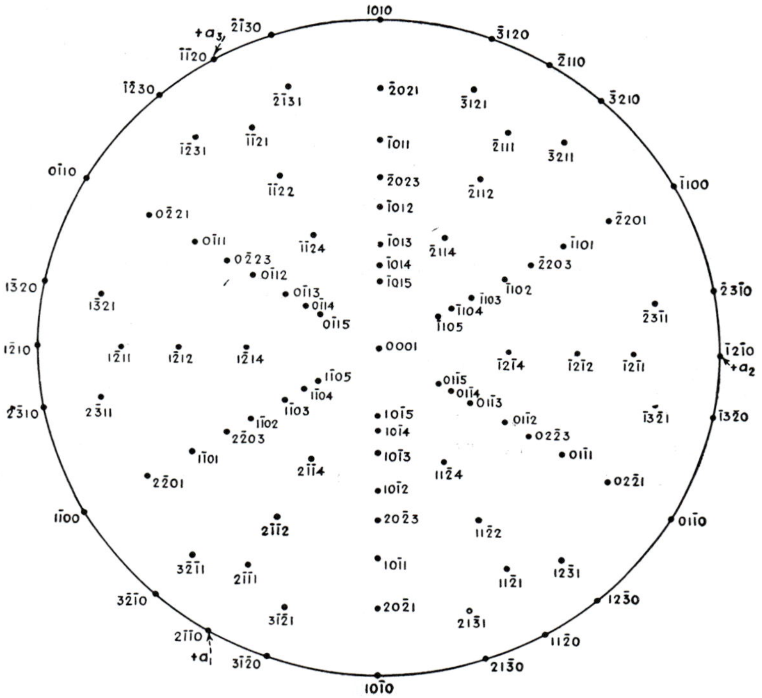

Fig. 2-38 Standard (0001) projection for zinc (hexagonal, $c/a = 1.86$) after Barrett [1.7].

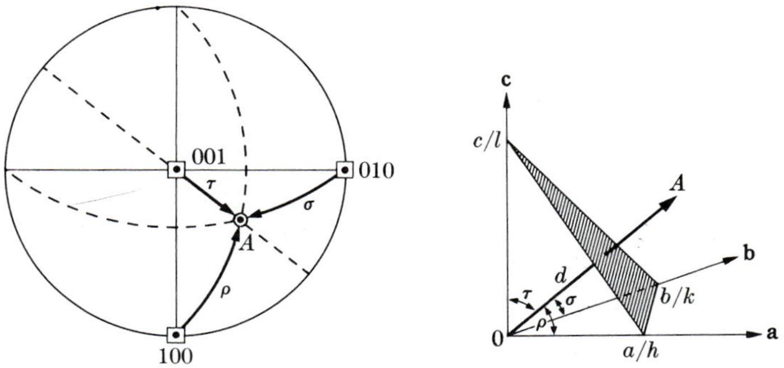

(a) (b)

Fig. 2-39 Determination of the Miller indices of a pole.

direction cosines of the line A be p, q, r. Therefore

$$p = \cos\rho = \frac{d}{a/h}, \qquad q = \cos\sigma = \frac{d}{b/k}, \qquad r = \cos\tau = \frac{d}{c/l},$$

$$h:k:l = pa:qb:rc. \tag{2-8}$$

For the cubic system we have the simple result that the Miller indices required are in the same ratio as the direction cosines.

The lattice reorientation caused by *twinning* can be clearly shown on the stereographic projection. In Fig. 2–40 the open symbols are the {100} poles of a cubic crystal projected on the (001) plane. If this crystal is FCC, then one of its possible twin planes is ($\bar{1}\bar{1}1$), represented on the projection both by its pole and its trace. The cube poles of the twin formed by reflection in this plane are shown as solid symbols; these poles are located by rotating the projection on a Wulff net until the pole of the twin plane lies on the equator, after which the cube poles of the crystal can be moved along latitude circles of the net to their final position.

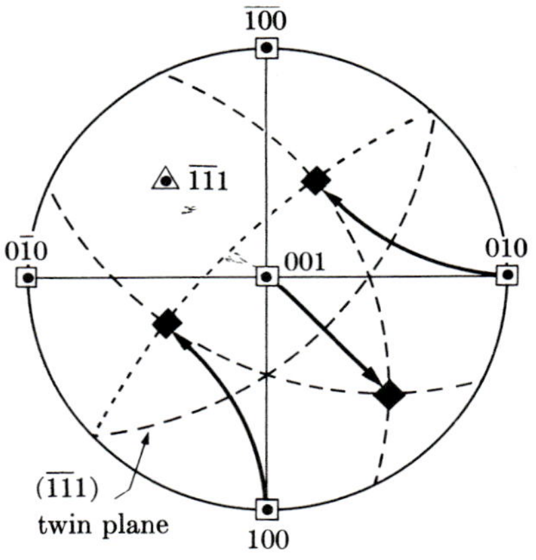

Fig. 2–40 Stereographic projection of an FCC crystal and its twin.

The main principles of the stereographic projection have now been presented, and we will have occasion to use them later in dealing with various practical problems in x-ray crystallography. The student is reminded, however, that a mere reading of this section is not sufficient preparation for such problems. In order to gain real familiarity with the stereographic projection, he must practice, with Wulff net and tracing paper, the operations described above and solve problems of the kind given below. Only in this way will he be able to read and manipulate the stereographic projection with facility and think in three dimensions of what is represented in two.

PROBLEMS

2–1 Draw the following planes and directions in a tetragonal unit cell: (001), (011), (113), [110], [201], [$\bar{1}$01]. Show cell axes.

2–2 Show by means of a (1$\bar{1}$0) sectional drawing that [111] is perpendicular to (111) in the cubic system, but not, in general, in the tetragonal system.

2–3 In a drawing of a hexagonal prism, indicate the following planes and directions: (1$\bar{2}$10), (10$\bar{1}$2), ($\bar{1}$011), [110], [11$\bar{1}$], [021]. Show cell axes.

2–4 Derive Eq. (2–2) of the text.

2–5 Show that the planes (1$\bar{1}$0), (1$\bar{2}$1), and ($\bar{3}$12) belong to the zone [111].

2–6 Do the following planes all belong to the same zone: ($\bar{1}$10), ($\bar{3}$11), ($\bar{1}\bar{3}$2)? If so, what is the zone axis? Give the indices of any other plane belonging to this zone.

* **2–7** Prepare a cross-sectional drawing of an HCP structure which will show that all atoms do not have identical surroundings and therefore do not lie on a point lattice.

2–8 Show that c/a for hexagonal close packing of spheres is 1.633.

2–9 Show that the HCP structure (with $c/a = 1.633$) and the FCC structure are equally close-packed, and that the BCC structure is less closely packed than either of the former.

2–10 The unit cells of several orthorhombic crystals are described below. What is the Bravais lattice of each and how do you know? Do not change axes. (In solving this kind of problem, examining the given atom positions for the existence or nonexistence of centering translations is generally more helpful than making a drawing of the structure.)

a) Two atoms of the same kind per unit cell located at $0\ \frac{1}{2}\ 0$, $\frac{1}{2}\ 0\ \frac{1}{2}$.
b) Four atoms of the same kind per unit cell located at $0\ 0\ z$, $0\ \frac{1}{2}\ z$, $0\ \frac{1}{2}\ (\frac{1}{2} + z)$, $0\ 0\ (\frac{1}{2} + z)$.
c) Four atoms of the same kind per unit cell located at $x\ y\ z$, $\bar{x}\ \bar{y}\ z$, $(\frac{1}{2} + x)\ (\frac{1}{2} - y)\ \bar{z}$, $(\frac{1}{2} - x)\ (\frac{1}{2} + y)\ \bar{z}$.
d) Two atoms of one kind A located at $\frac{1}{2}\ 0\ 0$, $0\ \frac{1}{2}\ \frac{1}{2}$; and two atoms of another kind B located at $0\ 0\ \frac{1}{2}$, $\frac{1}{2}\ \frac{1}{2}\ 0$.

***2–11** Make a drawing, similar to Fig. 2–23, of a (112) twin in a BCC lattice and show the shear responsible for its formation. Obtain the magnitude of the shear strain graphically.

2–12 Construct a Wulff net, 18 cm in diameter and graduated at 30° intervals, by the use of compass, dividers, and straightedge only. Show all construction lines.

In some of the following problems, the coordinates of a point on a stereographic projection are given in terms of its latitude and longitude, measured from the center of the projection. Thus, the N pole is 90°N, 0°E, the E pole is 0°N, 90°E, etc.

2–13 Plane A is represented on a stereographic projection by a great circle passing through the N and S poles and the point 0°N, 70°W. The pole of plane B is located at 30°N, 50°W.

a) Find the angle between the two planes by measuring the angle between the poles of A and B.
b) Draw the great circle of plane B and demonstrate that the stereographic projection is angle-true by measuring with a protractor the angle between the great circles of A and B.

***2–14** Pole A, whose coordinates are 20°N, 50°E, is to be rotated about the axes described below. In each case, find the coordinates of the final position of pole A and show the path traced out during its rotation.

a) 100° rotation about the NS axis, counterclockwise looking from N to S.
b) 60° rotation about an axis normal to the plane of projection, clockwise to the observer.
c) 60° rotation about an inclined axis B, whose coordinates are 10°S, 30°W, clockwise to the observer.

2–15 Draw a standard (111) projection of a cubic crystal, showing all poles of the form {100}, {110}, {111} and the important zone circles between them. Compare with Figs. 2–36(a) and (b).

2–16 Draw a standard (001) projection of white tin (tetragonal, $c/a = 0.545$), showing all poles of the form {001}, {100}, {110}, {011}, {111} and the important zone circles between them. Compare with Fig. 2–36(a).

2–17 Draw a standard (0001) projection of beryllium (hexagonal, $c/a = 1.57$), showing all poles of the form $\{2\bar{1}\bar{1}0\}$, $\{10\bar{1}0\}$, $\{2\bar{1}\bar{1}1\}$, $\{10\bar{1}1\}$ and the important zone circles between them. Compare with Fig. 2–38.

2–18 On a standard (001) projection of a cubic crystal, in the orientation of Fig. 2–36(a), the pole of a certain plane has coordinates 53.3°S, 26.6°E. What are its Miller indices? Verify your answer by comparison of measured angles with those given in Table 2–3.

***2–19** Duplicate the operations shown in Fig. 2–40 and thus find the locations of the cube poles of a $(\bar{1}\bar{1}1)$ reflection twin in a cubic crystal. What are their coordinates?

2–20 Show that the twin orientation found in Prob. 2–19 can also be obtained by

a) Reflection in a {112} plane. Which one?
b) 180° rotation about a ⟨111⟩ axis. Which one?
c) 60° rotation about a ⟨111⟩ axis. Which one?

In (c), show the paths traced out by the cube poles during their rotation.

***2–21** Plot the great-circle route from Washington, D.C. (39°N, 77°W) to Moscow (56°N, 38°E).

a) What is the distance between the two cities? (Radius of the earth = 6360 km.)
b) What is the true bearing of an airplane flying from Washington to Moscow at the beginning, midpoint, and end of the trip? (The bearing is the angle measured clockwise from north to the flight direction. Thus east is 90° and west is 270°.)

3

Diffraction I: Directions of Diffracted Beams

3-1 INTRODUCTION

After our preliminary survey of the physics of x-rays and the geometry of crystals, we can now proceed to fit the two together and discuss the phenomenon of x-ray diffraction, which is an interaction of the two. Historically, this is exactly the way this field of science developed. For many years, mineralogists and crystallographers had accumulated knowledge about crystals, chiefly by measurement of interfacial angles, chemical analysis, and determination of physical properties. There was little knowledge of interior structure, however, although some very shrewd guesses had been made, namely, that crystals were built up by periodic repetition of some unit, probably an atom or molecule, and that these units were situated some 1 or 2 Å apart. On the other hand, there were indications, but only indications, that x-rays might be electromagnetic waves about 1 or 2 Å in wavelength. In addition, the phenomenon of diffraction was well understood, and it was known that diffraction, as of visible light by a ruled grating, occurred whenever wave motion encountered a set of regularly spaced scattering objects, provided that the wavelength of the wave motion was of the same order of magnitude as the repeat distance between the scattering centers.

Such was the state of knowledge in 1912 when the German physicist von Laue (1879–1960) took up the problem. He reasoned that, *if* crystals were composed of regularly spaced atoms which might act as scattering centers for x-rays, and *if* x-rays were electromagnetic waves of wavelength about equal to the interatomic distance in crystals, then it should be possible to diffract x-rays by means of crystals. Under his direction, experiments to test this hypothesis were carried out: a crystal of copper sulfate was set up in the path of a narrow beam of x-rays and a photographic plate was arranged to record the presence of diffracted beams, if any. The second attempt was successful and showed without doubt that x-rays *were* diffracted by the crystal out of the primary beam to form a pattern of spots on the photographic plate. These experiments proved, at one and the same time, the wave nature of x-rays and the periodicity of the arrangement of atoms within a crystal. Hindsight is always easy and these ideas appear quite simple to us now, when viewed from the vantage point of more than sixty years' development of the subject, but they were not at all obvious in 1912, and von Laue's hypothesis and its experimental verification must stand as a great intellectual achievement [3.1].

The account of these experiments was read with great interest by two English physicists, W. H. Bragg (1862–1942) and his son W. L. Bragg (1890–1971). The latter, although only a young student at the time—it was still the year 1912—successfully analyzed the Laue experiment and was able to express the necessary conditions for diffraction in a considerably simpler mathematical form than that used by von Laue. He also attacked the problem of crystal structure with the new tool of x-ray diffraction and, in the following year, solved the structures of NaCl, KCl, KBr, and KI, all of which have the NaCl structure; these were the first complete crystal-structure determinations ever made [3.2]. The simpler structures of metals like iron and copper were not determined until later.

3–2 DIFFRACTION

Diffraction is due essentially to the existence of certain phase relations between two or more waves, and it is advisable, at the start, to get a clear notion of what is meant by phase relations. Consider a beam of x-rays, such as beam **1** in Fig. 3–1, proceeding from left to right. For convenience only, this beam is assumed to be plane-polarized in order that we may draw the electric field vector **E** always in one plane. We may imagine this beam to be composed of two equal parts, ray **2** and ray **3**, each of half the amplitude of beam **1**. These two rays, on the wave front AA', are said to be completely *in phase* or in step; i.e., their electric-field vectors have the same magnitude and direction at the same instant at any point x measured along the direction of propagation of the wave. A *wave front* is a surface perpendicular to this direction of propagation.

Now consider an imaginary experiment, in which ray **3** is allowed to continue in a straight line but ray **2** is diverted by some means into a curved path before

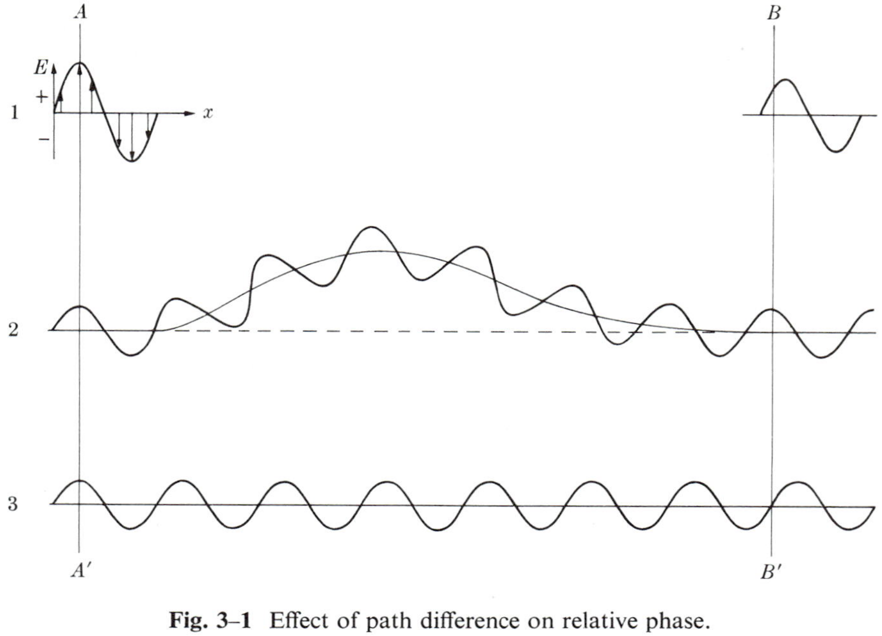

Fig. 3–1 Effect of path difference on relative phase.

rejoining ray **3**. What is the situation on the wave front BB' where both rays are proceeding in the original direction? On this front, the electric vector of ray **2** has its maximum value at the instant shown, but that of ray **3** is zero. The two rays are therefore *out of phase*. If we add these two imaginary components of the beam together, we find that beam **1** now has the form shown in the upper right of the drawing. If the amplitudes of rays **2** and **3** are each 1 unit, then the amplitude of beam **1** at the left is 2 units and that of beam **1** at the right is 1.4 units, if a sinusoidal variation of **E** with x is assumed.

Two conclusions may be drawn from this illustration:

1. Differences in the length of the path traveled lead to differences in phase.
2. The introduction of phase differences produces a change in amplitude.

The greater the path difference, the greater the difference in phase, since the path difference, measured in wavelengths, exactly equals the phase difference, also measured in wavelengths. If the diverted path of ray **2** in Fig. 3–1 were a quarter wavelength longer than shown, the phase difference would be a half wavelength. The two rays would then be completely out of phase on the wave front BB' and beyond, and they would therefore annul each other, since at any point their electric vectors would be either both zero or of the same magnitude and opposite in direction. If the difference in path length were made three quarters of a wavelength greater than shown, the two rays would be one complete wavelength out of phase, a condition indistinguishable from being completely in phase since in both cases the two waves would combine to form a beam of amplitude 2 units, just like the original beam. We may conclude that two rays are completely in phase whenever their path lengths differ either by zero or by a whole number of wavelengths.

Differences in the path length of various rays arise quite naturally when we consider how a crystal diffracts x-rays. Figure 3–2 shows a section of a crystal, its atoms arranged on a set of parallel planes A, B, C, D, . . . , normal to the plane of the drawing and spaced a distance d' apart. Assume that a beam of perfectly parallel, perfectly monochromatic x-rays of wavelength λ is incident on this crystal at an angle θ, called the Bragg angle, where θ is measured between the incident beam and the particular crystal planes under consideration.

We wish to know whether this incident beam of x-rays will be diffracted by the crystal and, if so, under what conditions. *A diffracted beam may be defined as a beam composed of a large number of scattered rays mutually reinforcing one another.* Diffraction is, therefore, essentially a scattering phenomenon and not one involving any "new" kind of interaction between x-rays and atoms. We saw in Sec. 1–5 that atoms scatter incident x-rays in all directions, and we shall see presently that in some of these directions the scattered beams will be completely in phase and so reinforce each other to form diffracted beams.

For the particular conditions described by Fig. 3–2 the only diffracted beam formed is that shown, namely one making an angle θ of reflection* equal to the

* Note that these angles are defined differently in x-ray diffraction and in general optics. In the latter, the angles of incidence and reflection are the angles which the incident and reflected beams make with the *normal* to the reflecting surface.

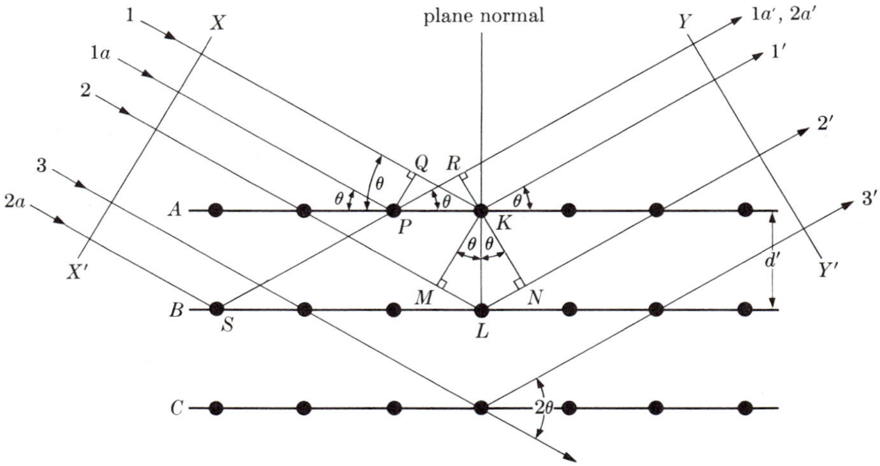

Fig. 3–2 Diffraction of x-rays by a crystal.

angle θ of incidence. We will show this, first, for one plane of atoms and, second, for all the atoms making up the crystal. Consider rays **1** and **1a** in the incident beam; they strike atoms K and P in the first plane of atoms and are scattered in all directions. Only in the directions **1'** and **1a'**, however, are these scattered beams completely in phase and so capable of reinforcing one another; they do so because the difference in their length of path between the wave fronts XX' and YY' is equal to

$$QK - PR = PK \cos\theta - PK \cos\theta = 0.$$

Similarly, the rays scattered by all the atoms in the first plane in a direction parallel to **1'** are in phase and add their contributions to the diffracted beam. This will be true of all the planes separately, and it remains to find the condition for reinforcement of rays scattered by atoms in different planes. Rays **1** and **2**, for example, are scattered by atoms K and L, and the path difference for rays **1**K**1'** and **2**L**2'** is

$$ML + LN = d' \sin\theta + d' \sin\theta.$$

This is also the path difference for the overlapping rays scattered by S and P in the direction shown, since in this direction there is no path difference between rays scattered by S and L or P and K. Scattered rays **1'** and **2'** will be completely in phase if this path difference is equal to a whole number n of wavelengths, or if

$$n\lambda = 2d' \sin\theta. \tag{3-1}$$

This relation was first formulated by W. L. Bragg and is known as the Bragg law. It states the essential condition which must be met if diffraction is to occur. n is called the order of reflection; it may take on any integral value consistent with $\sin\theta$ not exceeding unity and is equal to the number of wavelengths in the path difference between rays scattered by *adjacent* planes. Therefore, for fixed values of λ and d', there may be several angles of incidence $\theta_1, \theta_2, \theta_3 \ldots$ at which diffraction may occur, corresponding to $n = 1, 2, 3, \ldots$. In a first-order reflection

($n = 1$), the scattered rays **1′** and **2′** of Fig. 3–2 would differ in length of path (and in phase) by one wavelength, rays **1′** and **3′** by two wavelengths, rays **1′** and **4′** by three wavelengths, and so on throughout the crystal. The rays scattered by all the atoms in all the planes are therefore completely in phase and reinforce one another (constructive interference) to form a diffracted beam in the direction shown. In all other directions of space the scattered beams are out of phase and annul one another (destructive interference). The diffracted beam is rather strong compared to the sum of all the rays scattered in the same direction, simply because of the reinforcement which occurs,* but extremely weak compared to the incident beam since the atoms of a crystal scatter only a small fraction of the energy incident on them.

It is helpful to distinguish three scattering modes:

1. By atoms arranged randomly in space, as in a monatomic gas. This scattering occurs in *all* directions and is weak. Intensities add.

2. By atoms arranged periodically in space, as in a perfect crystal:

a) In a very few directions, those satisfying the Bragg law, the scattering is strong and is called diffraction. Amplitudes add.

b) In most directions, those not satisfying the Bragg law, there is *no* scattering because the scattered rays cancel one another.

At first glance, the diffraction of x-rays by crystals and the reflection of visible light by mirrors appear very similar, since in both phenomena the angle of incidence is equal to the angle of reflection. It seems that we might regard the planes of atoms as little mirrors which "reflect" the x-rays. Diffraction and reflection, however, differ fundamentally in at least three aspects:

1. The diffracted beam from a crystal is built up of rays scattered by all the atoms of the crystal which lie in the path of the incident beam. The reflection of visible light takes place in a thin surface layer only.

2. The diffraction of monochromatic x-rays takes place only at those particular angles of incidence which satisfy the Bragg law. The reflection of visible light takes place at any angle of incidence.

---

* If the scattering atoms were not arranged in a regular, periodic fashion but in some independent manner, then the rays scattered by them would have a random phase relationship to one another. In other words, there would be an equal probability of the phase difference between any two scattered rays having any value between zero and one wavelength. Neither constructive nor destructive interference takes place under these conditions, and the intensity of the beam scattered in a particular direction is simply *the sum of the intensities* of all the rays scattered in that direction. If there are N scattered rays each of amplitude A and therefore of intensity A^2 in arbitrary units, then the intensity of the scattered beam is NA^2. On the other hand, if the rays are scattered by the atoms of a crystal in a direction satisfying the Bragg law, then they are all in phase and the amplitude of the scattered beam is *the sum of the amplitudes* of the scattered rays. The total amplitude is then N times the amplitude A of each scattered ray, or NA. The intensity of the scattered beam is therefore N^2A^2, or N times as large as if reinforcement had not occurred. Since N is very large for the scattering of x-rays from even a small bit of crystal, ($N = 1.1 \times 10^{19}$ atoms for 1 mg of iron), the role of reinforcement in producing a strong diffracted beam is considerable.

3. The reflection of visible light by a good mirror is almost 100 percent efficient. The intensity of a diffracted x-ray beam is extremely small compared to that of the incident beam.

Despite these differences, we often speak of "reflecting planes" and "reflected beams" when we really mean diffracting planes and diffracted beams. This is common usage and, from now on, we will frequently use these terms without quotation marks but with the tacit understanding that we really mean diffraction and not reflection.*

To sum up, diffraction is essentially a scattering phenomenon in which a large number of atoms cooperate. Since the atoms are arranged periodically on a lattice, the rays scattered by them have definite phase relations between them; these phase relations are such that destructive interference occurs in most directions of scattering, but in a few directions constructive interference takes place and diffracted beams are formed. The two essentials are a wave motion capable of interference (x-rays) and a set of periodically arranged scattering centers (the atoms of a crystal).

3-3 THE BRAGG LAW

Two geometrical facts are worth remembering: (1) The incident beam, the normal to the reflecting plane, and the diffracted beam are always coplanar. (2) The angle between the diffracted beam and the transmitted beam is always 2θ. This is known as the diffraction angle, and it is this angle, rather than θ, which is usually measured experimentally.

As previously stated, diffraction in general occurs only when the wavelength of the wave motion is of the same order of magnitude as the repeat distance between scattering centers. This requirement follows from the Bragg law. Since $\sin\theta$ cannot exceed unity, we may write

$$\frac{n\lambda}{2d'} = \sin\theta < 1. \tag{3-2}$$

Therefore, $n\lambda$ must be less than $2d'$. For diffraction, the smallest value of n is 1. ($n = 0$ corresponds to the beam diffracted in the same direction as the transmitted beam. It cannot be observed.) Therefore the condition for diffraction at any observable angle 2θ is

$$\lambda < 2d'. \tag{3-3}$$

For most sets of crystal planes d' is of the order of 3 Å or less, which means that λ cannot exceed about 6 Å. A crystal could not possibly diffract ultraviolet radiation, for example, of wavelength about 500 Å. On the other hand, if λ is very small, the diffraction angles are too small to be conveniently measured.

* For the sake of completeness, it should be mentioned that x-rays *can* be totally reflected by a solid surface, just as visible light is by a mirror, but only at very small angles of incidence (below about one degree).

The Bragg law may be written in the form

$$\lambda = 2\frac{d'}{n}\sin\theta. \tag{3-4}$$

Since the coefficient of λ is now unity, we can consider a reflection of any order as a first-order reflection from planes, real or fictitious, spaced at a distance $1/n$ of the previous spacing. This turns out to be a real convenience, so we set $d = d'/n$ and write the Bragg law in the form

$$\boxed{\lambda = 2d\sin\theta}. \tag{3-5}$$

This form will be used throughout this book.

This usage is illustrated by Fig. 3-3. Consider the second-order 100 reflection* shown in (a). Since it is second-order, the path difference ABC between rays scattered by adjacent (100) planes must be two whole wavelengths. If there is no real plane of atoms between the (100) planes, we can always imagine one as in Fig. 3-3(b), where the dotted plane midway between the (100) planes forms part of the (200) set of planes. For the same reflection as in (a), the path difference DEF between rays scattered by adjacent (200) planes is now only one whole wavelength, so that this reflection can properly be called a first-order 200 reflection. Similarly, 300, 400, etc., reflections are equivalent to reflections of the third, fourth, etc., orders from the (100) planes. In general, an nth-order reflection from (hkl) planes of spacing d' may be considered as a first-order reflection from the $(nh\ nk\ nl)$ planes of spacing $d = d'/n$. Note that this convention is in accord with the definition of Miller indices since $(nh\ nk\ nl)$ are the Miller indices of planes parallel to the (hkl) planes but with $1/n$th the spacing of the latter.

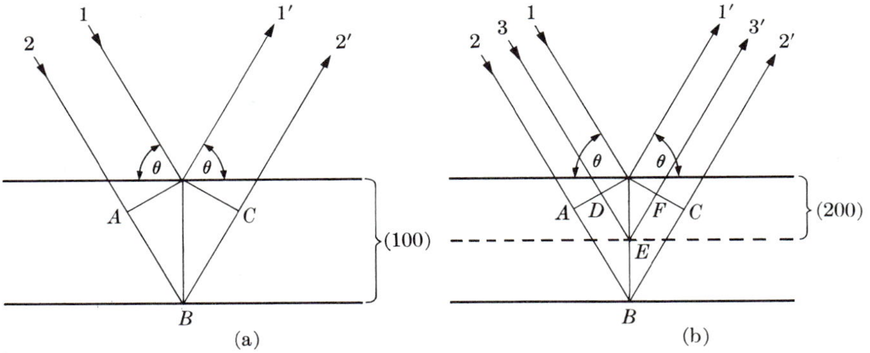

Fig. 3-3 Equivalence of (a) a second-order 100 reflection and (b) a first-order 200 reflection.

* This means the reflection from the (100) planes. Conventionally, the Miller indices of a reflecting plane hkl, written without parentheses, stand for the reflected beam from the plane (hkl).

3-4 X-RAY SPECTROSCOPY

Experimentally, the Bragg law can be applied in two ways. By using x-rays of known wavelength λ and measuring θ, we can determine the spacing d of various planes in a crystal: this is *structure analysis* and is the subject, in one way or another, of the greater part of this book. Alternatively, we can use a crystal with planes of known spacing d, measure θ, and thus determine the wavelength λ of the radiation used: this is *x-ray spectroscopy.*

The essential features of an x-ray spectrometer are shown in Fig. 3-4. X-rays from the tube T are incident on a crystal C which may be set at any desired angle to the incident beam by rotation about an axis through O, the center of the spectrometer circle. D is a counter which measures the intensity of the diffracted x-rays; it can also be rotated about O and set at any desired angular position. The crystal is usually cut or cleaved so that a particular set of reflecting planes of known spacing is parallel to its surface, as suggested by the drawing. In use, the crystal is positioned so that its reflecting planes make some particular angle θ with the incident beam, and D is set at the corresponding angle 2θ. The intensity of the diffracted beam is then measured and its wavelength calculated from the Bragg law, this procedure being repeated for various angles θ. It is in this way that curves such as Fig. 1-5 and the characteristic wavelengths tabulated in Appendix 7 were obtained. W. H. Bragg designed and used the first x-ray spectrometer, and the Swedish physicist Siegbahn developed it into an instrument of very high precision.

Except for one application, the subject of chemical analysis described in Chap. 15, we are here concerned with x-ray spectroscopy only insofar as it concerns certain units of wavelength. Wavelength measurements made in the way just described are obviously relative, and their accuracy is no greater than the accuracy with which the plane spacing of the crystal is known.

Before considering how the first plane spacing was determined, we must digress to consider the subject of *x-ray density*. Normally the density of a solid is found by measuring the volume, usually of the order of a few cubic centimeters, and the weight of a particular specimen. But x-ray diffraction allows us to determine the lattice parameters of a crystal's unit cell, and therefore its volume, together with the number of atoms in the cell. We can therefore base a density

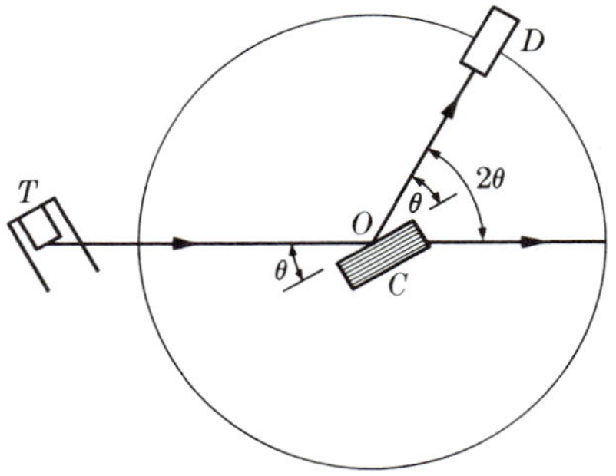

Fig. 3-4 The x-ray spectrometer.

determination, not on a few cubic centimeters but on the volume of a single unit cell, by defining the

$$\text{x-ray density} = \frac{\text{weight of atoms in unit cell}}{\text{volume of unit cell}}.$$

$$\rho = \frac{\sum A/N}{V} \tag{3-6}$$

where ρ = density (gm/cm^3), $\sum A$ = sum of the atomic weights of all the atoms in the unit cell, N = Avogadro's number, and V = volume of unit cell (cm^3). Inserting the current value of N, we have

$$\rho = \frac{\sum A}{NV} = \frac{\sum A}{(6.02257 \times 10^{23})(V' \times 10^{-24})} = \frac{1.66042 \sum A}{V'} \tag{3-7}$$

where ρ is in gm/cm^3 and V' is the unit-cell volume in Å^3.

The macroscopic density of a particular specimen, determined from the weight and volume of that specimen, is usually less than, and cannot exceed, the x-ray density, because the macroscopic specimen will usually contain minute cracks and pores. The x-ray density is therefore a useful quantity to know. By comparing it to the macroscopic density of, for example, a pressed and sintered metal or ceramic compact, one can calculate the percent porosity in the compact. X-ray densities are sometimes loosely called "theoretical densities"; they are not theoretical because they are determined experimentally.

To return to the problem of wavelength determination, it is an interesting and crucial fact that Bragg was able to solve the cyrstal structure of NaCl without knowing the wavelength of the x-rays being diffracted. All he knew—all he needed to know—was that there was one *single*, strong wavelength in the radiation from the x-ray tube, namely, the strong $K\alpha$ line of the tube target. Once the NaCl structure is known [Fig. 2-18(b)], it follows that there are four sodium and four chlorine atoms per unit cell, and that

$$\sum A = 4(\text{at. wt. Na}) + 4(\text{at. wt. Cl}).$$

If this value is inserted into Eq. (3-7) together with the macroscopic density ρ, the volume V' of the unit cell can be found. Because NaCl is cubic, the lattice parameter a is given simply by the cube root of V'. From this value of a and the cubic plane-spacing equation (Eq. 2-5), the spacing of any set of planes can be found.

In this way, Siegbahn obtained a value of 2.814 Å for the spacing of the (200) planes of rock salt (NaCl), which he could use as a basis for wavelength measurements. This spacing was known to only four significant figures, because it was derived from a macroscopic density of that precision. However, Siegbahn was able to measure wavelengths in terms of this spacing much more accurately, namely, to six significant figures. Not wishing to throw away the high relative precision he could attain, he wisely decided to arbitrarily define a new unit in which relative wavelengths could be expressed. This was the X unit (XU), so called because its true value in absolute units (angstroms) was unknown. By defining the

(200) spacing of rock salt to six significant figures as 2814.00 XU, the new unit was made as nearly as possible equal to 0.001 Å.

Once a particular wavelength was determined in terms of this spacing, the spacing of a given set of planes in any other crystal could be measured. Siegbahn thus measured the (211) spacing of calcite ($CaCO_3$), which he found more suitable as a standard crystal, and thereafter based all his wavelength measurements on this spacing. Its value is 3029.45 XU. Later on, the kilo X unit (kX) was introduced, a thousand times as large as the X unit and nearly equal to an angstrom. The kX unit is therefore *defined* by the relation

$$1 \text{ kX} = \frac{\text{(211) plane spacing of calcite}}{3.02945}. \tag{3-8}$$

On this basis, Siegbahn and his associates made very accurate measurements of wavelength in relative (kX) units and these measurements form the basis of most published wavelength tables.

It was found later that x-rays could be diffracted by a ruled grating such as is used in the spectroscopy of visible light, provided that the angle of incidence (the angle between the incident beam and the plane of the grating) is kept below the critical angle for total reflection. Gratings thus offer a means of making absolute wavelength measurements, independent of any knowledge of crystal structure. By a comparison of values so obtained with those found by Siegbahn from crystal diffraction, it was possible to calculate the following relation between the relative and absolute units:

$$1 \text{ kX} = 1.00202 \text{ Å}.$$

This conversion factor was decided on in 1946 by international agreement. Later work improved the accuracy of this factor, and the relation is now believed to be

$$\boxed{1 \text{ kX} = 1.002056 \text{ Å*}}. \tag{3-9}$$

Note that this relation is stated in terms of still another unit, the Å* unit, which was introduced because of the still remaining uncertainty in the conversion factor. The difference between Å and Å* is only some five parts per million, and the distinction between the two units is negligible except in work of the very highest accuracy.

The present situation is not entirely clear, but the wavelength tables published in 1974 by the International Union of Crystallography [Vol. 4, G.11], which are reproduced in Appendix 7, are based on Eq. (3–9).

The distinction between kX and Å is unimportant if no more than about three significant figures are involved, because the kX unit is only about 0.2 percent larger than the angstrom. In precise work, on the other hand, units must be correctly stated, and on this point there has been considerable confusion in the past. Some wavelength values published prior to about 1946 are stated to be in angstrom units but are actually in kX units. Some crystallographers have used such a value as the basis for a precise measurement of the lattice parameter of a crystal and the result has been stated, again incorrectly, in angstrom units. Many published parameters are therefore in error, and it is unfortunately not always easy to determine which ones are and which ones are not. The only safe rule to follow, in stating a precise parameter, is to give the wavelength of the radiation used in its determination. Similarly, any published table of wavelengths can be tested for the correctness of its units by noting the wavelength given for a particular characteristic line, Cu $K\alpha_1$ for example. The wavelength of this line is 1.540562 Å* (1974 value, 1.002056 as conversion factor), 1.54051 Å (1946 value, 1.00202 factor), or 1.53740 kX. See Appendix 7 for the estimated accuracy of the wavelengths listed there.

3-5 DIFFRACTION DIRECTIONS

What determines the possible directions, i.e., the possible angles 2θ, in which a given crystal can diffract a beam of monochromatic x-rays? Referring to Fig. 3-3, we see that various diffraction angles $2\theta_1, 2\theta_2, 2\theta_3, \ldots$ can be obtained from the (100) planes by using a beam incident at the correct angle $\theta_1, \theta_2, \theta_3, \ldots$ and producing first-, second-, third-, ... order reflections. But diffraction can also be produced by the (110) planes, the (111) planes, the (213) planes, and so on. We obviously need a general relation which will predict the diffraction angle for *any* set of planes. This relation is obtained by combining the Bragg law and the plane-spacing equation (Appendix 3) applicable to the particular crystal involved.

For example, if the crystal is cubic, then

$$\lambda = 2d \sin \theta$$

and

$$\frac{1}{d^2} = \frac{(h^2 + k^2 + l^2)}{a^2}.$$

Combining these equations, we have

$$\sin^2 \theta = \frac{\lambda^2}{4a^2}(h^2 + k^2 + l^2). \tag{3-10}$$

This equation predicts, for a particular incident wavelength λ and a particular cubic crystal of unit cell size a, all the possible Bragg angles at which diffraction can occur from the planes (hkl). For (110) planes, for example, Eq. (3-10) becomes

$$\sin^2 \theta_{110} = \frac{\lambda^2}{2a^2}.$$

If the crystal is tetragonal, with axes a and c, then the corresponding general equation is

$$\sin^2 \theta = \frac{\lambda^2}{4}\left(\frac{h^2 + k^2}{a^2} + \frac{l^2}{c^2}\right), \tag{3-11}$$

and similar equations can readily be obtained for the other crystal systems.

These examples show that the directions in which a beam of given wavelength is diffracted by a given set of lattice planes are determined by the crystal system to which the crystal belongs and its lattice parameters. In short, *diffraction directions are determined solely by the shape and size of the unit cell.* This is an important point and so is its converse: all we can possibly determine about an unknown crystal by measurements of the *directions* of diffracted beams are the shape and size of its unit cell. We will find, in the next chapter, that the *intensities* of diffracted beams are determined by the positions of the atoms within the unit cell, and it follows that we must measure intensities if we are to obtain any information at all about atom positions. We will find, for many crystals, that there are particular atomic arrangements which reduce the intensities of some diffracted beams to zero. In such a case, there is simply no diffracted beam at the angle predicted by an equation of the type of Eqs. (3–10) and (3–11). It is in this sense that equations of this kind predict all *possible* diffracted beams.

3-6 DIFFRACTION METHODS

Diffraction can occur whenever the Bragg law, $\lambda = 2d \sin \theta$, is satisfied. This equation puts very stringent conditions on λ and θ for any given crystal. With monochromatic radiation, an arbitrary setting of a single crystal in a beam of x-rays will not in general produce *any* diffracted beams. Some way of satisfying the Bragg law must be devised, and this can be done by continuously varying either λ or θ during the experiment. The ways in which these quantities are varied distinguish the three main diffraction methods:

	λ	θ
Laue method	Variable	Fixed
Rotating-crystal method	Fixed	Variable (in part)
Powder method	Fixed	Variable

Laue Method

The Laue method was the first diffraction method ever used, and it reproduces von Laue's original experiment. A beam of white radiation, the continuous spectrum from an x-ray tube, is allowed to fall on a fixed single crystal. The Bragg angle θ is therefore fixed for every set of planes in the crystal, and each set picks out and diffracts that particular wavelength which satisfies the Bragg law for the particular values of d and θ involved. Each diffracted beam thus has a different wavelength.

There are two variations of the Laue method, depending on the relative positions of source, crystal, and film (Fig. 3–5). In each, the film is flat and placed perpendicular to the incident beam. The film in the *transmission Laue method* (the original Laue method) is placed behind the crystal so as to record the beams diffracted in the forward direction. This method is so called because the diffracted beams are partially transmitted through the crystal. In the *back-reflection Laue method* the film is placed between the crystal and the x-ray source, the incident beam passing through a hole in the film, and the beams diffracted in a backward direction are recorded.

In either method, the diffracted beams form an array of spots on the film as shown in Fig. 3–6. This array of spots is commonly called a *pattern*, but the term is not used in any strict sense and does not imply any periodic arrangement of the

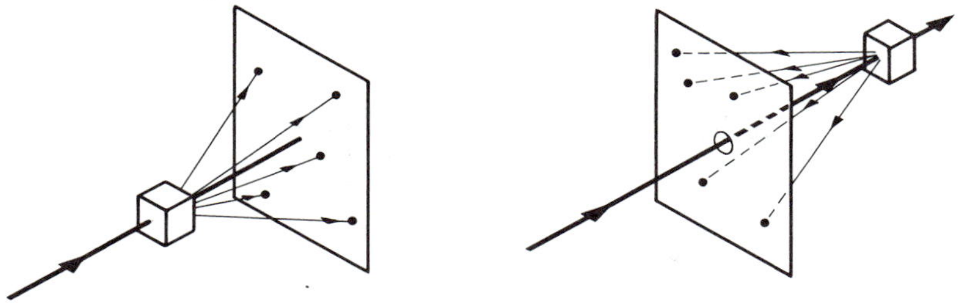

(a) (b)

Fig. 3–5 (a) Transmission and (b) back-reflection Laue methods.

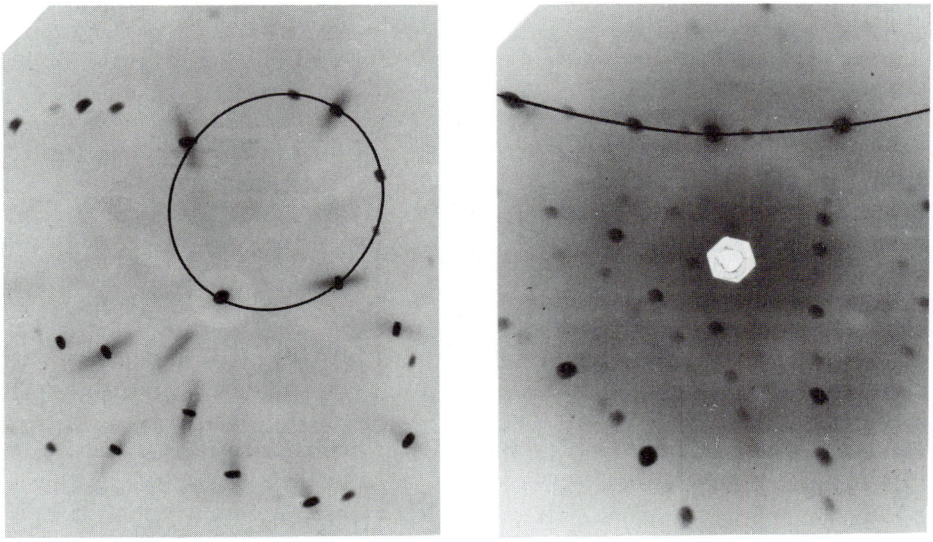

(a) (b)

Fig. 3–6 (a) Transmission and (b) back-reflection Laue patterns of an aluminum crystal (cubic). Tungsten radiation, 30 kV, 19 mA.

spots. On the contrary, the spots are seen to lie on certain curves, as shown by the lines drawn on the photographs. These curves are generally ellipses or hyperbolas for transmission patterns [Fig. 3–6(a)] and hyperbolas for back-reflection patterns [Fig. 3–6(b)].

The spots lying on any one curve are reflections from planes belonging to one zone. This is due to the fact that the Laue reflections from planes of a zone all lie on the surface of an imaginary cone whose axis is the zone axis. As shown in Fig. 3–7(a), one side of the cone is tangent to the transmitted beam, and the angle of inclination ϕ of the zone axis (Z.A.) to the transmitted beam is equal to the semi-apex angle of the cone. A film placed as shown intersects the cone in an

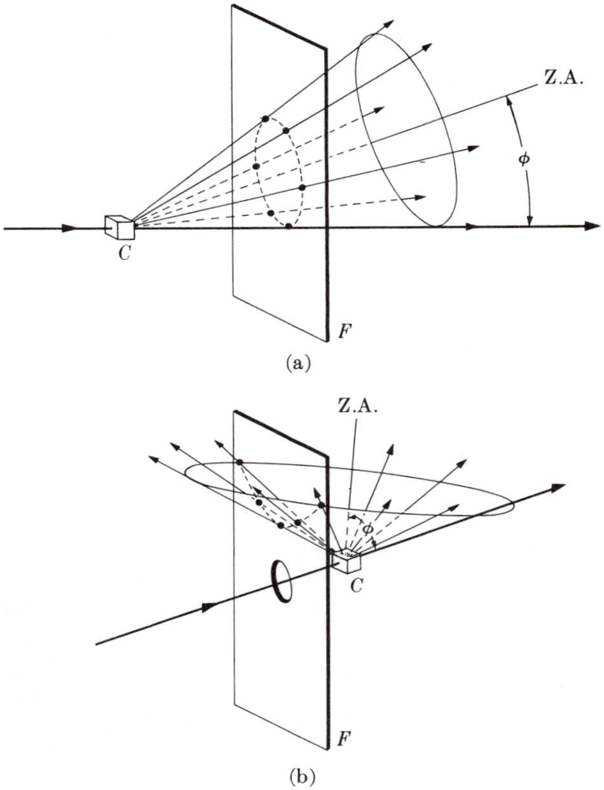

Fig. 3–7 Location of Laue spots (a) on ellipses in transmission method and (b) on hyperbolas in back-reflection method. (C = crystal, F = film, Z.A. = zone axis.)

imaginary ellipse passing through the center of the film, the diffraction spots from planes of a zone being arranged on this ellipse. When the angle ϕ exceeds 45°, a film placed between the crystal and the x-ray source to record the back-reflection pattern will intersect the cone in a hyperbola, as shown in Fig. 3–7(b).

The fact that the Laue reflections from planes of a zone lie on the surface of a cone can be nicely demonstrated with the stereographic projection. In Fig. 3–8, the crystal is at the center of the reference sphere, the incident beam I enters at the left, and the transmitted beam T leaves at the right. The point representing the zone axis lies on the circumference of the basic circle and the poles of five planes belonging to this zone, P_1 to P_5, lie on the great circle shown. The direction of the beam diffracted by any one of these planes, for example the plane P_2, can be found as follows. I, P_2, D_2 (the diffraction direction required), and T are all coplanar. Therefore D_2 lies on the great circle through I, P_2, and T. The angle between I and P_2 is $(90° - \theta)$, and D_2 must lie at an equal angular distance on the other side of P_2, as shown. The diffracted beams so found, D_1 to D_5, are seen

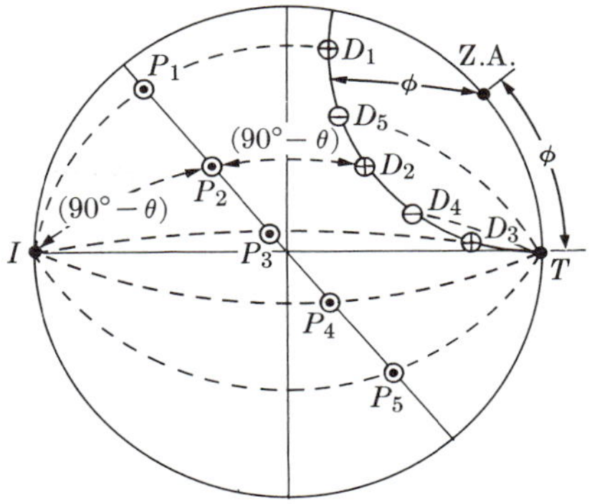

Fig. 3-8 Stereographic projection of transmission Laue method.

to lie on a small circle, the intersection with the reference sphere of a cone whose axis is the zone axis.

The positions of the spots on the film, for both the transmission and the back-reflection method, depend on the orientation of the crystal relative to the incident beam, and the spots themselves become distorted and smeared out if the crystal has been bent or twisted in any way. These facts account for the two main uses of the Laue methods: the determination of crystal orientation and the assessment of crystal quality.

Rotating-crystal Method

In the rotating-crystal method a single crystal is mounted with one of its axes, or some important crystallographic direction, normal to a monochromatic x-ray beam. A cylindrical film is placed around it and the crystal is rotated about the chosen direction, the axis of the film coinciding with the axis of rotation of the crystal (Fig. 3-9). As the crystal rotates, a particular set of lattice planes will, for an instant, make the correct Bragg angle for reflection of the monochromatic incident beam, and at that instant a reflected beam will be formed. The reflected beams are again located on imaginary cones but now the cone axes coincide with the rotation axis. The result is that the spots on the film, when the film is laid out flat, lie on imaginary horizontal lines, as shown in Fig. 3-10. Since the crystal is rotated about only one axis, the Bragg angle does not take on all possible values between 0° and 90° for every set of planes. Not every set, therefore, is able to produce a diffracted beam; sets perpendicular or almost perpendicular to the rotation axis are examples.

The chief use of the rotating-crystal method and its variations is in the determination of unknown crystal structures, and for this purpose it is the most powerful tool the x-ray crystallographer has at his disposal. However, the complete determination of complex crystal structures is a subject beyond the scope of this book and outside the province of the average metallurgist who uses x-ray diffraction as a laboratory tool. For this reason the rotating-crystal method will not be described in any further detail, except for a brief discussion in Appendix 1.

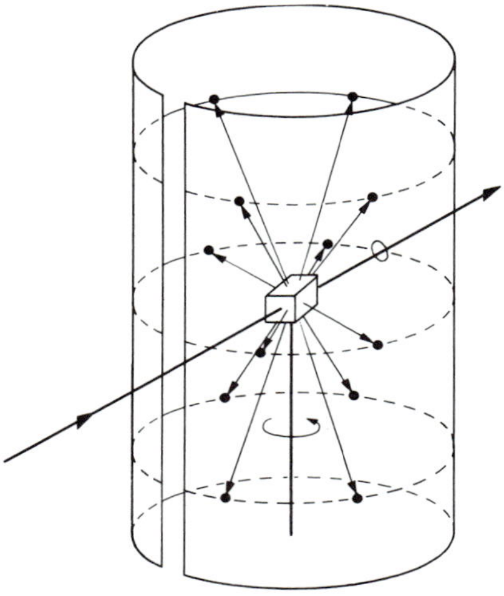

Fig. 3–9 Rotating-crystal method.

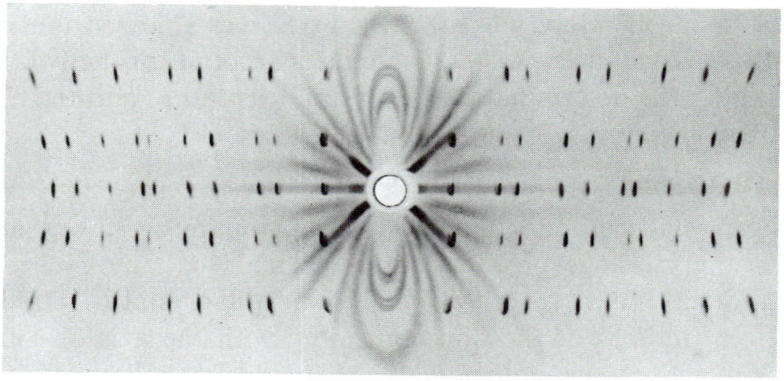

Fig. 3–10 Rotating-crystal pattern of a quartz crystal (hexagonal) rotated about its **c** axis. Filtered copper radiation. (The streaks are due to the white radiation not removed by the filter.) (Courtesy of B. E. Warren.)

Powder Method

In the powder method, the crystal to be examined is reduced to a very fine powder and placed in a beam of monochromatic x-rays. Each particle of the powder is a tiny crystal, or assemblage of smaller crystals, oriented at random with respect to the incident beam. Just by chance, some of the crystals will be correctly oriented so that their (100) planes, for example, can reflect the incident beam. Other crystals will be correctly oriented for (110) reflections, and so on. The result is that every set of lattice planes will be capable of reflection. The mass of powder is equivalent, in fact, to a single crystal rotated, not about one axis, but about all possible axes.

Consider one particular *hkl* reflection. One or more little crystals will, by

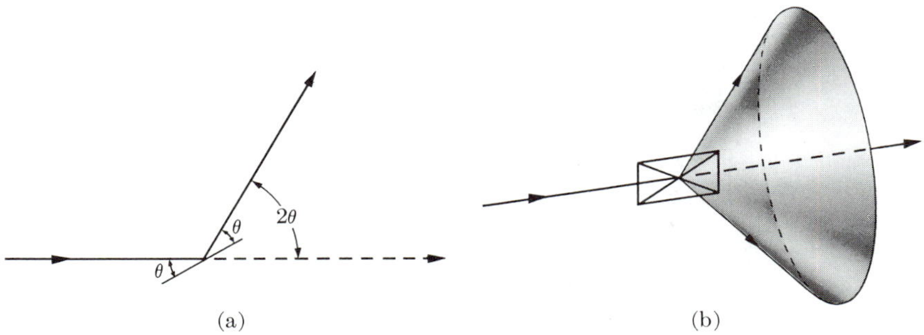

Fig. 3-11 Formation of a diffracted cone of radiation in the powder method.

chance, be so oriented that their (hkl) planes make the correct Bragg angle for reflection; Fig. 3-11(a) shows one plane in this set and the diffracted beam formed. If this plane is now rotated about the incident beam as axis in such a way that θ is kept constant, then the reflected beam will travel over the surface of a cone as shown in Fig. 3-11(b), the axis of the cone coinciding with the transmitted beam. This rotation does not actually occur in the powder method, but the presence of a large number of crystal particles having all possible orientations is equivalent to this rotation, since among these particles there will be a certain fraction whose (hkl) planes make the right Bragg angle with the incident beam and which at the same time lie in all possible rotational positions about the axis of the incident beam. The hkl reflection from a stationary mass of powder thus has the form of a conical sheet of diffracted radiation, and a separate cone is formed for each set of differently spaced lattice planes.

Figure 3-12 shows three such cones and also illustrates the most common powder-diffraction method. In this, the Debye-Scherrer method, a narrow strip of film is curved into a short cylinder with the specimen placed on its axis and the incident beam directed at right angles to this axis. The cones of diffracted radiation intersect the cylindrical strip of film in lines and, when the strip is unrolled and laid out flat, the resulting pattern has the appearance of the one illustrated in Fig. 3-12(b). Actual patterns, produced by various metal powders, are shown in Fig. 3-13. Each diffraction line is made up of a large number of small spots, each from a separate crystal particle, the spots lying so close together that they appear as a continuous line. The lines are generally curved, unless they occur exactly at $2\theta = 90°$ when they will be straight. From the measured position of a given diffraction line on the film, θ can be determined, and, knowing λ, we can calculate the spacing d of the reflecting lattice planes which produced the line.

Conversely, if the shape and size of the unit cell of the crystal are known, we can predict the position of all possible diffraction lines on the film. The line of lowest 2θ value is produced by reflection from planes of the greatest spacing. In the cubic system, for example, d is a maximum when $(h^2 + k^2 + l^2)$ is a minimum, and the minimum value of this term is 1, corresponding to (hkl) equal to (100). The 100 reflection is accordingly the one of lowest 2θ value. The next possible

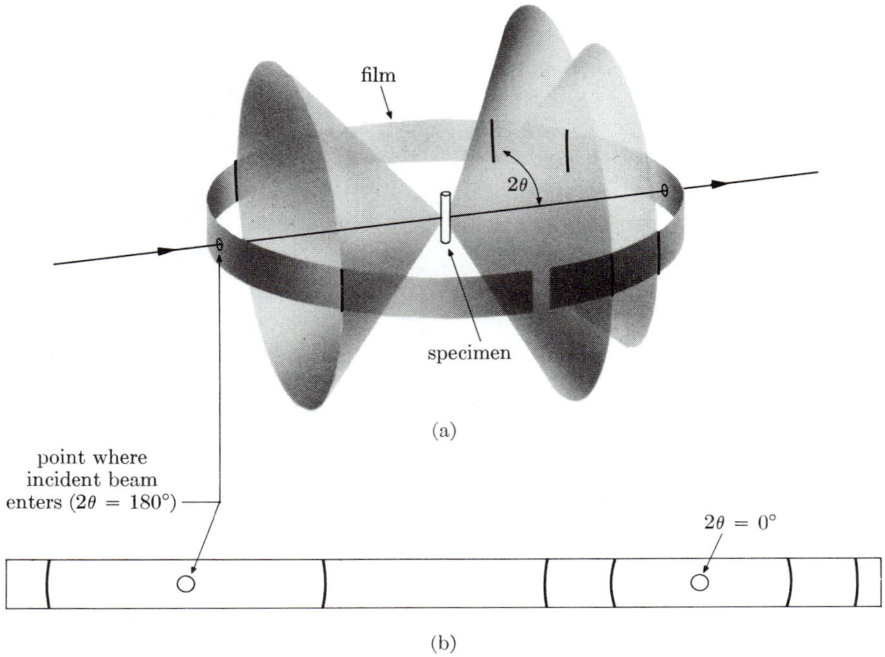

Fig. 3–12 Debye–Scherrer powder method: (a) relation of film to specimen and incident beam; (b) appearance of film when laid out flat.

reflection will have indices *hkl* corresponding to the next higher value of $(h^2 + k^2 + l^2)$, namely 2, in which case (hkl) equals (110), and so on.

The Debye–Scherrer and other variations of the powder method are very widely used, especially in metallurgy. The powder method is, of course, the only method that can be employed when a single-crystal specimen is not available, and this is the case more often than not in metallurgical work. The method is especially suited for determining lattice parameters with high precision and for the identification of phases, whether they occur alone or in mixtures such as polyphase alloys, corrosion products, refractories, and rocks. These and other uses of the powder method will be fully described in later chapters.

Diffractometer Method

The x-ray spectrometer can also be used as a tool in diffraction analysis. This instrument is known as a *diffractometer* when it is used with x-rays of *known* wavelength to determine the *unknown* spacing of crystal planes, and as a spectrometer in the reverse case, when crystal planes of known spacing are used to determine unknown wavelengths. The diffractometer is always used with monochromatic radiation and measurements may be made on either single crystals or polycrystalline specimens; in the latter case, it functions much like a Debye–Scherrer camera

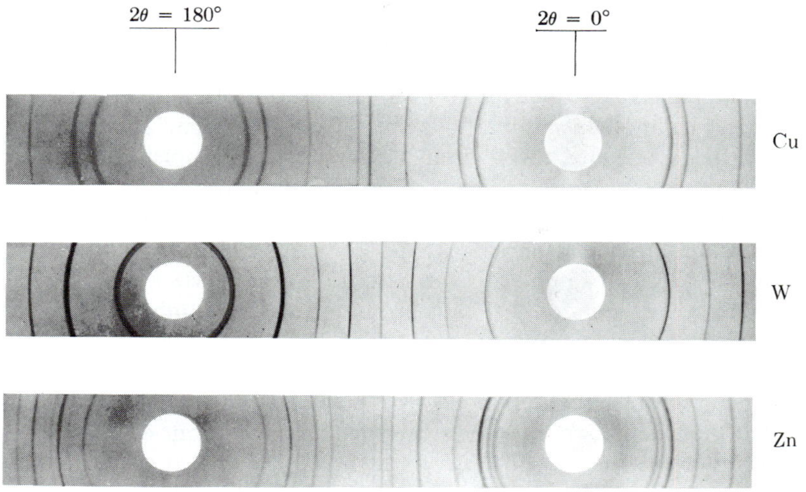

Fig. 3–13 Debye–Scherrer powder patterns of copper (FCC), tungsten (BCC), and zinc (HCP). Filtered copper radiation, camera diameter = 5.73 cm.

in that the counter intercepts and measures only a short arc of any one cone of diffracted rays.

3–7 DIFFRACTION UNDER NONIDEAL CONDITIONS

Before going any further, it is important to stop and consider with some care the derivation of the Bragg law given in Sec. 3–2 in order to understand precisely under what conditions it is strictly valid. In our derivation we assumed certain ideal conditions, namely a perfect crystal and an incident beam composed of perfectly parallel and strictly monochromatic radiation. These conditions never actually exist, so we must determine the effect on diffraction of various kinds of departure from the ideal.

In particular, the way in which destructive interference is produced in all directions except those of the diffracted beams is worth examining in some detail, both because it is fundamental to the theory of diffraction and because it will lead us to a method for estimating the size of very small crystals. We will find that only the infinite crystal is really perfect and that small size alone, of an otherwise perfect crystal, can be considered a crystal imperfection.

The condition for reinforcement used in Sec. 3–2 is that the waves involved must differ in path length, that is, in phase, by exactly an integral number of wavelengths. But suppose that the angle θ in Fig. 3–2 is such that the path difference for rays scattered by the first and second planes is only a quarter wavelength. These rays do not annul one another but, as we saw in Fig. 3–1, simply unite to form a beam of smaller amplitude than that formed by two rays which are completely in phase. How then does destructive interference take place? The answer lies in the contributions from planes deeper in the crystal. Under the assumed conditions, the rays scattered by the second and third planes would also be a quarter wavelength out of phase. But this means that the rays scattered by the first and third planes are exactly half a wavelength out of phase and would completely cancel one another. Similarly, the rays from the second and fourth planes, third and fifth planes, etc., throughout the crystal, are completely out of phase; the result is destructive interference and no diffracted beam. *Destructive interference is therefore just as much a consequence of the periodicity of atom arrangement as is constructive interference.*

This is an extreme example. If the path difference between rays scattered by the first two planes differs only slightly from an integral number of wavelengths, then the plane scattering a ray exactly out of phase with the ray from the first plane will lie deep within the crystal. If the crystal is so small that this plane does not exist, then complete cancellation of all the scattered rays will not result. It follows that there is a connection between the amount of "out-of-phaseness" that can be tolerated and the size of the crystal. We will find that very small crystals cause broadening (a small angular divergence) of the diffracted beam, i.e., diffraction (scattering) at angles near to, but not equal to, the exact Bragg angle. We must therefore consider the scattering of rays incident on the crystal planes at angles deviating slightly from the exact Bragg angle.

Suppose, for example, that the crystal has a thickness t measured in a direction perpendicular to a particular set of reflecting planes (Fig. 3–14). Let there be $(m + 1)$ planes in this set. We will regard the Bragg angle θ as a variable and call θ_B the angle which exactly satisfies the Bragg law for the particular values of λ and d involved, or

$$\lambda = 2d \sin \theta_B.$$

In Fig. 3–14, rays **A**, **D**, . . . , **M** make exactly this angle θ_B with the reflecting planes. Ray **D′**, scattered by the first plane below the surface, is therefore one

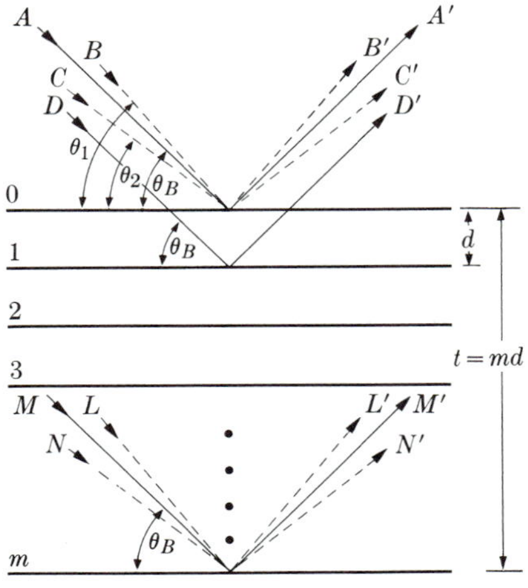

Fig. 3–14 Effect of crystal size on diffraction.

wavelength out of phase with **A**′; and ray **M**′, scattered by the *m*th plane below the surface, is *m* wavelengths out of phase with **A**′. Therefore, at a diffraction angle $2\theta_B$, rays **A**′, **D**′, . . . , **M**′ are completely in phase and unite to form a diffracted beam of maximum amplitude, i.e., a beam of maximum intensity, since the intensity is proportional to the square of the amplitude.

When we consider incident rays that make Bragg angles only slightly different from θ_B, we find that destructive interference is not complete. Ray **B**, for example, makes a slightly larger angle θ_1, such that ray **L**′ from the *m*th plane below the surface is $(m + 1)$ wavelengths out of phase with **B**′, the ray from the surface plane. This means that midway in the crystal there is a plane scattering a ray which is one-half (actually, an integer plus one-half) wavelength out of phase with ray **B**′ from the surface plane. These rays cancel one another, and so do the other rays from similar pairs of planes throughout the crystal, the net effect being that rays scattered by the top half of the crystal annul those scattered by the bottom half. The intensity of the beam diffracted at an angle $2\theta_1$ is therefore zero. It is also zero at an angle $2\theta_2$ where θ_2 is such that ray **N**′ from the *m*th plane below the surface is $(m - 1)$ wavelengths out of phase with ray **C**′ from the surface plane. We have therefore found two limiting angles, $2\theta_1$ and $2\theta_2$, at which the diffracted intensity must drop to zero. It follows that the diffracted intensity at angles near $2\theta_B$, but not greater than $2\theta_1$ or less than $2\theta_2$, is *not zero* but has a value intermediate between zero and the maximum intensity of the beam diffracted at an angle $2\theta_B$. The curve of diffracted intensity vs. 2θ will thus have the form of Fig. 3–15(a) in contrast to Fig. 3–15(b), which illustrates the hypothetical case of diffraction occurring only at the exact Bragg angle.

The width of the diffraction curve of Fig. 3–15(a) increases as the thickness of the crystal decreases, because the angular range $(2\theta_1 - 2\theta_2)$ increases as *m*

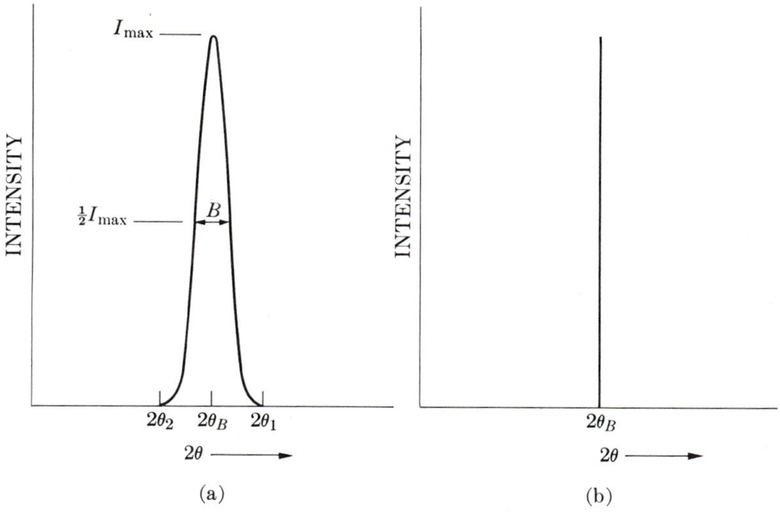

Fig. 3–15 Effect of fine particle size on diffraction curves (schematic).

decreases. The width B is usually measured, in radians, at an intensity equal to half the maximum intensity. [Note that B is an angular width, in terms of 2θ (not θ), and not a linear width.] As a rough measure of B, we can take half the difference between the two extreme angles at which the intensity is zero, which amounts to assuming that the diffraction line is triangular in shape. Therefore,

$$B = \tfrac{1}{2}(2\theta_1 - 2\theta_2) = \theta_1 - \theta_2.$$

We now write path-difference equations for these two angles, similar to Eq. (3–1) but related to the entire thickness of the crystal rather than to the distance between adjacent planes:

$$2t \sin \theta_1 = (m + 1)\lambda,$$

$$2t \sin \theta_2 = (m - 1)\lambda.$$

By subtraction we find

$$t(\sin \theta_1 - \sin \theta_2) = \lambda,$$

$$2t \cos \left(\frac{\theta_1 + \theta_2}{2}\right) \sin \left(\frac{\theta_1 - \theta_2}{2}\right) = \lambda.$$

But θ_1 and θ_2 are both very nearly equal to θ_B, so that

$$\theta_1 + \theta_2 = 2\theta_B \quad \text{(approx.)}$$

and

$$\sin \left(\frac{\theta_1 - \theta_2}{2}\right) = \left(\frac{\theta_1 - \theta_2}{2}\right) \quad \text{(approx.).}$$

Therefore

$$2t \left(\frac{\theta_1 - \theta_2}{2}\right) \cos \theta_B = \lambda,$$

$$t = \frac{\lambda}{B \cos \theta_B}. \tag{3–12}$$

A more exact treatment of the problem gives

$$t = \frac{0.9\lambda}{B \cos \theta_B}, \tag{3–13}$$

which is known as the Scherrer formula. It is used to estimate the *particle size* of very small crystals from the measured width of their diffraction curves. What is the order of magnitude of this effect? Suppose $\lambda = 1.5$ Å, $d = 1.0$ Å, and $\theta = 49°$. Then for a crystal 1 mm in diameter the breadth B, due to the small crystal effect alone, would be about 2×10^{-7} radian (10^{-5} degree), or too small to be observable. Such a crystal would contain some 10^7 parallel lattice planes of the spacing assumed above. However, if the crystal were only 500 Å thick, it would contain only 500 planes, and the diffraction curve would be relatively broad, namely about 4×10^{-3} radian (0.2°), which is easily measurable.

Nonparallel incident rays, such as **B** and **C** in Fig. 3–14, actually exist in any real diffraction experiment, since the "perfectly parallel beam" assumed in Fig. 3–2 has never been produced in the laboratory. As will be shown in Sec. 5–4, any actual beam of x-rays contains divergent and convergent rays as well as parallel rays, so that the phenomenon of diffraction at angles not exactly satisfying the Bragg law actually takes place.

Neither is any real beam ever strictly monochromatic. The usual "monochromatic" beam is simply one containing the strong $K\alpha$ component superimposed on the continuous spectrum. But the $K\alpha$ line itself has a width of about 0.001 Å and this narrow range of wavelengths in the nominally monochromatic beam is a further cause of line broadening, i.e., of measurable diffraction at angles close, but not equal, to $2\theta_B$, since for each value of λ there is a corresponding value of θ. (Translated into terms of diffraction line width, a range of wavelengths extending over 0.001 Å leads to an increase in line width, for $\lambda = 1.5$ Å and $\theta = 45°$, of about 0.08° over the width one would expect if the incident beam were strictly monochromatic.) Line broadening due to this natural "spectral width" is proportional to tan θ and becomes quite noticeable as θ approaches 90°.

Finally, there is a kind of crystal imperfection known as *mosaic structure* which is possessed by all real crystals to a greater or lesser degree and which has a decided effect on diffraction phenomena. It is a kind of substructure into which a "single" crystal is broken up and is illustrated in Fig. 3-16 in an enormously exaggerated fashion. A crystal with mosaic structure does not have its atoms arranged on a perfectly regular lattice extending from one side of the crystal to the other; instead, the lattice is broken up into a number of tiny blocks, each slightly disoriented one from another. The size of these blocks is of the order of 1000 Å, while the maximum angle of disorientation between them may vary from a very small value to as much as one degree, depending on the crystal. If this angle is ε, then diffraction of a parallel monochromatic beam from a "single" crystal will occur not only at an angle of incidence θ_B but at all angles between θ_B and $\theta_B + \varepsilon$. Another effect of mosaic structure is to increase the integrated intensity of the reflected beam relative to that theoretically calculated for an ideally perfect crystal (Sec. 4-12).

The notion of the mosaic crystal dates from the early years of x-ray diffraction and depends on much indirect evidence, both theoretical and experimental. In the 1960s the electron microscope provided direct evidence. It showed that real crystals, whether single crystals or individual grains in a polycrystalline aggregate,

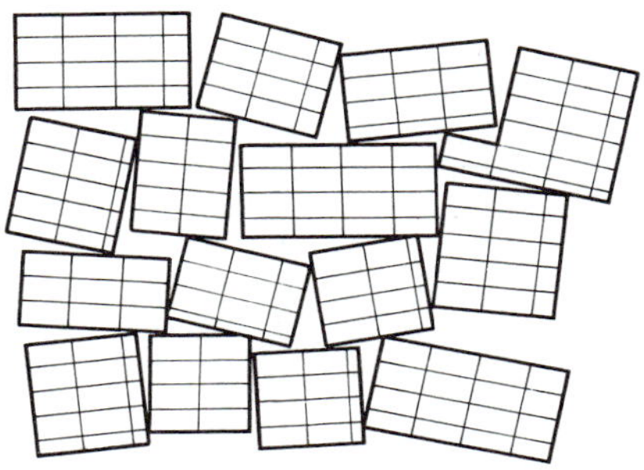

Fig. 3-16 The mosaic structure of a real crystal.

had a substructure defined by the dislocations present. The density of these dislocations is not uniform; they tend to group themselves into walls (sub-grain boundaries) surrounding small volumes having a low dislocation density (sub-grains or cells). Today the term "mosaic structure" is seldom used, but the little blocks of Fig. 3–16 are identical with sub-grains and the regions between the blocks are the dislocation walls.

These, then, are some examples of diffraction under nonideal conditions, that is, of diffraction as it actually occurs. We should not regard these as "deviations" from the Bragg law, and we will not as long as we remember that this law is derived for certain ideal conditions and that diffraction is only a special kind of scattering. This latter point cannot be too strongly emphasized. A single atom scatters an incident beam of x-rays in all directions in space, but a large number of atoms arranged in a perfectly periodic array in three dimensions to form a crystal scatters (diffracts) x-rays in relatively few directions, as illustrated schematically in Fig. 3–17. It does so precisely because the periodic arrangement of atoms causes destructive interference of the scattered rays in all directions *except* those predicted by the Bragg law, and in these directions constructive interference (reinforcement) occurs. It is not surprising, therefore, that measurable diffraction (scattering) occurs at non-Bragg angles whenever any crystal imperfection results in the partial

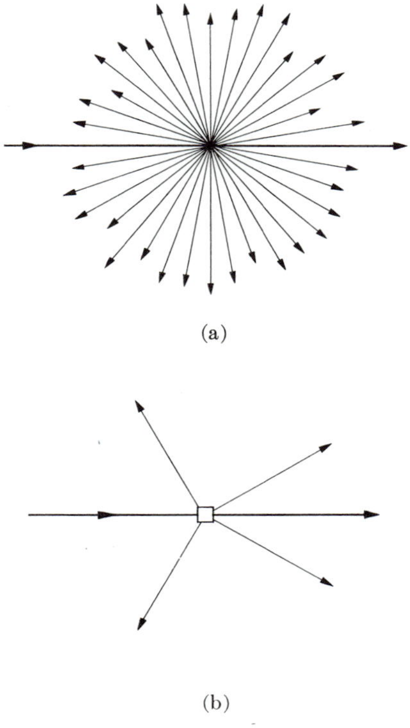

Fig. 3–17 (a) Scattering by an atom. (b) Diffraction by a crystal.

absence of one or more of the necessary conditions for perfect destructive interference at these angles. These imperfections are generally slight compared to the over-all regularity of the lattice, with the result that diffracted beams are confined to very narrow angular ranges centered on the angles predicted by the Bragg law for ideal conditions.

This relation between destructive interference and structural periodicity can be further illustrated by a comparison of x-ray scattering by solids, liquids, and gases (Fig. 3-18). The curve of scattered intensity vs. 2θ for a crystalline solid is almost zero everywhere except at certain angles where high sharp maxima occur: these are the diffracted beams. Both amorphous solids and liquids have structures characterized by an almost complete lack of periodicity and a tendency to "order" only in the sense that the atoms are fairly tightly packed together and show a statistical preference for a particular interatomic distance; the result is an x-ray scattering curve showing nothing more than one or two broad maxima. Finally, there are the monatomic gases, which have no structural periodicity whatever; in such gases, the atoms are arranged perfectly at random and their relative positions

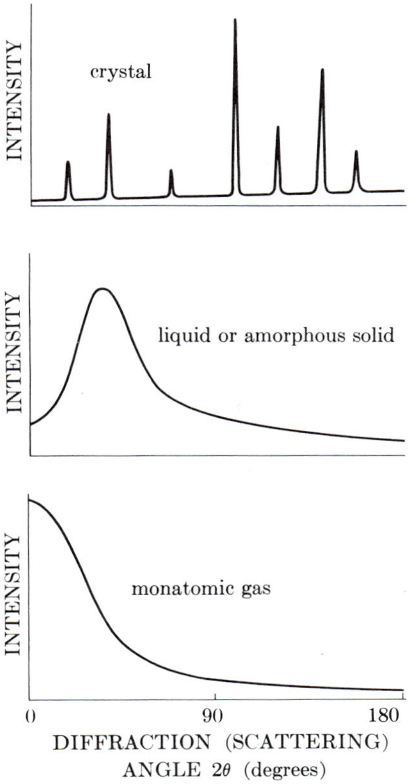

Fig. 3-18 Comparative x-ray scattering by crystalline solids, amorphous solids, liquids, and monatomic gases (schematic). The three vertical scales are not equal.

change constantly with time. The corresponding scattering curve shows no maxima, merely a regular decrease of intensity with increase in scattering angle. This curve would be entirely featureless, i.e., horizontal, if it were not for the fact that isolated atoms scatter x-rays more intensely at low 2θ angles than at high (Sec. 4–3).

PROBLEMS

3–1 A transmission Laue pattern is made of a cubic crystal having a lattice parameter of 4.00 Å. The x-ray beam is horizontal. The $[0\bar{1}0]$ axis of the crystal points along the beam towards the x-ray tube, the $[\bar{1}00]$ axis points vertically upward, and the $[001]$ axis is horizontal and parallel to the photographic film. The film is 5.00 cm from the crystal.

a) What is the wavelength of the radiation diffracted from the $(\bar{3}\bar{1}0)$ planes?
b) Where will the $\bar{3}\bar{1}0$ reflection strike the film?

***3–2** A transmission Laue pattern is made of a cubic crystal in the orientation of Prob. 3–1. By means of a stereographic projection similar to Fig. 3–8, show that the beams diffracted by the planes $(\bar{2}\bar{1}0)$, $(\bar{2}\bar{1}3)$, and (211), all of which belong to the zone $[\bar{1}20]$, lie on the surface of a cone whose axis is the zone axis. What is the angle ϕ between the zone axis and the transmitted beam?

3–3 Determine, and list in order of increasing angle, the values of 2θ and (hkl) for the first three lines (those of lowest 2θ values) on the powder patterns of substances with the following structures, the incident radiation being Cu $K\alpha$:

a) simple cubic (a = 3.00 Å),
b) simple tetragonal (a = 2.00 Å, c = 3.00 Å),
c) simple tetragonal (a = 3.00 Å, c = 2.00 Å),
d) simple rhombohedral (a = 3.00 Å, $\alpha = 80°$).

3–4 In Fig. 3–14, put $m = 10$. (a) Write down a complete list of the path differences, in wavelengths λ, between the ray scattered by each plane below the surface and the ray scattered by the surface plane, for a scattering angle of $2\theta_1$. What plane scatters a ray exactly out of phase with the ray scattered by the third plane below the surface? What is the path difference for these two rays? (b) Write down a similar list of path differences for rays scattered at an angle halfway between $2\theta_B$ and $2\theta_1$ in order to convince yourself that these rays do *not* cancel one another.

***3–5** In Fig. 3–14, assume that the incident beam is perfectly parallel, instead of convergent, and incident at the angle θ_B. Does broadening of the diffracted beam still occur? If so, derive the relation between t and B.

***3–6** Calculate the breadth B (in degrees of 2θ), due to the small crystal effect alone, of the powder pattern lines of particles of diameter 1000, 750, 500, and 250 Å. Assume $\theta = 45°$ and $\lambda = 1.5$ Å. For particles 250 Å in diameter, calculate the breadth B for $\theta = 10$, 45, and 80°.

3–7 Check the value given in Sec. 3–7 for the increase in breadth of a diffraction line due to the natural width of the $K\alpha$ emission line. [*Hint:* Differentiate the Bragg law and find an expression for the rate of change of 2θ with λ.]

4

Diffraction II: Intensities of Diffracted Beams

4-1 INTRODUCTION

As stated earlier, the positions of the atoms in the unit cell affect the intensities but not the directions of the diffracted beams. That this must be so may be seen by considering the two structures shown in Fig. 4–1. Both are orthorhombic with two atoms of the same kind per unit cell, but the one on the left is base-centered and the one on the right body-centered. Either is derivable from the other by a simple shift of one atom by the vector $\frac{1}{2}\mathbf{c}$.

Consider reflections from the (001) planes which are shown in profile in Fig. 4–2. For the base-centered lattice shown in (a), suppose that the Bragg law is satisfied for the particular values of λ and θ employed. This means that the path difference ABC between rays **1′** and **2′** is one wavelength, so that rays **1′** and **2′** are in phase and diffraction occurs in the direction shown. Similarly, in the body-centered lattice shown in (b), rays **1′** and **2′** are in phase, since their path difference ABC is one wavelength. However, in this case, there is another plane of atoms midway between the (001) planes, and the path difference DEF between rays **1′** and **3′** is exactly half of ABC, or one-half wavelength. Thus rays **1′** and **3′** are completely out of phase and annul each other. Similarly, ray **4′** from the next plane down (not shown) annuls ray **2′**, and so on throughout the crystal. There is no 001 reflection from the body-centered lattice.

This example shows how a simple rearrangement of atoms within the unit cell can eliminate a reflection completely. More generally, the intensity of a diffracted beam is changed, not necessarily to zero, by any change in atomic positions, and, conversely, we can determine atomic positions only by observations of diffracted intensities. To establish an exact relation between atom position and intensity is the main purpose of this chapter. The problem is complex because of the many variables involved, and we will have to proceed step by step: we will consider how x-rays are scattered first by a single electron, then by an atom, and finally by all the atoms in the unit cell. We will apply these results to the powder method of x-ray diffraction only, and, to obtain an expression for the intensity of a powder pattern line, we will have to consider a number of other factors which affect the way in which a crystalline powder diffracts x-rays.

4-2 SCATTERING BY AN ELECTRON

We have seen in Chap. 1 that an x-ray beam is an electromagnetic wave characterized by an electric field whose strength varies sinusoidally with time at any one

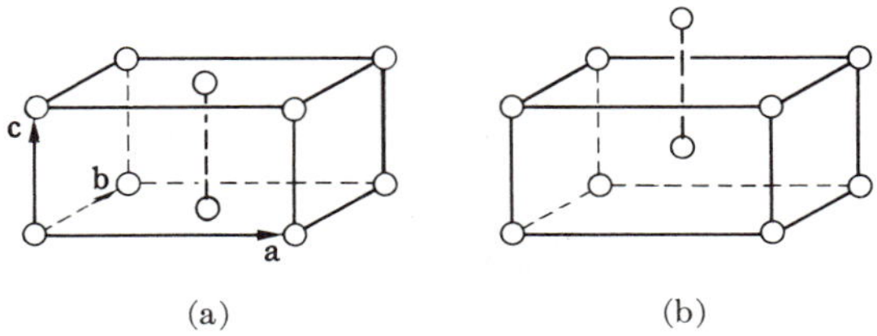

Fig. 4–1 (a) Base-centered and (b) body-centered orthorhombic unit cells.

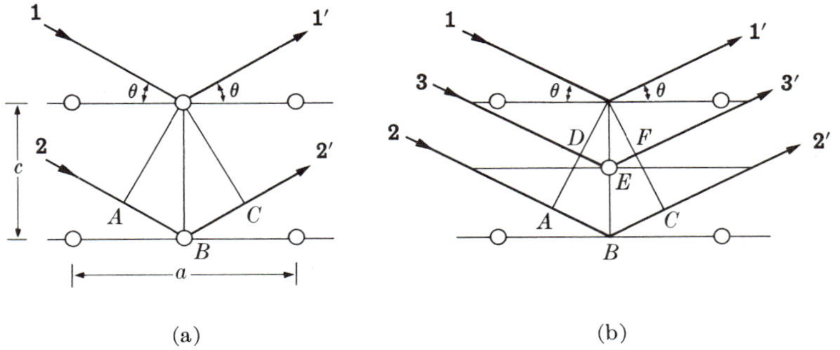

Fig. 4–2 Diffraction from the (001) planes of (a) base-centered and (b) body-centered orthorhombic lattices.

point in the beam. Since an electric field exerts a force on a charged particle such as an electron, the oscillating electric field of an x-ray beam will set any electron it encounters into oscillatory motion about its mean position.

Now an accelerating or decelerating electron emits an electromagnetic wave. We have already seen an example of this phenomenon in the x-ray tube, where x-rays are emitted because of the rapid deceleration of the electrons striking the target. Similarly, an electron which has been set into oscillation by an x-ray beam is continuously accelerating and decelerating during its motion and therefore emits an electromagnetic wave. In this sense, an electron is said to *scatter* x-rays, the scattered beam being simply the beam radiated by the electron under the action of the incident beam. The scattered beam has the same wavelength and frequency as the incident beam and is said to be *coherent* with it, since there is a definite relationship between the phase of the scattered beam and that of the incident beam which produced it. (The phase change on scattering from an electron is $\lambda/2$. Because it is exactly the same for all the electrons in a crystal, it cancels out in any consideration of phase differences between rays scattered by different atoms, as in Fig. 3–2, and so does not affect the derivation of the Bragg law given in Sec. 3–2.)

Although x-rays are scattered in all directions by an electron, the intensity of the scattered beam depends on the angle of scattering, in a way which was first

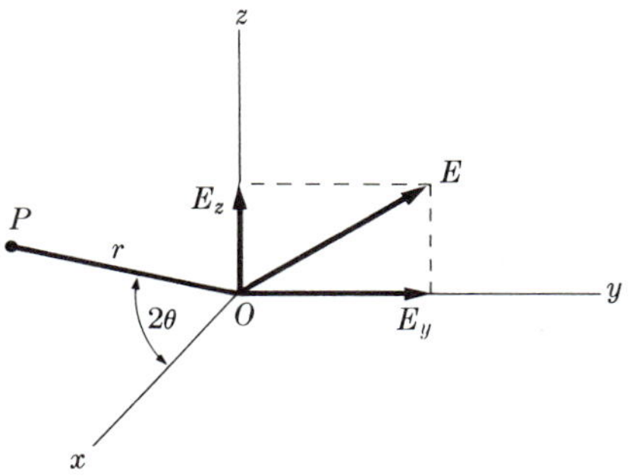

Fig. 4–3 Coherent scattering of x-rays by a single electron.

worked out by J. J. Thomson. He found that the intensity I of the beam scattered by a single electron of charge e coulombs (C) and mass m kg, at a distance r meters from the electron, is given by

$$I = I_0 \left(\frac{\mu_0}{4\pi}\right)^2 \left(\frac{e^4}{m^2 r^2}\right) \sin^2 \alpha = I_0 \frac{K}{r^2} \sin^2 \alpha \qquad (4\text{–}1)$$

where I_0 = intensity of the incident beam, $\mu_0 = 4\pi \times 10^{-7}$ m kg C^{-2}, K = constant, and α = angle between the scattering direction and the direction of acceleration of the electron. Suppose the incident beam is traveling in the direction Ox (Fig. 4–3) and encounters an electron at O. We wish to know the scattered intensity at P in the xz plane where OP is inclined at a scattering angle of 2θ to the incident beam. An unpolarized incident beam, such as that issuing from an x-ray tube, has its electric vector **E** in a random direction in the yz plane. This beam may be resolved into two plane-polarized components, having electric vectors $\mathbf{E}_y$ and $\mathbf{E}_z$ where

$$\mathbf{E}^2 = \mathbf{E}_y^2 + \mathbf{E}_z^2.$$

On the average, $\mathbf{E}_y$ will be equal to $\mathbf{E}_z$, since the direction of **E** is perfectly random. Therefore

$$\mathbf{E}_y^2 = \mathbf{E}_z^2 = \tfrac{1}{2}\mathbf{E}^2.$$

The intensity of these two components of the incident beam is proportional to the square of their electric vectors, since **E** measures the amplitude of the wave and the intensity of a wave is proportional to the square of its amplitude. Therefore

$$I_{Oy} = I_{Oz} = \tfrac{1}{2} I_0.$$

The y component of the incident beam accelerates the electron in the direction Oy. It therefore gives rise to a scattered beam whose intensity at P is found from Eq. (4–1) to be

$$I_{Py} = I_{Oy} \frac{K}{r^2},$$

since $\alpha = \measuredangle yOP = \pi/2$. Similarly, the intensity of the scattered z component is given by

$$I_{Pz} = I_{0z} \frac{K}{r^2} \cos^2 2\theta,$$

since $\alpha = \pi/2 - 2\theta$. The total scattered intensity at P is obtained by summing the intensities of these two scattered components:

$$\begin{aligned} I_P &= I_{Py} + I_{Pz} \\ &= \frac{K}{r^2}(I_{0y} + I_{0z} \cos^2 2\theta) \\ &= \frac{K}{r^2}\left(\frac{I_0}{2} + \frac{I_0}{2} \cos^2 2\theta\right) \\ &= I_0 \frac{K}{r^2}\left(\frac{1 + \cos^2 2\theta}{2}\right). \end{aligned} \tag{4–2}$$

This is the Thomson equation for the scattering of an x-ray beam by a single electron. The intensity of the scattered beam is only a minute fraction of the intensity of the incident beam; the value of K is 7.94×10^{-30} m^2, so that I_P/I_0 is only 7.94×10^{-26} in the forward direction at 1 cm from the electron. The equation also shows that the scattered intensity decreases as the inverse square of the distance from the scattering electron, as one would expect, and that the scattered beam is stronger in forward or backward directions than in a direction at right angles to the incident beam.

The Thomson equation gives the absolute intensity (in ergs/sq cm/sec) of the scattered beam in terms of the absolute intensity of the incident beam. These absolute intensities are both difficult to measure and difficult to calculate, so it is fortunate that relative values are sufficient for our purposes in practically all diffraction problems. In most cases, all factors in Eq. (4–2) except the last are constant during the experiment and can be omitted. This last factor, $\frac{1}{2}(1 + \cos^2 2\theta)$, is called the *polarization factor*; this is a rather unfortunate term because, as we have just seen, this factor enters the equation simply because the incident beam is unpolarized. The polarization factor is common to all intensity calculations, and we will use it later in our equation for the intensity of a beam diffracted by a crystalline powder.

There is another and quite different way in which an electron can scatter x-rays, and that is manifested in the *Compton effect*. This effect, discovered by A. H. Compton in 1923, occurs whenever x-rays encounter loosely bound or free electrons and can be understood only by considering the incident beam not as a wave motion, but as a stream of x-ray quanta or photons, each of energy $h\nu_1$. When such a photon strikes a loosely bound electron, the collision is an elastic one like that of two billiard balls (Fig. 4–4). The electron is knocked aside and the photon is deviated through an angle 2θ. Since some of the energy of the incident photon is used in providing kinetic energy for the electron, the energy $h\nu_2$ of the photon after impact is less than its energy $h\nu_1$ before impact. The wavelength λ_2 of the

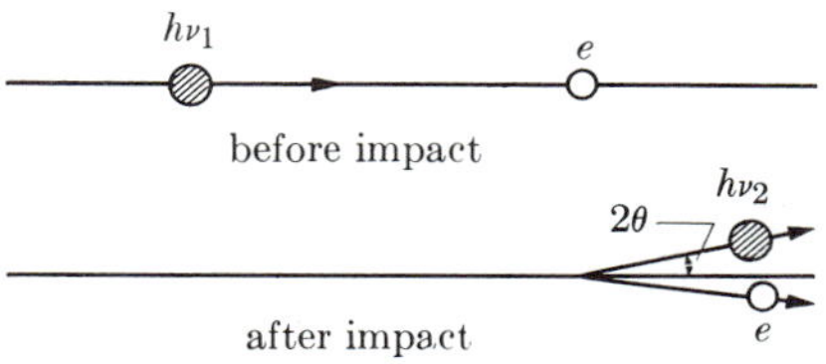

Fig. 4-4 Elastic collision of photon and electron (Compton effect).

scattered radiation is thus slightly greater than the wavelength λ_1 of the incident beam, the magnitude of the change being given by the equation

$$\Delta\lambda(\text{Å}) = \lambda_2 - \lambda_1 = 0.0486 \sin^2 \theta. \tag{4-3}$$

The increase in wavelength depends only on the scattering angle, and it varies from zero in the forward direction ($2\theta = 0$) to 0.05 Å in the extreme backward direction ($2\theta = 180°$).

Radiation so scattered is called *Compton modified radiation*, and, besides having its wavelength increased, it has the important characteristic that *its phase has no fixed relation to the phase of the incident beam.* For this reason it is also known as incoherent radiation. It cannot take part in diffraction because its phase is only randomly related to that of the incident beam and cannot therefore produce any interference effects. Compton modified scattering cannot be prevented, however, and it has the undesirable effect of darkening the background of diffraction patterns.

(It should be noted that the quantum theory can account for both the coherent and the incoherent scattering, whereas the wave theory is applicable only to the former. In terms of the quantum theory, coherent scattering occurs when an incident photon bounces off an electron which is so tightly bound that the electron receives no momentum from the impact. The scattered photon therefore has the same energy, and hence wavelength, as it had before.)

4-3 SCATTERING BY AN ATOM

When an x-ray beam encounters an atom, each electron in it scatters part of the radiation coherently in accordance with the Thomson equation. One might also expect the nucleus to take part in the coherent scattering, since it also bears a charge and should be capable of oscillating under the influence of the incident beam. However, the nucleus has an extremely large mass relative to that of the electron and cannot be made to oscillate to any appreciable extent; in fact, the Thomson equation shows that the intensity of coherent scattering is inversely proportional to the square of the mass of the scattering particle. The net effect is that coherent scattering by an atom is due only to the electrons contained in that atom.

The following question then arises: is the wave scattered by an atom simply the sum of the waves scattered by its component electrons? More precisely, does an atom of atomic number Z, i.e., an atom containing Z electrons, scatter a wave whose amplitude is Z times the amplitude of the wave scattered by a single electron? The answer is yes, if the scattering is in the forward direction ($2\theta = 0$), because the waves scattered by all the electrons of the atom are then in phase and the amplitudes of all the scattered waves can be added directly.

This is not true for other directions of scattering. The fact that the electrons of an atom are situated at different points in space introduces differences in phase between the waves scattered by different electrons. Consider Fig. 4–5, in which, for simplicity, the electrons are shown as points arranged around the central nucleus. The waves scattered in the forward direction by electrons A and B are exactly in phase on a wave front such as XX', because each wave has traveled the same distance before and after scattering. The other scattered waves shown in the figure, however, have a path difference equal to $(CB - AD)$ and are thus somewhat out of phase along a wave front such as YY', the path difference being less than one wavelength. Partial interference occurs between the waves scattered by A and B, with the result that the net amplitude of the wave scattered in this direction is less than that of the wave scattered by the same electrons in the forward direction.

A quantity f, *the atomic scattering factor*, is used to describe the "efficiency" of scattering of a given atom in a given direction. It is defined as a ratio of amplitudes:

$$f = \frac{\text{amplitude of the wave scattered by an atom}}{\text{amplitude of the wave scattered by one electron}}.$$

From what has been said already, it is clear that $f = Z$ for any atom scattering in the forward direction. As θ increases, however, the waves scattered by individual electrons become more and more out of phase and f decreases. The atomic scattering factor depends also on the wavelength of the incident beam: at a fixed

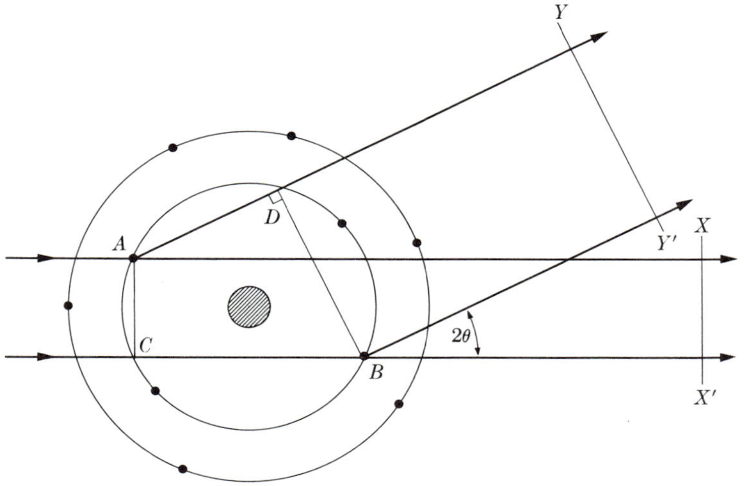

Fig. 4–5 X-ray scattering by an atom.

value of θ, f will be smaller the shorter the wavelength, since the path differences will be larger relative to the wavelength, leading to greater interference between the scattered beams. The actual calculation of f involves $\sin\theta$ rather than θ, so that the net effect is that f decreases as the quantity $(\sin\theta)/\lambda$ increases. The scattering factor f is sometimes called the *form factor*, because it depends on the way in which the electrons are distributed around the nucleus.

Calculated values of f for various atoms and various values of $(\sin\theta)/\lambda$ are tabulated in Appendix 12, and a curve showing the typical variation of f, in this case for copper, is given in Fig. 4-6. Note again that the curve begins at the atomic number of copper, 29, and decreases to very low values for scattering in the backward direction (θ near 90°) or for very short wavelengths. Since the intensity of a wave is proportional to the square of its amplitude, a curve of scattered intensity from an atom can be obtained simply by squaring the ordinates of a curve such as Fig. 4-6. (The resulting curve closely approximates the observed scattered intensity per atom of a monatomic gas, as shown in Fig. 3-18.)

Strictly, the scattering factors f tabulated in Appendix 12 apply only when the scattered radiation has a wavelength much shorter than that of an absorption edge of the scattering atom. When these two wavelengths are nearly the same, a small correction to f must be applied in precise work. An example is given in Sec. 13-4. Ordinarily we neglect this effect, called *anomalous dispersion.*

The scattering just discussed, whose amplitude is expressed in terms of the atomic scattering factor, is coherent, or unmodified, scattering, which is the only kind capable of being diffracted. On the other hand, incoherent, or Compton modified, scattering is occurring at the same time. Since the latter is due to collisions of quanta with loosely bound electrons, its intensity relative to that of the

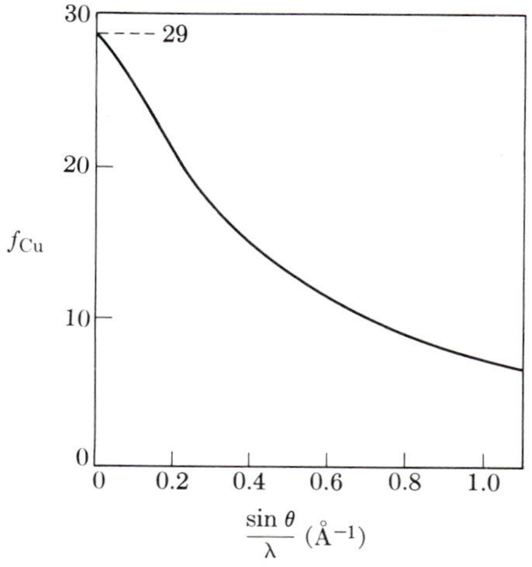

Fig. 4-6 The atomic scattering factor of copper.

unmodified radiation increases as the proportion of loosely bound electrons increases. The intensity of Compton modified radiation thus increases as the atomic number Z decreases. It is for this reason that it is difficult to obtain good diffraction photographs of organic materials, which contain light elements such as carbon, oxygen, and hydrogen, since the strong Compton modified scattering from these substances darkens the background of the photograph and makes it difficult to see the diffraction lines formed by the unmodified radiation. It is also found that the intensity of the modified radiation increases as the quantity $(\sin \theta)/\lambda$ increases. The intensities of modified scattering and of unmodified scattering therefore vary in opposite ways with Z and with $(\sin \theta)/\lambda$.

To summarize, when a monochromatic beam of x-rays strikes an atom, two scattering processes occur. Tightly bound electrons are set into oscillation and radiate x-rays of the same wavelength as that of the incident beam. More loosely bound electrons scatter part of the incident beam and slightly increase its wavelength in the process, the exact amount of increase depending on the scattering angle. The former is called coherent or unmodified scattering and the latter incoherent or modified; both kinds occur simultaneously and in all directions. If the atom is a part of a large group of atoms arranged in space in a regular periodic fashion as in a crystal, then another phenomenon occurs. The coherently scattered radiation from all the atoms undergoes reinforcement in certain directions and cancellation in other directions, thus producing diffracted beams. Diffraction is, essentially, reinforced coherent scattering.

We are now in a position to summarize, from the preceding sections and from Chap. 1, the chief effects associated with the passage of x-rays through matter. This is done schematically in Fig. 4–7. The incident x-rays are assumed to be of high enough energy, i.e., of short enough wavelength, to cause the emission of photoelectrons and characteristic fluorescent radiation. The Compton recoil electrons shown in the diagram are the loosely bound electrons knocked out of

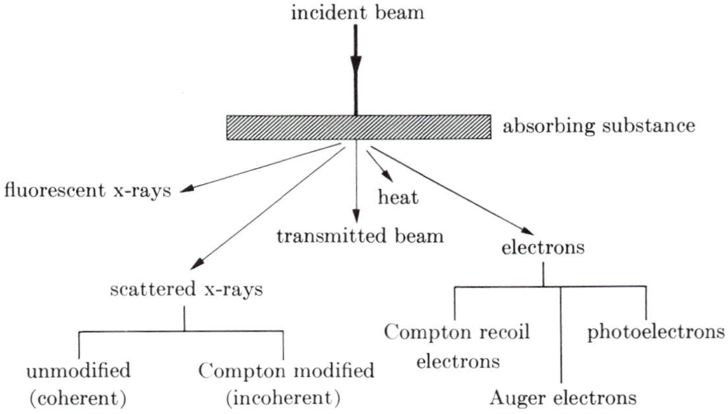

Fig. 4–7 Effects produced by the passage of x-rays through matter, after Henry, Lipson, and Wooster [G.8].

the atom by x-ray quanta, the interaction giving rise to Compton modified radiation. Auger electrons are those ejected from an atom by characteristic x-rays produced within the atom.

4-4 SCATTERING BY A UNIT CELL

To arrive at an expression for the intensity of a diffracted beam, we must now restrict ourselves to a consideration of the coherent scattering, not from an isolated atom but from all the atoms making up the crystal. The mere fact that the atoms are arranged in a periodic fashion in space means that the scattered radiation is now severely limited to certain definite directions and is now referred to as a set of diffracted beams. The directions of these beams are fixed by the Bragg law, which is, in a sense, a negative law. If the Bragg law is not satisfied, no diffracted beam can occur; however, the Bragg law may be satisfied for a certain set of atomic planes and yet no diffraction may occur, as in the example given at the beginning of this chapter, because of a particular arrangement of atoms within the unit cell [Fig. 4-2(b)].

Assuming that the Bragg law is satisfied, we wish to find the intensity of the beam diffracted by a crystal as a function of atom position. Since the crystal is merely a repetition of the fundamental unit cell, it is enough to consider the way in which the arrangement of atoms within a single unit cell affects the diffracted intensity.

Qualitatively, the effect is similar to the scattering from an atom, discussed in the previous section. There we found that phase differences occur in the waves scattered by the individual electrons, for any direction of scattering except the extreme forward direction. Similarly, the waves scattered by the individual atoms of a unit cell are not necessarily in phase except in the forward direction, and we must now determine how the phase difference depends on the arrangement of the atoms.

This problem is most simply approached by finding the phase difference between waves scattered by an atom at the origin and another atom whose position is variable in the x direction only. For convenience, consider an orthogonal unit cell, a section of which is shown in Fig. 4-8. Take atom A as the origin and let diffraction occur from the ($h00$) planes shown as heavy lines in the drawing. This means that the Bragg law is satisfied for this reflection and that $\delta_{2'1'}$, the path difference between ray **2'** and ray **1'**, is given by

$$\delta_{2'1'} = MCN = 2d_{h00} \sin\theta = \lambda.$$

From the definition of Miller indices,

$$d_{h00} = AC = \frac{a}{h}.$$

How is this reflection affected by x-rays scattered in the same direction by atom B, located at a distance x from A? Note that only this direction need be considered since only in this direction is the Bragg law satisfied for the $h00$ reflection. Clearly, the path difference between ray **3'** and ray **1'**, $\delta_{3'1'}$, will be less

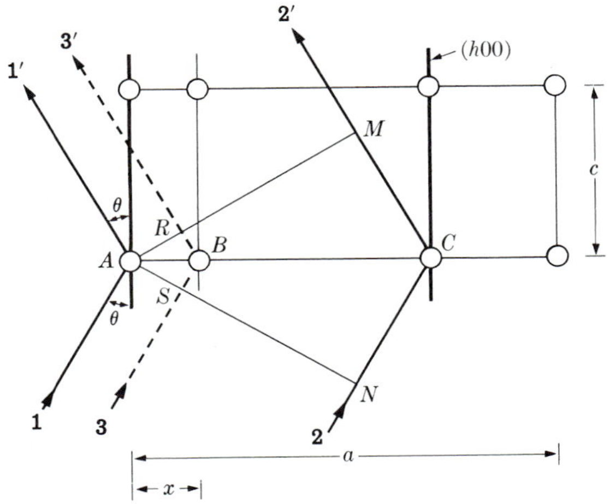

Fig. 4–8 The effect of atom position on the phase difference between diffracted rays.

than λ; by simple proportion it is found to be

$$\delta_{3'1'} = RBS = \frac{AB}{AC}(\lambda) = \frac{x}{a/h}(\lambda).$$

Phase differences may be expressed in angular measure as well as in wavelength: two rays, differing in path length by one whole wavelength, are said to differ in phase by 360°, or 2π radians. If the path difference is δ, then the phase difference ϕ in radians is given by

$$\phi = \frac{\delta}{\lambda}(2\pi).$$

The use of angular measure is convenient because it makes the expression of phase differences independent of wavelength, whereas the use of a path difference to describe a phase difference is meaningless unless the wavelength is specified.

The phase difference, then, between the wave scattered by atom B and that scattered by atom A at the origin is given by

$$\phi_{3'1'} = \frac{\delta_{3'1'}}{\lambda}(2\pi) = \frac{2\pi hx}{a}.$$

If the position of atom B is specified by its fractional coordinate $u = x/a$, then the phase difference becomes

$$\phi_{3'1'} = 2\pi hu.$$

This reasoning may be extended to three dimensions, as in Fig. 4–9, in which atom B has actual coordinates $x\ y\ z$ or fractional coordinates $\frac{x}{a}\ \frac{y}{b}\ \frac{z}{c}$ equal to $u\ v\ w$, respectively. We then arrive at the following important relation for the

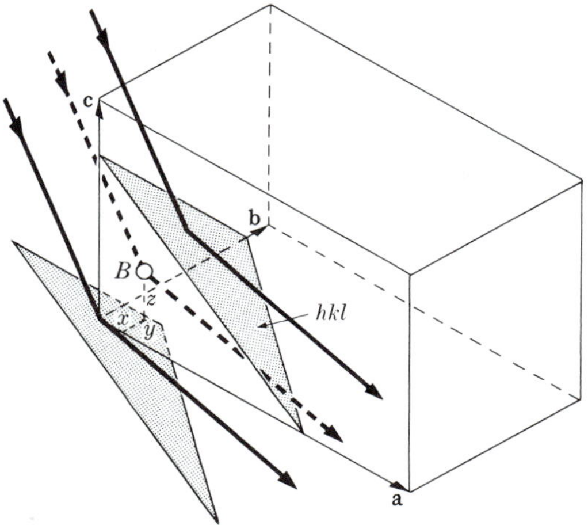

Fig. 4-9 The three-dimensional analogue of Fig. 4-8.

phase difference between the wave scattered by atom B and that scattered by atom A at the origin, for the *hkl* reflection:

$$\phi = 2\pi(hu + kv + lw). \tag{4-4}$$

This relation is general and applicable to a unit cell of any shape.

These two waves may differ, not only in phase, but also in amplitude if atom B and the atom at the origin are of different kinds. In that case, the amplitudes of these waves are given, relative to the amplitude of the wave scattered by a single electron, by the appropriate values of f, the atomic scattering factor.

We now see that the problem of scattering from a unit cell resolves itself into one of adding waves of different phase and amplitude in order to find the resultant wave. Waves scattered by all the atoms of the unit cell, including the one at the origin, must be added. The most convenient way of carrying out this summation is by expressing each wave as a complex exponential function.

The two waves shown as full lines in Fig. 4-10 represent the variations in electric field intensity $\mathbf{E}$ with time t of two rays on any given wave front in a diffracted x-ray beam. Their equations may be written

$$\mathbf{E}_1 = A_1 \sin(2\pi\nu t - \phi_1), \tag{4-5}$$

$$\mathbf{E}_2 = A_2 \sin(2\pi\nu t - \phi_2). \tag{4-6}$$

These waves are of the same frequency ν and therefore of the same wavelength λ, but differ in amplitude A and in phase ϕ. The dotted curve shows their sum $\mathbf{E}_3$, which is also a sine wave, but of different amplitude and phase.

Waves differing in amplitude and phase may also be added by representing them as vectors. In Fig. 4-11, each component wave is represented by a vector whose length is equal to the amplitude of the wave and which is inclined to the x-axis at an angle equal to the phase angle. The amplitude and phase of the

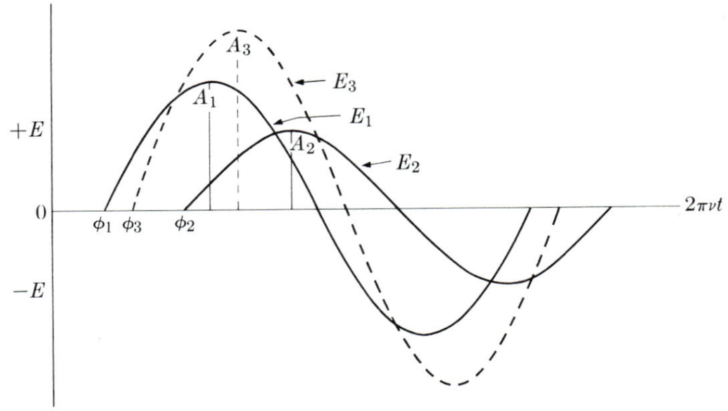

Fig. 4–10 The addition of sine waves of different phase and amplitude.

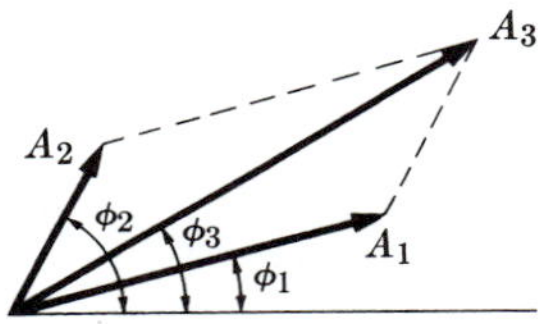

Fig. 4–11 Vector addition of waves.

resultant wave are then found simply by adding the vectors by the parallelogram law.

This geometrical construction may be avoided by use of the following analytical treatment, in which complex numbers are used to represent the vectors. A complex number is the sum of a real and an imaginary number, such as $(a + bi)$, where a and b are real and $i = \sqrt{-1}$ is imaginary. Such numbers may be plotted in the "complex plane," in which real numbers are plotted as abscissae and imaginary numbers as ordinates. Any point in this plane, or the vector drawn from the origin to this point, then represents a particular complex number $(a + bi)$.

To find an analytical expression for a vector representing a wave, we draw the wave vector in the complex plane as in Fig. 4–12. Here again the amplitude and phase of the wave are given by A, the length of the vector, and ϕ, the angle between the vector and the axis of real numbers. The analytical expression for the wave is now the complex number $(A \cos \phi + iA \sin \phi)$, since these two terms are the horizontal and vertical components OM and ON of the vector. Note that multiplication of a vector by i rotates it counterclockwise by 90°; thus multiplication by i converts the horizontal vector 2 into the vertical vector $2i$. Multiplication twice by i, that is, by $i^2 = -1$, rotates a vector through 180° or reverses its sense; thus multiplication twice by i converts the horizontal vector 2 into the horizontal vector -2 pointing in the opposite direction.

If we write down the power-series expansions of e^{ix}, $\cos x$, and $\sin x$ and

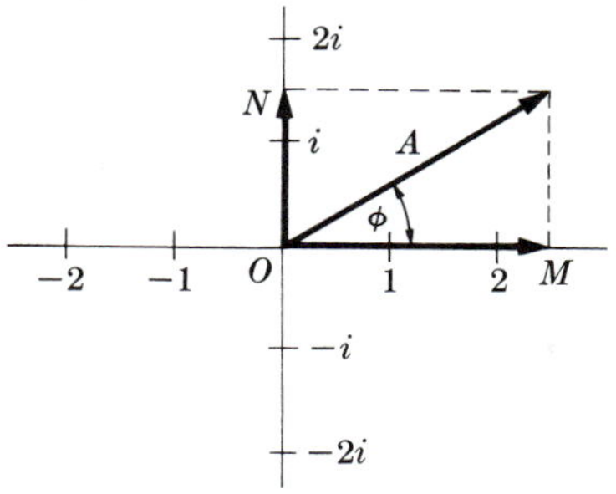

Fig. 4-12 A wave vector in the complex plane.

compare them, we find that

$$e^{ix} = \cos x + i \sin x \tag{4-7}$$

or

$$Ae^{i\phi} = A \cos \phi + Ai \sin \phi. \tag{4-8}$$

Thus the wave vector may be expressed analytically by either side of Eq. (4-8). The expression on the left is called a complex exponential function.

Since the intensity of a wave is proportional to the square of its amplitude, we now need an expression for A^2, the square of the absolute value of the wave vector. When a wave is expressed in complex form, this quantity is obtained by multiplying the complex expression for the wave by its complex conjugate, which is obtained simply by replacing i by $-i$. Thus, the complex conjugate of $Ae^{i\phi}$ is $Ae^{-i\phi}$. We have

$$|Ae^{i\phi}|^2 = Ae^{i\phi}Ae^{-i\phi} = A^2, \tag{4-9}$$

which is the quantity desired. Or, using the other form given by Eq. (4-8), we have

$$A(\cos \phi + i \sin \phi)A(\cos \phi - i \sin \phi) = A^2(\cos^2 \phi + \sin^2 \phi) = A^2.$$

We return now to the problem of adding the scattered waves from each of the atoms in the unit cell. The amplitude of each wave is given by the appropriate value of f for the scattering atom considered and the value of $(\sin \theta)/\lambda$ involved in the reflection. The phase of each wave is given by Eq. (4-4) in terms of the hkl reflection considered and the uvw coordinates of the atom. Using our previous relations, we can then express any scattered wave in the complex exponential form

$$Ae^{i\phi} = fe^{2\pi i(hu+kv+lw)}. \tag{4-10}$$

The resultant wave scattered by all the atoms of the unit cell is called the *structure factor*, because it describes how the atom arrangement, given by uvw for each atom, affects the scattered beam. The structure factor, designated by the symbol F, is obtained by simply adding together all the waves scattered by the individual atoms. If a unit cell contains atoms 1, 2, 3, ..., N, with fractional coordinates $u_1\ v_1\ w_1$, $u_2\ v_2\ w_2$, $u_3\ v_3\ w_3$, ... and atomic scattering factors $f_1, f_2, f_3, \ldots$, then the structure factor for the hkl reflection is given by

$$F = f_1 e^{2\pi i(hu_1+kv_1+lw_1)} + f_2 e^{2\pi i(hu_2+kv_2+lw_2)} + f_3 e^{2\pi i(hu_3+kv_3+lw_3)} + \cdots.$$

This equation may be written more compactly as

$$F_{hkl} = \sum_{1}^{N} f_n e^{2\pi i(hu_n + kv_n + lw_n)}, \tag{4–11}$$

where the summation extends over all the N atoms of the unit cell.

F is, in general, a complex number, and it expresses both the amplitude and phase of the resultant wave. Its absolute value $|F|$ gives the amplitude of the resultant wave in terms of the amplitude of the wave scattered by a single electron. Like the atomic scattering factor f, $|F|$ is defined as a ratio of amplitudes:

$$|F| = \frac{\text{amplitude of the wave scattered by all the atoms of a unit cell}}{\text{amplitude of the wave scattered by one electron}}.$$

The intensity of the beam diffracted by all the atoms of the unit cell in a direction predicted by the Bragg law is proportional simply to $|F|^2$, the square of the amplitude of the resultant beam, and $|F|^2$ is obtained by multiplying the expression given for F in Eq. (4–11) by its complex conjugate. Equation (4–11) is therefore a very important relation in x-ray crystallography, since it permits a calculation of the intensity of any *hkl* reflection from a knowledge of the atomic positions.

We have found the resultant scattered wave by adding together waves, differing in phase, scattered by individual atoms in the unit cell. Note that the phase difference between rays scattered by any two atoms, such as A and B in Fig. 4–8, is *constant* for every unit cell. There is no question here of these rays becoming increasingly out of phase as we go deeper in the crystal as there was when we considered diffraction at angles not exactly equal to the Bragg angle θ_B. In the direction predicted by the Bragg law, the rays scattered by all the atoms A in the crystal are exactly in phase and so are the rays scattered by all the atoms B, but between these two sets of rays there is a definite phase difference which depends on the relative positions of atoms A and B in the unit cell and which is given by Eq. (4–4).

Although it is more unwieldy, the following trigonometric equation may be used instead of Eq. (4–11):

$$F = \sum_{1}^{N} f_n[\cos 2\pi(hu_n + kv_n + lw_n) + i \sin 2\pi(hu_n + kv_n + lw_n)].$$

One such term must be written down for each atom in the unit cell. In general, the summation will be a complex number of the form

$$F = a + ib,$$

where

$$a = \sum_1^N f_n \cos 2\pi(hu_n + kv_n + lw_n),$$

$$b = \sum_1^N f_n \sin 2\pi(hu_n + kv_n + lw_n),$$

$$|F|^2 = (a + ib)(a - ib) = a^2 + b^2.$$

Substitution for a and b gives the final form of the equation:

$$|F|^2 = [f_1 \cos 2\pi(hu_1 + kv_1 + lw_1) + f_2 \cos 2\pi(hu_2 + kv_2 + lw_2) + \cdots]^2 + [f_1 \sin 2\pi(hu_1 + kv_1 + lw_1) + f_2 \sin 2\pi(hu_2 + kv_2 + lw_2) + \cdots]^2.$$

Equation (4–11) is much easier to manipulate, compared to this trigonometric form, particularly if the structure is at all complicated, since the exponential form is more compact.

4-5 SOME USEFUL RELATIONS

In calculating structure factors by complex exponential functions, many particular relations occur often enough to be worthwhile stating here. They may be verified by means of Eq. (4–7).

a) $e^{\pi i} = e^{3\pi i} = e^{5\pi i} = -1,$

b) $e^{2\pi i} = e^{4\pi i} = e^{6\pi i} = +1,$

c) in general, $e^{n\pi i} = (-1)^n,$ where n is any integer,

d) $e^{n\pi i} = e^{-n\pi i},$ where n is any integer,

e) $e^{ix} + e^{-ix} = 2 \cos x.$

4-6 STRUCTURE-FACTOR CALCULATIONS

Facility in the use of Eq. (4–11) can be gained only by working out some actual examples, and we shall consider a few such problems here and again in Chap. 10.

a) The simplest case is that of a unit cell containing only one atom at the origin, i.e., having fractional coordinates 0 0 0. Its structure factor is

$$F = fe^{2\pi i(0)} = f$$

and

$$F^2 = f^2.$$

F^2 is thus independent of h, k, and l and is the same for all reflections.

b) Consider now the base-centered cell discussed at the beginning of this chapter and shown in Fig. 4–1(a). It has two atoms of the same kind per unit cell located at 0 0 0 and $\frac{1}{2}\ \frac{1}{2}$ 0.

$$F = fe^{2\pi i(0)} + fe^{2\pi i(h/2 + k/2)}$$
$$= f[1 + e^{\pi i(h+k)}].$$

This expression may be evaluated without multiplication by the complex conjugate, since $(h + k)$ is always integral, and the expression for F is thus real and not complex. If h and k are both even or both odd, i.e., "unmixed," then their sum is always even and $e^{\pi i(h+k)}$ has the value 1. Therefore

$$F = 2f \qquad \text{for } h \text{ and } k \text{ unmixed;}$$
$$F^2 = 4f^2.$$

On the other hand, if h and k are one even and one odd, i.e., "mixed," then their sum is odd and $e^{\pi i(h+k)}$ has the value -1. Therefore

$$F = 0 \qquad \text{for } h \text{ and } k \text{ mixed;}$$
$$F^2 = 0.$$

Note that, in either case, the value of the l index has no effect on the structure factor. For example, the reflections 111, 112, 113, and 021, 022, 023 all have the same value of F, namely $2f$. Similarly, the reflections 011, 012, 013, and 101, 102, 103 all have a zero structure factor.

c) The structure factor of the body-centered cell shown in Fig. 4–1(b) may also be calculated. This cell has two atoms of the same kind located at 0 0 0 and $\frac{1}{2}\,\frac{1}{2}\,\frac{1}{2}$.

$$F = fe^{2\pi i(0)} + fe^{2\pi i(h/2+k/2+l/2)}$$
$$= f[1 + e^{\pi i(h+k+l)}].$$
$$F = 2f \qquad \text{when } (h + k + l) \text{ is even;}$$
$$F^2 = 4f^2.$$
$$F = 0 \qquad \text{when } (h + k + l) \text{ is odd;}$$
$$F^2 = 0.$$

We had previously concluded from geometrical considerations that the base-centered cell would produce a 001 reflection but that the body-centered cell would not. This result is in agreement with the structure-factor equations for these two cells. A detailed examination of the geometry of all possible reflections, however, would be a very laborious process compared to the straightforward calculation of the structure factor, a calculation that yields a set of rules governing the value of F^2 for all possible values of plane indices.

d) A face-centered cubic cell, such as that shown in Fig. 2–14, may now be considered. Assume it to contain four atoms of the same kind, located at 0 0 0, $\frac{1}{2}\,\frac{1}{2}\,0$, $\frac{1}{2}\,0\,\frac{1}{2}$, and $0\,\frac{1}{2}\,\frac{1}{2}$.

$$F = fe^{2\pi i(0)} + fe^{2\pi i(h/2+k/2)} + fe^{2\pi i(h/2+l/2)} + fe^{2\pi i(k/2+l/2)}$$
$$= f[1 + e^{\pi i(h+k)} + e^{\pi i(h+l)} + e^{\pi i(k+l)}].$$

If h, k, and l are unmixed, then all three sums $(h + k)$, $(h + l)$, and $(k + l)$ are even integers, and each term in the above equation has the value 1.

$$F = 4f \qquad \text{for unmixed indices;}$$
$$F^2 = 16f^2.$$

If h, k, and l are mixed, then the sum of the three exponentials is -1, whether two of the indices are odd and one even, or two even and one odd. Suppose, for example, that h and l are even and k is odd, e.g., 012. Then $F = f(1 - 1 + 1 - 1) = 0$, and no reflection occurs.

$$F = 0 \quad \text{for mixed indices;}$$
$$F^2 = 0$$

Thus, reflections will occur for such planes as (111), (200), and (220) but not for the planes (100), (210), (112), etc.

The reader may have noticed in the previous examples that some of the information given was not used in the calculations. In (a), for example, the cell was said to contain only one atom, but the shape of the cell was not specified; in (b) and (c), the cells were described as orthorhombic and in (d) as cubic, but this information did not enter into the structure-factor calculations. This illustrates the important point that *the structure factor is independent of the shape and size of the unit cell.* For example, *any* body-centered cell will have missing reflections for those planes which have $(h + k + l)$ equal to an odd number, whether the cell is cubic, tetragonal, or orthorhombic. The rules we have derived in the above examples are therefore of wider applicability than would at first appear and demonstrate the close connection between the Bravais lattice of a substance and its diffraction pattern. They are summarized in Table 4-1. These rules are subject to

Table 4-1

Bravais lattice	Reflections possibly present	Reflections necessarily absent
Simple	all	none
Base-centered	h and k unmixed*	h and k mixed*
Body-centered	$(h + k + l)$ even	$(h + k + l)$ odd
Face-centered	h, k, and l unmixed	h, k, and l mixed

* These relations apply to a cell centered on the C face. If reflections are present only when h and l are unmixed, or when k and l are unmixed, then the cell is centered on the B or A face, respectively.

some qualification, since some cells may contain more atoms than the ones given in examples (a) through (d), and these atoms may be in such positions that reflections normally present are now missing. For example, diamond has a face-centered cubic lattice, but it contains eight carbon atoms per unit cell. All the reflections present have unmixed indices, but reflections such as 200, 222, 420, etc., are missing. The fact that the only reflections *present* have unmixed indices proves that the lattice is face-centered, while the extra missing reflections are a clue to the actual atom arrangement in this crystal.

e) This point may be further illustrated by the structure of NaCl (Fig. 2-18). This crystal has a cubic lattice with 4 Na and 4 Cl atoms per unit cell, located as follows:

Na	0 0 0	$\frac{1}{2}\ \frac{1}{2}\ 0$	$\frac{1}{2}\ 0\ \frac{1}{2}$	$0\ \frac{1}{2}\ \frac{1}{2}$
Cl	$\frac{1}{2}\ \frac{1}{2}\ \frac{1}{2}$	$0\ 0\ \frac{1}{2}$	$0\ \frac{1}{2}\ 0$	$\frac{1}{2}\ 0\ 0$

In this case, the proper atomic scattering factors for each atom* must be inserted in the structure-factor equation, which will have eight terms:

$$F = f_{Na}e^{2\pi i(0)} + f_{Na}e^{2\pi i(h/2+k/2)} + f_{Na}e^{2\pi i(h/2+l/2)} + f_{Na}e^{2\pi i(k/2+l/2)}$$
$$+ f_{Cl}e^{2\pi i(h/2+k/2+l/2)} + f_{Cl}e^{2\pi i(l/2)} + f_{Cl}e^{2\pi i(k/2)} + f_{Cl}e^{2\pi i(h/2)},$$
$$F = f_{Na}[1 + e^{\pi i(h+k)} + e^{\pi i(h+l)} + e^{\pi i(k+l)}]$$
$$+ f_{Cl}[e^{\pi i(h+k+l)} + e^{\pi il} + e^{\pi ik} + e^{\pi ih}].$$

As discussed in Sec. 2–7, the sodium-atom positions are related by the face-centering translations and so are the chlorine-atom positions. *Whenever a lattice contains common translations, the corresponding terms in the structure-factor equation can always be factored out,* leading to considerable simplification. In this case we proceed as follows:

$$F = f_{Na}[1 + e^{\pi i(h+k)} + e^{\pi i(h+l)} + e^{\pi i(k+l)}]$$
$$+ f_{Cl}e^{\pi i(h+k+l)}[1 + e^{\pi i(-h-k)} + e^{\pi i(-h-l)} + e^{\pi i(-k-l)}].$$

The signs of the exponents in the second bracket may be changed, by relation (d) of Sec. 4–5. Therefore

$$F = [1 + e^{\pi i(h+k)} + e^{\pi i(h+l)} + e^{\pi i(k+l)}][f_{Na} + f_{Cl}e^{\pi i(h+k+l)}].$$

Here the terms corresponding to the face-centering translations appear in the first factor; the second factor contains the terms that describe the "basis" of the unit cell, namely, the Na atom at 0 0 0 and the Cl atom at $\frac{1}{2}\ \frac{1}{2}\ \frac{1}{2}$. The terms in the first bracket, describing the face-centering translations, have already appeared in example (d), and they were found to have a total value of zero for mixed indices and 4 for unmixed indices. This shows at once that NaCl has a face-centered lattice and that

$$F = 0 \qquad \text{for mixed indices;}$$
$$F^2 = 0.$$

For unmixed indices,

$$F = 4[f_{Na} + f_{Cl}e^{\pi i(h+k+l)}].$$
$$F = 4(f_{Na} + f_{Cl}) \qquad \text{if } (h + k + l) \text{ is even;}$$
$$F^2 = 16(f_{Na} + f_{Cl})^2.$$
$$F = 4(f_{Na} - f_{Cl}) \qquad \text{if } (h + k + l) \text{ is odd;}$$
$$F^2 = 16(f_{Na} - f_{Cl})^2.$$

In this case, there are more than four atoms per unit cell, but the lattice is still face-centered. The introduction of additional atoms has not eliminated any reflections present in the case of the four-atom cell, but it has decreased some in intensity. For example, the 111 reflection now involves the difference, rather than the sum, of the scattering powers of the two atoms.

---

* Strictly, and if the calculation of F is to be made to the highest accuracy, scattering factors f for the ions Na^+ and Cl^- must be used, rather than the f values for the neutral atoms Na and Cl, because NaCl is ionized.

The student should carefully note that a lot of algebra can be eliminated, whenever a lattice is known to be centered in any way, by factoring common translations out of the structure-factor equation and inserting immediately the known values of the terms representing these translations. This shortcut procedure is illustrated for NaCl:

1. Write down the atom positions in abbreviated form:

4 Na at 0 0 0 + face-centering translations,

4 Cl at $\frac{1}{2}\ \frac{1}{2}\ \frac{1}{2}$ + face-centering translations.

2. Write down the equation for F as a product of two factors. The first is the value of the terms representing the common translations; the second has terms corresponding to the "basis" atoms of the cell. The equation is

$$F = \begin{bmatrix} 4 \\ 0 \end{bmatrix} [f_{\text{Na}} + f_{\text{Cl}} e^{\pi i(h+k+l)}] \quad \begin{matrix} \text{unmixed indices} \\ \text{mixed indices.} \end{matrix}$$

3. Simplify further, as necessary. In all structure-factor calculations the aim is to obtain a set of *general* equations that will give the value of F for *any* value of hkl.

This shortcut procedure is illustrated again, for the ZnS structure, in Sec. 4–13.

f) One other example of structure factor calculation will be given here. The close-packed hexagonal cell shown in Fig. 2–15 has two atoms of the same kind located at 0 0 0 and $\frac{1}{3}\ \frac{2}{3}\ \frac{1}{2}$.

$$F = fe^{2\pi i(0)} + fe^{2\pi i(h/3 + 2k/3 + l/2)}$$
$$= f[1 + e^{2\pi i[(h+2k)/3 + l/2]}].$$

For convenience, put $[(h + 2k)/3 + l/2] = g$.

$$F = f(1 + e^{2\pi i g}).$$

Since g may have fractional values, such as $\frac{1}{3}$, $\frac{2}{3}$, $\frac{5}{6}$, etc., this expression is still complex. Multiplication by the complex conjugate, however, will give the square of the absolute value of the resultant wave amplitude F.

$$|F|^2 = f^2(1 + e^{2\pi i g})(1 + e^{-2\pi i g})$$
$$= f^2(2 + e^{2\pi i g} + e^{-2\pi i g}).$$

By relation (e) of Sec. 4–5, this becomes

$$|F|^2 = f^2(2 + 2\cos 2\pi g)$$
$$= f^2[2 + 2(2\cos^2 \pi g - 1)]$$
$$= f^2(4\cos^2 \pi g)$$
$$= 4f^2 \cos^2 \pi \left(\frac{h + 2k}{3} + \frac{l}{2}\right)$$
$$= 0 \quad \text{when } (h + 2k) \text{ is a multiple of 3 and } l \text{ is odd.}$$

It is by these missing reflections, such as $11 \cdot 1$, $11 \cdot 3$, $22 \cdot 1$, $22 \cdot 3$, that a hexagonal structure is recognized as being close-packed. Not all the reflections present have the same structure factor. For example, if $(h + 2k)$ is a multiple of 3 and l is even, then

$$\left(\frac{h + 2k}{3} + \frac{l}{2}\right) = n, \qquad \text{where } n \text{ is an integer;}$$

$$\cos \pi n = \pm 1,$$

$$\cos^2 \pi n = 1,$$

$$|F|^2 = 4f^2.$$

When all possible values of h, k, and l are considered, the results may be summarized as follows, where m is an integer:

$h + 2k$	l	$\lvert F\rvert^2$
$3m$	odd	0
$3m$	even	$4f^2$
$3m \pm 1$	odd	$3f^2$
$3m \pm 1$	even	f^2

4-7 APPLICATION TO POWDER METHOD

Any calculation of the intensity of a diffracted beam must always begin with the structure factor. The remainder of the calculation, however, varies with the particular diffraction method involved. For the Laue method, intensity calculations are so difficult that they are rarely made, since each diffracted beam has a different wavelength and blackens the film by a variable amount, depending on both the intensity and the film sensitivity for that particular wavelength. The factors governing diffracted intensity in the rotating-crystal and powder methods are somewhat similar, in that monochromatic radiation is used in each, but they differ in detail. The remainder of this chapter will be devoted to the powder method, since it is of most general utility in metallurgical work.

There are six factors affecting the relative intensity of the diffraction lines on a powder pattern:

1. polarization factor,
2. structure factor,
3. multiplicity factor,
4. Lorentz factor,
5. absorption factor,
6. temperature factor.

The first two of these have already been described, and the others will be discussed in the following sections.

4-8 MULTIPLICITY FACTOR

Consider the 100 reflection from a cubic lattice. In the powder specimen, some of the crystals will be so oriented that reflection can occur from their (100) planes. Other crystals of different orientation may be in such a position that reflection can occur from their (010) or (001) planes. Since all these planes have the same spacing, the beams diffracted by them all form part of the same cone. Now consider the 111 reflection. There are four sets of planes of the form {111} which have the same spacing but different orientation, namely, (111), $(11\bar{1})$, $(1\bar{1}\bar{1})$, and $(1\bar{1}1)$, whereas there are only three sets of the form {100}. Therefore, the probability that {111} planes will be correctly oriented for reflection is $\frac{4}{3}$ the probability that {100} planes will be correctly oriented. It follows that the intensity of the 111 reflection will be $\frac{4}{3}$ that of the 100 reflection, other things being equal.

This relative proportion of planes contributing to the same reflection enters the intensity equation as the quantity p, the *multiplicity factor*, which may be defined as the number of different planes in a form having the same spacing. Parallel planes with different Miller indices, such as (100) and $(\bar{1}00)$, are counted separately as different planes, yielding numbers which are double those given in the preceding paragraph. Thus the multiplicity factor for the {100} planes of a cubic crystal is 6 and for the {111} planes 8.

The value of p depends on the crystal system: in a tetragonal crystal, the (100) and (001) planes do not have the same spacing, so that the value of p for {100} planes is reduced to 4 and the value for {001} planes to 2. Values of the multiplicity factor as a function of hkl and crystal system are given in Appendix 13.

4-9 LORENTZ FACTOR

We must now consider certain trigonometrical factors which influence the intensity of the reflected beam. Suppose there is incident on a crystal [Fig. 4-13(a)] a narrow beam of parallel monochromatic rays, and let the crystal be rotated at a uniform angular velocity about an axis through O and normal to the drawing, so that a particular set of reflecting planes, assumed for convenience to be parallel to the crystal surface, passes through the angle θ_B, at which the Bragg law is exactly satisfied. As mentioned in Sec. 3-7, the intensity of reflection is greatest at the exact Bragg angle but still appreciable at angles deviating slightly from the Bragg angle, so that a curve of intensity vs. 2θ is of the form shown in Fig. 4-13(b). If all the diffracted beams sent out by the crystal as it rotates through the Bragg angle are received on a photographic film or in a counter, the total energy of the diffracted beam can be measured. This energy is called the *integrated intensity* of the reflection and is given by the area under the curve of Fig. 4-13(b). The integrated intensity is of much more interest than the maximum intensity, since the former is characteristic of the specimen while the latter is influenced by slight adjustments of the experimental apparatus. Moreover, in the visual comparison of the intensities of diffraction lines, it is the integrated intensity of the line rather than the maximum intensity which the eye evaluates.

The integrated intensity of a reflection depends on the particular value of θ_B

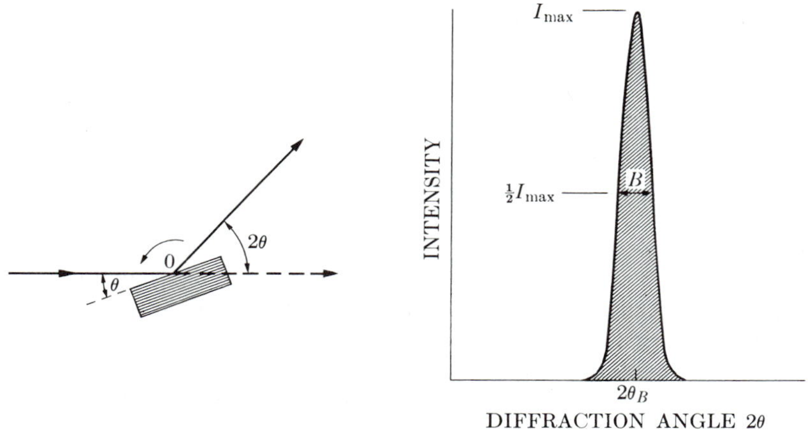

(a) (b)

Fig. 4–13 Diffraction by a crystal rotated through the Bragg angle.

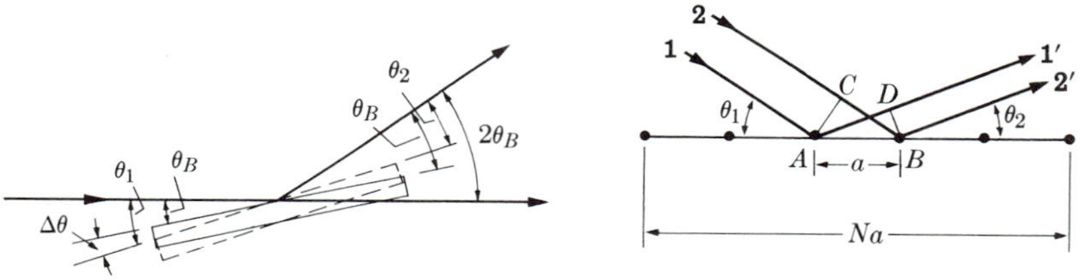

(a) (b)

Fig. 4–14 Scattering in a fixed direction during crystal rotation.

involved, even though all other variables are held constant. We can find this dependence by considering, separately, two aspects of the diffraction curve: the maximum intensity and the breadth. When the reflecting planes make an angle θ_B with the incident beam, the Bragg law is exactly satisfied and the intensity diffracted in the direction $2\theta_B$ is a maximum. But some energy is still diffracted in this direction when the angle of incidence differs slightly from θ_B, and the total energy diffracted in the direction $2\theta_B$ as the crystal is rotated through the Bragg angle is given by the value of I_{max} of the curve of Fig. 4–13(b). The value of I_{max} therefore depends on the angular range of crystal rotation over which the energy diffracted in the direction $2\theta_B$ is appreciable. In Fig. 4–14(a), the dashed lines show the position of the crystal after rotation through a small angle $\Delta\theta$ from the Bragg position. The incident beam and the diffracted beam under consideration now make unequal angles with the reflecting planes, the former making an angle $\theta_1 = \theta_B + \Delta\theta$ and the latter an angle $\theta_2 = \theta_B - \Delta\theta$. The situation on an atomic scale is shown in Fig. 4–14(b). Here we need only consider a single plane of atoms, since the rays scattered by all other planes are in phase with the corresponding rays scattered by the first plane. Let a equal the atom spacing in the plane and Na the total length* of the plane. The difference in path length for rays **1′** and **2′** scattered by adjacent atoms is given by

$$\begin{aligned}\delta_{1'2'} &= AD - CB \\ &= a\cos\theta_2 - a\cos\theta_1 \\ &= a[\cos(\theta_B - \Delta\theta) - \cos(\theta_B + \Delta\theta)].\end{aligned}$$

By expanding the cosine terms and setting sin $\Delta\theta$ equal to $\Delta\theta$, since the latter is small, we find:

$$\delta_{1'2'} = 2a\,\Delta\theta\sin\theta_B,$$

and the path difference between the rays scattered by atoms at either end of the plane is simply N times this quantity. When the rays scattered by the two end atoms are one wavelength out of phase, the diffracted intensity will be zero. (The argument here is exactly analogous to that used in Sec. 3–7.) The condition for zero diffracted intensity is therefore

$$2Na\,\Delta\theta\sin\theta_B = \lambda,$$

or

$$\Delta\theta = \frac{\lambda}{2Na\sin\theta_B}.$$

This equation gives the maximum angular range of crystal rotation over which appreciable energy will be diffracted in the direction $2\theta_B$. Since I_{max} depends on this range, we can conclude that I_{max} is proportional to $1/\sin\theta_B$. Other things being equal, I_{max} is therefore large at low scattering angles and small in the back-reflection region.

The breadth of the diffraction curve varies in the opposite way, being larger at large values of $2\theta_B$, as was shown in Sec. 3–7, where the half-maximum breadth B was found to be proportional to $1/\cos\theta_B$. The integrated intensity of the reflection is given by the area under the diffraction curve and is therefore proportional to the product $I_{max}\,B$, which is in turn proportional to $(1/\sin\theta_B)(1/\cos\theta_B)$ or to $1/\sin 2\theta_B$. Thus, as a crystal is rotated through the Bragg angle, the integrated intensity of a reflection, which is the quantity of most experimental interest, turns out to be greater for large and small values of $2\theta_B$ than for intermediate values, other things being equal.

The preceding remarks apply just as well to the powder method as they do to the case of a rotating crystal, since the range of orientations available among the powder particles, some satisfying the Bragg law exactly, some not so exactly, are the equivalent of single-crystal rotation.

However, in the powder method, a second geometrical factor arises when we consider that the integrated intensity of a reflection at any particular Bragg angle depends on the number of crystals oriented at or near that angle. This number is

---

* If the crystal is larger than the incident beam, then Na is the irradiated length of the plane; if it is smaller, Na is the actual length of the plane.

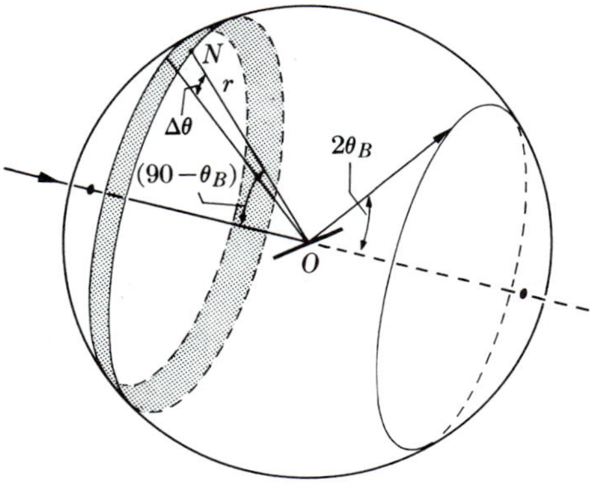

Fig. 4–15 The distribution of plane normals for a particular cone of reflected rays.

not constant even though the crystals are oriented completely at random. In Fig. 4–15 a reference sphere of radius r is drawn around the powder specimen located at O. For the particular *hkl* reflection shown, ON is the normal to this set of planes in one crystal of the powder. Suppose that the range of angles near the Bragg angle over which reflection is appreciable is $\Delta\theta$. Then, for this particular reflection, only those crystals will be in a reflecting position which have the ends of their plane normals lying in a band of width $r\,\Delta\theta$ on the surface of the sphere. Since the crystals are assumed to be oriented at random, the ends of their plane normals will be uniformly distributed over the surface of the sphere; the fraction favorably oriented for a reflection will be given by the ratio of the area of the strip to that of the whole sphere. If ΔN is the number of such crystals and N the total number, then

$$\frac{\Delta N}{N} = \frac{r\,\Delta\theta \cdot 2\pi r \sin(90^\circ - \theta_B)}{4\pi r^2} = \frac{\Delta\theta \cos\theta_B}{2}.$$

The number of crystals favorably oriented for reflection is thus proportional to $\cos\theta_B$ and is quite small for reflections in the backward direction.

In assessing relative intensities, we do not compare the total diffracted energy in one cone of rays with that in another but rather the integrated intensity per unit length of one diffraction line with that of another. For example, in the most common arrangement of specimen and film, the Debye-Scherrer method, shown in Fig. 4–16, the film obviously receives a greater proportion of a diffraction cone when the reflection is in the forward or backward direction than it does near $2\theta = 90^\circ$. Inclusion of this effect thus leads to a third geometrical factor affecting the intensity of a reflection. The length of any diffraction line being $2\pi R \sin 2\theta_B$, where R is the radius of the camera, the relative intensity per unit length of line is proportional to $1/\sin 2\theta_B$.

In intensity calculations, the three factors just discussed are combined into one and called the Lorentz factor. Dropping the subscript on the Bragg angle, we have:

$$\text{Lorentz factor} = \left(\frac{1}{\sin 2\theta}\right)(\cos\theta)\left(\frac{1}{\sin 2\theta}\right) = \frac{\cos\theta}{\sin^2 2\theta} = \frac{1}{4\sin^2\theta\cos\theta}.$$

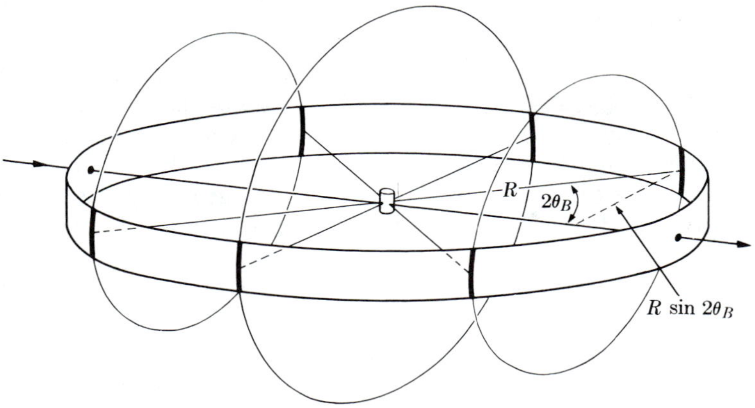

Fig. 4-16 Intersection of cones of diffracted rays with Debye-Scherrer film.

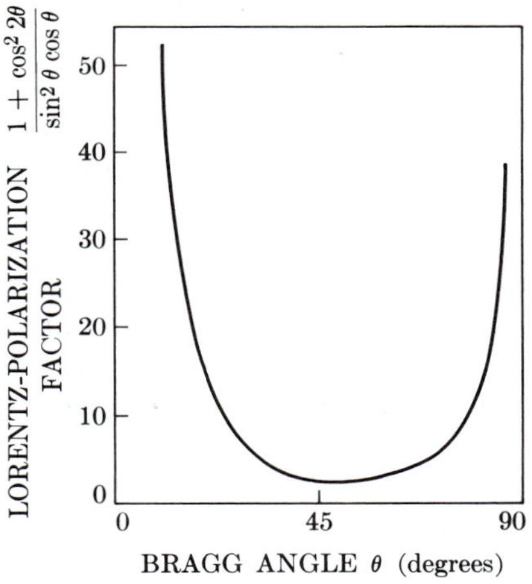

Fig. 4-17 Lorentz-polarization factor.

This in turn is combined with the polarization factor $\frac{1}{2}(1 + \cos^2 2\theta)$ of Sec. 4-2 to give the combined Lorentz-polarization factor which, with a constant factor of $\frac{1}{8}$ omitted, is given by

$$\text{Lorentz-polarization factor} = \frac{1 + \cos^2 2\theta}{\sin^2 \theta \cos \theta}.$$

Values of this factor are given in Appendix 14 and plotted in Fig. 4-17 as a function of θ. The overall effect of these geometrical factors is to decrease the intensity of reflections at intermediate angles compared to those in forward or backward directions.

4–10 ABSORPTION FACTOR

Still another factor affecting the intensities of the diffracted rays must be considered, and that is the absorption which takes place in the specimen itself. We allow for this effect in intensity calculations by introducing the *absorption factor A*, which is a number by which the calculated intensity is to be multiplied to allow for absorption. The calculation of A depends on the geometry of the diffraction method involved, and we will consider below the two most-used methods.

Debye–Scherrer Camera

The specimen in the Debye–Scherrer method has the form of a very thin cylinder of powder placed on the camera axis, and Fig. 4–18(a) shows the cross section of such a specimen. For the low-angle reflection shown, absorption of a particular ray in the incident beam occurs along a path such as AB; at B a small fraction of the incident energy is diffracted by a powder particle, and absorption of this diffracted beam occurs along the path BC. Similarly, for a high-angle reflection, absorption of both the incident and diffracted beams occurs along a path such as $(DE + EF)$. The net result is that the diffracted beam is of lower intensity than one would expect for a specimen of no absorption.

A calculation of this effect shows that the relative absorption increases as θ decreases, for any given cylindrical specimen. That this must be so can be seen from Fig. 4–18(b) which applies to a specimen (for example, tungsten) of very high absorption. The incident beam is very rapidly absorbed, and most of the diffracted beams originate in the thin surface layer on the left side of the specimen; backward-reflected beams then undergo very little absorption, but forward-reflected beams have to pass through the whole specimen and are greatly absorbed. Actually, the forward-reflected beams in this case come almost entirely from the top and bottom

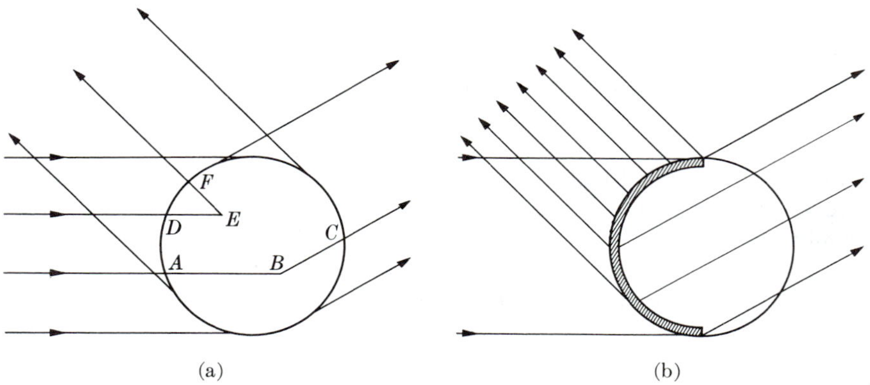

Fig. 4–18 Absorption in Debye–Scherrer specimens: (a) general case, (b) highly absorbing specimen.

edges of the specimen.* This difference in absorption between high-θ and low-θ reflections decreases as the linear absorption coefficient decreases, but the absorption is always greater for the low-θ reflections. We therefore write the Debye–Scherrer absorption factor as $A(\theta)$ to emphasize the fact that it varies with θ. Qualitatively, we conclude that $A(\theta)$ for any specimen increases as 2θ increases.

Exact calculation of the absorption factor for a cylindrical specimen is often difficult, so it is fortunate that this effect can usually be neglected in the calculation of diffracted intensities, when the Debye–Scherrer method is used. Justification of this omission will be found in Sec. 4–11.

The calculation of $A(\theta)$ for a cylindrical specimen proceeds as follows. In Fig. 4–18(a) the path length $(AB + BC)$, for a given value of θ, is expressed as a function of the position x, y of the point B relative to coordinate axes fixed relative to the specimen. The absorption factor $A(\theta)$ is then given by the function $e^{-\mu(AB+BC)}$ integrated over the entire cross-sectional area of the specimen. This integration can only be performed numerically. The result is a table of values of $A(\theta)$ as a function of θ and of the product μr, where μ is the linear absorption coefficient of the specimen and r is its radius. The specimen is usually a powder compact, with an absorption coefficient given by

$$\mu_{\text{compact}} = \mu_{\text{solid}} \left(\frac{\rho_{\text{compact}}}{\rho_{\text{solid}}} \right) \tag{4–12}$$

where ρ is density.

Values of $A(\theta)$ have been calculated and tabulated by Bradley [4.1]. Tables of values can also be found in [G.11, Vol. 2, p. 295–299] and in [G.13, p. 663–666].

Diffractometer

A diffractometer specimen usually has the form of a flat plate making equal angles with the incident and diffracted beams as in Fig. 3–4, if one imagines a polycrystalline plate substituted for the single crystal indicated there. It is shown below that the absorption factor A is equal to $1/2\mu$, *independent of* θ. This independence of θ is due to the exact balancing of two opposing effects. When θ is small, the specimen area irradiated by an incident beam of fixed cross section is large, but the effective depth of x-ray penetration is small; when θ is large, the irradiated area is small, but the penetration depth is relatively large. The net effect is that the effective irradiated volume is constant and independent of θ. Absorption occurs in any case, however, and the larger the absorption coefficient of the specimen, the lower the intensity of the diffracted beams, other things being equal. The important fact to note is that absorption decreases the intensities of all diffracted beams by the same factor and therefore does not enter into the calculation of *relative* intensities.

---

* The powder patterns reproduced in Fig. 3–13 show this effect, at least on the original films. The lowest-angle line in each pattern is split in two, because the beam diffracted through the center of the specimen is so highly absorbed. It is important to keep the possibility of this phenomenon in mind when examining Debye–Scherrer photographs, or split low-angle lines may be incorrectly interpreted as separate diffraction lines from two different sets of planes.

The calculation of A proceeds as follows. The incident beam in the diffractometer is actually divergent (Sec. 7–2), but we will assume here that the beam is composed of parallel rays, because the divergence angle is very small (3° or less). We will calculate the effect of absorption in the specimen on the intensity of the diffracted beam, and, since this effect will come up again in later parts of this book, we will make our calculation quite general. In Fig. 4–19, the incident beam has intensity I_0 (ergs/cm^2/sec), is 1 cm square in cross section, and is incident on the powder plate at an angle γ. We consider the energy diffracted from this beam by a layer of the powder of length l and thickness dx, located at a depth x below the surface. Since the incident beam undergoes absorption by the specimen over the path length AB, the energy incident per second on the layer considered is $I_0 e^{-\mu(AB)}$ (ergs/sec), where μ is the linear absorption coefficient of the powder compact, given by Eq. (4–12). Let a be the volume fraction of the specimen containing particles having the correct orientation for reflection of the incident beam, and b the fraction of the incident energy which is diffracted by unit volume. Then the energy diffracted per second by the layer considered, which has a volume l dx, is given by $ablI_0 e^{-\mu(AB)}\,\mathrm{d}x$. But this diffracted energy is also decreased by absorption, by a factor of $e^{-\mu(BC)}$, since the diffracted rays have a path length of BC in the specimen. The energy flux per second in the diffracted beam outside the specimen, i.e., the integrated intensity, is therefore given by

$$\mathrm{d}I_D = ablI_0 e^{-\mu(AB+BC)}\,\mathrm{d}x \quad \text{(ergs/sec)}. \tag{4–13}$$

But

$$l = \frac{1}{\sin\gamma}, \qquad AB = \frac{x}{\sin\gamma}, \qquad BC = \frac{x}{\sin\beta}.$$

Therefore,

$$\mathrm{d}I_D = \frac{I_0 ab}{\sin\gamma}\, e^{-\mu x(1/\sin\gamma + 1/\sin\beta)}\,\mathrm{d}x. \tag{4–14}$$

(The reader might note that the analogous absorption effect in transmission, rather than reflection, is given later as Eq. (9–7.)

For the particular specimen arrangement used in the diffractometer, $\gamma = \beta = \theta$, and the above equation becomes

$$\mathrm{d}I_D = \frac{I_0 ab}{\sin\theta}\, e^{-2\mu x/\sin\theta}\,\mathrm{d}x. \tag{4–15}$$

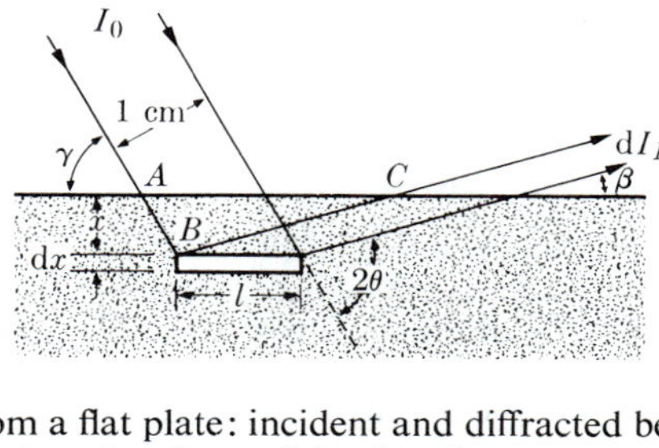

Fig. 4–19 Diffraction from a flat plate: incident and diffracted beams have a thickness of 1 cm in a direction normal to the plane of the drawing.

The total diffracted intensity is obtained by integrating over an infinitely thick specimen:

$$I_D = \int_{x=0}^{x=\infty} dI_D = \frac{I_0 ab}{2\mu}. \tag{4-16}$$

Here I_0, b, and μ are constant for all reflections (independent of θ) and we may also regard a as constant. Actually, a varies with θ, but this variation is already taken care of by the $\cos\theta$ portion of the Lorentz factor (see Sec. 4-9) and need not concern us here. We conclude that the absorption factor $1/2\mu$ is independent of θ for a flat specimen making equal angles with the incident and diffracted beams, provided the specimen fills the incident beam at all angles and is effectively of infinite thickness.

The criterion adopted for "infinite thickness" depends on the sensitivity of our intensity measurements or on what we regard as negligible diffracted intensity. For example, we might arbitrarily but quite reasonably define infinite thickness as that thickness t which a specimen must have in order that the intensity diffracted by a thin layer on the back side be $\frac{1}{1000}$ of the intensity diffracted by a thin layer on the front side. Then, from Eq. (4-15) we have

$$\frac{dI_D \text{ (at } x = 0)}{dI_D \text{ (at } x = t)} = e^{2\mu t/\sin\theta} = 1000,$$

from which

$$t = \frac{3.45 \sin\theta}{\mu}.$$

This expression shows that "infinite thickness," for a metal specimen, is very small indeed. For example, suppose a specimen of nickel powder is being examined with Cu $K\alpha$ radiation at θ values approaching 90°. The density of the powder compact may be taken as about 0.6 the density of bulk nickel, which is 8.9 gm/cm^3, leading to a value of μ for the compact of 261 cm^{-1}. The value of t is therefore 1.32×10^{-2} cm, or about five thousandths of an inch.

4-11 TEMPERATURE FACTOR

So far we have considered a crystal as a collection of atoms located at fixed points in the lattice. Actually, the atoms undergo thermal vibration about their mean positions even at the absolute zero of temperature, and the amplitude of this vibration increases as the temperature increases. In aluminum at room temperature, the average displacement of an atom from its mean position is about 0.17 Å, which is by no means negligible, being about 6 percent of the distance of closest approach of the mean atom positions in this crystal.

Increased thermal vibration of the atoms, as the result of an increase in temperature, has three main effects:

1. The unit cell expands, causing changes in plane spacings d and therefore in the 2θ positions of the diffraction lines. If the positions of one or more lines are measured as a function of temperature (Sections 6–5 and 7–2), the thermal expansion coefficient of the specimen can be determined by x-ray diffraction.

2. The intensities of the diffraction lines decrease.

3. The intensity of the background scattering between lines increases.

The second and third effects are described below. Here we are usually interested not in intensity changes with temperature, but in variations in intensity with 2θ at constant temperature (usually room temperature).

Thermal agitation decreases the intensity of a diffracted beam because it has the effect of smearing out the lattice planes; atoms can be regarded as lying no longer on mathematical planes but rather in platelike regions of ill-defined thickness. Thus the reinforcement of waves scattered at the Bragg angle by various parallel planes, the reinforcement which is called a diffracted beam, is not as perfect as it is for a crystal with fixed atoms. This reinforcement requires that the path difference, which is a function of the plane spacing d, between waves scattered by adjacent planes be an integral number of wavelengths. Now the thickness of the platelike "planes" in which the vibrating atoms lie is, on the average, $2u$, where u is the average displacement of an atom from its mean position. Under these conditions reinforcement is no longer perfect, and it becomes more imperfect as the ratio u/d increases, i.e., as the temperature increases, since that increases u, or as θ increases, since high-θ reflections involve planes of low d value. Thus the intensity of a diffracted beam decreases as the temperature is raised, and, for a constant temperature, thermal vibration causes a greater decrease in the reflected intensity at high angles than at low angles. In intensity calculations we allow for this effect by introducing the *temperature factor* e^{-2M}, which is a number by which the calculated intensity is to be multiplied to allow for thermal vibration of the atoms. Qualitatively, we conclude that e^{-2M} decreases as 2θ increases. A method of calculating e^{-2M} when it is needed is outlined later, and Fig. 4–20 shows the result of such a calculation for iron.

The temperature effect and the previously discussed absorption effect in cylindrical specimens depend on angle in opposite ways and, to a first approximation, cancel each other in the Debye-Scherrer method. In back reflection, for

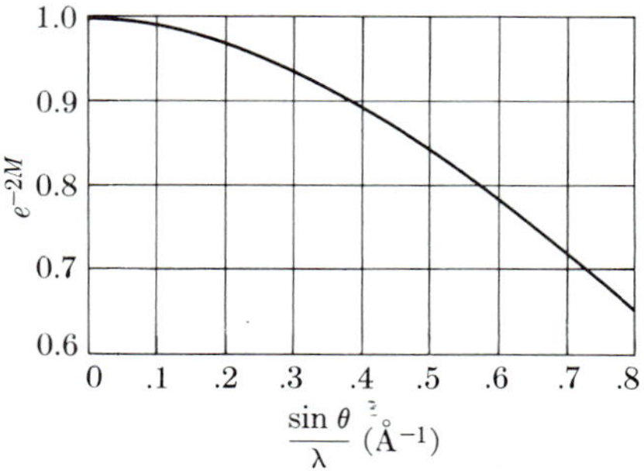

Fig. 4–20 Temperature factor e^{-2M} of iron at 20°C as a function of $(\sin\theta)/\lambda$.

example, the intensity of a diffracted beam is decreased very little by absorption but very greatly by thermal agitation, while in the forward direction the reverse is true. The two effects do not exactly cancel one another at all angles; however, if the comparison of line intensities is restricted to lines not differing too greatly in θ values, the absorption and temperature effects can be safely ignored in the Debye-Scherrer method. This is a fortunate circumstance, since both of these effects are rather difficult to calculate exactly.

Theoretically, thermal vibration of the atoms causes a very slight increase in the breadth B, measured at half-maximum intensity, of the diffraction lines. However, this expected effect has never been detected [4.2], and diffraction lines are observed to be sharp right up to the melting point, but their maximum intensity gradually decreases.

It is also worth noting that the mean amplitude of atomic vibration is not a function of the temperature alone but depends also on the elastic constants of the crystal. At any given temperature, the less "stiff" the crystal, the greater the vibration amplitude u. This means that u is much greater at any one temperature for a soft, low-melting-point metal like lead than it is for, say, tungsten. Substances with low melting points have quite large values of u even at room temperature and therefore yield rather poor back-reflection photographs. For example, thermal atomic vibration in lead at 20°C reduces the intensity of the highest-angle line observed with Cu $K\alpha$ radiation (at about 161° 2θ) to only 18 percent ($e^{-2M} = 0.18$) of the value for atoms at rest.

In only one application described in this book (Sec. 14-10) will we need any quantitative information about the temperature factor e^{-2M}, but it is convenient to describe the calculation here. Formally, we allow for the effect by defining f as the atomic scattering factor of an atom undergoing thermal vibration, f_0 as the same quantity for an atom at rest, and relating the two by

$$f = f_0 e^{-M}.$$

(The quantity f_0 is then the scattering factor as usually tabulated, for example in Appendix 12.) Because the intensity of any line depends on f^2, calculated intensities must be multiplied by e^{-2M} to allow for thermal vibration. The quantity M depends on both the amplitude u of thermal vibration and the scattering angle 2θ:

$$M = 2\pi^2 \left(\frac{\overline{u^2}}{d^2}\right) = 8\pi^2\, \overline{u^2} \left(\frac{\sin\theta}{\lambda}\right)^2 = B\left(\frac{\sin\theta}{\lambda}\right)^2 \tag{4-17}$$

where $\overline{u^2}$ is the mean square displacement of the atom in a direction normal to the reflecting planes. The exact calculation of $\overline{u^2}$ as a function of temperature is extremely difficult, which means that M or B is hard to determine accurately. Debye has given the following expression:

$$M = \frac{6h^2 T}{mk\Theta^2}\left[\phi(x) + \frac{x}{4}\right]\left(\frac{\sin\theta}{\lambda}\right)^2, \tag{4-18}$$

where h is Planck's constant, T the absolute temperature, m the mass of the vibrating atom, k Boltzmann's constant, Θ the Debye characteristic temperature of the substance in °K, $x = \Theta/T$, and $\phi(x)$ is a function tabulated, along with values of Θ, in Appendix 15. Because $m = A/N$, where A = atomic weight and N = Avogadro's number, the coefficient of the bracketed terms above becomes

$$\frac{6h^2T}{mk\Theta^2} = \frac{(6)(6.02 \times 10^{26})(6.63 \times 10^{-34})^2 T}{A\Theta^2(1.38 \times 10^{-23})(10^{-20})} = \frac{1.15 \times 10^4 T}{A\Theta^2}$$

if λ is in angstroms. Equation (4–18) is approximate and applies only to elements with cubic crystal structure.

For thorough treatments of the effect of thermal vibration on the diffraction pattern, see James [G.7] and Warren [G.30].

The thermal vibration of atoms has another effect on diffraction patterns. Besides decreasing the intensity of diffraction lines, it causes some general coherent scattering in all directions. This is called *temperature-diffuse scattering*; it contributes only to the general background of the pattern and its intensity gradually increases with 2θ. Contrast between lines and background naturally suffers, so this effect is a very undesirable one, leading in extreme cases to diffraction lines in the back-reflection region scarcely distinguishable from the background. Figure 4–21 illustrates this effect. In (a) is shown an extremely hypothetical pattern (only three lines, equally spaced, equally strong, with no background whatever) for atoms at rest; in (b) the lines, decreased in intensity by the factor e^{-2M}, are superimposed on a background of thermal diffuse scattering.

In the phenomenon of temperature-diffuse scattering we have another example, beyond those alluded to in Sec. 3–7, of scattering at non-Bragg angles. Here again it is not surprising that such scattering should occur, since the displacement of atoms from their mean positions constitutes a kind of crystal imperfection and leads to a partial breakdown of the conditions necessary for perfect destructive interference between rays scattered at non-Bragg angles.

The effect of thermal vibration also illustrates what has been called "the approximate law of conservation of diffracted energy." This law states that the total energy diffracted by a particular specimen under particular experimental conditions is roughly constant. Therefore, anything done to alter the physical condition of the specimen does not alter the total amount of diffracted energy but

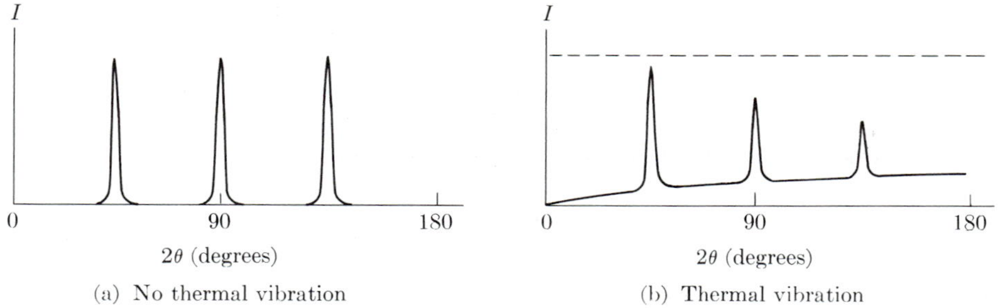

Fig. 4–21 Effect of thermal vibration of the atoms on a powder pattern. Very schematic, see text.

only its distribution in space. This "law" is not at all rigorous, but it does prove helpful in considering many diffraction phenomena. For example, at low temperatures there is very little background scattering due to thermal agitation and the diffraction lines are relatively intense; if the specimen is now heated to a high temperature, the lines will become quite weak and the energy which is lost from the lines will appear in a spread-out form as temperature-diffuse scattering.

4-12 INTENSITIES OF POWDER PATTERN LINES

We are now in a position to gather together the factors discussed in preceding sections into an equation for the relative intensity of powder pattern lines.

Debye–Scherrer Camera

$$\text{(Approximate)} \qquad \boxed{I = |F|^2 p \left(\frac{1 + \cos^2 2\theta}{\sin^2 \theta \cos \theta}\right)}, \qquad (4\text{–}19)$$

where I = relative integrated intensity (arbitrary units), F = structure factor, p = multiplicity factor, and θ = Bragg angle. The trigonometric terms in parentheses are the Lorentz-polarization factor. In arriving at this equation, we have omitted factors which are constant for all lines of the pattern. For example, all that is retained of the Thomson equation (Eq. 4–2) is the polarization factor $(1 + \cos^2 2\theta)$, with constant factors, such as the intensity of the incident beam and the charge and mass of the electron, omitted. The intensity of a diffraction line is also directly proportional to the irradiated volume of the specimen and inversely proportional to the camera radius, but these factors are again constant for all diffraction lines and may be neglected. Omission of the temperature and absorption factors means that Eq. (4–19) is valid only for lines fairly close together on the pattern; this latter restriction is not as serious as it may sound. Finally, it should be remembered that this equation gives the relative *integrated* intensity, i.e., the relative area under the curve of intensity vs. 2θ.

If an exact expression is required, the absorption factor $A(\theta)$ and the temperature factor e^{-2M} must be inserted:

$$\text{(Exact)} \qquad I = |F|^2 p \left(\frac{1 + \cos^2 2\theta}{\sin^2 \theta \cos \theta}\right) A(\theta) e^{-2M}. \qquad (4\text{–}20)$$

Diffractometer

Here the absorption factor is independent of θ and so does not enter into the calculation of relative intensities. Equation (4–19) becomes still less precise, because there is no longer any approximate cancellation of the absorption and temperature factors. Equation (4–19) may still be used, for adjacent lines on the pattern, but the calculated intensity of the higher-angle line, relative to that of the lower-angle line, will always be somewhat too large because of the omission of the temperature factor.

The exact equation for the diffractometer is

$$\text{(Exact)} \qquad I = |F|^2 p \left(\frac{1 + \cos^2 2\theta}{\sin^2 \theta \cos \theta}\right) e^{-2M}. \qquad (4\text{–}21)$$

Qualifications

The two following effects can make the above intensity equations invalid:

1) *Preferred orientation.* From the way in which the cos θ portion of the Lorentz factor was determined in Sec. 4–9, it follows that Eqs. (4–19) through (4–21) are valid only when the crystals making up the specimen are randomly oriented in space. Preferred orientation of the crystal grains causes radical disagreement between calculated and observed intensities and, when such disagreement exists, preferred orientation should be the first possible cause to be suspected. It is relatively easy to prepare powder-compact specimens from ground powders or metal filings so that the ideal of perfect randomness of orientation is closely approached, but virtually all polycrystalline specimens of metal wire, metal sheet, manufactured ceramics, and even natural rocks or minerals will exhibit more or less preferential orientation of the grains.

2) *Extinction* [G.7, G.30]. As mentioned in Sec. 3–7, all real crystals are imperfect, in the sense that they have a mosaic structure, and the degree of imperfection can vary greatly from one crystal to another. Equations (4–19) through (4–21) are derived on the basis of the so-called "ideally imperfect" crystal, one in which the mosaic blocks are quite small (of the order of 10^{-4} cm to 10^{-5} cm in thickness) and so disoriented that they are all essentially nonparallel. Such a crystal has maximum reflecting power. A crystal made up of large mosaic blocks, some or all of which are accurately parallel to one another, is more nearly perfect and has a lower reflecting power. This *decrease* in the integrated intensity of the diffracted beam as the crystal becomes more nearly perfect is called *extinction.* Extinction is absent for the ideally imperfect crystal, and the presence of extinction invalidates Eqs. (4–19) through (4–21). Any treatment that will make a crystal more imperfect will reduce extinction and, for this reason alone, powder specimens should be ground as fine as possible. Grinding not only reduces the crystal size but also tends to decrease the mosaic block size, disorient the mosaic blocks, and strain them nonuniformly. (The theory of the extinction effect is difficult. To prove that imperfections in a crystal increase its reflecting power would take us too far afield, but an experimental proof is given in Sec. 8–7.)

The extinction effect can operate, not only in single-crystal specimens, but also in the individual grains of polycrystalline specimens. Extinction may be assumed to be absent in ground or filed powders and is usually negligible in fine-grained polycrystalline specimens. If its presence is suspected in the latter, the specimen can always be reduced to powder by grinding or filing.

4–13 EXAMPLES OF INTENSITY CALCULATIONS

The use of Eq. (4–19) will be illustrated by the calculation of the position and relative intensities of the diffraction lines on a Debye–Scherrer pattern of copper, made with Cu $K\alpha$ radiation. The calculations are most readily carried out in tabular form, as in Table 4–2.

Remarks

Column 2: Since copper is face-centered cubic, F is equal to $4f_{\text{Cu}}$ for lines of unmixed indices and zero for lines of mixed indices. The reflecting plane indices, all unmixed, are

Table 4-2

1	2	3	4	5	6	7	8
Line	hkl	$h^2+k^2+l^2$	$\sin^2\theta$	$\sin\theta$	θ	$\frac{\sin\theta}{\lambda}$ (Å^{-1})	f_{Cu}
1	111	3	0.1365	0.369	21.7°	0.24	22.1
2	200	4	0.1820	0.427	25.3	0.27	20.9
3	220	8	0.364	0.603	37.1	0.39	16.8
4	311	11	0.500	0.707	45.0	0.46	14.8
5	222	12	0.546	0.739	47.6	0.48	14.2
6	400	16	0.728	0.853	58.5	0.55	12.5
7	331	19	0.865	0.930	68.4	0.60	11.5
8	420	20	0.910	0.954	72.6	0.62	11.1

1	9	10	11	12	13	14
Line	F^2	p	$\frac{1+\cos^2 2\theta}{\sin^2\theta\cos\theta}$	Relative integrated intensity		
				Calc.	Calc.	Obs.
1	7810	8	12.03	7.52×10^5	10.0	vs
2	6990	6	8.50	3.56	4.7	s
3	4520	12	3.70	2.01	2.7	s
4	3500	24	2.83	2.38	3.2	s
5	3230	8	2.74	0.71	0.9	m
6	2500	6	3.18	0.48	0.6	w
7	2120	24	4.81	2.45	3.3	s
8	1970	24	6.15	2.91	3.9	s

written down in this column in order of increasing values of $(h^2 + k^2 + l^2)$, from Appendix 10.

Column 4: For a cubic crystal, values of $\sin^2\theta$ are given by Eq. (3-10):

$$\sin^2\theta = \frac{\lambda^2}{4a^2}(h^2 + k^2 + l^2).$$

In this case, $\lambda = 1.542$ Å (Cu $K\alpha$) and $a = 3.615$ Å (lattice parameter of copper). Therefore, multiplication of the integers in column 3 by $\lambda^2/4a^2 = 0.0455$ gives the values of $\sin^2\theta$ listed in column 4. In this and similar calculations, three-figure accuracy is ample.

Column 6: Needed to determine the Lorentz-polarization factor and $(\sin\theta)/\lambda$.

Column 7: Obtained from Appendix 11. Needed to determine f_{Cu}.

Column 8: Obtained from Appendix 12.

Column 9: Obtained from the relation $F^2 = 16f_{Cu}^2$.

Column 10: Obtained from Appendix 13.

Column 11: Obtained from Appendix 14.

Column 12: These values are the product of the values in columns 9, 10, and 11, according to Eq. (4-19).

Column 13: Values from column 12 recalculated to give the first line an arbitrary intensity of 10, i.e., "normalized" to 10 for the first line.

Column 14: These entries give the observed intensities, visually estimated according to the following simple scale, from the original film for copper in Fig. 3-13 (vs = very strong, s = strong, m = medium, w = weak).

The agreement obtained here between observed and calculated intensities is satisfactory. Note how the value of the multiplicity p exerts a strong control over the line intensity. The values of $|F|^2$ and of the Lorentz-polarization factor vary smoothly with θ, but the values of p, and therefore of I, vary quite irregularly.

A more complicated structure may now be considered, namely that of the zinc-blende form of ZnS, shown in Fig. 2–19(b). This form of ZnS is cubic and has a lattice parameter of 5.41 Å. We will calculate the relative intensities of the first six lines on a Debye–Scherrer pattern made with Cu $K\alpha$ radiation.

As always, the first step is to work out the structure factor. ZnS has four zinc and four sulfur atoms per unit cell, located in the following positions:

$$\text{Zn}: \tfrac{1}{4}\,\tfrac{1}{4}\,\tfrac{1}{4} + \text{face-centering translations,}$$

$$\text{S}: 0\,0\,0 + \text{face-centering translations.}$$

Since the structure is face-centered, we know that the structure factor will be zero for planes of mixed indices. We also know, from example (e) of Sec. 4–6, that the terms in the structure-factor equation corresponding to the face-centering translations can be factored out and the equation for unmixed indices written down at once:

$$F = 4[f_S + f_{Zn}e^{(\pi i/2)(h+k+l)}].$$

$|F|^2$ is obtained by multiplication of the above by its complex conjugate:

$$|F|^2 = 16[f_S + f_{Zn}e^{(\pi i/2)(h+k+l)}][f_S + f_{Zn}e^{-(\pi i/2)(h+k+l)}].$$

This equation reduces to the following form:

$$|F|^2 = 16\left[f_S^2 + f_{Zn}^2 + 2f_S f_{Zn} \cos\frac{\pi}{2}(h + k + l)\right].$$

Further simplification is possible for various special cases:

$$|F|^2 = 16(f_S^2 + f_{Zn}^2) \qquad \text{when } (h + k + l) \text{ is odd;} \qquad (4\text{–}22)$$

$$|F|^2 = 16(f_S - f_{Zn})^2 \qquad \text{when } (h + k + l) \text{ is an odd multiple of 2;} \qquad (4\text{–}23)$$

$$|F|^2 = 16(f_S + f_{Zn})^2 \qquad \text{when } (h + k + l) \text{ is an even multiple of 2.} \qquad (4\text{–}24)$$

The intensity calculations are carried out in Table 4–3, with some columns omitted for the sake of brevity.

Remarks

Columns 5 and 6: These values are read from scattering-factor curves plotted from the data of Appendix 12.

Column 7: $|F|^2$ is obtained by the use of Eq. (4–22), (4–23), or (4–24), depending on the particular values of hkl involved. Thus, Eq. (4–22) is used for the 111 reflection and Eq. (4–24) for the 220 reflection.

Columns 10 and 11: The agreement obtained here between calculated and observed intensities is again satisfactory. In this case, both the values of $|F|^2$ and of p vary irregularly with θ, leading to an irregular variation of I.

One further remark on intensity calculations is necessary. In the powder method, two sets of planes with different Miller indices can reflect to the same

Table 4-3

1	2	3	4	5	6
Line	hkl	θ	$\frac{\sin\theta}{\lambda}$ (Å^{-1})	f_S	f_{Zn}
1	111	14.3°	0.16	12.3	25.8
2	200	16.6	0.19	11.4	24.6
3	220	23.8	0.26	9.7	22.1
4	311	28.2	0.30	9.0	20.7
5	222	29.6	0.32	8.8	20.0
6	400	34.8	0.37	8.2	18.4

1	7	8	9	10	11
Line	$\lvert F\rvert^2$	p	$\frac{1+\cos^2 2\theta}{\sin^2\theta\cos\theta}$	Relative intensity	
				Calc.	Obs.
1	13070	8	30.0	10.0	vs
2	2790	6	21.7	1.2	w
3	16180	12	9.76	6.1	vs
4	8150	24	6.64	4.1	vs
5	2010	8	5.95	0.3	vw
6	11320	6	4.19	0.9	w

point on the film: for example, the planes (411) and (330) in the cubic system, since they have the same value of $(h^2 + k^2 + l^2)$ and hence the same spacing, or the planes (501) and (431) of the tetragonal system, since they have the same values of $(h^2 + k^2)$ and l^2. In such a case, the intensity of each reflection must be calculated separately, since in general the two will have different multiplicity and structure factors, and then added to find the total intensity of the line.

4-14 MEASUREMENT OF X-RAY INTENSITY

In the examples just given, the observed intensity was estimated simply by visual comparison of one line with another. Although this simple procedure is satisfactory in a surprisingly large number of cases, there are problems in which a more precise measurement of diffracted intensity is necessary. Two methods are available for making such measurements, one dependent on the photographic effect of x-rays and the other on the ability of x-rays to activate an electronic counter. These methods have already been mentioned briefly in Sec. 1-8 and will be described more fully in Chaps. 6 and 7, respectively.

PROBLEMS

* **4-1** Derive an expression for the absorption factor of a diffractometer specimen in the form of a flat plate of finite thickness t. (Note that the absorption factor now depends on θ.)

* **4–2** Consider the highest-angle line on the diffraction pattern of Cu and of Pb, measured at 20°C with Cu $K\alpha$ radiation. By what percentage is the intensity of each of these lines reduced by thermal vibration of the atoms?

4–3 Consider a hypothetical element whose structure can be based on either of the following:

a) Cell A, base-centered tetragonal containing two atoms per cell, at 0 0 0 and $\frac{1}{2}\,\frac{1}{2}\,0$, for which $a = 2$ Å and $c = 3$ Å;
b) Cell B, simple tetragonal with one atom per cell at 0 0 0.

Determine simplified structure-factor equations for each cell and the positions (2θ values) of the first four lines that would be *observed* ($F^2 \neq 0$) on a powder pattern made with Cu $K\alpha$ radiation. Plot the 2θ values of these lines in the manner of Fig. 10–2, and label each line with its indices relative to Cell A and Cell B. Draw the two cells in the proper relation to one another, and show that the indices of any one observed line, other than obvious ones of the form $0\,0\,l$, refer to the same plane of atoms.

[This problem illustrates several points: (1) we can choose any unit cell we wish, (2) the Miller indices of any given plane of atoms depend on the choice of cell, and (3) the diffraction pattern of the material is independent of the choice of cell.]

* **4–4** Derive simplified expressions for F^2 for diamond, including the rules governing observed reflections. This crystal is cubic and contains 8 carbon atoms per unit cell, located in the following positions:

$$0\,0\,0 \qquad \tfrac{1}{2}\,\tfrac{1}{2}\,0 \qquad \tfrac{1}{2}\,0\,\tfrac{1}{2} \qquad 0\,\tfrac{1}{2}\,\tfrac{1}{2}$$
$$\tfrac{1}{4}\,\tfrac{1}{4}\,\tfrac{1}{4} \qquad \tfrac{3}{4}\,\tfrac{3}{4}\,\tfrac{1}{4} \qquad \tfrac{3}{4}\,\tfrac{1}{4}\,\tfrac{3}{4} \qquad \tfrac{1}{4}\,\tfrac{3}{4}\,\tfrac{3}{4}$$

4–5 A certain tetragonal crystal has four atoms of the same kind per unit cell, located at $0\,\tfrac{1}{2}\,\tfrac{1}{4}$, $\tfrac{1}{2}\,0\,\tfrac{1}{4}$, $\tfrac{1}{2}\,0\,\tfrac{3}{4}$, $0\,\tfrac{1}{2}\,\tfrac{3}{4}$. (Do not change axes.)

a) Derive simplified expressions for F^2.
b) What is the Bravais lattice of this crystal?
c) What are the values of F^2 for the 100, 002, 111, and 011 reflections?

* **4–6** Derive simplified expressions for F^2 for the wurtzite form of ZnS, including the rules governing observed reflections. This crystal is hexagonal and contains 2 ZnS per unit cell, located in the following positions:

$$\text{Zn}: 0\,0\,0,\ \tfrac{1}{3}\,\tfrac{2}{3}\,\tfrac{1}{2},$$
$$\text{S}: 0\,0\,\tfrac{3}{8},\ \tfrac{1}{3}\,\tfrac{2}{3}\,\tfrac{7}{8}.$$

Note that these positions involve a common translation, which may be factored out of the structure-factor equation.

Ignore the absorption and temperature factors in all of the following problems.

* **4–7** A Debye–Scherrer pattern of tungsten (BCC) is made with Cu $K\alpha$ radiation. The first four lines on this pattern were observed to have the following θ values:

Line	θ
1	20.3°
2	29.2
3	36.7
4	43.6

Index these lines (i.e., determine the Miller indices of each reflection by the use of Eq. (3–10) and Appendix 10) and calculate their relative integrated intensities.

4–8 A Debye–Scherrer pattern is made of silicon, which has the same structure as diamond, with Cu $K\alpha$ radiation. What are the indices of the first two lines on the pattern, and what is the ratio of the integrated intensity of the first to that of the second?

* **4–9** A Debye–Scherrer pattern is made of the intermediate phase InSb with Cu $K\alpha$ radiation. This phase has the zinc-blende structure and a lattice parameter of 6.46 Å. What are the indices of the first two lines on the pattern, and what is the ratio of the integrated intensity of the first to the second?

4–10 Calculate the relative integrated intensities of the first six lines of the Debye–Scherrer pattern of zinc, made with Cu $K\alpha$ radiation. The indices and observed θ values of these lines are:

Line	*hkl*	θ
1	00 · 2	18.8°
2	10 · 0	20.2
3	10 · 1	22.3
4	10 · 2	27.9
5	11 · 0, 10 · 3	36.0
6	00 · 4	39.4

(Line 5 is made up of two unresolved lines from planes of very nearly the same spacing.) Compare your results with the intensities observed in the pattern shown in Fig. 3–13.

Experimental Methods

5

Laue Photographs

5-1 INTRODUCTION

This chapter begins a section of the book devoted to experimental methods. A large range of laboratory equipment is available for x-ray diffraction and spectroscopy, and the International Union of Crystallography has published a useful, worldwide directory of suppliers [5.1], which shows what apparatus and supplies are available and where to find them. A laboratory manual by Azaroff and Donahue [G.27] describes twenty-one experiments in x-ray crystallography.

The present chapter deals with the Laue method only from the experimental side. Its main applications will be examined in Chap. 8. Both aspects are described in a book by Amoros *et al* [G.40].

Laue photographs are the easiest kind of diffraction pattern to make and require only the simplest kind of apparatus. White radiation is necessary, and the best source is a tube with a heavy-metal target, such as tungsten, since the intensity of the continuous spectrum is proportional to the atomic number of the target metal. Good patterns can also be obtained with radiation from other metals, such as molybdenum or copper. Ordinarily, the presence of strong characteristic components, such as W $L\alpha_1$, Cu $K\alpha$, Mo $K\alpha$, etc., in the radiation used, does not complicate the diffraction pattern in any way or introduce difficulties in its interpretation. Such a component will be reflected only if a set of planes in the crystal happens to be oriented in just such a way that the Bragg law is satisfied for that component, and then the only effect will be the formation of a Laue spot of exceptionally high intensity.

The specimen used in the Laue method is a single crystal. This may mean an isolated single crystal or one particular crystal grain, not too small, in a polycrystalline aggregate. The only restriction on the size of a crystal in a polycrystalline mass is that it must be no smaller than the incident x-ray beam, if the pattern obtained is to correspond to that crystal alone.

Each diffracted beam in the Laue method has a different wavelength, selected out of the incident beam of white radiation by the d spacing and θ value of the crystal planes producing the reflection. If x-rays were visible like light, we would see each reflected beam as a different color. Although we cannot see these colors, colored Laue patterns can be made by special photographic procedures. Blum [5-2] made such patterns by preparing a special film, in which the three color-producing layers of emulsion in color film were separated by layers containing metal atoms that acted as filters for the incident x-rays. Ting *et al.* [5–3] achieved the same result with commercially available materials.

The position of any Laue spot is unaltered by a change in plane spacing, since the only effect of such a change is to alter the wavelength of the diffracted beam. It follows that two crystals of the same orientation and crystal structure, but of different lattice parameter, will produce identical Laue patterns.

5–2 CAMERAS

Any device designed to hold a specimen and photographic film to record diffracted beams is called an x-ray *camera*, even though it bears little resemblance to cameras used for photography by visible light. Laue cameras are so simple to construct that homemade models are not uncommon.

Transmission Camera

Figure 5–1 shows a transmission camera, in this case a dual-purpose instrument, and Fig. 5–2 illustrates its essential parts. *A* is the collimator, a device used to produce a narrow incident beam made up of rays as nearly parallel as possible; it usually consists of two pinholes in line, one in each of two lead disks set into the ends of the collimator tube. *C* is the single-crystal specimen supported on the holder *B*. *F* is the light-tight film holder, or *cassette*, made of a frame, a removable metal back, and a sheet of opaque paper; the film, usually 4 × 5 in. (10 × 13 cm) in size, is sandwiched between the metal back and the paper. *S* is the beam stop, designed to prevent the transmitted beam from striking the film and causing excessive blackening. A small copper disk, about 0.5 mm thick, cemented on the paper film cover serves very well for this purpose: it stops all but a small fraction of the beam transmitted through the crystal, while this small fraction serves to record the position of this beam on the film. The shadow of a beam stop of this kind can be seen in Fig. 3–6(a).

The Bragg angle θ corresponding to any transmission Laue spot is found very simply from the relation

$$\tan 2\theta = \frac{r_1}{D}, \qquad (5\text{–}1)$$

where r_1 = distance of spot from center of film (point of incidence of transmitted beam) and D = specimen-to-film distance (usually 5 cm). Adjustment of the specimen-to-film distance is best made by using a feeler gauge of the correct length.

The voltage applied to the x-ray tube has a decided effect on the appearance of a transmission Laue pattern. It is of course true that the higher the tube voltage, the more intense the spots, other variables, such as tube current and exposure time, being held constant. But there is still another effect due to the fact that the continuous spectrum is cut off sharply on the short-wavelength side at a value of the wavelength which varies inversely as the tube voltage [Eq. (1–4)]. Laue spots near the center of a transmission pattern are caused by first-order reflections from planes inclined at very small Bragg angles to the incident beam. Only short-wave-

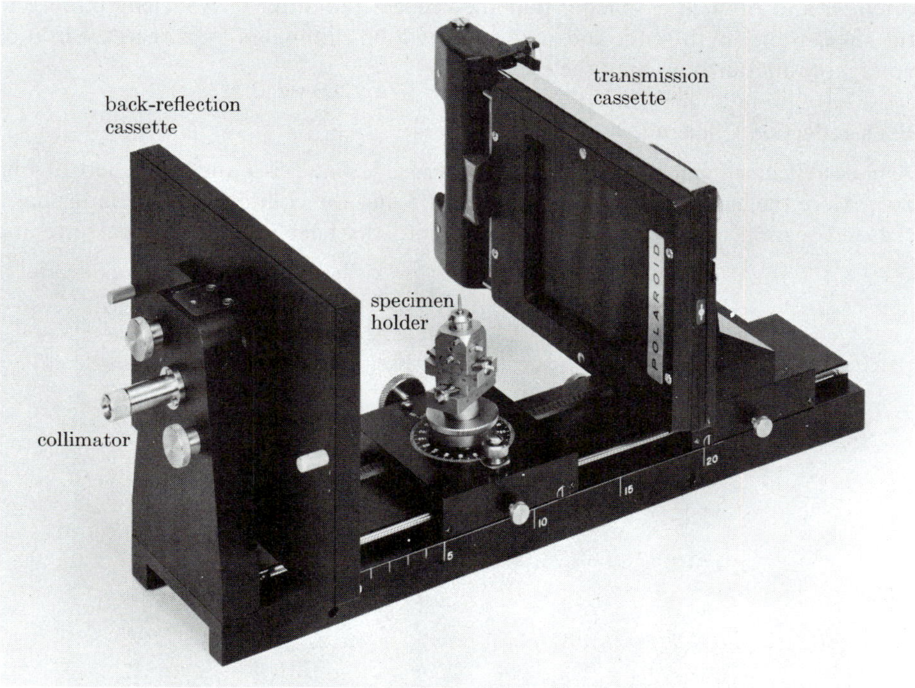

Fig. 5-1 Combination transmission and back-reflection Laue camera. In this camera the Polaroid cassette (at right) and the cassette for ordinary film (at left) are interchangeable; either can be used for transmission or for back reflection. (Courtesy of Blake Industries, Inc.)

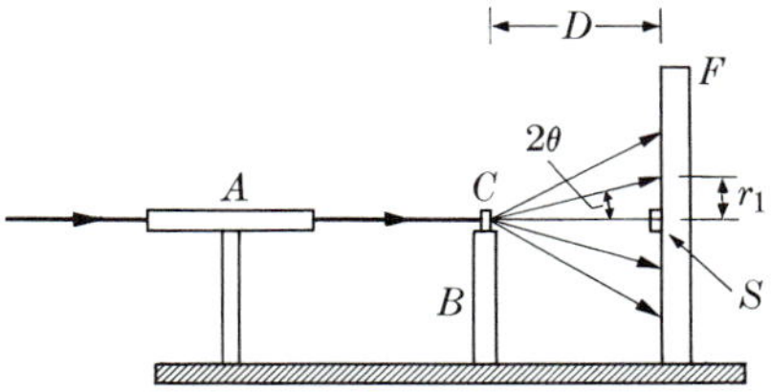

Fig. 5-2 Transmission Laue camera.

length radiation can satisfy the Bragg law for such planes, but if the tube voltage is too low to produce the wavelength required, the corresponding Laue spot will not appear on the pattern. It therefore follows that there is a region near the center of the pattern which is devoid of Laue spots and that the size of this region increases as the tube voltage decreases. The tube voltage therefore affects not only the intensity of each spot, but also the number of spots. This is true also of spots far removed from the center of the pattern; some of these are due to planes so oriented and of such a spacing that they reflect radiation of wavelength close to the short-wavelength limit, and such spots will be eliminated by a decrease in tube voltage no matter how long the exposure.

Back-reflection Camera

A back-reflection camera is illustrated in Figs. 5–3 and 5–4 and at the left of Fig. 5–1. Here the cassette supports both the film and the collimator. The latter has a reduced section at one end which screws into the back plate of the cassette and

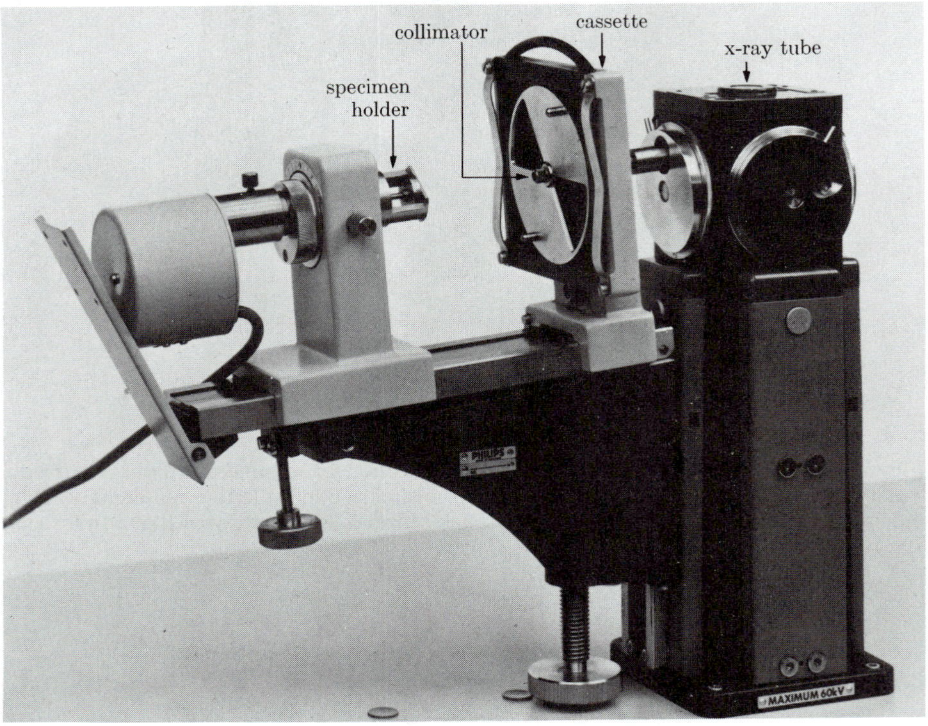

Fig. 5–3 Back-reflection Laue camera. The sectored disc on the front of the cassette would be removed for a Laue photograph; the purpose of this disc and of the drive motor at left is described in Sec. 6–9. (Courtesy of Philips Electronic Instruments, Inc.)

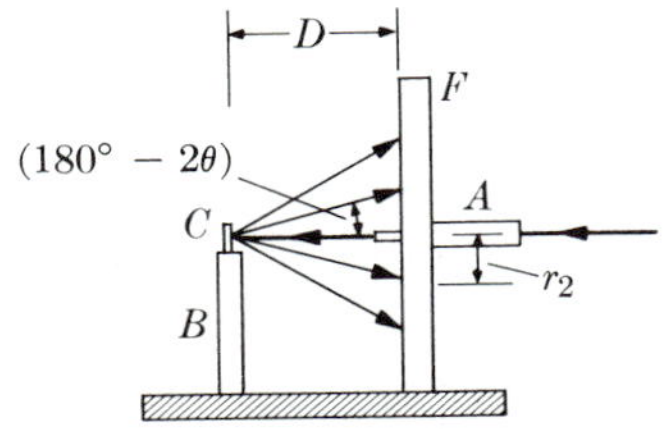

Fig. 5–4 Back-reflection Laue camera (schematic).

projects a short distance in front of the cassette through holes punched in the film and its paper cover.

The Bragg angle θ for any spot on a back-reflection pattern may be found from the relation

$$\tan (180^\circ - 2\theta) = \frac{r_2}{D}, \tag{5-2}$$

where r_2 = distance of spot from center of film and D = specimen-to-film distance (usually 3 cm). In contrast to transmission patterns, back-reflection patterns may have spots as close to the center of the film as the size of the collimator permits. Such spots are caused by high-order overlapping reflections from planes almost perpendicular to the incident beam. Since each diffracted beam is formed of a number of wavelengths, the only effect of a decrease in tube voltage is to remove one or more short-wavelength components from some of the diffracted beams. The longer wavelengths will still be diffracted, and the decrease in voltage will not, in general, remove any spots from the pattern.

General

To obtain a diffraction pattern a Laue camera must be correctly oriented with respect to the x-ray tube. This alignment requires that the collimator axis point directly at the focal spot on the tube target and make an angle of about 6° with the face of the target. The camera is moved relative to the tube until the primary beam, observed on a small fluorescent screen held near the collimator exit, is of maximum intensity and circular, not elliptical, in section.

Transmission patterns can usually be obtained with much shorter exposures than back-reflection patterns. For example, with a tungsten-target tube operating at 30 kV and 20 mA and an aluminum crystal about 1 mm thick, the required exposure is about 5 min in transmission and 30 min in back reflection. This difference is due to the fact that the atomic scattering factor f decreases as the quantity $(\sin \theta)/\lambda$ increases, and this quantity is much larger in back reflection than in transmission. Transmission patterns are also clearer, in the sense of having greater contrast between the diffraction spots and the background, since the coherent scattering, which forms the spots, and the incoherent (Compton modified) scattering, which contributes to the background, vary in opposite ways with $(\sin \theta)/\lambda$. The incoherent scattering reaches its maximum value in the back-reflection region, as shown in Fig. 3-6(a) and (b); it is in this region also that the temperature-diffuse scattering is most intense. In both Laue methods, the short-wavelength radiation in the incident beam will cause most specimens to emit K fluorescent radiation. If this becomes troublesome in back reflection, it may be minimized by placing a filter of aluminum sheet 0.01 in. (0.25 mm) thick in front of the film.

If necessary, the intensity of a Laue spot may be increased by means of an *intensifying screen*, as used in radiography. This resembles a fluorescent screen in having an active material coated on an inert backing such as cardboard, the active material having the ability to fluoresce in the visible region under the action of

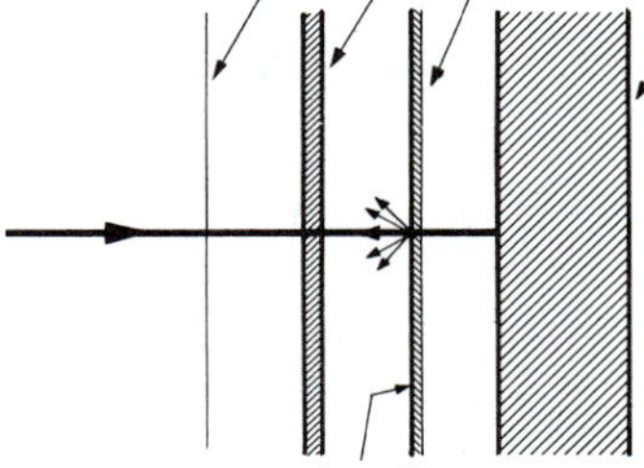

Fig. 5–5 Arrangement of film and intensifying screen (exploded view).

x-rays. When such a screen is placed with its active face in contact with the film (Fig. 5–5), the film is blackened not only by the incident x-ray beam but also by the visible light which the screen emits under the action of the beam. Whereas fluorescent screens emit yellow light, intensifying screens are designed to emit blue light, which is more effective than yellow in blackening the film. Two kinds of intensifying screens are in use today, one containing calcium tungstate and the other zinc sulfide with a trace of silver; the former is most effective at short x-ray wavelengths (about 0.5 Å or less), while the latter can be used at longer wavelengths.

An intensifying screen should not be used if it is important to record fine detail in the Laue spots, as in some studies of crystal distortion, since the presence of the screen will cause the spots to become more diffuse than they would ordinarily be. Each particle of the screen which is struck by x-rays emits light in *all* directions and therefore blackens the film outside the region blackened by the diffracted beam itself, as suggested in Fig. 5–5. This effect is aggravated by the fact that most x-ray film is double-coated, the two layers of emulsion being separated by an appreciable thickness of film base. Even when an intensifying screen is not used, double-coated film causes the size of a diffraction spot formed by an obliquely incident beam to

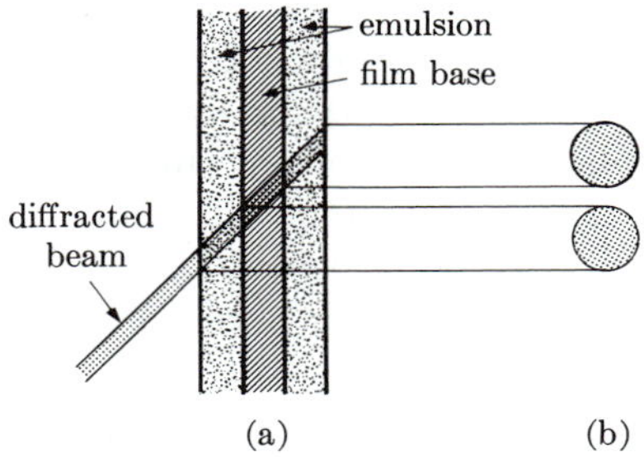

Fig. 5–6 Effect of double-coated film on appearance of Laue spot: (a) section through diffracted beam and film; (b) front view of doubled spot on film.

be larger than the cross section of the beam itself; in extreme cases, an apparent doubling of the diffraction spot results, as shown in Fig. 5-6.

Intensifying screens of one kind or another are routinely used for x-ray radiography but not for diffraction. Most film manufacturers make one or more kinds of film designed for radiography and another kind for diffraction, whether for the Laue or powder method (Chap. 6). Diffraction films are faster and have coarser grain. They are designed for use without an intensifying screen and are often named "no screen" to emphasize this fact.

5-3 SPECIMENS AND HOLDERS

Obviously, a specimen for the transmission method must have low enough absorption to transmit the diffracted beams; in practice, this means that relatively thick specimens of a light element like aluminum may be used but that the thickness of a fairly heavy element like copper must be reduced, by etching, for example, to a few thousandths of an inch. On the other hand, the specimen must not be too thin or the diffracted intensity will be too low, since the intensity of a diffracted beam is proportional to the volume of diffracting material. In the back-reflection method, there is no restriction on the specimen thickness and quite massive specimens may be examined, since the diffracted beams originate in only a thin surface layer of the specimen. This difference between the two methods may be stated in another way and one which is well worth remembering: any information about a thick specimen obtained by the back-reflection method applies only to a thin surface layer of that specimen, whereas information recorded on a transmission pattern is representative of the complete thickness of the specimen, simply because the transmission specimen must necessarily be thin enough to transmit diffracted beams from all parts of its cross section. See Sec. 9-5 for further discussion of this point.

There is a large variety of specimen holders in use, each suited to some particular purpose. The simplest consists of a fixed post to which the specimen is attached with wax or plasticine. A more elaborate holder is required when it is necessary to set a crystal in some particular orientation relative to the x-ray beam. In this case, a three-circle goniometer is used (Fig. 5-7); it has three mutually perpendicular axes of rotation, two horizontal and one vertical, and is so constructed that the crystal, cemented to the tip of the short metal rod at the top, is not displaced in space by any of the three possible rotations.

After the orientation of a crystal has been determined by the Laue method, it is sometimes necessary to cut the crystal along some selected plane. A more massive goniometer-holder than that of Fig. 5-7 is then required; such a holder can be removed from the track of the Laue camera and transferred to a similar track on the cutting device without disturbing the orientation of the crystal.

In the examination of sheet specimens, it is frequently necessary to obtain diffraction patterns from various points on the surface, and this requires movement of the specimen, between exposures, in two directions at right angles in the plane of the specimen surface, this surface being perpendicular to the incident x-ray

Fig. 5–7 Goniometer with three rotation axes. (Courtesy of Charles Supper Co.)

beam. The mechanical stage from a microscope can be easily converted to this purpose.

It is often necessary to know exactly where the incident x-ray beam strikes the specimen, as, for example, when one wants to obtain a pattern from a particular grain, or a particular part of a grain, in a polycrystalline mass. This is sometimes a rather difficult matter in a back-reflection camera because of the short distance between the film and the specimen. One method is to project a light beam through the collimator and observe its point of incidence on the specimen with a mirror or prism held near the collimator. An even simpler method is to push a stiff straight wire through the collimator and observe where it touches the specimen with a small mirror, of the kind used by dentists, fixed at an angle to the end of a rod.

5–4 COLLIMATORS

Collimators of one kind or another are used in all varieties of x-ray cameras, and it is therefore important to understand their function and to know what they can and cannot do. To "collimate" means, literally, to "render parallel," and the perfect collimator would produce a beam composed of perfectly parallel rays. Such a collimator does not exist, and the reason, essentially, lies in the source of the radiation, since every source emits radiation in all possible directions.

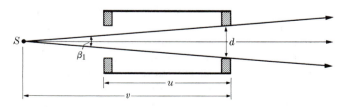

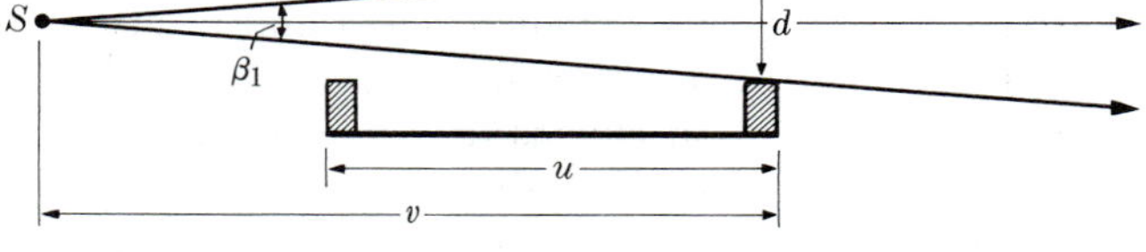

Fig. 5–8 Pinhole collimator and small source.

Consider the simplest kind of collimator (Fig. 5-8), consisting of two circular apertures of diameter d separated by a distance u, where u is large compared to d. If there is a point source of radiation at S, then all the rays in the beam from the collimator are nonparallel, and the beam is conical in shape with a maximum angle of divergence β_1 given by the equation

$$\tan \frac{\beta_1}{2} = \frac{d/2}{v},$$

where v is the distance of the exit pinhole from the source. Since β_1 is always very small, this relation can be closely approximated by the equation

$$\beta_1 = \frac{d}{v} \text{ radian.} \tag{5-3}$$

Whatever we do to decrease β_1 and therefore render the beam more nearly parallel will at the same time decrease the energy of the beam. We note also that the entrance pinhole serves no function when the source is very small, and may be omitted.

No actual source is a mathematical point, and, in practice, we usually have to deal with x-ray tubes which have focal spots of finite size, usually rectangular in shape. The projected shape of such a spot, at a small target-to-beam angle, is either a small square or a very narrow line (Fig. 1-17), depending on the direction of projection. Such sources produce beams having parallel, divergent, and convergent rays.

Figure 5-9 illustrates the case when the projected source shape is square and of such a height h that convergent rays from the edges of the source cross at the center of the collimator and then diverge. The maximum divergence angle is now given by

$$\beta_2 = \frac{2d}{u} \text{ radian,} \tag{5-4}$$

and the center of the collimator may be considered as the virtual source of these divergent rays. The beam issuing from the collimator contains not only parallel

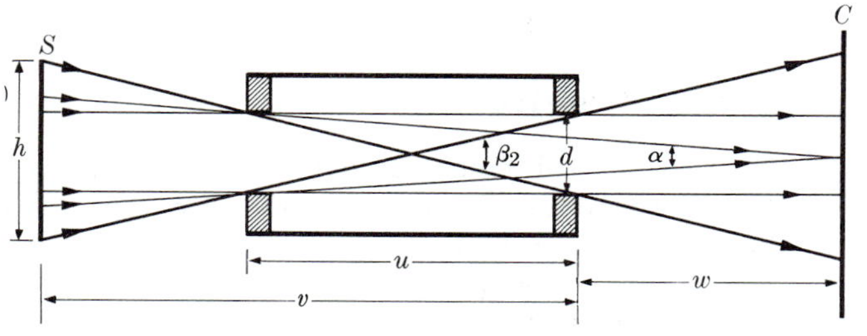

Fig. 5-9 Pinhole collimator and large source. S = source, C = crystal.

and divergent rays but also convergent ones, the maximum angle of convergence being given by

$$\alpha = \frac{d}{u + w} \text{ radian,} \tag{5-5}$$

where w is the distance of the crystal from the exit pinhole. The size of the source shown in Fig. 5–9 is given by

$$h = d\left(\frac{2v}{u} - 1\right). \tag{5-6}$$

In practice, v is very often about twice as large as u, which means that the conditions illustrated in Fig. 5–9 are achieved when the pinholes are about one-third the size of the projected source. If the value of h is smaller than that given by Eq. (5–6), then conditions will be intermediate between those shown in Figs. 5–8 and 5–9; as h approaches zero, the maximum divergence angle decreases from the value given by Eq. (5–4) to that given by Eq. (5–3) and the proportion of parallel rays in the beam and the maximum convergence angle both approach zero. When h exceeds the value given by Eq. (5–6), none of the conditions depicted in Fig. 5–9 is changed, and the increase in the size of the source merely represents wasted energy.

When the shape of the projected source is a fine line, the geometry of the beam varies between two extremes in two mutually perpendicular planes. In a plane at right angles to the line source, the shape is given by Fig. 5–8, and in a plane parallel to the source by Fig. 5–9. Aside from the component which diverges in the plane of the source, the resulting beam is shaped somewhat like a wedge. Since the length of the line source greatly exceeds the value given by Eq. (5–6), a large fraction of the x-ray energy is wasted with this arrangement of source and collimator.

The extent of the nonparallelism of actual x-ray beams may be illustrated by taking, as typical values, $d = 0.5$ mm, $u = 5$ cm, and $w = 3$ cm. Then Eq. (5–4) gives $\beta_2 = 1.15°$ and Eq. (5–5) gives $\alpha = 0.36°$. These values may of course be reduced by decreasing the size of the pinholes, for example, but this reduction will be obtained at the expense of decreased energy in the beam and increased exposure time.

The collimators of most cameras, both for Laue and powder photographs, produce beams about 1 mm in diameter. Much smaller beams are used in *microcameras* for special purposes (Sec. 6–10).

5-5 THE SHAPES OF LAUE SPOTS

We will see later that Laue spots become smeared out if the reflecting crystal is distorted. Here, however, we are concerned with the shapes of spots obtained from perfect, undistorted crystals. These shapes are greatly influenced by the nature of the incident beam, i.e., by its convergence or divergence, and it is important to realize this fact, or Laue spots of "unusual" shape may be erroneously taken as evidence of crystal distortion.

Consider the transmission case first, and assume that the crystal is thin and larger than the cross section of the primary beam at the point of incidence. If this beam is mainly divergent, which is the usual case in practice (Fig. 5–8 or 5–9), then a focusing action takes place on diffraction. Figure 5–10 is a section through the incident beam and any diffracted beam; the incident beam, whose cross section at any point is circular, is shown issuing from a small source, real or virtual. Each ray of the incident beam which lies in the plane of the drawing strikes the reflecting lattice planes of the crystal at a slightly different Bragg angle, this angle being a maximum at A and decreasing progressively toward B. The lowermost rays are therefore deviated through a greater angle 2θ than the upper ones, with the result that the diffracted beam converges to a focus at F. This is true only of the rays in the plane of the drawing; those in a plane at right angles continue to diverge after diffraction, with the result that the diffracted beam is elliptical in cross section. The film intersects different diffracted beams at different distances from the crystal, so elliptical spots of various sizes are observed, as shown in Fig. 5–11. This is not a sketch of a Laue pattern but an illustration of spot size and shape as a function of spot position in one quadrant of the film. Note that the spots

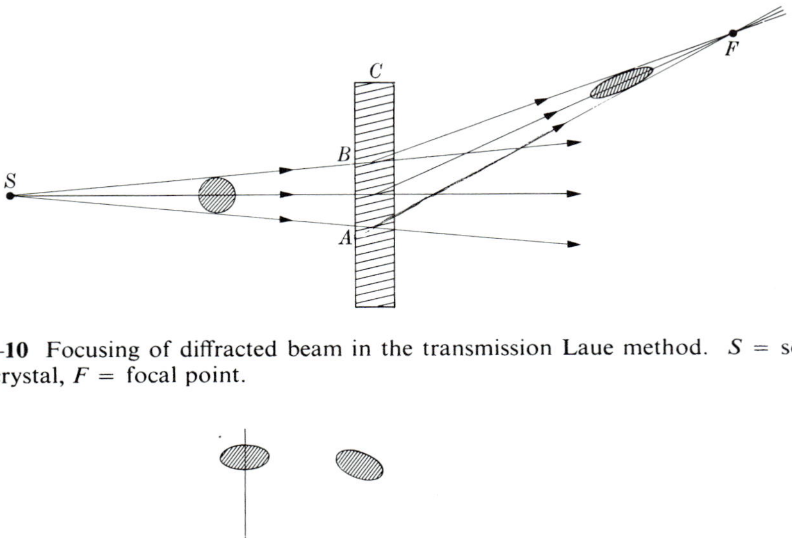

Fig. 5–10 Focusing of diffracted beam in the transmission Laue method. S = source, C = crystal, F = focal point.

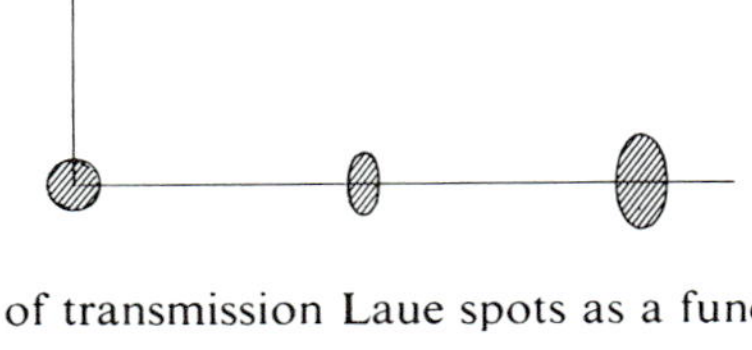

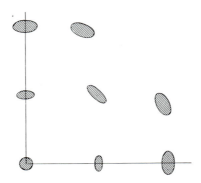

Fig. 5–11 Shape of transmission Laue spots as a function of position.

are all elliptical with their minor axes aligned in a radial direction and that spots near the center and edge of the pattern are thicker than those in intermediate positions, the latter being formed by beams near their focal point. Spots having the shapes illustrated are fairly common, and Fig. 3–6(a) is an example.

In back reflection, no focusing occurs and a divergent incident beam continues to diverge in all directions after diffraction. Back-reflection Laue spots are therefore more or less circular near the center of the pattern, and they become increasingly elliptical toward the edge, due to the oblique incidence of the rays on the film, the major axes of the ellipses being approximately radial. Figure 3–6(b) is typical.

PROBLEMS

***5–1** A transmission Laue pattern is made of an aluminum crystal with 40-kV tungsten radiation. The film is 5 cm from the crystal. How close to the center of the pattern can Laue spots be formed by reflecting planes of maximum spacing, namely (111), and those of next largest spacing, namely (200)?

5–2 A transmission Laue pattern is made of an aluminum crystal with a specimen-to-film distance of 5 cm. The (111) planes of the crystal make an angle of 3° with the incident beam. What minimum tube voltage is required to produce a 111 reflection?

***5–3** (a) A back-reflection Laue pattern is made of an aluminum crystal at 50 kV. The (111) planes make an angle of 88° with the incident beam. What orders of reflection are present in the beam diffracted by these planes? (Assume that wavelengths larger than 2.0 Å are too weak and too easily absorbed by air to register on the film.)

b) What orders of the 111 reflection are present if the tube voltage is reduced to 40 kV?

6

Powder Photographs

6-1 INTRODUCTION

The powder method of x-ray diffraction was devised independently in 1916 by Debye and Scherrer in Germany and in 1917 by Hull in the United States. It is the most generally useful of all diffraction methods and, when properly employed, can yield a great deal of structural information about the material under investigation. Basically, this method involves the diffraction of monochromatic x-rays by a powder specimen. In this connection, "monochromatic" usually means the strong K characteristic component of the filtered radiation from an x-ray tube operated above the K excitation potential of the target material. "Powder" can mean either an actual, physical powder held together with a suitable binder or any specimen in polycrystalline form. The method is thus eminently suited for metallurgical work, since single crystals are not always available to the metallurgist and such materials as polycrystalline wire, sheet, rod, etc., may be examined nondestructively without any special preparation.

There are three main powder methods in use, differentiated by the relative position of the specimen and film:

1. *Debye–Scherrer method.* The film is placed on the surface of a cylinder and the specimen on the axis of the cylinder.

2. *Focusing method.* The film, specimen, and x-ray source are all placed on the surface of a cylinder.

3. *Pinhole method.* The film is flat, perpendicular to the incident x-ray beam, and located at any convenient distance from the specimen.

In all these methods, the diffracted beams lie on the surfaces of cones whose axes lie along the incident beam or its extension; each cone of rays is diffracted from a particular set of lattice planes. In the Debye–Scherrer and focusing methods, only a narrow strip of film is used and the recorded diffraction pattern consists of short lines formed by the intersections of the cones of radiation with the film. In the pinhole method, the whole cone intersects the film to form a circular diffraction ring.

The various powder cameras are in many respects competitive with the diffractometer (Chap. 7). The diffractometer has the great advantage of being able to measure the positions *and* intensities of diffraction lines simultaneously and quickly. Cameras are very much cheaper than a diffractometer, but photographic methods are generally slow (Polaroid Land cassettes are available only for flat-film cameras), and the measurement of line intensities requires an additional operation. However, exact line intensities are by no means needed in many investigations, and there are at least three circumstances when a powder camera is definitely superior to a diffractometer:

1. When only a very small amount of specimen is available. (Specimens weighing as little as one milligram, or even less, can be successfully examined in a powder camera; the diffractometer ordinarily requires a specimen of the order of half a gram or more.)

2. When an entire diffraction ring must be recorded, as in the rapid estimation of grain size and preferred orientation.

3. When the specimen is so large, heavy, or immovable that the diffraction-recording equipment must be brought to the specimen rather than vice versa.

Anyone setting up a diffraction laboratory for the first time, and not blessed with a large budget, can begin with the basic necessities (a high-voltage power supply and an x-ray tube), a few cameras, and a darkroom, leaving open the option of adding a diffractometer later.

Powder photographic methods are described in great and useful detail by Klug and Alexander [G.39], and the book by Azaroff and Buerger [G.17] is devoted solely to such methods.

6-2 DEBYE-SCHERRER METHOD

A typical Debye camera is shown in Fig. 6-1. It consists essentially of a cylindrical chamber with a light-tight cover, a collimator to admit and define the incident beam, a beam stop to confine and stop the transmitted beam, a means for holding the film tightly against the inside circumference of the camera, and a specimen holder that can be rotated.

Camera diameters vary from about 5 to about 20 cm. The greater the diameter, the greater the resolution or separation of a particular pair of lines on the film. In spectroscopy, resolving power is the power of distinguishing between two components of radiation which have wavelengths very close together and is given by $\lambda/\Delta\lambda$, where $\Delta\lambda$ is the difference between the two wavelengths and λ is their mean value; in crystal-structure analysis, we may take resolving power as the ability to separate diffraction lines from sets of planes of very nearly the same spacing, or as the value of $d/\Delta d$.* Thus, if S is the distance measured on the film from a particular diffraction line to the point where the transmitted beam would strike the film (Fig. 6-2), then

$$S = 2\theta R$$

* Resolving power is often defined by the quantity $\Delta\lambda/\lambda$, which is the reciprocal of that given above. However, the *power* of resolving two wavelengths which are nearly alike is a quantity which should logically *increase* as $\Delta\lambda$, the difference between the two wavelengths to be separated, decreases. This is the reason for the definition given in the text. The same argument applies to interplanar spacings d.

Fig. 6-1 Debye-Scherrer camera, with cover plate removed. (Courtesy of Philips Electronic Instruments, Inc.)

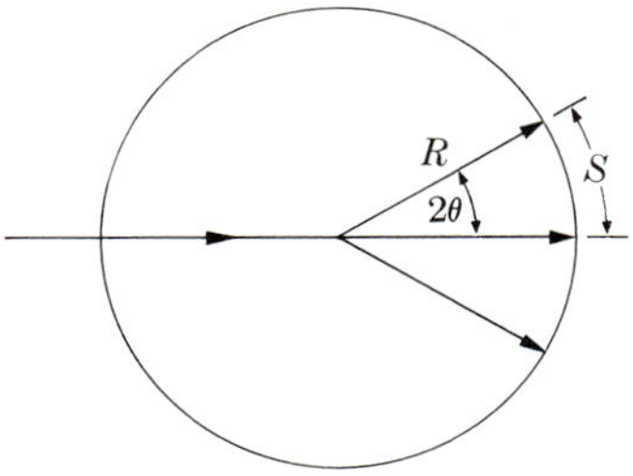

Fig. 6-2 Geometry of the Debye-Scherrer method. Section through film and one diffraction cone.

and

$$\Delta S = R\Delta 2\theta, \tag{6-1}$$

where R is the radius of the camera. Two sets of planes of very nearly the same spacing will give rise to two diffracted beams separated by a small angle $\Delta 2\theta$; for a given value of $\Delta 2\theta$, Eq. (6-1) shows that ΔS, the separation of the lines on the film, increases with R. The resolving power may be obtained by differentiating the

Bragg law:*

$$\lambda = 2d \sin \theta$$

$$\frac{\mathrm{d}\theta}{\mathrm{d}d} = \frac{-1}{d} \tan \theta. \tag{6–2}$$

But

$$\mathrm{d}\theta = \frac{\mathrm{d}S}{2R}.$$

Therefore

$$\frac{\mathrm{d}S}{\mathrm{d}d} = \frac{-2R}{d} \tan \theta$$

$$\text{Resolving power} = \frac{d}{\Delta d} = \frac{-2R}{\Delta S} \tan \theta, \tag{6–3}$$

where d is the mean spacing of the two sets of planes, Δd the difference in their spacings, and ΔS the separation of two diffraction lines which appear just resolved on the film. Equation (6–3) shows that the resolving power increases with the size of the camera; this increased resolution is obtained, however, at the cost of increased exposure time, and the smaller cameras are usually preferred for all but the most complicated patterns. A camera diameter of 5.73 cm is often used and will be found suitable for most work. This particular diameter, equal to 1/10 the number of degrees in a radian, facilitates calculation, since θ (in degrees) is obtained simply by multiplication of S (in cm) by 10, except for certain corrections necessary in precise work. Equation (6–3) also shows that the resolving power of a given camera increases with θ, being directly proportional to tan θ.

The increased exposure time required by an increase in camera diameter is due not only to the decrease in intensity of the diffracted beam with increased distance from the specimen, but also to the partial absorption of both the incident and diffracted beams by the air in the camera. For example, Prob. 1–8 and the curves of Fig. 6–3 show that, in a camera of 19 cm diameter (about the largest in common use), the decrease in intensity due to air absorption is about 18 percent for Cu $K\alpha$ radiation and about 48 percent for Cr $K\alpha$ radiation. This decrease in intensity may be avoided by evacuating the camera or by filling it with a light gas such as hydrogen or helium during the exposure.

Correct design of the pinhole system which collimates the incident beam is important, especially when weak diffracted beams must be recorded. The exit pinhole scatters x-rays in all directions, and these scattered rays, if not prevented from striking the film, can seriously increase the intensity of the background. A "guarded-pinhole" assembly which practically eliminates this effect is shown in Fig. 6–4, where the divergent and convergent rays in the incident beam are ignored

---

* A lower-case roman d is used throughout this book for differentials in order to avoid confusion with the symbol d for distance between atomic planes.

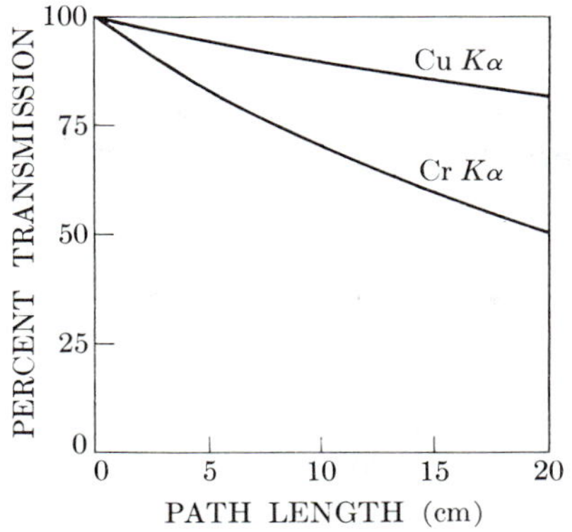

Fig. 6-3 Absorption of Cu $K\alpha$ and Cr $K\alpha$ radiation by air.

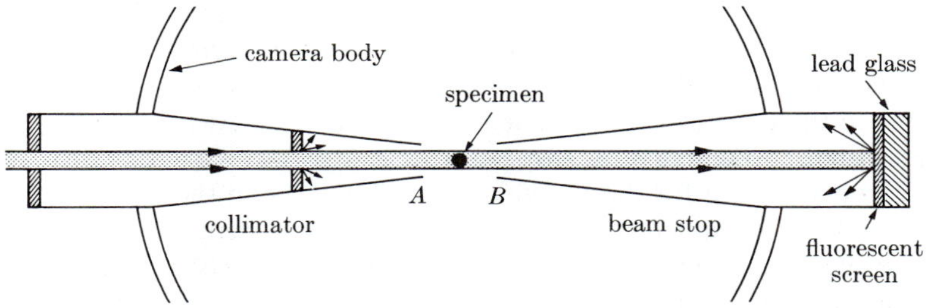

Fig. 6-4 Design of collimator and beam stop (schematic).

and only the parallel component is shown. The collimator tube is extended a considerable distance beyond the exit pinhole and constricted so that the end A is close enough to the main beam to confine the radiation scattered by the exit pinhole to a very narrow angular range and yet not close enough to touch the main beam and be itself a cause of further scattering. The beam stop is usually a thick piece of lead glass placed behind a fluorescent screen, the combination allowing the transmitted beam to be viewed with safety when adjusting the camera in front of the x-ray tube. Back scatter from the stop is minimized by extending the beam-stop tube backward and constricting its end B. Another reason for extending the collimator and beam-stop tubes as close to the specimen as possible is to minimize the extent to which rays scattered by air from the primary beam can reach the film. Both tubes are tapered to interfere as little as possible with low-angle and high-angle diffracted beams.

Some cameras have rectangular slits rather than pinholes to define the beam, the long edges of the slits being parallel to the axis of the specimen. The use of slits instead of pinholes decreases exposure time by increasing the irradiated volume of the specimen, but requires more accurate positioning of the camera relative to the source and produces diffraction lines which are sharp only along the median line of the film.

6-3 SPECIMEN PREPARATION

Metals and alloys may be converted to powder by filing or, if they are sufficiently brittle, by grinding in a small agate mortar. In either case, the powder should be filed or ground as fine as possible, preferably to pass a 325-mesh screen, in order to produce smooth, continuous diffraction lines. The screened powder is usually annealed in evacuated glass or quartz capsules in order to relieve the strains due to filing or grinding. These strains are so low in extremely brittle solids (most minerals and ceramics) that annealing may be omitted.

Special precautions are necessary in screening two-phase alloys. If a small, representative sample is selected from an ingot for x-ray analysis, then that entire sample must be ground or filed to pass through the screen. The common method of grinding until an amount sufficient for the x-ray specimen has passed the screen, the oversize being rejected, may lead to very erroneous results. One phase of the alloy is usually more brittle than the other, and that phase will more easily be ground into fine particles; if the grinding and screening are interrupted at any point, then the material remaining on the screen will contain less of the more brittle phase than the original sample while the undersize will contain more, and neither will be representative.

The final specimen for the Debye camera should be in the form of a thin rod, 0.5 mm or less in diameter and about 1 cm long. There are various ways of preparing such a specimen, one of the simplest being to coat the powder on the surface of a fine glass fiber with a small amount of glue or petroleum jelly. Other methods consist in packing the powder into a thin-walled capillary tube made of a weakly absorbing substance such as cellophane or lithium borate glass, or in extruding a mixture of powder and binder through a small hole. Polycrystalline wires may be used directly, but since they usually exhibit some preferred orientation, the resulting diffraction pattern must be interpreted with that fact in mind (Chap. 9). Strongly absorbing substances may produce split low-angle lines (see Sec. 4–10); if this effect becomes troublesome, it may be eliminated by diluting the substance involved with some weakly absorbing substance, so that the absorption coefficient of the composite specimen is low. Both flour and cornstarch have been used for this purpose. The diluent chosen should not produce any strong diffraction lines of its own and too much of it should not be used, or the lines from the substance being examined will become spotty. These and other details of specimen preparation are described on pp. 21–34 of Vol. 3 of [G.11].

After the specimen rod is prepared, it is mounted in its holder so that it will lie accurately along the rotation axis of the camera when the specimen holder is rotated. This adjustment is made by viewing the specimen through a short-focus lens or low-power microscope temporarily inserted into the camera in place of the beam-stop tube; the specimen holder is then adjusted so that the specimen does not appear to wobble when the holder is rotated. Rotation of the specimen during the exposure is common practice but not an intrinsic part of the powder method; its only purpose is to produce continuous, rather than spotty, diffraction lines by increasing the number of powder particles in reflecting positions.

6-4 FILM LOADING

The film for the Debye method is a narrow strip punched with one or two holes. Devices are available for cutting strip of the proper size from sheet film and punching properly located holes; alternately, one can buy a roll of 35-mm film, which has the correct width for many Debye cameras, and punch and cut it to the right length.

Figure 6-5 illustrates three methods of arranging the film strip. The small sketches on the right show the loaded film in relation to the incident beam, while the films laid out flat are indicated on the left. In (a), a hole is punched in the center of the film so that the film may be slipped over the beam stop; the transmitted beam thus *leaves* through the hole in the film. The pattern is symmetrical about the center, and the θ value of a particular reflection is obtained by measuring U, the distance apart of two diffraction lines formed by the same cone of radiation, and using the relation

$$4\theta R = U.$$

Photographic film always shrinks slightly during processing and drying, and this shrinkage effectively changes the camera radius. The film-shrinkage error may

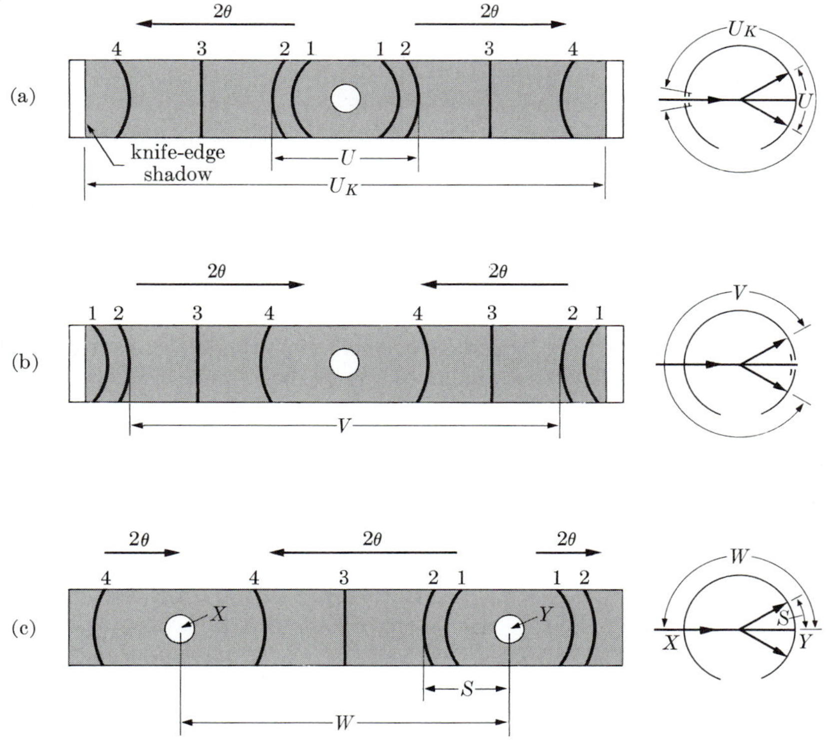

Fig. 6-5 Methods of film loading in Debye cameras. Corresponding lines have the same numbers in all films.

be allowed for by slipping the ends of the film under metal knife-edges which cast a sharp shadow near each end of the film. In this way, a standard distance is impressed on the film which will shrink in the same proportion as the distance between a given pair of diffraction lines. If the angular separation $4\theta_K$ of the knife-edges in the camera is known, either by direct measurement or by calibration with a substance of known lattice parameter, then the value of θ for a particular reflection may be obtained by simple proportion:

$$\frac{\theta}{\theta_K} = \frac{U}{U_K},$$

where U_K is the distance apart of the knife-edge shadows on the film.

Figure 6–5(b) illustrates a method of loading the film which is just the reverse of the previous one. Here the incident beam *enters* through the hole in the film, and θ is obtained from the relation

$$(2\pi - 4\theta)R = V.$$

Knife-edges may also be used in this case as a basis for film-shrinkage corrections.

The unsymmetrical, or Straumanis, method of film loading is shown in Fig. 6–5(c) and Fig. 3–12. Two holes are punched in the film so that it may be slipped over both the entrance collimator and the beam stop. Since it is possible to determine from measurements on the film where the incident beam entered the film circle and where the transmitted beam left it, no knife-edges are required to make the film-shrinkage correction. The point X ($2\theta = 180°$), where the incident beam entered, is halfway between the measured positions of lines 4,4; similarly, the point Y ($2\theta = 0°$), where the transmitted beam left, is halfway between lines 1,1. The difference between the positions of X and Y gives W, and θ is found by proportion:

$$\frac{2\theta}{\pi} = \frac{S}{W}.$$

Unsymmetrical loading thus provides for the film-shrinkage correction without calibration of the camera or knowledge of any camera dimension. It is the most popular film arrangement.

The shapes of the diffraction lines in Fig. 6–5 should be noted. The low-angle lines are strongly curved because they are formed by cones of radiation which have a small apex angle 4θ. The same is true of the high-angle lines, although naturally they are curved in the opposite direction. Lines for which 4θ is nearly equal to 180° are practically straight. This change of line shape with change in θ may also be seen in the powder photographs shown in Fig. 3–13.

6-5 CAMERAS FOR SPECIAL CONDITIONS

Various kinds of special cameras have been devised for obtaining diffraction patterns from specimens subjected to unusual conditions of temperature or pressure [G.39]. These cameras, usually of the Debye–Scherrer type and usually homemade, have designs that vary almost from laboratory to laboratory.

High-Temperature Cameras

Metallurgical investigations frequently require that the crystal structure of a phase stable only at high temperature be determined. In many cases, this can be accomplished by quenching the specimen at a high enough rate to suppress the decomposition of the high-temperature phase and then examining the specimen in an ordinary camera at room temperature. In other cases, the transformation into the phases stable at room temperature cannot be suppressed, and a high-temperature camera is necessary in order that the specimen may be examined at the temperature at which the phase in question is stable. Such a camera may also be used to determine coefficients of thermal expansion from measurements of lattice parameters as a function of temperature. A bibliography on high-temperature techniques has been prepared by Goldschmidt [6.1].

High-temperature cameras all involve a small furnace, usually of the electric-resistance type, to heat the specimen and a thermocouple to measure its temperature. The main design problem is to keep the film cool without too great an increase in the camera diameter; this requires water-cooling of the body of the camera and/or the careful placing of radiation shields between the furnace and the film, shields so designed that they will not interfere with the diffracted x-ray beams. The furnace which surrounds the specimen must also be provided with a slot of some kind to permit the passage of the incident and diffracted beams. If the specimen is susceptible to oxidation at high temperatures, means of evacuating the camera or of filling it with an inert gas must be provided; or the powder specimen may be sealed in an evacuated thin-walled silica tube. Because of the small size of the furnace in a high-temperature camera, the temperature gradients in it are usually quite steep, and special care must be taken to ensure that the temperature recorded by the thermocouple is actually that of the specimen itself. Specimen temperatures as high as 3000°C have been attained. Since the intensity of any reflection is decreased by an increase in temperature, the exposure time required for a high-temperature diffraction pattern is normally rather long.

Low-Temperature Cameras

These cameras are usually designed for specimen temperatures down to about −150°C. Diffraction studies at still lower temperatures can be more easily made with the diffractometer.

The simplest cooling method is to run a thin stream of coolant, such as liquid air, over the specimen throughout the x-ray exposure. The diffraction pattern of the coolant will also be recorded but this is easily distinguished from that of a crystalline solid, because the typical pattern of a liquid contains only one or two very diffuse maxima in contrast to the sharp diffraction lines from a solid. Scattering from the liquid will, however, increase the background blackening of the photograph.

A better method is to cool the specimen with a stream of cold gas. Liquid nitrogen, for example, boils at 77°K (= −196°C) and can easily produce nitrogen gas at the specimen at about −150°C. Gas produces much less background scattering than a liquid coolant.

Post [6.2] has prepared a bibliography on low-temperature techniques.

High-Pressure Cameras

Many substances have quite different crystal structures at high pressure. Uusually these structures exist only under pressure, and their study necessarily requires high-pressure cameras. Maximum pressures reached in such cameras are of the order of 100 kilobars (1 bar $= 10^5$ N/m^2 $= 14.5$ lb/in^2 ≈ 1 atmosphere).

Camera design involves the difficult problem of applying large forces to the specimen and simultaneously getting x-ray beams into and out of it. Hydrostatic pressures up to about 5 kbars have been achieved in cameras operated under gas pressure. Higher pressures are obtained by compressing the specimen between anvils, either uniaxially or tetrahedrally (along directions, in cubic notation, of the form ⟨111⟩).

6-6 FOCUSING CAMERAS

Cameras in which diffracted rays originating from an extended region of the specimen all converge to one point on the film are called focusing cameras. The design of all such cameras is based on the following geometrical theorem (Fig. 6-6):

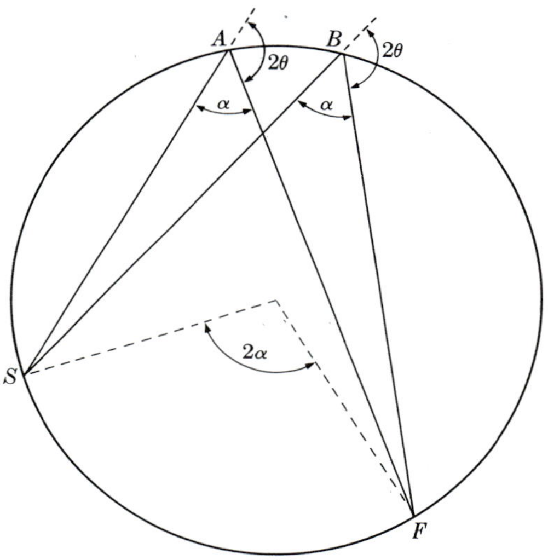

Fig. 6-6 Geometry of focusing cameras.

all angles inscribed in a circle and based on the same arc *SF* are equal to one another and equal to half the angle subtended at the center by the same arc. Suppose that x-rays proceeding in the directions *SA* and *SB* encounter a powder specimen located on the arc *AB*. Then the rays diffracted by the same (*hkl*) planes at points *A* and *B* will be deviated through the same angle 2θ. But these deviation angles 2θ are each equal to $(180° - \alpha)$, which means that the diffracted rays must proceed along *AF* and *BF*, and come to a focus at *F* on a film placed along the circumference of the circle.

6-7 SEEMANN-BOHLIN CAMERA

This focusing principle is exploited in the Seemann-Bohlin camera shown in Fig. 6-7. The slit *S* acts as a virtual line source of x-rays, the actual source being the extended focal spot on the target *T* of the x-ray tube. Only converging rays from

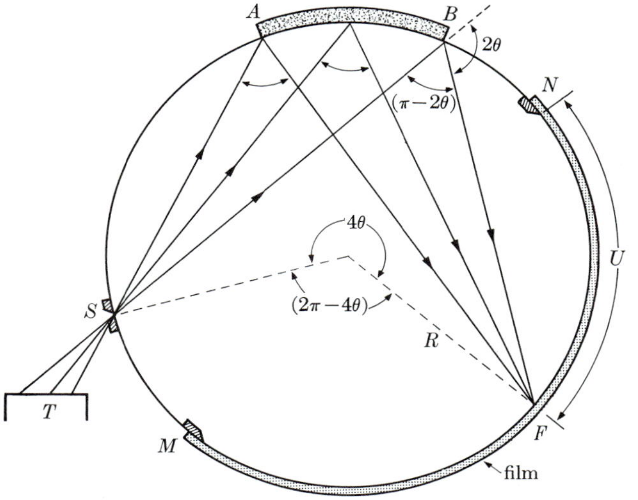

Fig. 6–7 Seemann–Bohlin focusing camera. Only one *hkl* reflection is shown.

the target can enter this slit and, after passing it, they diverge to the specimen *AB*. For a particular *hkl* reflection, each ray is then diffracted through the same angle 2θ, with the result that all diffracted rays from various parts of the specimen converge to a focus at *F*. As in any powder method, the diffracted beams lie on the surfaces of cones whose axes are coincident with the incident beam; in this case, a number of incident beams contribute to each reflection and a diffraction line is formed by the intersection of a number of cones with the film.

The ends of the film strip are covered by knife-edges *M* and *N*, which cast reference shadows on the film. The value of θ for any diffraction line may be found from the distance *U*, measured on the film, from the line to the shadow of the low-angle knife-edge *N*, by use of the relation

$$4\theta R = U + \text{arc } SABN. \tag{6–4}$$

In practice, θ is found by calibrating the camera with a standard substance of known lattice parameter, such as NaCl, rather than by the use of Eq. (6–4).

By differentiating Eq. (6–4), we obtain

$$d\theta = \frac{dU}{4R}.$$

This relation may be combined with Eq. (6–2) to give

$$\frac{dU}{dd} = -\frac{4R}{d}\tan\theta.$$

$$\text{Resolving power} = \frac{d}{\Delta d} = -\frac{4R}{\Delta U}\tan\theta. \tag{6–5}$$

The resolving power, or ability to separate diffraction lines from planes of almost the same spacing, is therefore twice that of a Debye–Scherrer camera of the same radius. In addition, the exposure time is much shorter, because a much larger specimen is used (the arc AB of Fig. 6–7 is of the order of 1 cm) and diffracted rays from a considerable volume of material are all brought to one focus. The Seemann–Bohlin camera is, therefore, useful in studying complex diffraction patterns, whether they are due to a single phase or to a mixture of phases such as occur in alloy systems.

On the debit side, the Seemann–Bohlin camera has the disadvantage that the lines registered on the film cover only a limited range of 2θ values, particularly on the low-angle side. The Seemann–Bohlin camera, in itself, is now virtually obsolete (the diffractometer has greater resolution), except in combination with a monochromator. The combination is then called a Guinier camera (Sec. 6–14).

6–8 BACK-REFLECTION FOCUSING CAMERAS

The most precise measurement of lattice parameter is made in the back-reflection region, as discussed in greater detail in Chap. 11. The most suitable camera for such measurements is the symmetrical back-reflection focusing camera illustrated in Fig. 6–8.

It employs the same focusing principle as the Seemann–Bohlin camera, but the film straddles the slit and the specimen is placed diametrically opposite the slit. Means are usually provided for slowly oscillating the specimen through a few degrees about the camera axis in order to produce smooth diffraction lines. A typical film, punched in the center to allow the passage of the incident beam, is shown in Fig. 6–9. The value of θ for any diffraction line may be calculated from the relation

$$(4\pi - 8\theta)R = V, \qquad (6\text{–}6)$$

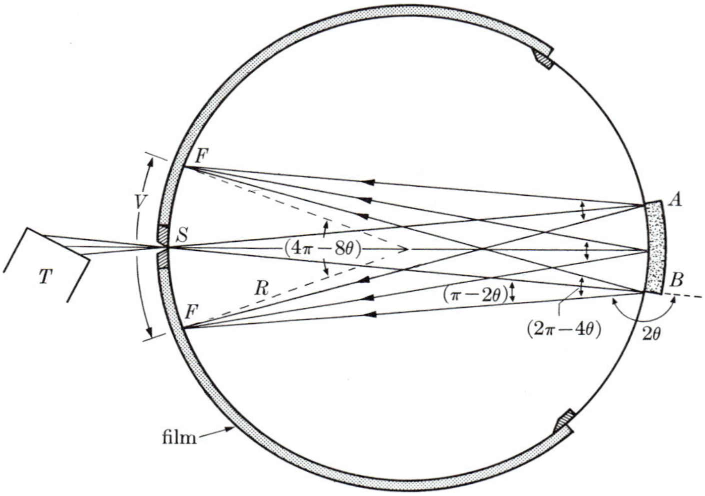

Fig. 6–8 Symmetrical back-reflection focusing camera. Only one *hkl* reflection is shown.

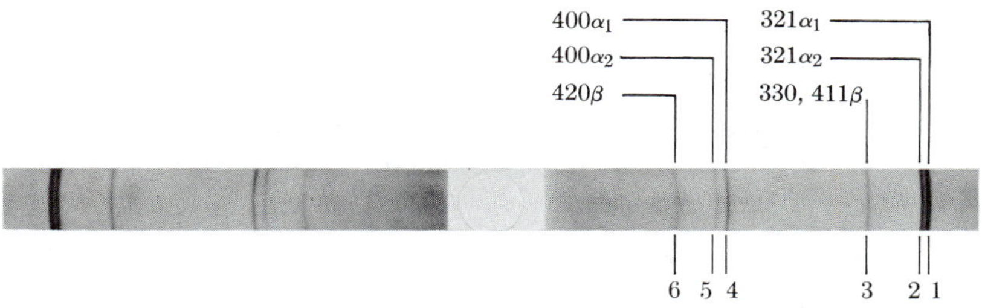

Fig. 6-9 Powder photograph of tungsten made in a symmetrical back-reflection focusing camera, 4.00 in. (10.16 cm) in diameter. Unfiltered copper radiation.

where V is the distance on the film between corresponding diffraction lines on either side of the entrance slit.

Differentiation of Eq. (6-6) gives

$$\Delta\theta = \frac{-1}{4R}\,\Delta\left(\frac{V}{2}\right), \tag{6-7}$$

where $\Delta(V/2)$ is the separation on the film of two reflections differing in Bragg angle by $\Delta\theta$. Combination of this equation with Eq. (6-2) shows that

$$\text{Resolving power} = \frac{d}{\Delta d} = \frac{4R}{\Delta(V/2)}\tan\theta.$$

The resolving power of this camera is therefore the same as that of a Seemann-Bohlin camera of the same diameter.

In the pattern shown in Fig. 6-9, two pairs of closely spaced lines can be seen, lines 1 and 2 and lines 4 and 5. Each pair is a doublet formed by reflection from one set of planes of the two components, $K\alpha_1$ and $K\alpha_2$, which make up $K\alpha$ radiation. These component lines are commonly found to be resolved, or separated, in the back-reflection region. (The β lines in this photograph are not resolved since $K\beta_1$ radiation consists only of a single wavelength.) To determine the conditions under which a given camera can separate two components of radiation which have almost the same wavelength, we must use the spectroscopic definition of resolving power, namely $\lambda/\Delta\lambda$, where $\Delta\lambda$ is the difference between the two wavelengths and λ is their mean value. For Cu $K\alpha$ radiation, these wavelengths are:

$$\lambda(\text{Cu } K\alpha_2) = 1.54439 \text{ Å*}$$

$$\lambda(\text{Cu } K\alpha_1) = \underline{1.54056}$$

$$\Delta\lambda = 0.00383$$

Therefore

$$\frac{\lambda}{\Delta\lambda} = \frac{1.542}{0.00383} = 403.$$

The resolving power of the camera must exceed this value, for the particular reflection considered, if the component lines are to be separated on the film.

By differentiating the Bragg law, we obtain

$$\lambda = 2d \sin \theta,$$

$$\frac{d\theta}{d\lambda} = \frac{1}{2d \cos \theta} = \frac{\tan \theta}{2d \sin \theta} = \frac{\tan \theta}{\lambda}.$$

$$\frac{\lambda}{\Delta\lambda} = \frac{\tan \theta}{\Delta\theta}. \tag{6–8}$$

Substitution of Eq. (6–7) gives

$$\text{Resolving power} = \frac{\lambda}{\Delta\lambda} = \frac{-4R \tan \theta}{\Delta(V/2)}. \tag{6–9}$$

The negative sign here can be disregarded; it merely means that an increase in λ causes a decrease in $V/2$, since the latter is measured from the center of the film. Equation (6–9) demonstrates that the resolving power increases with the camera radius and with θ, becoming very large near 90°. This latter point is clearly evident in Fig. 6–9, which shows a greater separation of the higher-angle 400 reflections as compared to the 321 reflections.

By use of Eq. (6–9), we can calculate the resolving power, for the 321 reflections, of the camera used to obtain Fig. 6–9. The camera radius is 2.00 in. (5.08 cm), and the mean θ value for these reflections is about 65.7°. The line breadth at half maximum intensity is about 0.04 cm. The two component lines of the doublet will be clearly resolved on the film if their separation is twice their breadth (Fig. 6–10). Therefore

$$\Delta\left(\frac{V}{2}\right) = 2(0.04) = 0.08 \text{ cm},$$

$$\frac{\lambda}{\Delta\lambda} = \frac{(4)(2.00)(2.54)(\tan 65.7°)}{(0.08)} = 563.$$

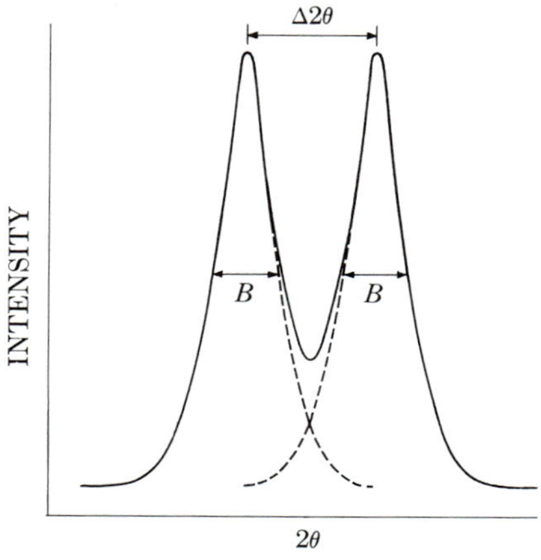

Fig. 6–10 Resolution of closely spaced lines. The lines shown have $\Delta 2\theta = 2B$. Any smaller separation might make the two lines appear as one.

Since this value exceeds the resolving power of 403, found above to be necessary for resolution of the Cu $K\alpha$ doublet, we would expect this doublet to be resolved for the 321 reflection, and such is seen to be the case in Fig. 6–9. At some lower angle, this would not be true and the two components would merge into a single, unresolved line. The fact that resolution of the $K\alpha$ doublet normally occurs only in the back-reflection region can be seen from the Debye photographs reproduced in Fig. 3–13.

6-9 PINHOLE PHOTOGRAPHS

When "monochromatic" radiation is used to examine a "powder" specimen in a Laue (flat-film) camera, the result is often called, for no particularly good reason, a pinhole photograph. (There is no general agreement on the name of this method. Klug and Alexander [G.39], for example, call it the "monochromatic-pinhole technique.") Either a transmission or a back-reflection camera may be used. A typical transmission photograph, made of fine-grained aluminum sheet, is shown in Fig. 6–11.

The pinhole method has the advantage that an entire Debye ring, and not just a part of it, is recorded on the film. On the other hand, the range of θ values which are recorded is rather limited: either low-angle or high-angle reflections may be obtained, but not those in the median range of θ (see Fig. 6–12). In the transmission method, the value of θ for a particular reflection is found from the relation

$$\tan 2\theta = \frac{U}{2D}, \tag{6–10}$$

where U = diameter of the Debye ring and D = specimen-to-film distance. The

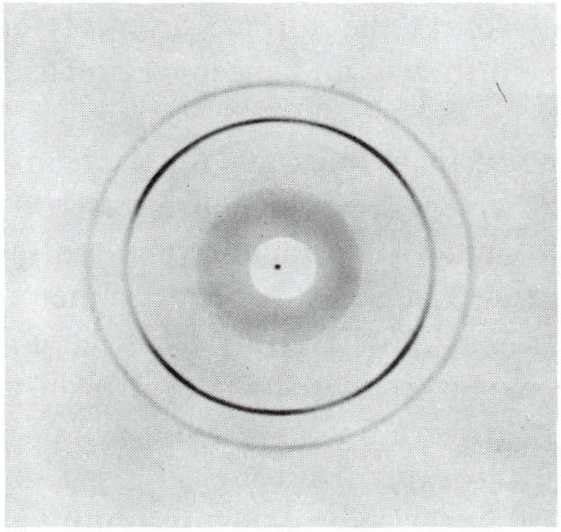

Fig. 6–11 Transmission pinhole photograph of an aluminum sheet specimen. Filtered copper radiation. (The diffuse circular band near the center is caused by white radiation. The nonuniform blackening of the Debye rings is due to preferred orientation in the specimen; see Chap. 9.)

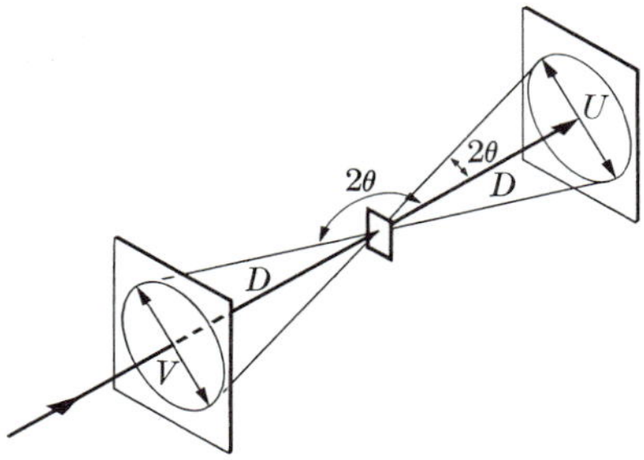

Fig. 6–12 Angular relationships in the pinhole method.

corresponding relation for the back-reflection method is

$$\tan(\pi - 2\theta) = \frac{V}{2D}, \tag{6–11}$$

where V = diameter of the Debye ring. The distance D is usually of the order of 3 to 5 cm.

Powder specimens may be prepared simply by spreading a bit of the powder mixed with a binder on any convenient support. For a transmission photograph a piece of paper or cellophane will do. If there is any doubt about diffraction from the support material, a control pattern without the specimen can be prepared.

However, the greatest utility of the pinhole method in metallurgical work lies in the fact that polycrystalline specimens can be examined directly. In back reflection, a metallographic specimen, mounted in the usual 1-in. (25-mm) diameter plastic mount for microscopic examination, can be positioned so that the primary beam falls on any selected area; the advantage of being able to examine the same area of the specimen both with the microscope and with x-rays is obvious; it is worth noting also that both methods of examination, the optical and the x-ray, provide information only about the surface layer of the specimen. The transmission method is restricted to wire and sheet specimens which are not too highly absorbing.

There is an optimum specimen thickness for the transmission method, because the diffracted beams will be very weak or entirely absent if the specimen is either too thin (insufficient volume of diffracting material) or too thick (excessive absorption). As will be shown in Sec. 9–8, the specimen thickness which produces the maximum diffracted intensity is given by $1/\mu$, where μ is the linear absorption coefficient of the specimen. Inspection of Eq. (1–10) shows that this condition can also be stated as follows: a transmission specimen is of optimum thickness when the intensity of the beam transmitted through the specimen is $1/e$, or about $\frac{1}{3}$, of the intensity of the incident beam. Normally this optimum thickness is of the order of a few thousandths of an inch (0.1 mm). There is one way, however, in which a partial transmission pattern can be obtained from a thick specimen and that is by diffraction from an edge (Fig. 6–13). Only the upper half of the pattern is recorded on the film, but that is all that is necessary in many applications. The same technique has also been used in some Debye–Scherrer cameras.

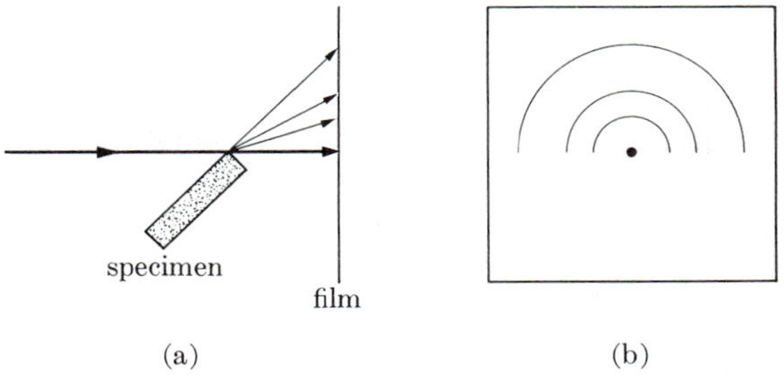

Fig. 6–13 Transmission pinhole method for thick specimens: (a) section through incident beam; (b) partial pattern obtained.

The pinhole method is used in studies of preferred orientation, grain size, and crystal quality. With a back-reflection camera, fairly precise parameter measurements can be made by this method. Precise knowledge of the specimen-to-film distance D is not necessary, provided the proper extrapolation equation is used (Chap. 11) or the camera is calibrated. The calibration is usually performed for each exposure, simply by smearing a thin layer of the calibrating powder over the surface of the specimen; in this way, reference lines of known θ value are formed on each film.

When the pinhole method is used for parameter measurements, the film or specimen, or both, is moved during the exposure to produce smooth, continuous diffraction lines. By rotating or oscillating the film about the axis of the incident beam, the reflections from each reflecting particle or grain are smeared out along the Debye ring. The specimen itself may be rotated about the incident beam axis or about any axis parallel to the incident beam, or translated back and forth in any direction in a plane parallel to the specimen surface. Such movements increase the number of grains in reflecting positions and allow a greater proportion of the total specimen surface to take part in diffraction, thus ensuring that the information recorded on the film is representative of the surface as a whole. Any camera in which the specimen can be so moved during the exposure that the incident beam traverses a large part of its surface is called an *integrating camera.*

The camera shown in Fig. 5–3 has a motor to rotate the specimen for such integration. The sectored disc on the cassette of this camera is designed for recording two or more partial patterns on one film for comparative purposes. After the first pattern is made, the specimen is changed, the disc is rotated about the collimator by an amount sufficient to cover the previously exposed portion of film and uncover an unexposed portion, and the second exposure is made. Or the disc may be removed, if one wishes to record complete Debye rings from a single specimen.

6–10 MICROBEAMS AND MICROCAMERAS

Sometimes one has only a minute amount of material from which a diffraction pattern has to be obtained. Or one may wish to obtain a pattern from only a very small region of a large specimen. In either case, a very small incident beam, called a *microbeam*, is needed, of the order of 100 μm ($=0.1$ mm $= 4 \times 10^{-3}$ in.) or less in diameter.

The central problem is the design and construction of the collimator and its placement relative to the focal spot of the x-ray tube. Hirsch [6.3] has discussed these matters. Collimators are often made of glass capillary tubing. If such collimators are used with ordinary x-ray tubes, the collimated beam will be weak and exposure times very long, because most of the x-rays coming from the tube are wasted. The proper procedure is to use a microfocus tube (Sec. 1–7).

While it may be possible to adapt an ordinary pinhole camera to some microbeam work simply by changing the collimator, better results will be obtained with a specially designed *microcamera* [6.3, G.39]. Such a camera will usually have a small specimen-to-focal-spot distance (to increase intensity and improve collimation), a small specimen-to-film distance (to reduce exposure time), and some arrangement for accurately positioning the specimen in the beam. Diffraction patterns of specimens amounting to as little as 10 micrograms have been obtained in such cameras.

6–11 CHOICE OF RADIATION

With any of the powder methods described above, the investigator must choose the radiation best suited to the problem at hand. In making this choice, the two most important considerations are:

1. The characteristic wavelength used should not be shorter than the K absorption edge of the specimen, or the fluorescent radiation produced will badly fog the film. In the case of alloys or compounds, it may be difficult or impossible to satisfy this condition for every element in the specimen.

2. The Bragg law shows that the shorter the wavelength, the smaller the Bragg angle for planes of a given spacing. Decreasing the wavelength will therefore shift every diffraction line to lower Bragg angles and increase the total number of lines on the film, while increasing the wavelength will have the opposite effect. The choice of a short or a long wavelength depends on the particular problem involved.

The characteristic radiations usually employed in x-ray diffraction are the following:

Mo $K\alpha$:	0.711 Å
Cu $K\alpha$:	1.542
Co $K\alpha$:	1.790
Fe $K\alpha$:	1.937
Cr $K\alpha$:	2.291

In each case, the appropriate filter is used to suppress the $K\beta$ component of the radiation. All in all, Cu $K\alpha$ radiation is generally the most useful. It cannot be employed with ferrous materials, however, since it will cause fluorescent radiation from the iron in the specimen; instead, Co $K\alpha$, Fe $K\alpha$, or Cr $K\alpha$ radiation should be used.

Precise lattice-parameter measurements require that there be a number of lines in the back-reflection region, while some specimens may yield only one or two. This difficulty may be avoided by using unfiltered radiation, in order to have $K\beta$ as well as $K\alpha$ lines present.

6–12 BACKGROUND RADIATION

A good powder photograph has sharp intense lines superimposed on a background of minimum intensity. However, the diffraction lines themselves vary in intensity, because of the structure of the crystal itself, and an appreciable background intensity may exist, due to a number of causes. The two effects together may cause the weakest diffraction line to be almost invisible in relation to the background.

This background intensity is due to the following causes:

1. *Fluorescent radiation emitted by the specimen.* It cannot be too strongly emphasized that the characteristic wavelength used should be longer than the K absorption edge of the specimen, in order to prevent the emission of fluorescent radiation. Incident radiation so chosen, however, will not completely eliminate fluorescence, since the short-wavelength components of the continuous spectrum will also excite K radiation in the specimen. For example, suppose a copper specimen is being examined with Cu $K\alpha$ radiation of wavelength 1.542 Å from a tube operated at 30 kV. Under these conditions the short-wavelength limit is 0.413 Å. The K absorption edge of copper is at 1.380 Å. The $K\alpha$ component of the incident radiation will not cause fluorescence, but all wavelengths between 0.413 and 1.380 Å will. If a nickel filter is used to suppress the $K\beta$ component of the incident beam, it will also have the desirable effect of reducing the intensity of some of the short wavelengths which cause fluorescence, but it will not eliminate them completely, particularly in the wavelength region near 0.6 Å, where the intensity of the continuous spectrum is high and the absorption coefficient of nickel rather low.

It is sometimes possible to filter part of the fluorescent radiation from the specimen by placing the proper filter over the *film.* For example, if a steel specimen is examined with copper radiation, which is not generally advisable, the situation may be improved by covering the film with aluminum foil, because aluminum has a greater absorption for the fluorescent Fe $K\alpha$ radiation contributing to the background than for the Cu $K\alpha$ radiation forming the diffraction lines. In fact, the following is a good general rule to follow: if it is impossible to use a wavelength longer than the K absorption edge of the specimen, choose one which is considerably shorter and cover the film with a filter. Sometimes the air itself will provide sufficient filtration. Thus excellent patterns of aluminum can be obtained with Cu $K\alpha$ radiation, even though this wavelength (1.54 Å) is much shorter than the K absorption edge of aluminum (6.74 Å), simply because the Al $K\alpha$ radiation excited has such a long wavelength (8.34 Å) that it is almost completely absorbed in a few centimeters of air.

2. *Diffraction of the continuous spectrum.* Each crystal in a powder specimen forms a weak Laue pattern, because of the continuous radiation component of the incident beam. This is true whether or not that particular crystal has the correct orientation to reflect the characteristic component into the Debye ring. Many crystals in the specimen are therefore contributing only to the background of the photograph and not to the diffraction ring, and the totality of the Laue patterns from all the crystals is a continuous distribution of background radiation. If the incident radiation has been so chosen that very little fluorescent radiation is emitted, then diffraction of the continuous spectrum is the largest single cause of high background intensity in powder photographs.

3. *Diffuse scattering from the specimen itself.*

a) Incoherent (Compton modified) scattering. This kind of scattering becomes more intense as the atomic number of the specimen decreases.

b) Coherent scattering.

(i) Temperature-diffuse scattering. This form is more intense with soft materials of low melting point.

(ii) Diffuse scattering due to various kinds of imperfection in the crystals. Any kind of randomness or strain will cause such scattering.

4. *Diffraction and scattering from other than the specimen material.*

a) Collimator and beam stop. This kind of scattering can be minimized by correct camera design, as discussed in Sec. 6–2.

b) Specimen binder, support, or enclosure. The glue or other adhesive used to compact the powder specimen, the glass fiber to which the powder is attached, or the glass or fused-quartz tube in which it is enclosed all contribute to the background of the photograph, since these are all amorphous substances. The amount of these materials should be kept to the absolute minimum.

c) Air. Diffuse scattering from the air may be avoided by evacuating the camera or filling it with a light gas such as hydrogen or helium.

6–13 CRYSTAL MONOCHROMATORS

The purest kind of radiation to use in a diffraction experiment is radiation which has itself been diffracted, since it is entirely monochromatic.* If a single crystal is set to reflect the strong $K\alpha$ component of the general radiation from an x-ray tube and this reflected beam is used as the incident beam in a diffraction camera, then the causes of background radiation listed under (1) and (2) above can be

* This statement requires some qualification. When a crystal monochromator is set to diffract radiation of wavelength λ from a particular set of planes, then these same planes will also diffract radiation of wavelength $\lambda/2$ and $\lambda/3$ in the second and third order, respectively, and at exactly the same angle 2θ. These components of submultiple wavelength are of relatively low intensity when the main component is $K\alpha$ characteristic radiation but, even so, their presence is undesirable whenever precise calculations of the intensity diffracted by the *specimen* must be made. The submultiple components may be eliminated from the beam from the monochromator by reducing the tube voltage to the point where these wavelengths are not produced. If the main component is Cu $K\alpha$ radiation, this procedure is usually impractical because of the decrease in intensity attendant on a reduction in tube voltage to 16 kV (necessary to eliminate the $\lambda/2$ and $\lambda/3$ components). Usually, a compromise is made by operating at a voltage just insufficient to generate the $\lambda/3$ component (24 kV for copper radiation) and by using a crystal which has, for a certain set of planes, a negligible reflecting power for the $\lambda/2$ component. Fluorite (CaF_2) is such a crystal, the structure factor for the 222 reflection being much less than for the 111. The diamond cubic crystals, silicon and germanium, are even better, since their structure factors for the 222 reflection are actually zero.

completely eliminated. Since the other causes of background scattering are less serious, the use of crystal-monochromated radiation produces diffraction photographs of remarkable clarity. There are two kinds of monochromators in use, depending on whether the reflecting crystal is unbent or bent and cut.

An unbent crystal is not a very efficient reflector, as can be seen from Fig. 6-14. The beam from an x-ray tube is never composed only of parallel rays, even when defined by a slit or collimator, but contains a large proportion of convergent and divergent radiation. When the crystal is set at the correct Bragg angle for the parallel component of the incident beam, it can reflect only that component and none of the other rays, with the result that the reflected beam is of very low intensity although it is itself perfectly parallel, at least in the plane of the drawing. In a plane at right angles, the reflected beam may contain both convergent and divergent radiation.

A large gain in intensity may be obtained by using a bent and cut crystal, which operates on the focusing principle illustrated in Fig. 6-15. A line source

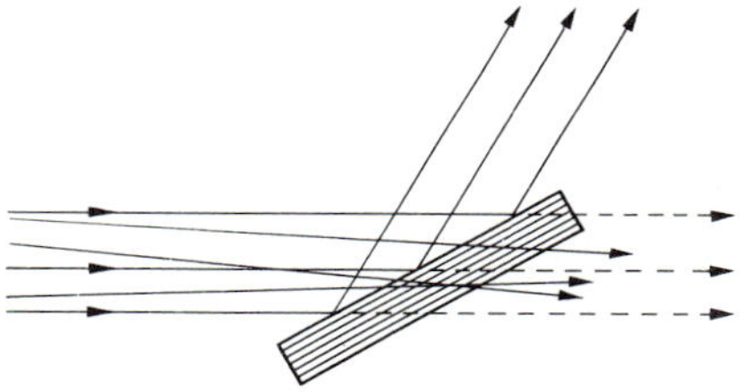

Fig. 6-14 Monochromatic reflection when the incident beam is nonparallel.

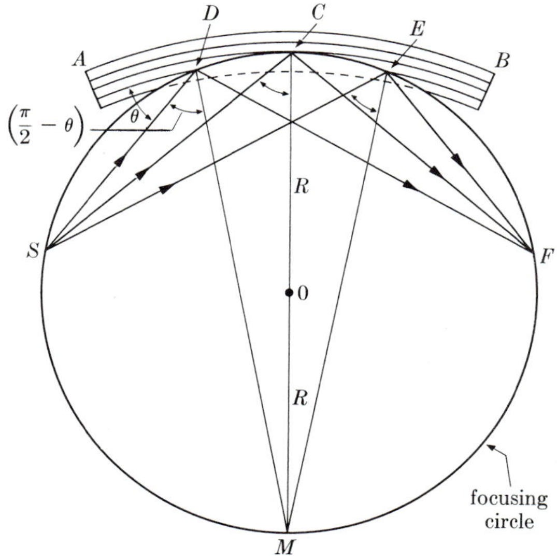

Fig. 6-15 Focusing monochromator (cut and bent).

of x-rays, the focal line on the tube target, is located at S perpendicular to the plane of the drawing. The crystal AB is in the form of a rectangular plate and has a set of reflecting planes parallel to its surface. It is elastically bent into a circular form so that the radius of curvature of the plane through C is $2R = CM$; in this way, all the plane normals are made to pass through M, which is located on the same circle, of radius R, as the source S. If the face of the crystal is then cut away behind the dotted line to a radius of R, then all rays diverging from the source S will encounter the lattice planes at the same Bragg angle, since the angles SDM, SCM, and SEM are all equal to one another, being inscribed on the same arc SM, and have the value $(\pi/2 - \theta)$.

When the Bragg angle is adjusted to that required for reflection of the $K\alpha$ component of the incident beam, then a strong monochromatic beam will be reflected by the crystal. Moreover, since the diffracted rays all originate on a circle passing through the source S, they will converge to a focus at F, located on the same circle as S and at the same distance from C, in much the same way as in the focusing cameras previously discussed. This kind of monochromator is called the Johansson type.

In practice the crystal is not bent and then cut as described above, but the unbent crystal, usually of quartz, is first cut to a radius of $2R$ and then bent against a circular form of radius R. This procedure will produce the same net result. The value of θ required for the diffraction of a particular wavelength λ from planes of spacing d is given by the Bragg law:

$$\lambda = 2d \sin \theta. \tag{6–12}$$

The source-to-crystal distance SC, which equals the crystal-to-focus distance CF, is given by

$$SC = 2R \cos\left(\frac{\pi}{2} - \theta\right). \tag{6–13}$$

By combining Eqs. (6–12) and (6–13), we obtain

$$SC = R\frac{\lambda}{d}. \tag{6–14}$$

For reflection of Cu $K\alpha$ radiation from the (10·1) planes of quartz, the distance SC is 14.2 cm for a value of R of 30 cm.

The chief value of the focusing monochromator lies in the fact that all the monochromatic rays in the incident beam are utilized and the diffracted rays from a considerable area of the crystal surface are all brought to a focus. This leads to a large concentration of energy and a considerable reduction in exposure time compared to the unbent-crystal monochromator first described. However, the latter does produce a semiparallel beam of radiation, and, even though it is of very low intensity, such a beam is required in some experiments.

If the monochromating crystal is bent but not cut, some concentration of energy will be achieved inasmuch as the reflected beam will be convergent, but it will not converge to a perfect focus.

Many crystals have been used as monochromators: NaCl, LiF, SiO_2 (quartz),

Al, Ge, graphite. For discussion of their relative merits and details of design and use, see Guinier [G.10, G.21], Brindley [6.4], pp. 79–88 of Vol. 3 of [G.11], and Klug and Alexander [G.39]. Graphite gives the strongest diffracted beam.

The use of a monochromator produces a change in the relative intensities of the beams diffracted by the specimen. Equation (4–19), for example, was derived for the completely unpolarized incident beam obtained from the x-ray tube. Any beam diffracted by a crystal, however, becomes partially polarized by the diffraction process itself, which means that the beam from a crystal monochromator is partially polarized before it reaches the specimen. Under these circumstances, the usual polarization factor $(1 + \cos^2 2\theta)/2$, which is included in Eqs. (4–19) through (4–21), must be replaced by the factor $(1 + \cos^2 2\alpha \cos^2 2\theta)/(1 + \cos^2 2\alpha)$, where 2α is the diffraction angle in the monochromator (Fig. 6–16). Since the denominator in this expression is independent of θ, it may be omitted; the combined Lorentz-polarization factor for crystal-monochromated radiation is therefore $(1 + \cos^2 2\alpha \cos^2 2\theta)/\sin^2 \theta \cos \theta$. This factor may be substituted into Eqs. (4–19) and (4–20), although a monochromator is not often used with a Debye–Scherrer camera, or into Eq. (4–21), when a monochromator is used with a diffractometer (Sec. 7–13). But note that Eq. (4–20) does not apply to the focusing cameras of the next section.

6-14 GUINIER CAMERAS [G.10, G.39]

The focusing monochromator is best used with powder cameras especially designed to take advantage of the particular property of the reflected beam, namely its focusing action. A cylindrical camera (Fig. 6–16) is used with the specimen and film arranged on the surface of the cylinder. Low-angle reflections are registered with the camera placed in position C, in which case the specimen D must be thin enough to be examined in transmission. High-angle reflections are obtained by back reflection with the camera in position C', shown dashed, and the specimen at D'; the geometry of the camera is then exactly similar to that of the Seemann–Bohlin camera, the focal point F of the monochromatic beam acting as a virtual source of divergent radiation. In either case, the diffracted rays from the specimen are focused on the film for all *hkl* reflections; the only requirement is that the film be located on a circle passing through the specimen and the point F.

The combination of a focusing monochromator and a focusing camera is known as a Guinier camera, pioneered by Guinier in the late 1930s. Later investigators produced

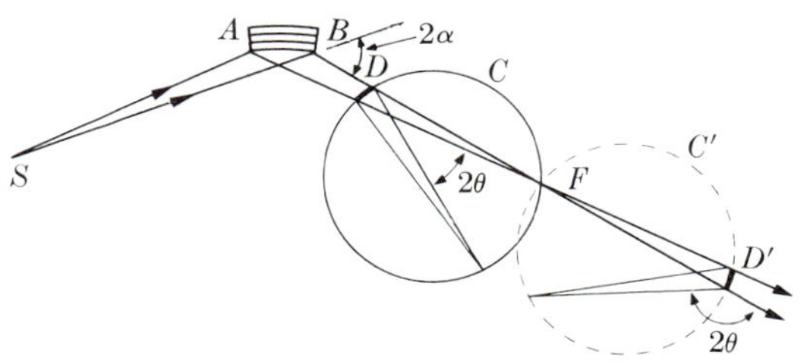

Fig. 6–16 Cameras used with focusing monochromators. Only one diffracted beam is shown in each case. After Guinier [G.10].

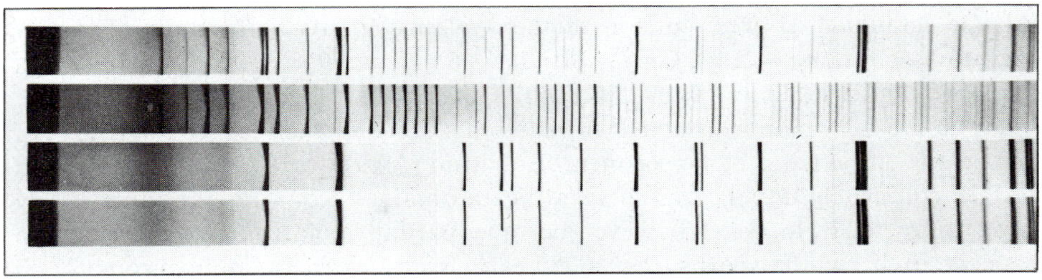

Fig. 6–17 Single film with four powder patterns, made with Guinier–de Wolff camera. The two bottom patterns are of quartz, the third from the bottom is of ammonium alum, and the top one is of a mixture of quartz and ammonium alum. The range of 2θ angles is 4° to 82°. (Courtesy of Enraf-Nonius Inc., Delft.)

variants known by hyphenated names. Thus, the Guinier–de Wolff "camera" [6.5] is a set of two, or four, Guinier cameras stacked one above the other and separated by baffles; patterns from two or four different specimens are registered simultaneously on one piece of film (Fig. 6–17); this arrangement is possible because the beam from the monochromator in Fig. 6–16 is physically wide enough normal to the plane of the drawing to be split into two or four beams. The Guinier–Jagodzinski "camera" [6.6] has two cameras on one support, arranged as in Fig. 6–16 but symmetrically with respect to the incident beam, one for transmission and one for back reflection; three specimens can be examined simultaneously in each camera.

Compared to a Debye–Scherrer camera of the same size, operated directly from the x-ray tube, a Guinier camera provides a much clearer pattern with twice the resolution and about the same exposure time, but any one Guinier camera covers only a limited range of 2θ. It is best suited to the examination of particular parts of complex patterns.

6–15 MEASUREMENT OF LINE POSITION

The solution of any powder photograph begins with the measurement of the positions of the diffraction lines on the film. A device of the kind shown in Fig. 6–18 is commonly used for this purpose. It is essentially a box with an opal-glass plate on top, illuminated from below, on which the film to be measured is placed. On top of the glass plate is a graduated scale carrying a slider equipped with a vernier and cross-hair; the cross-hair is moved over the illuminated film from one diffraction line to another and their positions noted. The film is usually measured without magnification. A low-power hand lens may be of occasional use, but magnification greater than 2 or 3 diameters usually causes the line to merge into the background and become invisible, because of the extreme graininess of x-ray film.

After the line positions are measured, their θ values are calculated from the appropriate equation for the camera involved. But if no great precision is needed, both of these steps can be avoided in routine work by making a cardboard scale marked with θ values; this scale is placed next to the film and the θ value of each line read off. A separate scale will be needed for each size and type of camera.

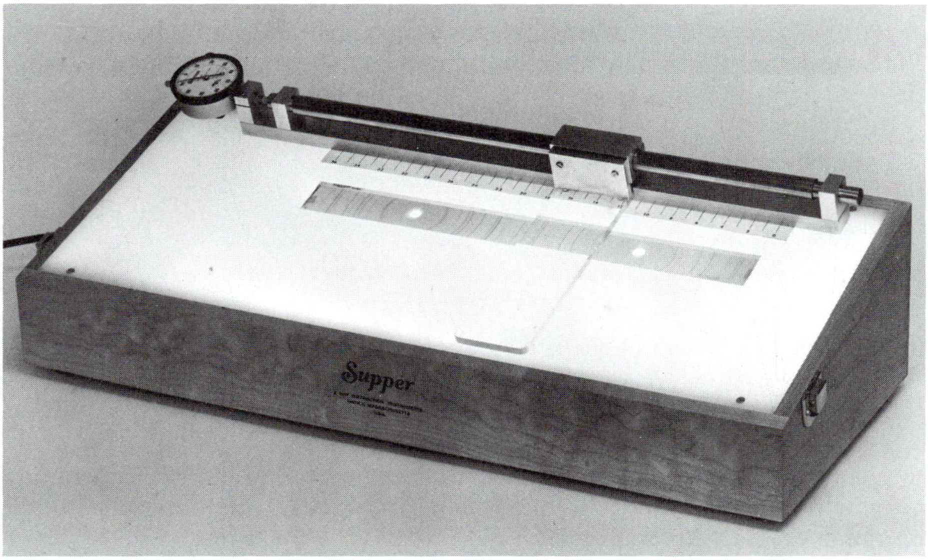

Fig. 6-18 Film-measuring device. (Courtesy of Charles Supper Company.)

Scales giving directly the spacing d of the reflecting planes causing each line may also be made for any particular wavelength, such as Cu $K\alpha$.

6-16 MEASUREMENT OF LINE INTENSITY

Many diffraction problems require an accurate measurement of the integrated intensity, or the breadth at half maximum intensity, of a diffraction line on a powder photograph. For this purpose it is necessary to obtain a curve of intensity vs. 2θ for the line in question.

The intensity of an x-ray beam may be measured by the amount of blackening it causes on a photographic film. The photographic density (blackening) of a film is in turn measured by the amount of visible light the film will transmit, with an instrument called a *microphotometer* or microdensitometer. A photometer is an instrument that measures light intensity with some kind of photocell, and a microphotometer is one that operates with an extremely thin light beam, about 0.1 mm wide, defined by narrow slits. The x-ray film to be measured is slowly traversed across this beam, and the intensity of the light passing through the film is continuously measured with a photocell connected to a recording galvanometer. The result is a plot like that of Fig. 6-19.

The microphotometer is little used today because the counter of a diffractometer is faster, more accurate, and more sensitive. However, when only a camera is available or when circumstances are such that a camera is required (Sec. 6-1),

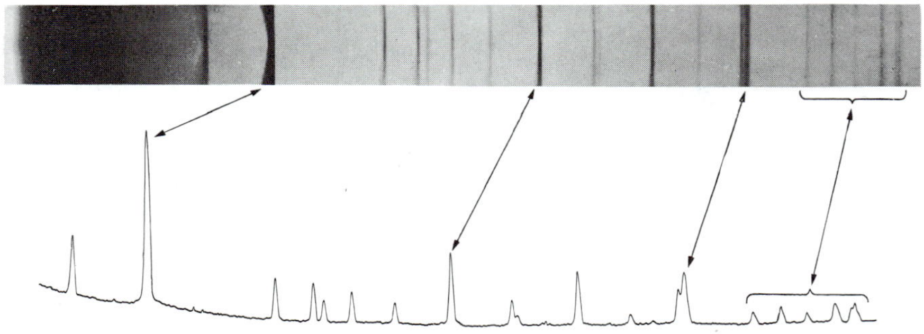

Fig. 6–19 Powder pattern of quartz (above) and corresponding microphotometer trace (below). (Courtesy of U.S. Bureau of Mines.) [6.7]

then a microphotometer is the only means of measuring diffracted intensities quantitatively. Readers who need more information on microphotometer operation can find it in sources such as [G.19], [G.39], and [6.8].

PROBLEMS

***6–1** Plot a curve similar to that of Fig. 6–3 showing the absorption of Fe $K\alpha$ radiation by air. Take the composition and density of air from Prob. 1–8. If a 1-hr exposure in air is required to produce a certain diffraction line intensity in a 19-cm-diameter camera with Fe $K\alpha$ radiation, what exposure is required to obtain the same line intensity with the camera evacuated, other conditions being equal?

***6–2** Derive an equation for the resolving power of a Debye–Scherrer camera for two wavelengths of nearly the same value, in terms of ΔS, where S is defined by Fig. 6–2.

6–3 For a Debye pattern made in a 5.73-cm-diameter camera with Cu $K\alpha$ radiation, calculate the separation of the components of the $K\alpha$ doublet in degrees and in centimeters for $\theta = 10, 35, 60$, and $85°$.

***6–4** What is the smallest value of θ at which the Cr $K\alpha$ doublet will be resolved in a 5.73-cm-diameter Debye camera? Assume that the line width is 0.03 cm and that the separation must be twice the width for resolution.

6–5 A powder pattern of zinc is made in a Debye–Scherrer camera 5.73 cm in diameter with Cu $K\alpha$ radiation.

a) Calculate the resolving power necessary to separate the 11·0 and 10·3 diffraction lines.
b) Calculate the resolving power of the camera used, for these lines. Assume that the line width is 0.03 cm.
c) What minimum camera diameter is required to produce resolution of these lines?

(See Fig. 3–13, which shows these lines unresolved from one another. They form the fifth line from the low-angle end.)

***6–6** A transmission pinhole photograph is made of copper with Cu $K\alpha$ radiation. The film measures 4 by 5 in. What is the maximum specimen-to-film distance which can be used and still have the first two Debye rings completely recorded on the film?

6–7 A powder pattern of iron is made with Cu $K\alpha$ radiation. Assume that the background is due entirely to fluorescent $K\alpha$ radiation from the specimen. The maximum intensity (measured above the background) of the weakest line on the pattern is found to be equal to the background intensity itself at that angle. If the film is covered with aluminum foil 0.0015 in. thick, what will be the ratio of I_{max} for this line to the background intensity?

7

Diffractometer and Spectrometer Measurements

7-1 INTRODUCTION

The x-ray spectrometer, briefly mentioned in Sec. 3–4, has had a long and uneven history in the field of x-ray diffraction. It was first used by W. H. and W. L. Bragg in their early work on x-ray spectra and crystal structure, but it then passed into a long period of relative disuse during which photographic recording in cameras was the most popular method of observing diffraction effects. The few spectrometers in use were all homemade and confined largely to the laboratories of research physicists. In the late 1940s, however, commercially made instruments became available; they rapidly became popular because they offered certain particular advantages over film techniques. Initially a research tool, the x-ray spectrometer has now become an instrument for control and analysis in a wide variety of industrial laboratories.

Depending solely on the way it is used, the basic x-ray spectrometer is really two instruments:

1. An instrument for measuring x-ray spectra by means of a crystal of known structure.
2. An instrument for studying crystalline (and noncrystalline) materials by measurements of the way in which they diffract x-rays of known wavelength.

The term *spectrometer* was originally used to describe both instruments, but, properly, it should be applied only to the first. The second instrument is aptly called a *diffractometer*: this name serves well to emphasize the particular use to which the instrument is being put, namely, diffraction analysis rather than spectrometry.

In this chapter, the design and operation of diffractometers will be described with particular reference to the commercial models available. Equipment suppliers are listed in [5.1], and Jenkins [G.37] gives worked-out examples of problems that arise in diffractometry. The best source of detailed information on diffractometer techniques is the book by Klug and Alexander [G.39].

Just as the emphasis in the present book is on diffraction rather than spectroscopy, the emphasis in this chapter is on the diffractometer. However, some experimental techniques used only, or mainly, in spectrometry are also described here, because they merge quite naturally with diffractometer techniques.

7-2 GENERAL FEATURES

In a diffraction camera, the intensity of a diffracted beam is measured through the amount of blackening it produces on a photographic film, a microphotometer measurement of the film being required to convert "amount of blackening" into x-ray intensity. In the diffractometer, the intensity of a diffracted beam is measured directly by an electronic *counter*. There are many types of counters, but they all convert incoming x-rays into surges or pulses of electric current in the circuit connected to the counter. This circuit counts the number of current pulses per unit of time, and this number is directly proportional to the intensity of the x-ray beam entering the counter. (Logically, the device known as a counter should be called a "detector," and the counting circuit a "counter," but the detecting, pulse-producing device is usually called a counter.)

Basically, a diffractometer is designed somewhat like a Debye-Scherrer camera, except that a movable counter replaces the strip of film. In both instruments, essentially monochromatic radiation is used and the x-ray detector (film or counter) is placed on the circumference of a circle centered on the powder specimen. The essential features of a diffractometer are shown in Fig. 7-1. A powder specimen C, in the form of a flat plate, is supported on a table H, which can be rotated about an axis O perpendicular to the plane of the drawing. The x-ray source is S, the line focal spot on the target T of the x-ray tube; S is also normal to the plane of the drawing and therefore parallel to the diffractometer axis O. X-rays diverge from this source and are diffracted by the specimen to form a convergent diffracted beam which comes to a focus at the slit F and then enters the counter G. A and

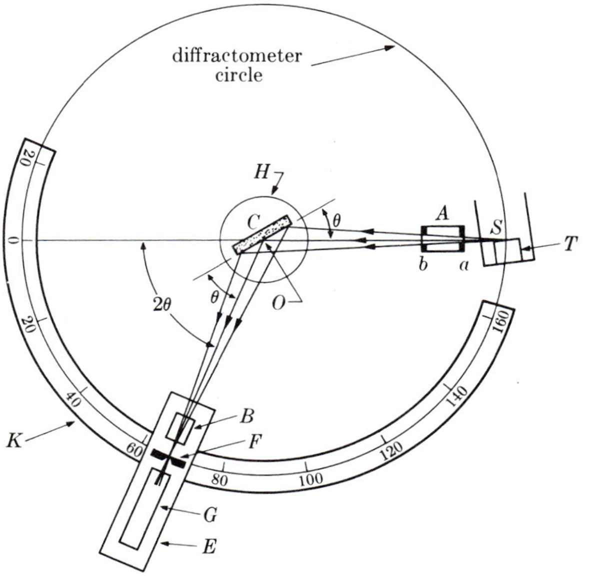

Fig. 7-1 X-ray diffractometer (schematic).

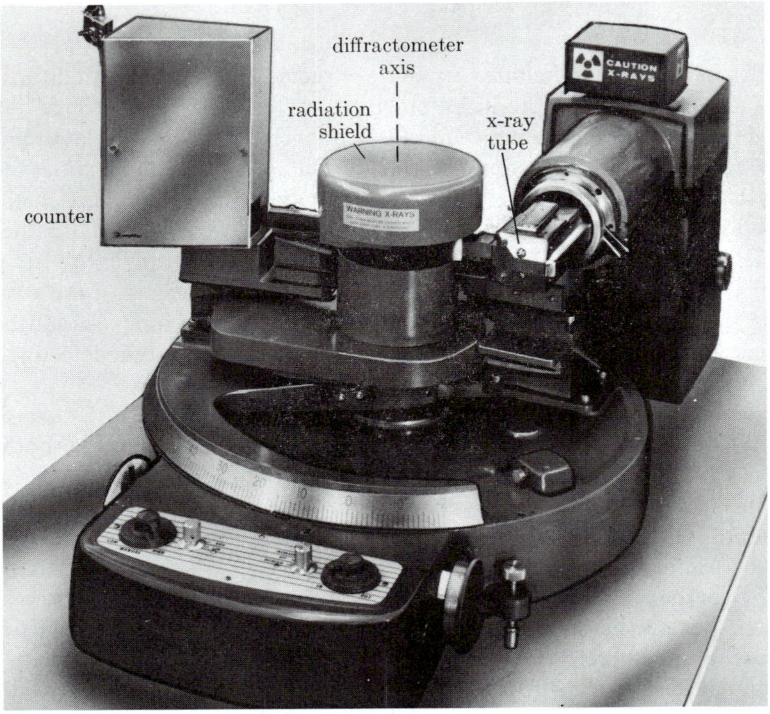

Fig. 7–2(a) Diano diffractometer. (Courtesy of Diano Corporation. This company has, since 1972, manufactured and sold the x-ray diffraction and spectroscopy equipment formerly made by the General Electric Company.)

B are special slits which define and collimate the incident and diffracted beams. The filter is usually placed in a special holder (not shown) in the diffracted, rather than the incident, beam; a filter in the diffracted beam not only serves its primary function (suppression of $K\beta$ radiation) but also decreases background radiation originating in the specimen.

The receiving slits and counter are supported on the carriage *E*, which may be rotated about the axis *O* and whose angular position 2θ may be read on the graduated scale *K*. The supports *E* and *H* are mechanically coupled so that a rotation of the counter through $2x$ degrees is automatically accompanied by rotation of the specimen through x degrees. This coupling ensures that the angles of incidence on, and reflection from, the flat specimen will always be equal to one another and equal to half the total angle of diffraction, an arrangement necessary to preserve focusing conditions. The counter may be power-driven at a constant angular velocity about the diffractometer axis or moved by hand to any desired angular position.

Figures 7–2 and 7–3 illustrate three commercial instruments. Basically, they all adhere to the design principles described above, but they differ in detail and

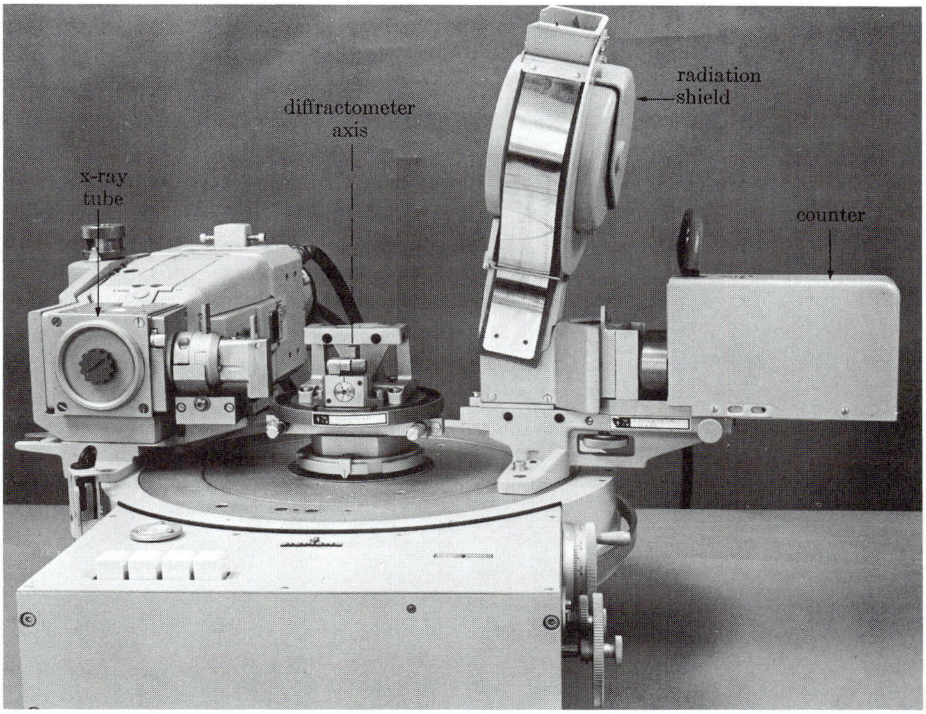

Fig. 7–2(b) Siemens diffractometer. The radiation shield has here been raised from its normal position covering the specimen. (Courtesy of Siemens Corporation.)

in positioning. In the Diano and Siemens instruments shown in Fig. 7–2(a) and (b), the diffractometer axis is vertical and the counter moves in a horizontal plane. The axis of the Philips diffractometer (Fig. 7–3) is horizontal and the counter moves in a vertical plane.

The way in which a diffractometer is used to measure a diffraction pattern depends on the kind of circuit used to measure the rate of production of pulses in the counter. The pulse rate may be measured in two different ways:

1. The succession of current pulses is converted into a steady current, which is measured on a meter called a *counting-rate meter*, calibrated in such units as counts (pulses) per second (c/s or cps). Such a circuit gives a continuous indication of x-ray intensity.

2. The pulses of current are counted electronically in a circuit called a *scaler*, and the average counting rate is obtained simply by dividing the number of pulses counted by the time spent in counting. This operation is essentially discontinuous because of the time spent in counting, and a scaling circuit cannot be used to follow continuous changes in x-ray intensity.

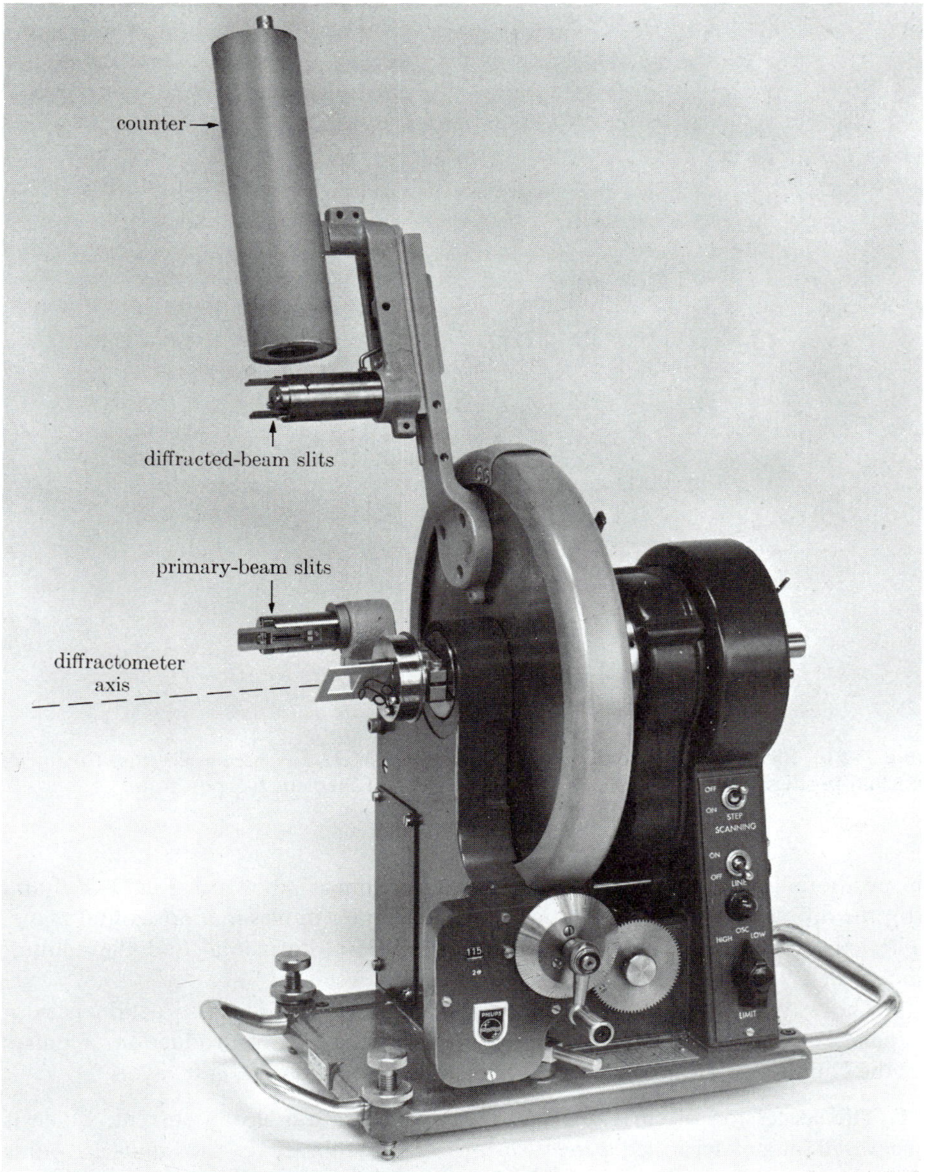

Fig. 7–3 Philips diffractometer. The window of a vertical x-ray tube, not shown, would be immediately behind the primary-beam slits. Radiation shield not shown. See also Fig. 15–10. (Courtesy of Philips Electronic Instruments, Inc.)

Corresponding to these two kinds of measuring circuits, there are two ways in which the diffraction pattern of an unknown substance may be obtained with a diffractometer (Fig. 7–4):

1. *Continuous.* The counter is set near $2\theta = 0°$ and connected to a counting-rate meter. The output of this circuit is fed to a strip-chart recorder. The counter is then driven at a constant angular velocity through increasing values of 2θ until the whole angular range is "scanned." At the same time, the paper chart on the recorder moves at a constant speed, so that distances along the length of the chart are proportional to 2θ. The result is a chart, such as Fig. 7–5, which gives a record of counts per second (proportional to diffracted intensity) vs. diffraction angle 2θ. A "high" scanning speed is typically $2°$ of 2θ per minute; at this rate a complete scan extending from, say, $10°$ to $160°$ 2θ, requires $150/2 = 75$ minutes. (The upper limit of counter motion, determined by contact between counter and x-ray tube, is about $165°$ 2θ.)

2. *Intermittent.* The counter is connected to a scaler and set at a fixed value of 2θ for a time sufficient to make an accurate count of the pulses obtained from the counter. The counter is then moved to a new angular position and the operation repeated. The range of 2θ of interest is covered in this fashion, and the curve of intensity vs. 2θ is finally plotted by hand. When the continuous background between diffraction lines is being measured, the counter may be moved in steps of several degrees, but determinations of line profile may require measurements of intensity at angular intervals as small as $0.01°$. This method of obtaining a diffraction

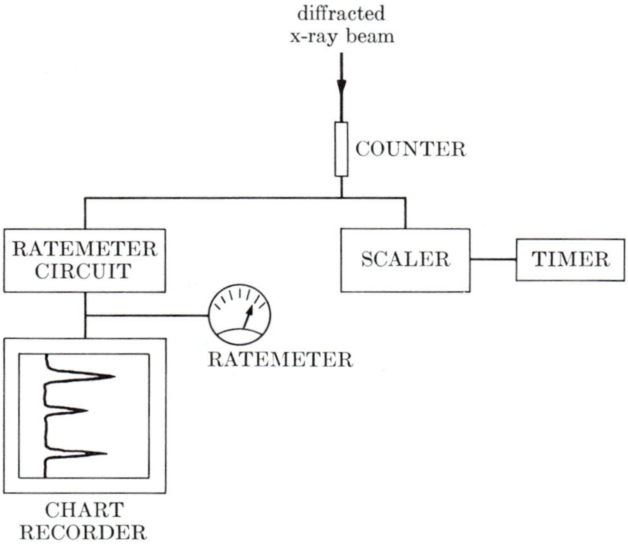

Fig. 7–4 Block diagram of detector circuits for a diffractometer. The ratemeter circuit actuates a meter, for a visual indication of the counting rate, and a chart recorder. The scalar and timer operate together.

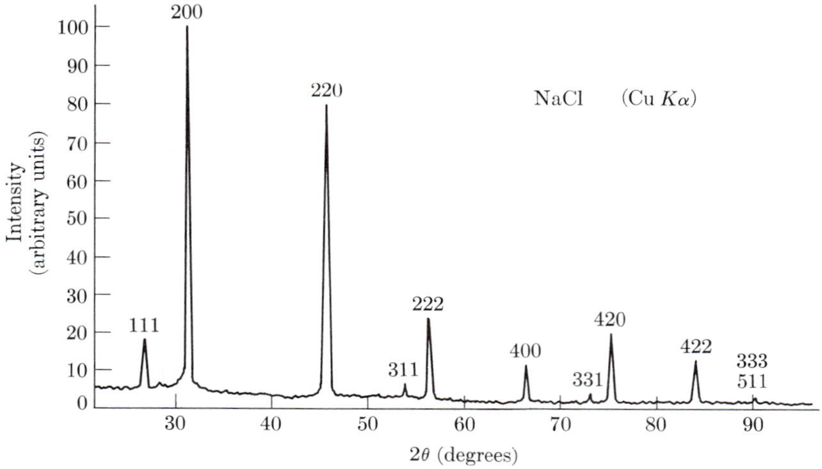

Fig. 7–5 Automatically recorded diffraction pattern of NaCl powder. Copper radiation, nickel filter. About half of the entire range of 2θ is shown here. The numbers on the ordinate are simply chart units, which may be converted to counts/sec if desired. Warren [G.30].

pattern is much slower than that involving a ratemeter and recorder but it yields more precise measurements of intensity. This operation may be performed automatically with some diffractometers (see below).

There is a fundamental difference between the operation of a powder camera and a diffractometer. In a camera, all diffraction lines are recorded simultaneously, and variations in the intensity of the incident x-ray beam during the exposure can have no effect on relative line intensities. On the other hand, with a diffractometer, diffraction lines are recorded one after the other, and it is therefore imperative to keep the incident-beam intensity constant when relative line intensities must be measured accurately. Since the usual variations in line voltage are quite appreciable, the x-ray tube circuit of a diffractometer must include a voltage stabilizer and a tube-current stabilizer.

The kind of specimen used depends on the kind of material available. Flat metal sheet or plate may be examined directly; however, such materials almost always exhibit preferred orientation and this fact must be kept in mind in assessing relative intensities. This is also true of wires, which are best examined by cementing a number of lengths side by side to a glass plate. This plate is then inserted in the specimen holder so that the wire axes are at right angles to the diffractometer axis. Powder specimens are best prepared by placing the powder in a recess in a glass or plastic plate, compacting it under just sufficient pressure to cause cohesion without use of a binder, and smoothing off the surface. Too much pressure causes preferred orientation of the powder particles. Alternatively, the powder may be mixed with a binder and smeared on the surface of a glass slide. The powder should be ground extremely fine, to a size of 10 microns or less, if relative line intensities are to be accurately reproducible; since the flat specimen is not rotated as a Debye–Scherrer specimen is, the only way of obtaining an adequate number of particles having the correct orientation for reflection is to reduce their average size. (Specimen "spinners" are available to continuously rotate the specimen in its own plane. But this kind of rotation is not nearly as effective, in bringing new orientations into the beam, as the kind that takes place in a Debye–Scherrer camera.) Surface roughness also has a marked effect on relative line intensities. If the surface is rough, as in the case of a coarse powder compact, and the linear absorption coefficient high, the intensities of low-angle reflections will be abnormally low because of the absorption of the diffracted rays in each projecting portion of the surface. The only way to avoid this effect is to use a flat-surfaced compact of very fine powders or a specimen with a polished surface.

Single-crystal specimens may also be examined in a diffractometer by mounting the crystal on a three-circle goniometer, such as that shown in Fig. 5–7, which will allow independent rotation of the specimen and counter about the diffractometer axis and two other axes passing through the specimen. (Incidentally, independent rotation of the specimen about the diffractometer axis is often called an ω, rather than a θ, rotation.) In fact, special *single-crystal diffractometers* are available, designed solely for the determination of complex crystal structures.

A diffractometer may be used for measurements at high or low temperatures by surrounding the specimen with the appropriate heating or cooling unit. Such an adaptation of the instrument is much easier with the diffractometer than with a camera because of the generally larger amount of free working space around the specimen in the former. An additional advantage is that dynamic measurements may be made. For example, the counter may be set to receive a prominent line of a phase stable only at room temperature; as the temperature is continuously increased, the ratemeter will continuously indicate and record the disappearance of that phase, in the form of a curve of line intensity vs. temperature.

Automatic diffractometers have been made by many investigators by modifying conventional instruments, and they are also available from manufacturers as original equipment [7.1 to 7.7]. Automation requires mechanical modifications to permit (1) automatic specimen changing and (2) step scanning. The latter allows the counter to be driven quickly from one angular position 2θ to another, pausing at each position long enough for the scaler to make an accurate count. Specimen and counter movements are controlled by a computer which permits, for example, a step scan over any selected range of 2θ in steps of any desired size. The scaler output (diffracted x-ray intensity) and the angle to which it refers may be printed by a teletypewriter, punched on paper tape, or stored on magnetic tape. A computer can then process the output data in various ways, depending on the problem involved. This kind of automation can result in continuous, unattended, day-and-night operation of a diffractometer.

In the succeeding sections, the various parts of the diffractometer will be described in greater detail. This summary of the general features of the instrument is enough to show its principal advantage over the powder camera: the quantitative measurement of line position and intensity is made in one operation with a diffractometer, whereas the same measurement with film technique requires at least two steps (recording the pattern on film and making a microphotometer record of the film) and leads to an over-all result which is generally of lower accuracy. This superiority of the diffractometer is reflected in the much higher cost of the instrument, a cost due not only to the precision machining necessary in its mechanical parts but also to the expensive circuits needed to stabilize the power supply and measure the intensity of diffracted beams.

The diffractometer is a superb instrument for dealing with many problems. But it is not without its faults. A counter "sees" only the radiation it is exposed to and is blind to all diffracted and scattered rays not lying in the plane of the diffractometer circle. A photographic film, with an area of several square inches, will intercept and record a great many rays, disclosing at a glance such conditions as coarse grains and preferred orientation (Chap. 9); a diffractometer does not immediately see these things. As Guinier [G.21] puts it, "The photographic method is admirably suited to the *qualitative exploration of an unknown pattern*, since one can then find totally unexpected phenomena. The counter is necessary for *quantitative measurements* on a pattern which is already known qualitatively."

7-3 X-RAY OPTICS

The chief reason for using a flat specimen is to take advantage of the focusing action described in Sec. 6-6 and so increase the intensity of weak diffracted beams to a point where they can be accurately measured. Figure 7-6 shows how this is done. For any position of the counter, the receiving slit F and the x-ray source S are always located on the diffractometer circle, which means that the face of the specimen, because of its mechanical coupling with the counter, is always tangent to a focusing circle centered on the normal to the specimen and passing through F and S. The focusing circle is not of constant size but increases in radius as the angle 2θ decreases, as indicated in Fig. 7-6. Perfect focusing at F requires that the specimen be curved to fit the focusing circle. Use of a flat specimen causes some broadening of the diffracted beam at F and a small shift in line position toward smaller angles, particularly at 2θ angles less than about 60°; both effects can be lessened by decreasing the divergence of the incident beam, at the expense

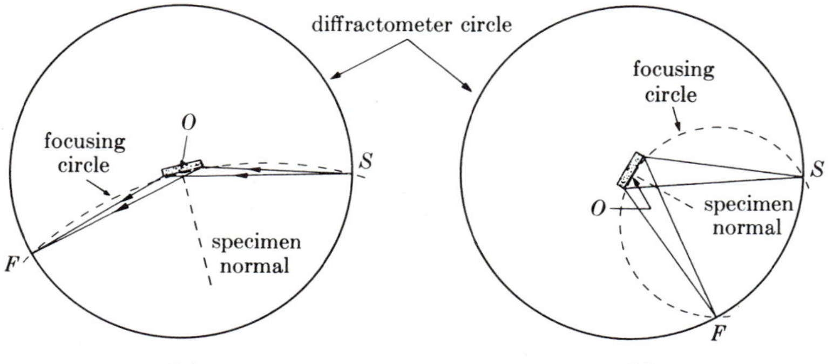

Fig. 7-6 Focusing geometry for flat specimens in (a) forward reflection and (b) back reflection.

of decreased intensity. (Neither focusing nor intensity has to be sacrificed if the specimen surface always conforms to the focusing circle. A device to do this is commercially available, based on a design by Ogilvie [7.24]. The powder specimen is mounted on a thin flexible strip, which is automatically bent to the proper curvature at each angle 2θ.)

The line source S extends considerably above and below the plane of the drawing of Fig. 7-6 and emits radiation in all directions, but the focusing described above requires that all rays in the incident beam be parallel to the plane of the drawing. This condition is realized as closely as possible experimentally by passing the incident beam through a *Soller slit* (Fig. 7-7), slit A in Fig. 7-1, which contains a set of closely spaced, thin metal plates parallel to the plane of the diffractometer circle. These plates remove a large proportion of rays inclined to the plane of the diffractometer circle and still allow the use of a line source of considerable length. Typical dimensions of a Soller slit are: length of plates 32 mm, thickness of plates 0.05 mm, clear distance between plates 0.43 mm. At either end of the slit assembly are rectangular slits a and b, the entrance slit a next to the source being narrower than the exit slit b. The combination of slits and plates breaks up the incident beam into a set of triangular wedges of radiation, as indicated in Fig. 7-7. There are, of course, some rays, not shown in the drawing, which diverge in planes perpendicular to the plane of the plates, and these rays cause the wedges of radiation to merge into one another a short distance away from the exit slit. However, the long, closely spaced plates do restrict this unwanted divergence to an angle of about 1.5°. Slits a and b define the divergence of the incident beam in the plane of the diffractometer circle. The slits commonly available have divergence angles ranging from very small values up to about 4°. In the forward-reflection region, a divergence angle of 1° is sufficient because of the low inclination of the specimen surface to the incident beam, but in back reflection an increase in divergence angle to 3 or 4° will increase the area irradiated and the diffracted intensity. But if line intensities are to be compared over the whole range of 2θ, the same divergence must be used throughout and the specimen must be wider than the beam at all angles.

The beam diffracted by the specimen passes through another Soller slit and the receiving slit F before entering the counter (Fig. 7-8). Since the receiving slit defines the width of the beam admitted to the counter, an increase in its width will increase the maximum intensity of any diffraction line being measured but at the expense of some loss of resolution. On the other hand, the relative *integrated intensity* of a diffraction line is independent of slit width, which is one reason for its greater fundamental importance.*

* A number of things besides slit width (e.g., x-ray tube current) will change the integrated intensity of a single diffraction line. The important thing to note, however, is that a change in any one of the operating variables changes the integrated intensities of all diffraction lines in the same ratio but can produce very unequal effects on maximum intensities. Thus, if I_1/I_2 is the ratio of the integrated intensities of two lines measured with a certain slit width and M_1/M_2 the ratio of their maximum intensities, then another measurement with a different slit width will result in the same ratio I_1/I_2 for the integrated intensities, but the ratio of the maximum intensities will now, in general, differ from M_1/M_2.

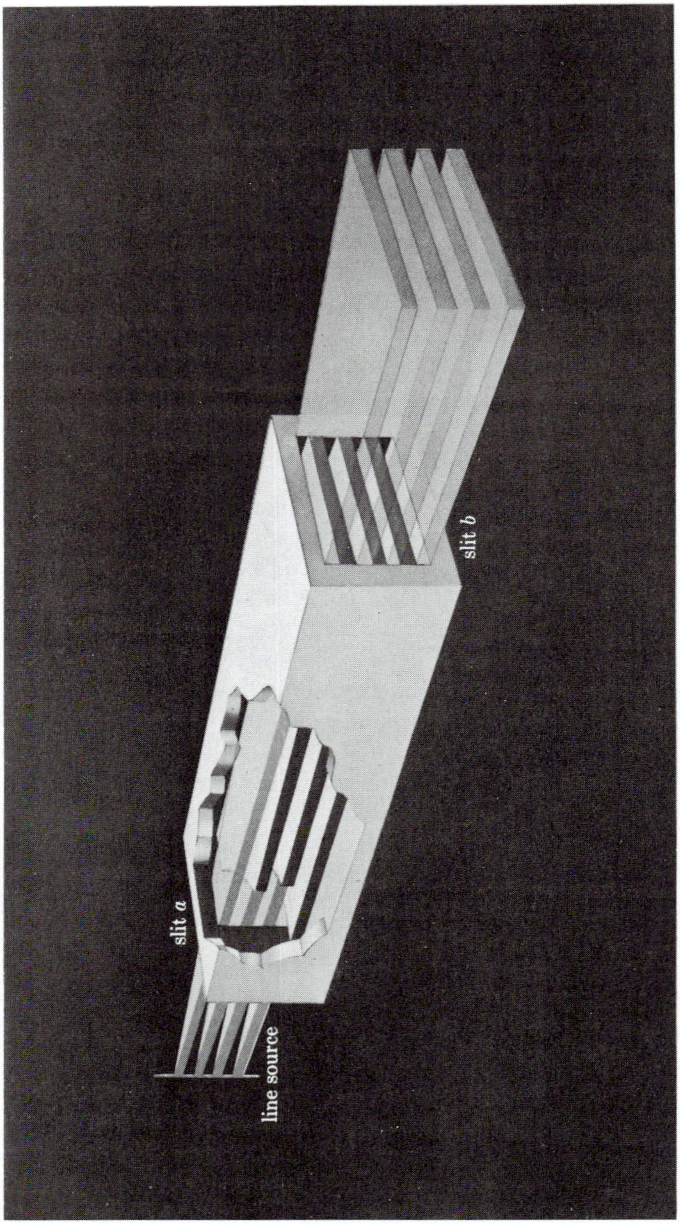

Fig. 7–7 Soller slit (schematic). For simplicity, only three metal plates are shown; actual Soller slits contain about a dozen.

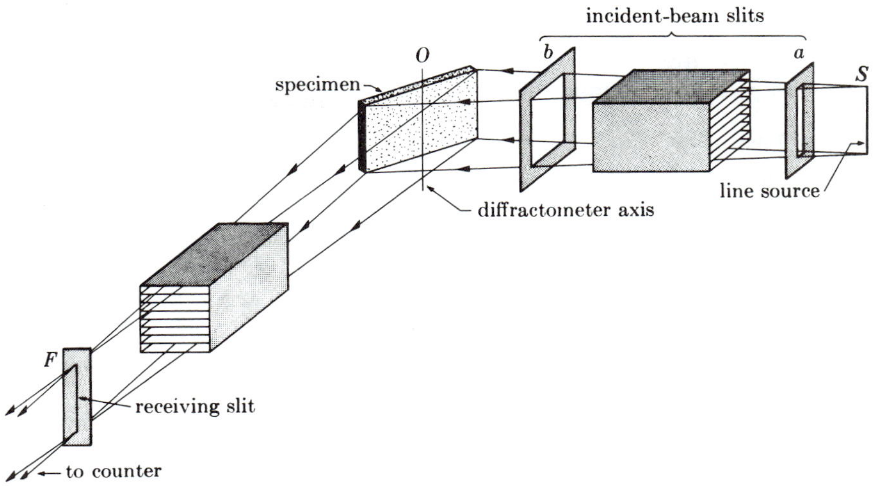

Fig. 7-8 Arrangement of slits in diffractometer.

Because of the focusing of the diffracted rays and the relatively large radius of the diffractometer circle, about 15 cm in commercial instruments, a diffractometer can resolve very closely spaced diffraction lines. Indicative of this is the fact that resolution of the Cu $K\alpha$ doublet can be obtained at 2θ angles as low as about 40°. Such resolution can only be achieved with a correctly adjusted instrument, and it is necessary to so align the component parts that the following conditions are satisfied for all diffraction angles:

1. line source, specimen surface, and receiving-slit axis are all parallel,
2. the specimen surface coincides with the diffractometer axis, and
3. the line source and receiving slit both lie on the diffractometer circle.

7-4 COUNTERS (GENERAL)

Without exception all electronic counters were developed by nuclear physicists for studies of radioactivity. They can detect not only x- and γ-radiation, but also charged particles such as electrons and α-particles, and the design of the counter and associated circuits depends to some extent on what is to be detected. Here we are concerned only with counters for the detection of x-rays of the wavelengths commonly encountered in diffraction.

Four types of counters are currently in use: proportional, Geiger, scintillation, and semiconductor. All depend on the power of x-rays to ionize atoms, whether they are atoms of a gas (proportional and Geiger counters) or atoms of a solid (scintillation and semiconductor counters). A general treatment of the first three types has been given by Parrish [7.8].

We will be interested in three aspects of counter behavior: losses, efficiency, and energy resolution. These are defined below and made more specific in later sections on particular counters.

Counting Losses

The absorption of a quantum (photon) of x-rays in the active volume of a counter causes a voltage pulse in the counter output. Pulses from the counter then enter some very complex electronic circuitry, consisting of one or more pulse amplifiers, pulse shapers, etc. and, at the end, a scaler or ratemeter and, possibly, a pulse-height analyzer (Sec. 7–9). Let us call all the circuitry beyond the counter simply the "electronics." Then we are interested not simply in the behavior of the counter alone, but in the behavior of the whole system, namely, the counter-electronics combination.

If the x-ray beam to be measured is strong, the rate of pulse production in the counter will be high, and the counting rate given by the ratemeter will be high. (Roughly speaking, several thousand counts per second is a "high" rate in powder diffractometry, and less than a hundred cps a "low" rate.) As the counting rate increases, the time interval between pulses decreases and may become so small that adjacent pulses merge with one another and are no longer resolved, or counted, as separate pulses. At this point counting loss has begun. The quantity that determines this point is the *resolving time* t_s of the counter-electronics system, defined as the minimum time between two resolvable pulses.

The arrival of x-ray quanta at the counter is random in time. Therefore pulse production in the counter is random in time, and a curve showing the change in voltage of the counter output would look like Fig. 7–9. If the arrival and absorption of entering quanta were absolutely periodic in time, the maximum counting rate without losses would be given simply by $1/t_s$. But even if their average rate of arrival is no greater than $1/t_s$, some successive quanta may be spaced less than t_s apart because of their randomness in time. It follows that counting losses will occur at rates less than $1/t_s$ and that losses will increase as the rate increases, as shown in Fig. 7–10. Here "quanta absorbed per second" are directly proportional to the x-ray intensity, so that this curve has an important bearing on diffractometer measurements, because it shows the point at which the observed counting rate is no longer proportional to the x-ray intensity. The straight line shows the ideal response that can be obtained with a proportional counter at the rates shown. This linear, no-loss behavior is fortunately typical of most counters used today in diffractometry; otherwise one would have the tedious task of correcting some observed counting rates for losses.

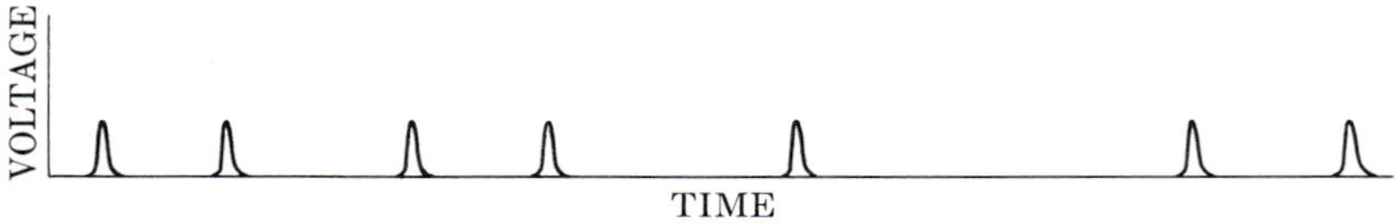

Fig. 7–9 Randomly spaced voltage pulses produced by a counter.

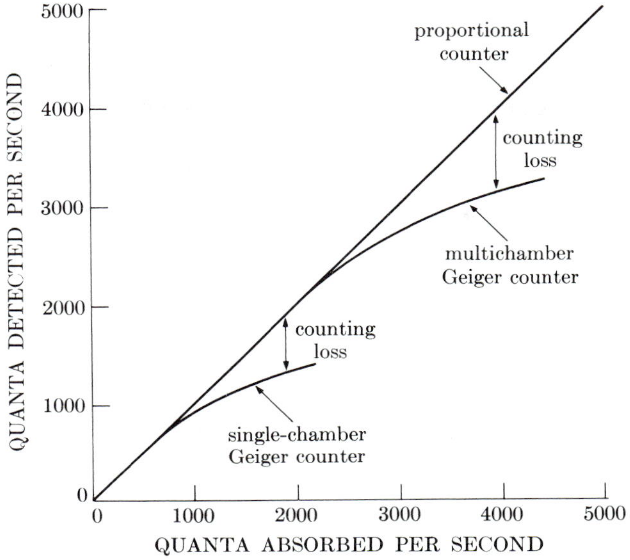

Fig. 7–10 The effect of counting rate on counting losses for three kinds of counter (schematic).

If the resolving time t_s of the counter-electronics is known, the point at which losses begin can be calculated by an easily remembered rule: a loss of one percent occurs at a rate of about one percent of $1/t_s$. Thus, if t_s is one microsecond, the counting rate should be linear to within one percent up to a rate of about 10,000 cps.

Ordinarily, the resolving time is unknown. But if nonlinear counting behavior is suspected, the counting rate at which losses begin can be determined experimentally by the following procedure. Position the counter to receive a strong diffracted beam, and insert in this beam a sufficient number of metal foils of uniform thickness to reduce the counting rate almost to the cosmic background. (Cosmic rays, because of their high penetrating power, pass right through the walls of the counter and continually produce a few counts per second.) Measure the counting rate, remove one foil, measure the counting rate, and continue in this manner until all the foils have been removed. Because each foil produces the same fractional absorption of the energy incident on it, a plot of observed counting rate (on a logarithmic scale) vs. number of foils removed from the beam (on a linear scale) will be linear up to the point where losses begin and will in fact resemble Fig. 7–10. A curve of this kind is shown in Fig. 7–11.

Counting Efficiency

The overall efficiency E of the counter-electronics system in detecting incident x-ray quanta as resolved pulses is the product of the absorption efficiency E_{abs} and the detection efficiency E_{det}.

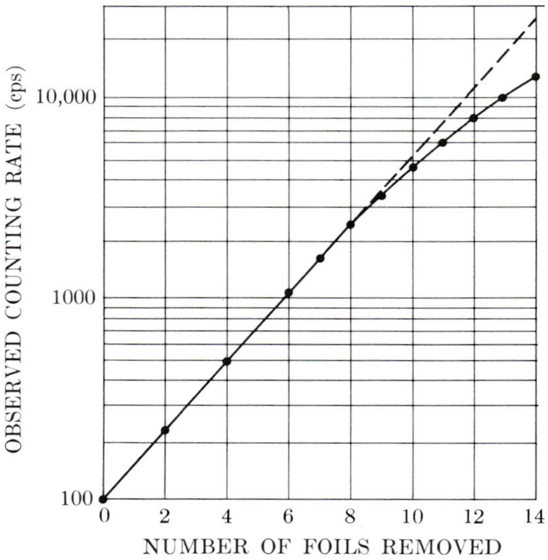

Fig. 7–11 Calibration curve of a multichamber Geiger counter. Cu $K\alpha$ radiation. Nickel foils, each 0.01 mm thick, used as absorbers.

All counters have a thin "window," usually of mica or beryllium, through which the x-rays must pass before reaching the active volume of the counter. The fraction of the incident radiation absorbed by the window $f_{\text{abs, w}}$ should be as small as possible, and the fraction absorbed by the counter itself $f_{\text{abs, c}}$ as large as possible. The absorption efficiency E_{abs}, expressed as a fraction, is given by $(1 - f_{\text{abs, w}})(f_{\text{abs, c}})$. The detection efficiency E_{det} is simply $(1 - f_{\text{losses}})$, where f_{losses} represents the fractional counting losses described above. The overall efficiency is then

$$E = E_{\text{abs}}E_{\text{det}} = [(1 - f_{\text{abs, w}})(f_{\text{abs, c}})][1 - f_{\text{losses}}]. \qquad (7\text{–}1)$$

As previously mentioned E_{det} is essentially 100 percent for most counters used in diffractometry. Therefore E is determined by E_{abs}, which can be calculated from the dimensions and absorption coefficients of the window and counter, and Fig. 7–12 shows the result. Note particularly the dependence of E_{abs} on wavelength, due to the dependence of absorption coefficients on wavelength. The efficiency of any counter is low for very short wavelengths, because most of these hard x-rays pass right through window and counter and are absorbed by neither; at long wavelengths E_{abs} decreases because of increasing absorption of soft x-rays by the window.

Energy Resolution

In most counters the size of the voltage pulse produced by the counter is proportional to the energy of the x-ray quantum absorbed. Thus, if absorption of a Cu $K\alpha$ quantum ($\lambda = 1.54$ Å, $h\nu = 9$ keV) produces a pulse of V volts, then absorption of a Mo $K\alpha$ quantum ($\lambda = 0.71$ Å, $h\nu = 20$ keV) will produce a pulse of $(20/9)V = 2.2\ V$.

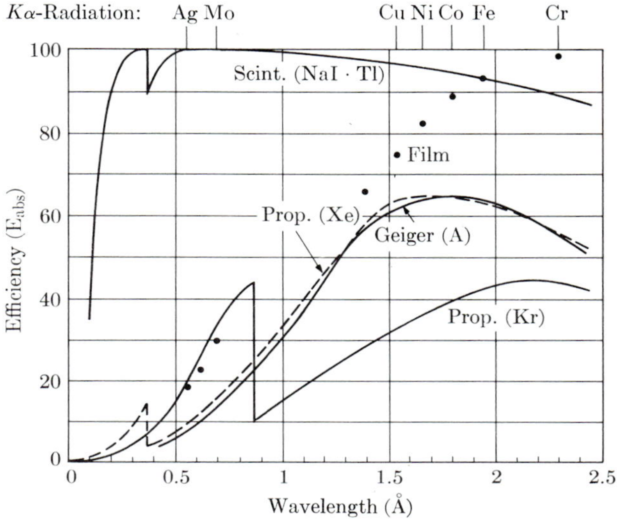

Fig. 7–12 Calculated values of absorption efficiency E_{abs} (in percent) of various kinds of counters and of photographic x-ray film (black dots). Parrish [7.8].

However, the size of a pulse is not sharply defined, even when the incident radiation is strictly monochromatic or, in energy terms, "monoenergetic." Instead of all pulses having exactly the same size V as suggested, for example, by Fig. 7–9, they have sizes distributed around V roughly as indicated in Fig. 7–13. Here the ordinate, "counting rate," is equivalent to the number of pulses having a particular size, so that this curve is a pulse-size distribution curve. If the width of the curve at half its maximum height is W and if V is the mean pulse size, then the resolution R of the counter is

$$R = \frac{W}{V}. \tag{7–2}$$

The smaller R, the better the resolution.

We will now examine the operation and performance of various counters.

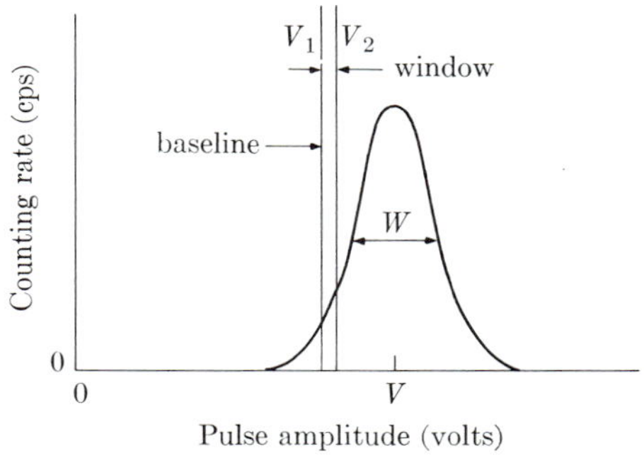

Fig. 7–13 Distribution curve of pulse size. (The "window" and "baseline" are explained in Sec. 7–9.)

7–5 PROPORTIONAL COUNTERS

Consider the device shown in Fig. 7–14, consisting of a cylindrical metal shell (the cathode), about 10 cm long and 2 cm in diameter, filled with a gas and containing a fine metal wire (the anode) running along its axis. Suppose there is a constant potential difference of about 200 volts between anode and cathode. One end of the cylinder is covered with a window of high transparency to x-rays. Of the x-rays which enter the cylinder, a small fraction passes right through, but the larger part is absorbed by the gas, and this absorption is accompanied by the ejection of photoelectrons and Compton recoil electrons from the atoms of the gas. The net result is ionization of the gas, producing electrons, which move under the influence of the electric field toward the wire anode, and positive gas ions, which move toward the cathode shell. At a potential difference of about 200 volts, all these electrons and ions will be collected on the electrodes, and, if the x-ray intensity is constant, there will be a small constant current of the order of 10^{-12} amp or less through the resistance R_1. This current is a measure of the x-ray intensity. When operated in this manner, this device is called an *ionization chamber*. It was used in the original Bragg spectrometer but is now obsolete for the measurement of diffracted x-rays because of its low sensitivity. It is still used in some radiation survey meters.

The same instrument, however, can be made to act as a *proportional counter* if the voltage is raised to the neighborhood of 1000 volts. A new phenomenon now occurs, namely, multiple ionization or "gas amplification." The electric-field intensity is now so high that the electrons produced by the primary ionization are rapidly accelerated toward the wire anode and at an ever-increasing rate of acceleration, since the field intensity increases as the wire is approached. The electrons thus acquire enough energy to knock electrons out of other gas atoms, and these in turn cause further ionization and so on, until the number of atoms ionized by the absorption of a single x-ray quantum is some 10^3 to 10^5 times as large as the number ionized in an ionization chamber. As a result of this amplification a veritable avalanche of electrons hits the wire and causes an easily detectable pulse of current in the external circuit. This pulse leaks away through

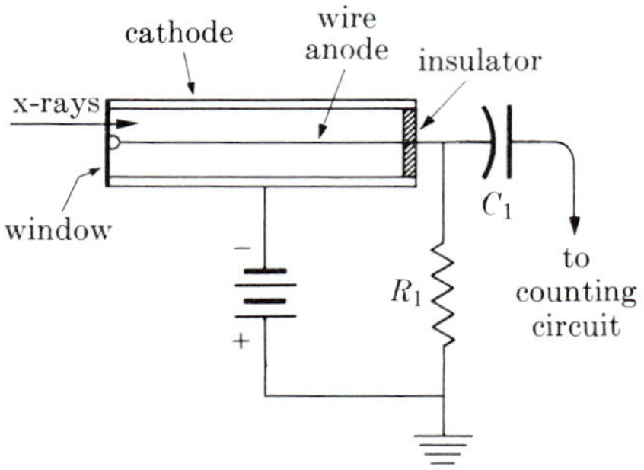

Fig. 7–14 Gas counter (proportional or Geiger) and basic circuit connections.

the large resistance R_1 but not before the charge momentarily added to the capacitor C_1 has been detected by the ratemeter or scaling circuit connected to C_1. At the same time the positive gas ions move to the cathode but at a much lower rate because of their larger mass. This whole process, which is extremely fast, is triggered by the absorption of one x-ray quantum.

We can define a gas amplification factor A as follows: if n is the number of atoms ionized by one x-ray quantum, then An is the total number ionized by the cumulative process described above. (For example, if the gas in the counter is argon, energy of about 26 eV is required to produce an ion pair, i.e., a positive ion and an electron. If the incident radiation is Cu $K\alpha$ of energy 8040 eV, then the number n of ion pairs formed is 8040/26 or 310.) Figure 7–15 shows schematically how the gas amplification factor varies with the applied voltage. At the voltages used in ionization chambers, $A = 1$; i.e., there is no gas amplification, since the electrons produced by the primary ionization do not acquire enough energy to ionize other atoms. But when the voltage is raised into the proportional counter region, A becomes of the order of 10^3 to 10^5, and a pulse of the order of a few millivolts is produced. Moreover, the size of this pulse is proportional to the energy of the x-ray quantum absorbed, which accounts for the name of this counter. This proportionality is important, because it allows us to distinguish (Sec. 7–9) between x-ray quanta of different energies (wavelengths). (Historically, this counter was the first kind to exhibit such proportionality. There are now others.) Pulses from the counter go to a preamplifier, mounted immediately adjacent to the counter; here they are amplified enough to be transmitted, without too much attenuation, along several feet of cable to the main amplifier and the rest of the electronics.

The correct voltage at which to operate the counter is found as follows. Position the counter to receive an x-ray beam of constant intensity. Measure the

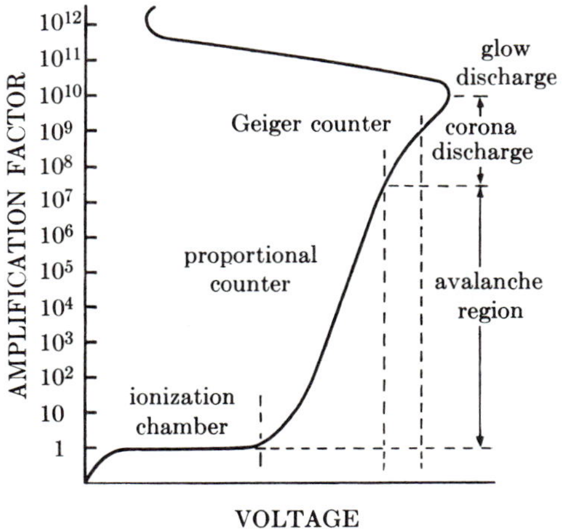

Fig. 7–15 Effect of voltage on the gas amplification factor. Friedman [7.9].

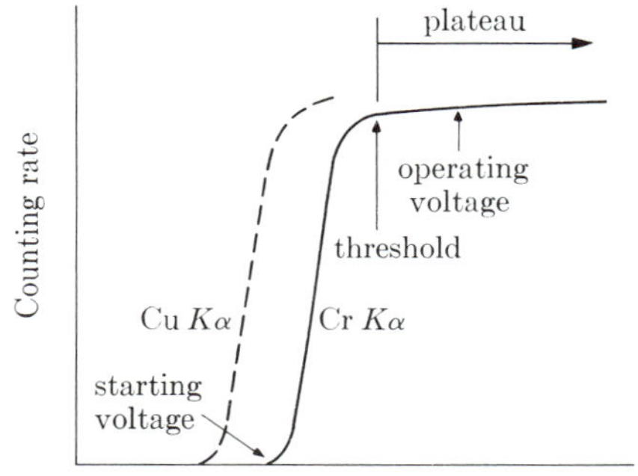

Fig. 7–16 Effect of voltage applied to proportional counter on observed counting rate at constant x-ray intensity (schematic).

counting rate with a ratemeter or scaler while slowly increasing the voltage applied to the counter from a low value. Figure 7–16 shows how the counting rate will vary with voltage. Below the starting voltage the pulse size is less than the input sensitivity of the counting circuit and no counts are observed. The pulse size and observed counting rate then increase rapidly with voltage up to the threshold of the plateau, where the counting rate is almost independent of voltage. The voltage is then fixed at about 100 volts above threshold. (Note that x-rays of longer wavelength require a higher counter voltage. This means that the counter voltage should be reset when the x-ray tube in the diffractometer is changed for one with a different target.)

The proportional counter is essentially a very fast counter and has a linear counting curve up to about 10,000 cps. This ability to separate closely spaced pulses is due to the fact that the avalanche triggered by the absorption of an x-ray quantum is confined to an extremely narrow region of the counter, 0.1 mm or less, and does not spread along the counter tube (Fig. 7–17). The rest of the counter volume is still sensitive to incoming x-rays.

The electric field near the end of the anode wire is not uniform. Most proportional counters are now made with a side window, rather than the end window shown in Fig. 7–17, so that x-ray absorption can take place in a region of uniform field.

The gas in the counter is usually xenon, argon, or krypton at a pressure somewhat less than atmospheric. Figure 7–12 shows that a krypton counter has about the same sensitivity for all the characteristic radiations normally used in diffraction. But an argon or xenon counter is much less sensitive to short wavelengths, an advantage in most cases. Thus, if a diffraction pattern is made with filtered radiation from a copper target, use of an argon counter will produce semi-monochromatic conditions, in that the counter will be highly sensitive to Cu $K\alpha$ radiation and relatively insensitive to the short wavelength radiation that forms the most intense part of the continuous spectrum. The diffraction background will therefore be lower than if a krypton counter had been used. (The

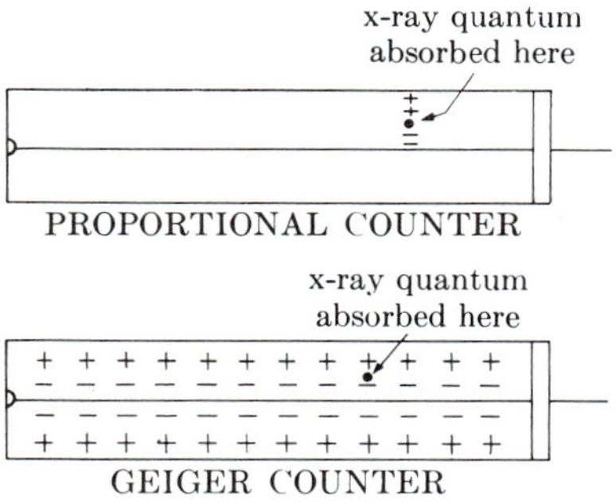

Fig. 7–17 Differences in the extent of ionization between proportional and Geiger counters. Each plus (or minus) symbol represents a large number of positive ions (or electrons).

student may wish to record the spectrum of the copper-target x-ray tube in a diffractometer. This can be done by operating it as a spectrometer, with a single crystal, such as quartz or rock salt, in the specimen holder and continuous scanning with a ratemeter and strip-chart recorder. With an argon-filled counter, the resulting spectrum will not look at all like what is expected (Fig. 1–5). Instead, only the Cu $K\alpha$ and Cu $K\beta$ lines will be visible on the chart, because the counter is insensitive to short wavelengths. The continuous spectrum can be observed only by effecting an opposite distortion: put several thicknesses of aluminum foil in the diffracted beam so as to absorb the Cu K lines more than the short wavelengths; at the same time, expand the ratemeter scale to allow for the decreased intensity of all wavelengths. With sufficiently heavy filtration by aluminum, the spectrum can be so distorted that the maximum in the continuous spectrum will be more intense, as recorded, than the characteristic lines. It is an instructive experiment.)

In x-ray spectroscopy (Chap. 15), but not in diffraction, there is a need to measure soft x-rays of wavelength about 5–20 Å. Because ordinary windows would almost totally absorb such radiation, thin sheet plastic is used as a window, so thin that it leaks. To allow for this, a stream of counter gas is continuously passed through the counter, which is then called a *gas-flow proportional counter* [G.29, 7–10].

Another special type, the *position-sensitive proportional counter* [7.11, 12, 13] may become important for certain applications. The diffracted x-ray beam enters the counter through a side window, striking the anode wire, which lies in the plane of the diffractometer circle, approximately at right angles. Because the electron avalanche is sharply localized (Fig. 7–17), the point where the electrons hit the wire can be determined by electronically measuring the time required for the pulse to travel from the point of impact to the end of the wire. Thus the angular position 2θ of a diffracted beam is found, not in the usual way by moving a counter with a narrow entrance slit to the position of the beam (Fig. 7–1), but by finding where the beam strikes the wire of a *fixed*, wide-window counter. The counting circuit must include a multichannel analyzer (Sec. 7–9) in order to determine the profile of the diffraction line. This new method is applicable only over a restricted range of 2θ values, but such a range is all that need be examined in some problems.

7-6 GEIGER COUNTERS

If the voltage on a proportional counter is increased to the neighborhood of 1500 volts, it will act as a Geiger counter. Historically, this was the first electronic counter; it is also called a Geiger–Müller or G-M counter.

The applied voltage is now so high that not only are some atoms ionized but others are raised to excited states and caused to emit ultraviolet radiation. These ultraviolet photons travel throughout the counter at high speed (light travels 10 cm in a third of a nanosecond), knocking electrons out of other gas atoms and out of the cathode shell. All the electrons so produced trigger other avalanches, and the net result is that one tremendous avalanche of electrons hits the whole length of the anode wire whenever an x-ray quantum is absorbed anywhere in the tube (Fig. 7–17). As a result the gas amplification factor A is now much larger, about 10^8 to 10^9, than in a proportional counter, and so is the size of the pulse produced, now some 1 to 10 volts. This means that no preamplifier is needed at the counter. On the debit side, all pulses have the same size, whatever the energy of the x-ray quanta.

The Geiger counter is also slow. Any one avalanche of electrons hits the anode wire in less than a microsecond, but the slowly moving positive ions require about 200 microseconds to reach the cathode. Thus the electron avalanche leaves behind it a cylindrical sheath of positive ions around the anode wire. The presence of this ion sheath reduces the electric field between it and the wire below the threshold value necessary to produce a Geiger pulse. Until this ion sheath has moved far enough away from the wire, the counter is insensitive to entering x-ray quanta. If these quanta are arriving at a very rapid rate, it follows that not every one will cause a separate pulse and the counter will become "choked." The resolving time is only about 10^{-4} sec, so that counting losses begin at a few hundred cps. Even the multichamber counter is not much better (Fig. 7–10); this counter has a number of chambers side by side, each with its own anode wire, and one chamber can therefore register a count while another one is in its insensitive period.

Because it cannot count at high rates without losses, the Geiger counter is now obsolete in diffractometry. It is still used in some radiation survey meters.

7-7 SCINTILLATION COUNTERS

This type of counter exploits the ability of x-rays to cause certain substances to fluoresce visible light, as in the fluorescent screens mentioned in Sec. 1–8. The amount of light emitted is proportional to the x-ray intensity and can be measured by means of a phototube. Since the amount of light emitted is small, a special kind of phototube called a *photomultiplier* has to be used in order to obtain a measurable current output.

The substance generally used to detect x-rays is a sodium iodide crystal activated with a small amount of thallium. It emits violet light under x-ray bombardment. (The details of this emission are roughly as follows. Absorbed x-rays ionize some atoms, i.e., raise some electrons from the valence to the conduction band of NaI. These electrons then transfer some of their energy to the Tl^+ ion. When the excited ion returns to its ground state, light is emitted.) The light-emitting crystal is cemented to the face of a photomultiplier tube, as indicated in Fig. 7–18, and shielded from external light by means of aluminum foil. A flash of light (scintillation) is produced in the crystal for every x-ray quantum absorbed, and this light passes into the photomultiplier tube and ejects a number of electrons from the photocathode, which is a photosensitive material generally made of a cesium-antimony intermetallic compound. (For simplicity, only one of these electrons is shown in Fig. 7–18.) The emitted electrons are then drawn to the first of several metal *dynodes*, each maintained at a potential about 100 volts more positive than the preceding one, the last one being connected to the measuring circuit. On reaching the first dynode, each electron from the photocathode knocks two electrons, say, out of the metal surface, as indicated in the drawing. These are drawn to the second dynode where each knocks out two more electrons and so on. Actually, the gain at each dynode may be 4 or 5 and there are usually at least 10 dynodes. If the gain per dynode is 5 and there are 10 dynodes, then the multiplication factor is $5^{10} = 10^7$. Thus the absorption of one x-ray quantum in the crystal results in the collection of a very large number of electrons at the final dynode, producing a pulse about as large as a Geiger pulse, i.e., of the order of volts. Furthermore, the whole process requires less than a microsecond, so that a scintillation counter can operate at rates as high as 10^5 counts per second without losses. The correct counter voltage is found by the method used for the proportional counter, by plotting counting rate vs. voltage (Fig. 7–16).

As in the proportional counter, the pulses produced in a scintillation counter have sizes proportional to the energy of the x-ray quanta absorbed. But the pulse size corresponding to a certain quantum energy is much less sharply defined, as shown in Fig. 7–19 for typical proportional and scintillation (NaI · Tl) counters. As a result, it is more difficult to discriminate, with a scintillation counter, between x-ray quanta of different wavelengths (energies) on the basis of pulse size.

The efficiency of a scintillation counter approaches 100 percent over the usual range of wavelengths (Fig. 7–12), because virtually all incident quanta are absorbed, even in a relatively thin crystal.

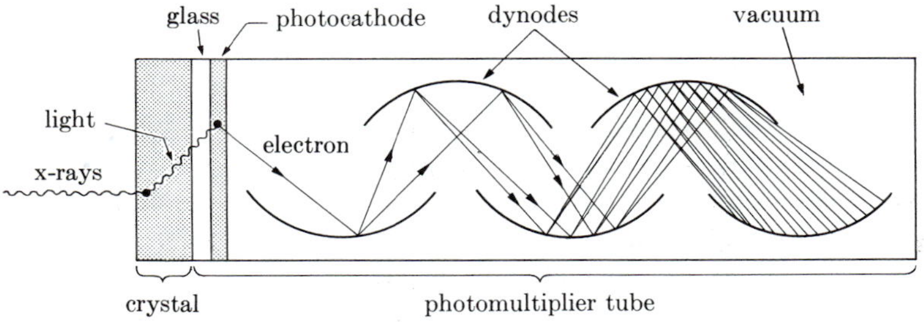

Fig. 7–18 Scintillation counter (schematic). Electrical connections not shown.

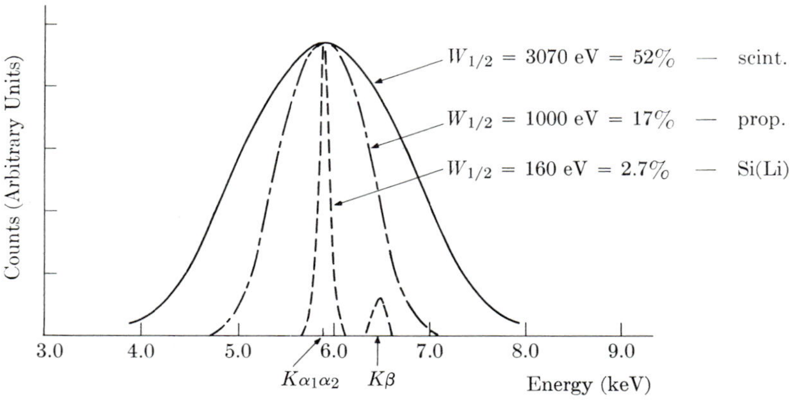

Fig. 7–19 Pulse-height distribution curves for three kinds of counter. Incident radiation is Mn $K\alpha$ (λ = 2.10 Å, $h\nu$ = 5.90 keV) and Mn $K\beta$ (λ = 1.91 Å, $h\nu$ = 6.50 keV). Frankel and Aitken [7.14].

7–8 SEMICONDUCTOR COUNTERS

Developed in the 1960s, semiconductors are the newest form of counter. They produce pulses proportional to the absorbed x-ray energy with better energy resolution than any other counter; this characteristic has made them of great importance in spectroscopy (Chap. 15). Although they have had little application in diffraction, it is convenient to describe them here along with the other counters.

Both silicon and germanium are used, germanium as a detector for gamma rays, because it is heavier and therefore a better absorber, and silicon for x-rays. Both contain a small amount of lithium, and they are designated Si(Li) and Ge(Li), inevitably referred to as "silly" and "jelly." Their properties have been reviewed by Dearnaley and Northrop [7.15], Heath [7.16], and Gedcke [7.17], among others, and in an ASTM symposium [7.18].

Pure silicon is an intrinsic semiconductor. It has very high electrical resistivity, especially at low temperatures, because few electrons are thermally excited across the energy gap into the conduction band. However, incident x-rays can cause excitation and thereby create a free electron in the conduction band and a free hole in the valence band. As shown later, the absorption of one x-ray quantum creates about a thousand electron-hole pairs. If a high voltage is maintained across opposite faces of the silicon crystal, the electrons and holes will be swept to these faces, creating a small pulse in the external circuit.

It is essential that the silicon be intrinsic (i). It must neither be n-type, containing free electrons from donor impurities, nor p-type, containing free holes from acceptor impurities; in either type, the free charge carriers, at their usual concentrations, would overwhelm the few carriers produced by x-rays. Production of a reasonably large intrinsic crystal, which is not easy, requires two operations:

1. The starting material is a cylindrical crystal, some 3–5 mm thick and 5–15 mm in diameter. It is p-type, having been lightly doped with boron. Lithium is applied to one face and diffused into the crystal at an elevated temperature, producing a gradient of lithium concentration from high to low through the thickness. The lithium exists as Li^+ ions, and the free electrons it provides convert the crystal into n-type on one side, where the lithium concentration is high, leaving the other side p-type.

2. A voltage is then applied, also at an elevated temperature, to opposite faces, positive on the n side and negative on the p side (called "reverse bias"). This causes the Li^+ ions to "drift" toward the p side, resulting in a wide central region of constant lithium concentration; this region is now intrinsic because it has equal lithium and boron concentrations.

The result is the lithium-drifted silicon counter sketched in Fig. 7–20. The crystal is virtually all intrinsic, with the p and n portions confined to thin surface layers, which are exaggerated in the drawing. The very small pulses from the counter are amplified to the millivolt level by a field-effect transistor, abbreviated FET. (There is no charge amplification, such as occurs in a gas counter. The pulse from the counter contains only the charge liberated by the absorbed x-rays.)

Putting aside all of the above details of semiconductor physics, we can regard a Si(Li) counter simply as a solid-state ionization chamber, with one difference. X-rays incident on a gas ionization chamber produce a *constant* current (Sec. 7–5). In a Si(Li) counter the current flows in discrete *pulses*, because the voltage is high enough to sweep the counter free of charge carriers (the electrons and holes are highly mobile) before the next incident photon creates new carriers.

A major disadvantage of the Si(Li) counter is that it must be operated at the temperature of liquid nitrogen (77°K = −196°C) in order to minimize (1) a constant current through the detector, even in the absence of x-rays, due to thermal excitation of electrons in the intrinsic region, and (2) thermal diffusion of lithium, which would destroy the even distribution attained by drifting. Even when not

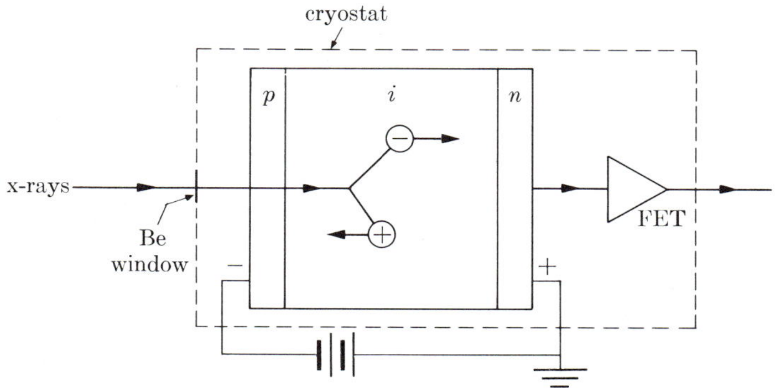

Fig. 7–20 Si(Li) counter and FET preamplifier, very schematic. Both are in a cooled evacuated space, and x-rays enter through a beryllium window. The counter is operated at about 1000 volts. ⊖ = electron, ⊕ = hole.

operating, the counter should not be allowed to warm too often to room temperature. Electronic "noise" in the FET increases with temperature and degrades the resolution. Thus both the counter and the FET have to be cooled, necessitating a bulky cryostat to hold several liters of liquid nitrogen (Fig. 15–10).

The efficiency of a Si(Li) counter resembles that of the other solid-state counter (scintillation), very high for intermediate wavelengths (Fig. 7–12). Very long wavelengths are partially absorbed by the counter window before they can reach the sensitive intrinsic layer. Very short wavelengths are partially transmitted by the entire counter.

The counting rate varies linearly with x-ray intensity up to rates of about 5,000–10,000 cps. Counting losses in the counter-electronics system occur in the electronics rather than the counter. The electronics are more complex than usual and include, besides the usual pulse amplifiers and shapers, a multichannel pulse-height analyzer (Sec. 7–9).

The excellent energy resolution of a Si(Li) counter is shown in Fig. 7–19. The width W of the pulse distribution is so small that the Si(Li) counter can resolve the $K\alpha$ and $K\beta$ lines of manganese, which the other two counters cannot do. Put another way, the resolution $R = W/V$ of the Si(Li) counter is 2.7 percent or some six times better than that of the proportional counter. For any kind of counter, both W and W/V vary with V, i.e., with the energy $h\nu$ of the incident x-rays. Therefore any description of counter performance must specify the x-ray energy at which it is measured; the 5.90 keV energy of the Mn $K\alpha$ line is the usual standard reference. The width W, incidentally, is often written as FWHM (full width at half maximum) in the literature of this subject.

To create an electron-hole pair in silicon at 77°K requires an average energy of 3.8 eV. The absorption of a Mn $K\alpha$ quantum should therefore create $5900/3.8 = 1550$ pairs. However, the actual number created by successive quanta might be 1540, 1560, 1555, . . . , leading to a corresponding variation in the size of the output pulse. This statistical variation in the number of charge carriers created by x-ray absorption is the basic reason for the finite width W of the pulse distribution, and the same is true of proportional and scintillation counters. In the Si(Li)–FET counter, there is an even larger contribution to W, namely, electronic noise in the FET preamplifier. At the energy of Mn $K\alpha$ more than half of the observed value of W is due to noise in the FET. Beyond a certain counting rate, the resolution of the system worsens (W becomes larger) as the count rate increases.

As stated earlier, silicon counters are usually preferred for x-rays and germanium counters for the more energetic, shorter wavelength gamma rays of interest to nuclear physicists. This situation may change. Germanium counters are usable over the energy range of about 3–100 keV (4–0.1 Å), and *intrinsic* germanium crystals are now being made that do not require lithium drifting. They may find application in x-ray spectroscopy for wavelengths less than 4 Å.

In any counter except the Geiger, the average number n of ion pairs or electron-hole pairs produced is proportional to the energy E of the absorbed quantum. The actual number has a Gaussian (normal) distribution about the mean, and the width at half maximum of this distribution is proportional to the standard deviation σ, which is equal to $\sqrt{n}$. Therefore the resolution R is

$$R = \frac{W}{V} = \frac{k_1\sqrt{n}}{k_2 n} = \frac{k_3}{\sqrt{n}}, \tag{7–3}$$

where the ks are constants. The superior resolution of a Si(Li) counter is simply due to the large value of n, which is 1550 for Mn $K\alpha$. By comparison, n is only $5600/26 = 230$ for an argon proportional counter and the same radiation, because 26 eV are needed to create an ion pair in argon. Actually, the inherent resolution of a Si(Li) counter (the resolution in the absence of preamplifier noise) is even better, for complex reasons, than the above statistical argument suggests.

Because n is proportional to E, we can find from Eq. (7–3) the energy or wavelength dependence of the resolution:

$$R = \frac{k_3}{\sqrt{n}} = \frac{k_4}{\sqrt{E}} = \frac{k_4}{\sqrt{h\nu}} = k_5\sqrt{\lambda}. \tag{7–4}$$

Although Eqs. (7–3) and (7–4) are useful for rough qualitative arguments, they do not include the substantial effect of electronic noise in Si(Li)–FET counters. A better estimate of resolution in such counters is given by

$$R = \frac{W}{V} = \frac{[(100)^2 + 2.62E]^{1/2}}{E}, \tag{7–5}$$

where E is the x-ray energy in eV and the term 100 eV in the numerator is the present level of the electronic noise. This relation is important in spectroscopy.

7-9 PULSE-HEIGHT ANALYSIS

All the counters in use today (proportional, scintillation, and semiconductor) are "proportional" in the sense that they produce pulses having a size (amplitude) that is proportional to the energy of the incident x-rays. Electrical circuits that can distinguish between pulses of different size can therefore distinguish between x-rays of different energies (wavelengths), and this ability is of great value in many experimental techniques. These circuits, in order of increasing complexity, are:

1. Pulse-height discriminator.
2. Single-channel pulse-height analyzer.
3. Multichannel pulse-height analyzer.

Circuits (1) and (2) may be used with ordinary diffractometers to increase the peak/background ratio of diffraction lines. They are by no means necessary; quite adequate diffraction patterns can be obtained from a wide variety of specimens with no other "discriminator" than a $K\beta$ filter. Circuit (3) is required only in x-ray spectroscopy (Chap. 15), in a very special kind of diffractometry (Sec. 7-10), and with a position-sensitive proportional counter. Any one of these circuits is more effective, the better the resolution of the counter with which it operates.

Pulse-Height Discriminator

Suppose that x-rays of three wavelengths $a > b > c$ are incident on a counter. Then the counter will produce pulses A,B,C of different size, C being the largest (Fig. 7–21). If a circuit is inserted, just ahead of the scaler or ratemeter, that will allow only pulses larger than a certain selected size (V_1 volts) to pass and discriminate against smaller ones, then the A pulses will not be counted and the system will be blind to wavelength a. Such a circuit is called a pulse-height discriminator. It may be of some value in decreasing diffractometer background due to the specimen's fluorescent radiation, when that radiation has a wavelength much longer than that which forms the diffraction lines, but it cannot weaken the short-wavelength components of the continuous spectrum.

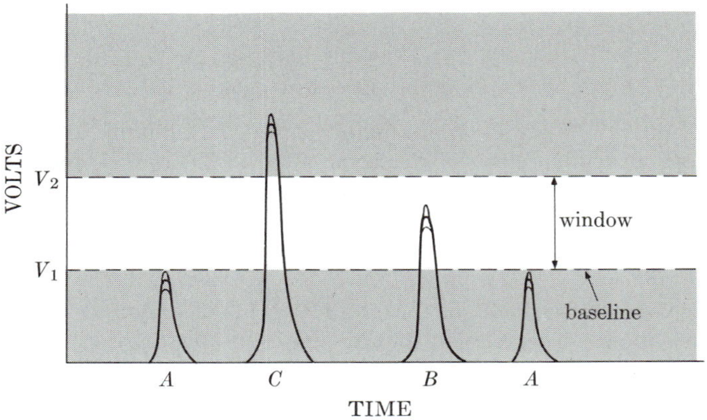

Fig. 7–21 Pulse-height discrimination and analysis. The statistical spread in pulse size, measured by the width W of Fig. 7–13, is suggested above by the variable heights of each pulse.

Single-Channel Pulse-Height Analyzer

This instrument discriminates against any pulses smaller than V_1 volts (Fig. 7–21). In addition, it contains an *anti-coincidence circuit* that rejects any pulse larger than V_2, because such a pulse will simultaneously trigger both the V_1 and V_2 levels. The net result is that only pulses having sizes between V_1 and V_2 volts are passed. Out of the mixture of A, B, and C pulses entering the analyzer, only B pulses are passed to the counting circuit. The level V_1 is called the *baseline*, and the range from V_1 to V_2 is the *window* or *channel*. (Here, then, is another sense of the word "window." It is not a physical window, as at the entrance to a counter, but a voltage window.)

Both V_1 and V_2 are adjustable. If the window ($V_2 - V_1$) is made quite small, say 0.5 volt, and the baseline V_1 continuously varied from low values to high values, with ($V_2 - V_1$) constant, then a narrow window can be traversed across the voltage range of pulse heights; if the counting rate is measured at each setting of V_1, the result is an analysis of the distribution of pulse heights. It is in this way that curves like that of Fig. 7–13 are measured.

An analyzer can markedly reduce the background of a diffraction pattern, chiefly by excluding short-wavelength white radiation. For example, examining the 1 1 1 line from silicon powder with copper radiation and a xenon-filled proportional counter, Parrish [7.8] found the peak/background ratio to be 57 without an analyzer and 146 with one. To achieve this almost three-fold improvement, the analyzer window was centered on the center V (Fig. 7–13) of the Cu $K\alpha$ distribution and made wide enough to accept about 90 percent of it.

This method works best when the wavelength to be passed and the wavelength to be rejected are far apart. If they are close together, or if the counter has poor resolution, it will be hard to pass one and reject the other. There will then be two pulse distributions like that of Fig. 7–13, side by side and partially overlapping. A window set to pass a reasonable proportion of one set of pulses will also pass some of the other set.

Problems of window settings can arise even when the incident x-rays are monochromatic, because pulses of *two* different sizes can be produced. This can occur when the incident radiation is energetic enough to cause x-ray fluorescence in the counter. The fluorescent radiation may escape from the counter without causing any ionization, carrying with it some of the energy that would normally be absorbed. An *escape peak* of smaller-than-normal pulses is then formed, corresponding to this less-than-normal energy absorption.

As a specific example, consider the absorption of Mo $K\alpha$ radiation ($\lambda = 0.71$ Å, energy $E_1 = 17.4$ keV) in a krypton-filled proportional counter. The K absorption edge of krypton is at 0.87 Å ($W_K = 14.3$ keV), so that Mo $K\alpha$ radiation can cause the emission of Kr $K\alpha$ radiation ($\lambda = 0.98$ Å, energy $E_2 = 12.6$ keV). Some of the Mo $K\alpha$ quanta are absorbed with a total loss of energy, causing normal pulses of average size $V_n = kE_1 = 17.4\ k$ volts, where k is a constant. Absorption of other Mo $K\alpha$ quanta will involve an energy loss in the counter of only $(E_1 - E_2)$, because of the escaping Kr $K\alpha$, causing escape-peak pulses of average size $V_{ep} = k(E_1 - E_2) = k(17.4 - 12.6) = 4.8\ k$ volts. These two processes are illustrated in Fig. 7–22(a), where k is evidently about 1.9. (The term *escape peak* may be misleading to some. We see from this example that the size of escape-peak pulses corresponds *not* to the escaping energy, but to the difference between the normal and escaping energies.) Shown in (b) are the normal and escape peaks for Cu $K\alpha$ radiation incident on a xenon counter; here the fluorescent radiation is the $L\alpha_1$ line.

The situation shown in Fig. 7–22(a) is unusual, in that there are more pulses in the escape peak than in the normal peak. If the window of a pulse-height analyzer is set to pass only the normal Mo $K\alpha$ pulses, then the observed counting rate (=observed x-ray intensity) would be less than half the value observed with no analyzer at all because without an analyzer all pulses would be counted. The number of pulses in the escape peak will be larger, relative to that in the normal peak, the greater the fluorescence yield (Sec. 1–5) of the counter material and the lower the absorption coefficient of the counter material for its own fluorescent radiation.

Escape pulses can occur in any "proportional" counter. In a NaI(Tl) scintillation

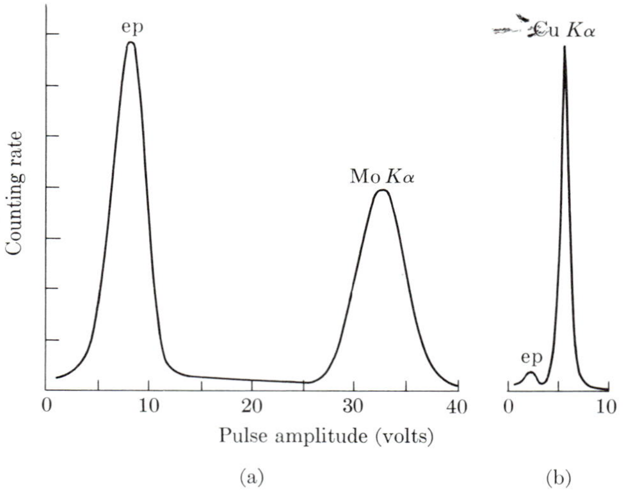

Fig. 7–22 Pulse-height distribution curves showing escape peaks (ep) in proportional counters for (a) Mo $K\alpha$ radiation incident on a krypton counter and (b) Cu $K\alpha$ radiation incident on a xenon counter. Parrish [7.8].

counter they are caused by fluorescent iodine K radiation; in a Si(Li) counter by fluorescent silicon K radiation.

Escape peaks can be troublesome in x-ray spectroscopy. When several wavelengths are incident on the counter, the escape peak for wavelength λ_1 may fall on or near the normal peak for wavelength λ_2, causing uncertainty in the identification of λ_2.

Multichannel Pulse-Height Analyzer

This remarkable instrument, called MCA for short, usually has not one but upward of a thousand channels. It is designed to separate pulses from a counter that is receiving incident radiation of many wavelengths, by sorting pulses according to their size (amplitude).

An MCA is actually a special-purpose computer with three functions [7.18]:

1. *Digitizing.* An analog-to-digital converter (ADC) converts the analog information contained in each pulse (amplitude in volts) into digital form, suitable for storage in a memory.

2. *Sorting and storage.* These operations are performed in the memory of the MCA. If the x-ray energy range to be examined extends from, say, 0 to 20 keV and the MCA has 1000 channels, then each channel spans an energy range of 20 eV. Channel number 295 would receive from the ADC information about any pulse activity in the range 5880–5900 eV. This channel would therefore get information (number of counts) about a vertical slice near the center of the Mn $K\alpha$ pulse distribution shown in Fig. 7–19; for the Si(Li) counter shown there, the base of the pulse distribution appears to be about 300 eV wide; information about Mn $K\alpha$ pulses would therefore be spread over 15 channels of the MCA.

3. *Display*. The contents of the MCA memory (total counts in each channel) may be displayed visually as counts vs. channel number (= counts vs. x-ray energy) on a TV screen or by an X–Y plotter, or they may be recorded on printed tape, punched tape, or magnetic tape.

The operation of multichannel and single-channel instruments in performing pulse-height analysis may be contrasted as follows. In a single-channel instrument the entire energy range is scanned serially in time by one moving channel, as suggested by Fig. 7–13. In the MCA, a large number of fixed channels covers the energy range and all channels simultaneously receive the count-rate information appropriate to each channel.

Applications of the MCA to specific problems are given in Chap. 15 and in the next section.

7–10 SPECIAL KINDS OF DIFFRACTOMETRY

The ability of the MCA to sort out signals of various magnitudes has permitted the development of new kinds of diffractometers, radically different from the conventional, moving-counter instrument to which this chapter is mainly devoted.

Energy-Dispersive Diffractometry

In the standard diffractometer, atomic planes of various spacings d in a powder specimen reflect a single wavelength in various directions 2θ; the diffraction pattern is observed by moving a counter to the position of each diffracted beam in turn. On the other hand, if the incident beam consists of white radiation and the angle θ is fixed for all planes, the different sets of planes will reflect a set of different wavelengths into a counter set at a fixed position 2θ. If the counter is "proportional" and connected to a multichannel analyzer, the reflected wavelengths can be sorted out on the basis of their energies. This procedure, called *energy-dispersive diffractometry*, was first developed by Giessen and Gordon [7.19, 7.20] and Buras *et al.* [7.21] independently.

It is now appropriate to write the Bragg law in terms of quantum energy E rather than wavelength λ:

$$E = h\nu = \frac{hc}{\lambda} = \frac{hc}{2d \sin \theta}. \tag{7–6}$$

For E in keV and d in Å, this becomes (see Eq. 1–4)

$$E = \frac{6.20}{d \sin \theta}. \tag{7–7}$$

At constant θ, measurement of E will give the spacing d of each set of planes, and their indices (hkl) are found by the methods of Chap. 10. The experimental arrangement and an example of the results obtained, from a plot of the contents of the various channels of the MCA, are shown in Fig. 7–23. The specimen was a sheet of polycrystalline platinum.

The fluorescent L lines of platinum appear at the left of Fig. 7–23(b). The energies of these lines are independent of 2θ. The diffraction lines, on the other hand, have energies that depend on 2θ according to Eq. (7–7). Therefore the whole diffraction pattern can be shifted to higher or lower energies by changing 2θ, if it is necessary to prevent an overlap of fluorescent and diffraction lines. (If the nature of the specimen is entirely unknown, the presence of the fluorescence lines can be advantageous. The fluorescence lines disclose the chemical elements present in the specimen, while the diffraction lines disclose its crystal structure. See Chaps. 14 and 15.)

Compared to the conventional method (single wavelength, moving counter), energy-dispersive diffractometry is much faster, because the diffraction pattern is acquired simultaneously rather than serially. Typically, the entire pattern can be recorded in 1 to 5 minutes, whereas the conventional technique requires over an hour. However, the resolution of closely spaced diffraction lines is inferior to that of the conventional technique. Also on the debit side are the added cost of an MCA and the inconvenience of cooling the Si (Li) counter.

Some specimens, notably steels, give diffraction patterns composed of rather widely separated lines. For such materials high resolution is not needed. Voskamp [7.22] has

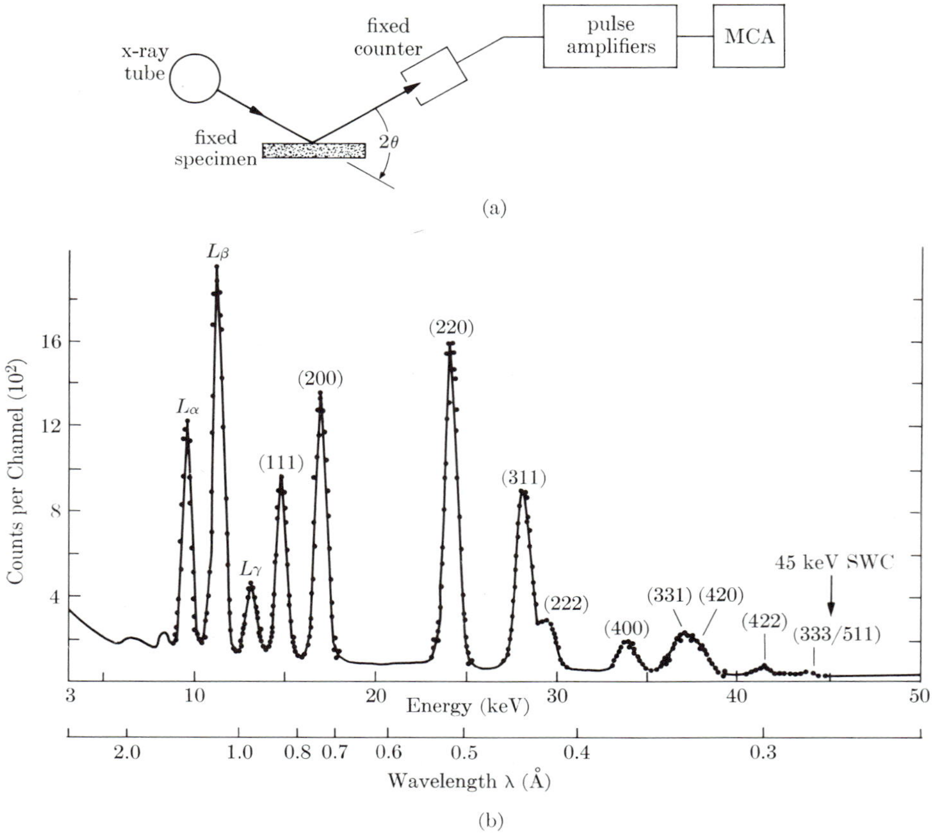

Fig. 7–23 Energy-dispersive diffractometry. (a) Experimental arrangement. The x-ray tube is seen end on. Diffracted-beam collimator not shown. (b) Diffraction pattern of polycrystalline platinum at $2\theta = 21.4°$ obtained with a Si(Li) counter and an iron-target x-ray tube operated at 45 kV and 8 mA. SWC = short wave cutoff = short-wavelength limit of incident beam. Giessen and Gordon [7.20].

described the use of energy-dispersive diffractometry for the examination of steels and gives references to other applications.

Energy-dispersive diffractometry has not yet been widely used. Prophecy is always risky, but it may turn out that this method will be most useful in process control, such as chemical analysis (Chap. 14), because it is fast, involves no counter movement, and is easily adapted to automation.

Time-Analysis Diffractometry

This is a very recent development. It involves a side-window position-sensitive proportional counter (Sec. 7–5), a multichannel analyzer, and the measurement of the angular positions of many diffraction lines simultaneously. The anode wire of the counter, which is long and *curved*, coincides with a segment of the diffractometer circle and is connected, through appropriate circuits, to an MCA. The powder specimen is in the form of a thin rod centered on the diffractometer axis. The geometry of the apparatus therefore resembles that of a Debye-Scherrer camera (Fig. 6–2), except that the curved film strip is replaced by a curved counter.

When "monochromatic" radiation is incident on the specimen, it sends out diffracted beams at particular 2θ angles. These beams enter the side window of the counter at particular points, causing pulse formation at those points. The times required for these pulses to travel to the end of the counter are converted to digital form, analyzed, and sorted by the MCA. The contents of the MCA memory are therefore number of pulses (counts) as a function of the position on the anode wire where the pulses originated (angle 2θ). A display of the contents of the MCA resembles the pattern recorded by a conventional diffractometer (Fig. 7–5).

This method is fast, because all diffracted beams incident on the counter are measured simultaneously. However, the arc of the present counter subtends a 2θ angle of only 60°. Although the counter is fixed in position for any one measurement, it may be rotated around the diffractometer axis to cover any desired 2θ range, for example, 20° to 80° or 90° to 150°. If the whole pattern is to be examined from near 0° to near 180°, the measurement is made in three overlapping steps. The present instrument has 1000 channels, so that each channel represents 0.06° 2θ.

This technique is still too new to have an agreed-on name, but perhaps "time-analysis diffractometry" is appropriate. The instrument itself has been aptly called an *electronic powder camera* by the manufacturer [7.23].

7–11 SCALERS

A scaler is an electronic device which counts each pulse produced by the counter. Once the number of pulses over a measured period of time is known, the average counting rate is obtained by simple division.

A scaler consists of a number of identical *stages* connected in series. Each stage is a circuit that divides, or scales down, the entering pulses by a constant factor before transmitting them to the next stage. This factor is 10 in a *decade scaler* and 2 in a *binary scaler*.

The first stage of a decade scaler transmits one pulse to the second stage for every ten pulses that enter the first stage. In a five-stage scaler the fifth stage will have transmitted one pulse when 10^5 pulses have entered the scaler, and the *scaling factor* is said to be 10^5. If, say, 12,327 pulses have entered the scaler when it is turned off, that count will be displayed as a decimal number on the front of the scaler, usually by ten-position glow tubes.

A binary scaler operates in a similar way, but now the scaling factor is 2^n, where n is the number of stages. Some binary scalers do not appear to be binary to the user, because the contents of the individual stages are not displayed; only the total count, in decimal form, appears at the front of the scaler.

A scaler can be operated in two ways to obtain the average counting rate:

1. *Counting for a fixed time.* The desired time t is selected by a switch, the START COUNT button is pushed, and the timer automatically stops the scaler after t seconds. The average counting rate is then N/t, where N is the displayed number of pulses (counts).

2. *Counting a fixed number of pulses.* The desired number of counts N is selected by a switch. If N is to be, say, 10,000 counts, the switch will connect the timer to the output of the fourth stage of a decade scaler. When 10,000 pulses have entered the scaler, the fourth stage will transmit its first pulse and that pulse will stop the timer.

Because the arrival of x-ray quanta in the counter is random in time, the accuracy of a counting rate measurement is governed by the laws of probability. Two counts of the same x-ray beam for identical periods of time will not be precisely the same because of the random spacing between pulses, even though the counter and scaler are functioning perfectly. Clearly, the accuracy of a rate measurement of this kind improves as the time of counting is prolonged, and it is therefore important to know how long to count in order to attain a specified degree of accuracy. The number of pulses N counted for a fixed time in repeated measurements of a constant x-ray intensity will have a Gaussian (normal) distribution about the true value N_t obtained by averaging many measurements, with a standard deviation σ of $\sqrt{N}$. The relative standard deviation to be expected in a single count of N pulses is then

$$\text{Relative } \sigma = \frac{\sqrt{N}}{N}(100) = \frac{100}{\sqrt{N}} \text{ percent.} \tag{7-8}$$

The probable error* in a single count is

$$\text{Probable error} = 0.67 \text{ (relative } \sigma) = \frac{67}{\sqrt{N}} \text{ percent.} \tag{7-9}$$

---

* The probable error is that which is just as likely to be exceeded as not. Three times the probable error is a somewhat more useful figure, as the probability that this will be exceeded is only 0.04. Thus, if a single measurement gives 1000 counts, then the probable error is $67/\sqrt{1000} = 2.1$ percent or 21 counts. Then the probability is 0.5 that this count lies in the range $N_t \pm 21$, where N_t is the true number of counts, while the probability is 0.96 that the measured value lies in the range $N_t \pm 63$.

For some of the total counts obtainable from a decade scaler, these expressions give the following errors:

Number of pulses counted	Relative standard deviation (percent)	Probable error (percent)
1,000	3.2	2.1
10,000	1.0	0.7
100,000	0.3	0.2

Note that the error depends only on the number of pulses counted and not on their rate, which means that high rates and low rates can be measured with the same accuracy, if the counting times are chosen to produce the same total number of counts in each measurement. It also follows that the second scaling method outlined above, in which the time is measured for a fixed number of counts, is generally preferable to the first, since it permits intensity measurements of the same accuracy of both high- and low-intensity beams.

The probable error in the measured intensity of a diffraction line *above* background increases as the background intensity increases. If N_P and N_B are the numbers of counts obtained in the same time at the peak of the diffraction line and in the background adjacent to the line, respectively, then we are more interested in the error in $(N_P - N_B)$ than in the error in N_P. When two quantities are combined, the result has a variance equal to the sum of the variances of the quantities involved. (Variance $= \sigma^2$, where $\sigma =$ standard deviation.) In this case,

$$\sigma_{P-B}^2 = \sigma_P^2 + \sigma_B^2 = N_P + N_B;$$

$$\sigma_{P-B} = (N_P + N_B)^{1/2}; \tag{7-10}$$

$$\text{Relative} \quad \sigma_{P-B} = \frac{(N_P + N_B)^{1/2}}{(N_P - N_B)}.$$

As indicated in Sec. 7-2, the integrated intensity of a diffraction line may be measured with a scaler by determining the average counting rate at several angular positions of the counter. The line profile, the curve of intensity vs. 2θ, is then plotted on graph paper, and the area under the curve, and above the continuous background, is measured with a planimeter. To obtain the same relative accuracy of both the line profile and the adjacent background, all measurements should be made by counting a fixed number of pulses.

A faster and more accurate method of measuring integrated intensity exploits the integrating property of the scaler. In Fig. 7-24 the shaded area P is the integrated intensity of the diffraction line shown. Select two counter positions $2\theta_1$ and $2\theta_2$, well into the background on either side of the line. Scan from $2\theta_1$ to $2\theta_2$, the scaler being started at the beginning of the scan and stopped at the end. Let the time required for this scan be t and the number of counts accumulated be N_{PB}; this number is proportional to the sum of the areas P and B. Then count for a time $t/2$ at $2\theta_1$ and for a time $t/2$ at $2\theta_2$. Let the total count accumulated

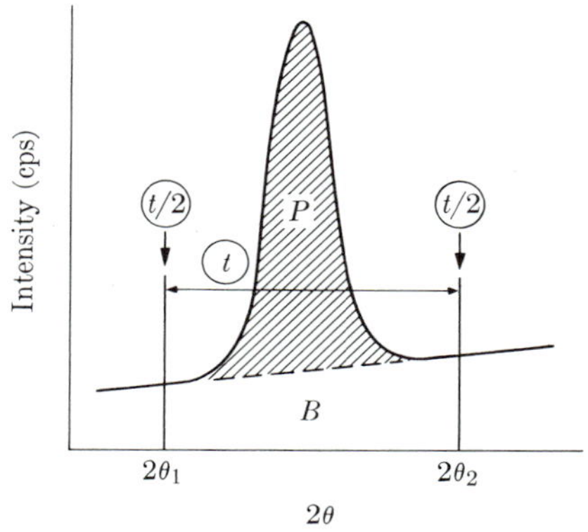

Fig. 7–24 Determination of integrated intensity by scanning and measuring counts with a scaler. The encircled symbols are the times spent in counting.

in these two fixed-position measurements be N_B. This number is proportional to the area B, because it is the count that would have been obtained in a scan for time t from $2\theta_1$ to $2\theta_2$ if the diffraction peak were absent. The integrated intensity of the peak is then

$$N_P = N_{PB} - N_B \tag{7–11}$$

When the integrated intensities of two diffraction lines are to be compared, both lines must be scanned with the same counter slit (receiving slit) at the same speed.

Because this method involves counting for a fixed time, the background and low-intensity portions of the diffraction line are measured with less accuracy than the high-intensity portions. The counting time should be chosen so that the low intensities are measured to the accuracy required by the particular problem involved; it will then follow that the high intensities are measured with unnecessarily high accuracy, but that is unavoidable in a fixed-time method.

7–12 RATEMETERS

The counting-rate meter, as its name implies, is a device which indicates the average counting rate directly without requiring, as in the scaler-timer combination, separate measurements of the number of counts and the time. It does this by a circuit which, in effect, smooths out the succession of randomly spaced pulses from the counter into a steady current, whose magnitude is proportional to the average rate of pulse production in the counter.

The heart of a ratemeter circuit is a series arrangement of a capacitor and resistor. To understand the action of a ratemeter, we must review some of the properties of such a circuit, notably the way in which the current and voltage vary with time. Consider the circuit shown in Fig. 7–25(a), in which the switch S can be used either to connect a to c and thus apply a voltage to the capacitor, or to connect b to c and thus short-circuit the capacitor and resistor. When a is suddenly connected to c, the voltage across the capacitor reaches its final value V not

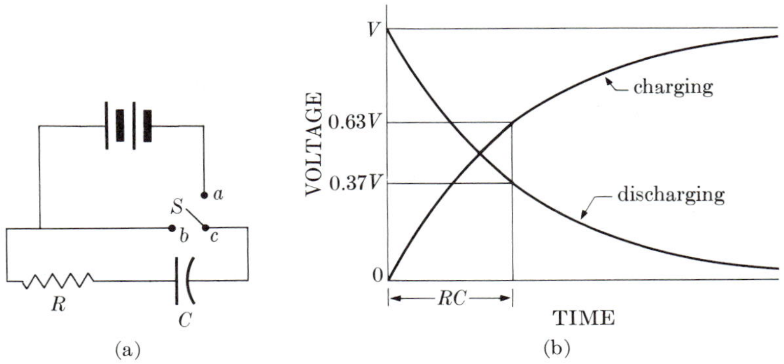

Fig. 7–25 The capacitor-resistor circuit.

instantaneously but only over a period of time, and at a rate which depends on the resistance R and the capacitance C, as shown in Fig. 7–25(b). The product of R and C has the dimensions of time (seconds, in fact, if R is in megohms and C in microfarads), and it may be shown that the voltage across the capacitor reaches 63 percent of its final value in a time given by RC, known as the *time constant* of the circuit. The time required to reach 99 percent of its final value is $4.6RC$. Conversely, if the fully charged capacitor, bearing a charge $Q = CV$, is suddenly shorted through the resistor by connecting b to c, the charge does not immediately disappear but leaks away at a rate dependent on the time constant. The charge drops to 37 percent of its initial value in a time equal to RC and to 1 percent in a time equal to $4.6RC$.

A complete ratemeter circuit consists of two parts. The first is a pulse-amplifying and pulse-shaping portion which electronically converts the counter pulses, which vary in amplitude and shape from counter to counter, into rectangular pulses of fixed dimensions in voltage and time. These pulses are then fed into the second portion, which is the measuring circuit shown in Fig. 7–26, a circuit basically similar to that of Fig. 7–25(a) and having a time constant R_2C_2. S, shown as a simple switch, is actually an electronic circuit which connects a to c each time a pulse arrives and then connects b to c immediately afterwards. A constant charge

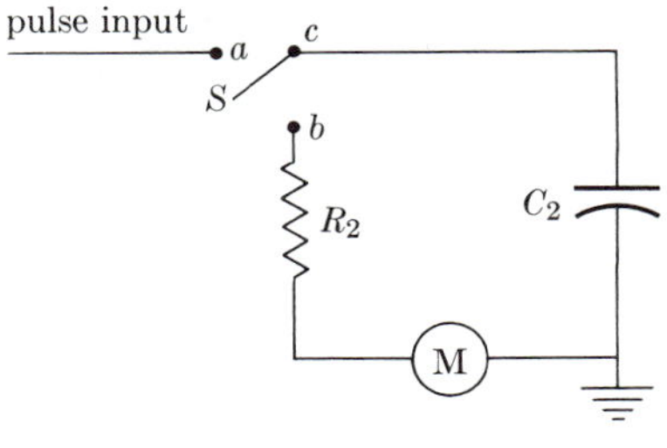

Fig. 7–26 Measuring portion of ratemeter circuit.

is thus added to the capacitor for each pulse received and this charge leaks away through the resistor until, at equilibrium, the rate of addition of charge is just balanced by the rate of leakage. The rate of charge leakage is simply the current through the microammeter M, which therefore indicates the rate of pulse production in the counter and, in turn, the x-ray intensity. The circuit usually contains, in addition to the indicating meter, a chart recorder which produces a continuous record of the intensity.

Even when the x-ray intensity is constant (constant average counting rate), the spacing of the counter pulses is random in time, which means that the counting rate actually varies with time over short periods. The ratemeter responds to these statistical fluctuations in the counting rate, and its response speed is greater the smaller the time constant. This follows from the discussion of the capacitor-resistor circuit: any change in the pulse rate causes a change in the current through the circuit, but the latter change always lags behind the former; the amount of lag is less for a small time constant than for a large one. Random fluctuations in the counting rate are therefore more evident with a small time constant, because the current in the circuit then follows the changes in counting rate more closely. This feature is illustrated in Fig. 7–27, which shows the automatically recorded output of a ratemeter when the counter is receiving a constant-intensity x-ray beam. The large fluctuations at the left have been reduced in magnitude by successive increases in the time constant, effected by changing the value of C_2. Evidently, a *single* reading of the position of the indicating meter needle or the recorder pen of a ratemeter may be seriously in error, and more so at low time constants than at high. In Sec. 7–11 we saw that the error in a counting-rate measurement decreased as the number of counts increased. Now it may be shown that a ratemeter acts as if it counted for a time $2R_2C_2$, in the sense that the accuracy of any single reading is equivalent to a count made with a scaler for a time $2R_2C_2$. Therefore, the relative probable error in any single ratemeter reading is given by the counterpart of Eq. (7–9), namely by

$$\text{Probable error} = \frac{67}{\sqrt{2nR_2C_2}} \text{ percent,} \tag{7–12}$$

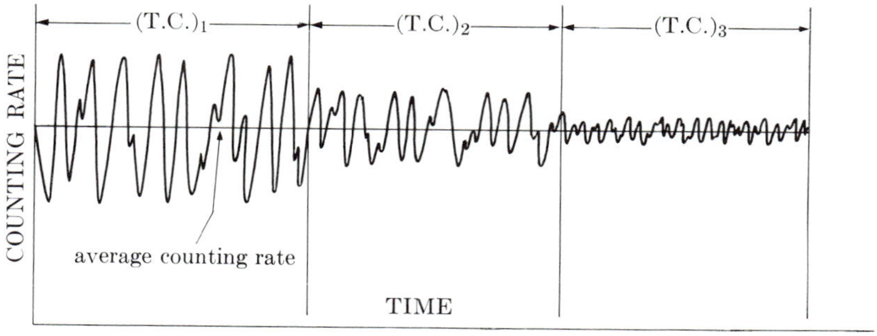

Fig. 7–27 Effect of time constant (T.C.) on recorded fluctuations in counting rate at constant x-ray intensity (schematic). Time constants changed abruptly at times shown. $(\text{T.C.})_1 < (\text{T.C.})_2 < (\text{T.C.})_3$.

where n is the average counting rate. This equation also shows that the probable error is less for high counting rates than for low, when the time constant remains the same; this effect is evident on the chart recording of any diffraction line, where the pen fluctuations are smaller at the top of the line than in the background.

The most useful feature of a ratemeter is its ability to follow *changes* in the average counting rate, a function which the scaler is totally unable to perform, since a change in the average counting rate occurring during the time a count is being made with a scaler will go entirely undetected. It is this feature of a ratemeter which is so useful in diffractometry. A diffraction pattern can be scanned from one end to the other, and the moving counter automatically transmits, through the ratemeter, a continuous record of the intensity it observes as the diffraction angle is changed. On the other hand, the ratemeter is less accurate than the scaler, both because of the unavoidable statistical fluctuations in its output and because of the errors inherent in its indicating or recording instruments.

As mentioned earlier, a large time constant smooths out fluctuations in the average counting rate by increasing the response time to changes in rate. But when a sharp diffraction line is being scanned, the average counting rate is changing rapidly and we would like the ratemeter to indicate this change as accurately as possible. From this point of view a short response time, produced by a small time constant, is required. A ratemeter must therefore be designed with these two conflicting factors in mind, and the time constant should be chosen large enough to smooth out most of the statistical fluctuations and yet small enough to give a reasonably short response time.

Most commercial ratemeters have several scales available to cover various ranges of x-ray intensity (100, 1000, and 10,000 cps for full-scale deflection of the recorder pen, for example). Smaller time constants are used with the higher scales, just as short counting times are used with a scaler when the counting rate is high. In some instruments, the time constant appropriate to each scale is fixed by the manufacturer, and in others the operator can select any one of several time constants, ranging from about 0.5 to 15 sec, by switches which insert the proper capacitance in the circuit. The proper time constant to use is, of course, not unrelated to the scanning speed, for a fast scan demands a fast response from the ratemeter and therefore a short time constant. A time constant which is too large for the scanning speed used will slightly shift the peaks of diffraction lines in the direction of the scan and lower their maximum intensity and, because of its excessive smoothing action, may actually obliterate weak diffraction lines and cause them to go unnoticed. In choosing a time constant, it is therefore better to err on the short side. A good rule to follow is to make the time constant less than half the *time width* of the receiving slit, where the time width is defined as the time required for the slit to travel its own width. For example, if a 0.2° slit is used at a scanning speed of 2°/min, then the time width of the slit is $(0.2/2)(60) = 6$ sec, and the time constant should therefore be less than 3 sec. The same rule can be used to find the proper slit width for a given scanning speed when the time constant is fixed.

The relation between the x-ray intensity, i.e., the average counting rate, and the deflection of the recorder pen may usually be made linear or logarithmic, at the turn of a switch. Logarithmic recording is useful when looking for weak lines or when exploring an unknown pattern; weak lines will be emphasized and the recording of strong lines will not go off-scale. If there is any doubt about the relation between x-ray intensity and pen deflection, it may be established experimentally by a procedure similar to that used for determining the counting rate at which losses begin, as outlined in Sec. 7–4. A number of identical metal foils are placed in a strong diffracted beam entering the counter and these are withdrawn one by one, with the counter in a fixed position. After each withdrawal, the counting rate is measured accurately with a scaler, and the ratemeter operated for a time at least equal to the scaling time, the recording chart speed being selected to give a trace of reasonable length. An average straight line is then drawn through each trace, in such a way as to make the positive and negative fluctuations as nearly equal as possible. Finally, the distances of these straight lines from the chart zero are plotted against the corresponding average counting rates as determined by the scaler, and the calibration curve so obtained is used as a basis for future intensity measurements with the ratemeter-recorder combination.

7–13 MONOCHROMATIC OPERATION

The background of a diffraction pattern obtained with a diffractometer may be reduced by means of a single-channel pulse-height analyzer, as mentioned in Sec. 7–9. An even better method is to use a crystal monochromator in the diffracted beam. Balanced filters present still another option.

Monochromating Crystal

With a diffractometer one has the option, which does not exist with a powder camera, of placing a crystal monochromator in the diffracted, rather than the incident, beam. Figure 7–28 shows such an arrangement. The diffracted beam from the specimen comes to a focus at the receiving slit S_1, diverges to the focusing monochromator M, and comes to a focus again at the counter slit S_2. Counter, crystal, and slits are mounted on one support and rotate as a unit about the diffractometer axis.

Even though intensity is decreased during diffraction by a monochromator, a $K\beta$ filter is not needed because the monochromator is set to diffract only $K\alpha$ radiation. As a result, and because of the focusing action of the monochromator, the intensity of a diffraction line at the counter can actually be higher with a monochromator than without, particularly if the monochromating crystal is graphite.

Placement of the monochromator in the diffracted beam has the advantage of suppressing background radiation originating in the *specimen*, such as fluorescent radiation and incoherent (Compton modified) scattered radiation. For example, if a steel specimen or any iron-rich material is examined with copper radiation in an ordinary diffractometer, the background due to fluorescent Fe K radiation will be unacceptably high. But if a monochromator is added and oriented to reflect only Cu $K\alpha$, the background is reduced practically to zero, because the fluoresced Fe $K\alpha$ and Fe $K\beta$ do not enter the counter. A monochromator may therefore

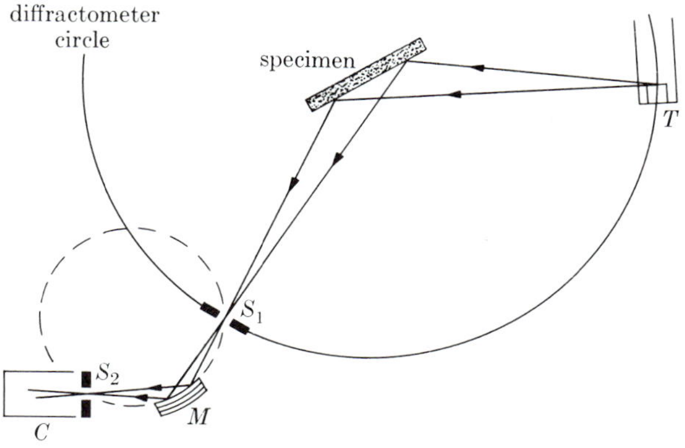

Fig. 7-28 Diffractometer with monochromating crystal M in diffracted beam. C = counter, T = x-ray tube.

eliminate the need for an extra tube, such as a Cr-target tube, for the examination of steel specimens.

The diffractometer in Fig. 15-10(a) is equipped with a diffracted-beam monochromator.

Balanced Filters

Another method of operating under essentially monochromatic conditions, a method peculiar to the diffractometer, is by the use of *Ross filters*, also called balanced filters [G.39, Vol. 3 of G.11]. This method depends on the fact that the absorption coefficients of all substances vary in the same way with wavelength; i.e., they are proportional to λ^3, as shown by Eq. (1-13). If filters are made of two substances differing in atomic number by one, and their thicknesses adjusted so that they produce the same absorption for a particular wavelength, then they will have the same absorption for all wavelengths *except* those lying in the narrow wavelength region between the K absorption edges of the two substances. This region is called the *pass band* of the filter combination. If these filters are placed alternately in a heterochromatic x-ray beam, i.e., a beam containing rays of different wavelengths, then the difference between the intensities transmitted in each case is due only to wavelengths lying in the pass band. When the pass band is chosen to include a strong characteristic component of the spectrum, then the net effect is that of a strong monochromatic beam.

The isolation of Cu $K\alpha$ radiation may be taken as an example. Its wavelength is 1.542 Å, which means that cobalt and nickel can be used as filter materials since their K absorption edges (1.608 and 1.488 Å, respectively) effectively bracket the Cu $K\alpha$ line. Their linear absorption coefficients μ are plotted in Fig. 7-29(a), which shows that balancing can be obtained by making the nickel filter somewhat thinner than the cobalt one. When their thicknesses x are adjusted to the correct ratio, then $\mu_{Ni}x_{Ni} = \mu_{Co}x_{Co}$ except in the pass band, and a plot of μx versus λ has

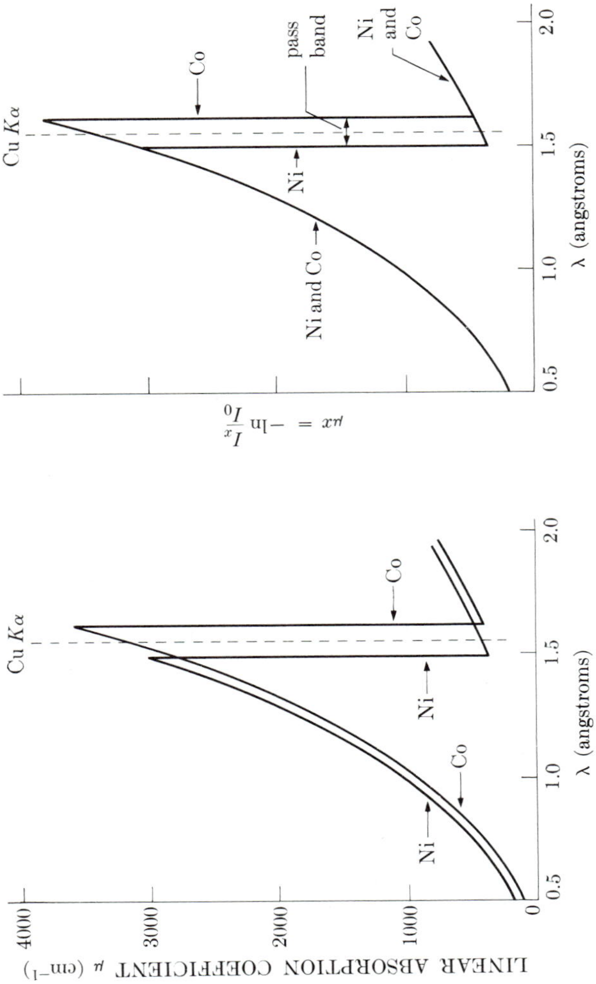

Fig. 7–29 Ross filters for Cu $K\alpha$ radiation: (a) absorption coefficients of filter materials; (b) μx values after balancing.

the appearance of Fig. 7–29(b). Since $\mu x = -\ln I_x/I_0$, the transmission factors I_x/I_0 (ratio of transmitted to incident intensity) of the two filters are now equal for all wavelengths except those in the pass band, which is only 0.12 Å wide. At each angle 2θ at which the intensity is to be measured with the diffractometer, first one filter and then the other is placed in the diffracted beam before it enters the counter. The intensity of the diffracted beam passing through each filter is then measured, and the difference in the measurements gives the diffracted intensity of only the Cu $K\alpha$ line and the relatively weak wavelengths immediately adjacent to it in the pass band.

It should be emphasized that the beam entering the counter is never physically monochromatic, as it is when a crystal monochromator is used. Radiation with a great many wavelengths enters the counter when either filter is in place, but every wavelength transmitted by one filter has the same intensity as that transmitted by the other filter, except those wavelengths lying in the pass band, and these are transmitted quite unequally by the two filters. Therefore, when the intensity measured with one filter is subtracted from that measured with the other filter, the difference is zero for every wavelength except those in the pass band.

In practice, balancing of the filters is carried out by inserting two foils of approximately the same thickness into suitable holders which can be slipped into place in the beam entering the counter. One foil is always perpendicular to the x-ray beam, while the other may be rotated about an axis at right angles to the beam; in this way the second foil may be inclined to the beam at such an angle that its effective thickness x equals the thickness required for balancing. Perfect balancing at all wavelengths outside the pass band is not possible, although it may be approached quite closely, because μ does not vary exactly as λ^3 and because the magnitude of the K absorption jump (ratio of absorption coefficients for wavelengths just shorter and just longer than the K edge) is not exactly the same for all elements.

Note also that balanced filters will not exclude Compton scattering, which differs in wavelength by only 0.05 Å or less from the wavelength of the diffracted beam (Eq. 4–3) and which will therefore generally fall inside the pass band. And if the radiation forming the diffracted beam is, for example, Cu $K\alpha$, then neither a monochromating crystal nor balanced filters will exclude diffusely scattered Cu $K\alpha$ caused, for example, by thermal vibration of the atoms.

PROBLEMS

***7–1** A powder specimen in the form of a rectangular plate has a width of 0.5 in., measured in the plane of the diffractometer circle, which has a radius of 5.73 in. If it is required that the specimen entirely fill the incident beam at all angles and that measurements must be made to angles as low as $2\theta = 10°$, what is the maximum divergence angle (measured in the plane of the diffractometer circle) that the incident beam may have?

***7–2** Even the weather can affect the long-term stability of the measured intensity of x-rays from a well-stabilized tube, because a change in barometric pressure or humidity changes the absorption of x-rays by air. What is the percent change in the measured intensity of Cr $K\alpha$ resulting from a 3-percent drop in pressure over a 12-hour period, a not uncommon event? (Assume a path length in air of 27 cm and take μ of air for Cr $K\alpha = 3.48 \times 10^{-2}$ cm^{-1}.)

7–3 Cu $K\alpha$ radiation is incident on a xenon-filled proportional counter. Calculate the ratio of the average pulse size in the escape peak to that in the normal peak. Compare your result with that of Fig. 7–22(b).

***7–4** If a count N_P of 30,000 is obtained at the peak of a diffraction line and, in the same time, a count N_B of 10,000 in the background adjacent to the line, calculate the percent probable error in (a) N_P and (b) $(N_P - N_B)$.

7–5 A diffraction pattern of polycrystalline platinum is obtained by energy-dispersive diffractometry at $\theta = 10.7°$. Calculate the energy (in keV) at which the 2 2 0 line will appear and compare your result with Fig. 7–23(b).

***7–6** (a) Calculate the ratio of the effective thicknesses of cobalt and nickel filters when they are balanced for all wavelengths except Cu $K\alpha$. (Obtain an average value applicable to a wavelength range extending from about 0.5 Å to about 2 Å.)

b) When the filters are balanced, calculate the ratio of the intensity of Cu $K\alpha$ radiation transmitted by the nickel filter to that transmitted by the cobalt filter, assuming the same incident intensity in each case. The effective thickness of the nickel filter is 0.00035 in.

Applications

8

Orientation and Quality of Single Crystals

8-1 INTRODUCTION

Much of our understanding of the properties of polycrystalline materials has been gained by studies of isolated single crystals, since such studies permit measurement of the properties of the individual building blocks in the composite mass. Because single crystals are anisotropic, research of this kind always requires accurate knowledge of the orientation of the single crystal test specimen in order that measurements may be made along known crystallographic directions or planes. By varying the crystal orientation, we can obtain data on the property measured (e.g., yield strength, electrical resistivity, corrosion rate) as a function of crystal orientation.

There is also an increasing production of single crystals, not for research studies, but for use as such in various devices, mainly electrical and magnetic. Examples of these applications are quartz crystals for oscillators and timing circuits, silicon and germanium crystals for transistors and other semiconductor products, and garnet crystals for magnetic memories. These crystals must all be produced with particular orientations.

Whether the crystal is grown for research purposes or for use in a device, we will also be interested in the *quality* of the crystal, i.e., in how closely it approaches perfection, because its intended use will set a limit on how much imperfection can be tolerated.

This chapter is therefore devoted to the two-fold problem of determining crystal orientation and assessing crystal quality.

CRYSTAL ORIENTATION

Described below are the three main methods of determining orientation: back-reflection Laue, transmission Laue, and diffractometer. Nor should the old etch-pit method be overlooked. This is an optical method, involving the reflection of visible light from the flat sides, of known Miller indices, of etch pits in crystal surfaces. Although not universally applicable, this method is fast and requires only simple apparatus [G.25].

8-2 BACK-REFLECTION LAUE METHOD

As mentioned in Sec. 3-6, the Laue pattern of a single crystal consists of a set of diffraction spots on the film and the positions of these spots depend on the orientation of the crystal. This is true of either Laue method, transmission or back-reflection, so either can be used to determine crystal orientation. However, the back-reflection method is the more widely used of the two because it requires no special preparation of the specimen, which may be of any thickness, whereas the transmission method requires relatively thin specimens of low absorption.

In either case, since the orientation of the specimen is to be determined from the location of the Laue spots on the film, it is necessary to orient the specimen relative to the film in some known manner. The single crystal specimens encountered in metallurgical work are usually in the form of wire, rod, sheet, or plate, but crystals of irregular shape must occasionally be dealt with. Wire or rod specimens are best mounted with their axis parallel to one edge of the square or rectangular film; a fiducial mark on the specimen surface, for example on the side nearest the film, then fixes the orientation of the specimen completely. It is convenient to mount sheet or plate specimens with their plane parallel to the plane of the film and one edge of the sheet or plate parallel to an edge of the film. Irregularly shaped crystals must have fiducial marks on their surface which will definitely fix their orientation relative to that of the film.

The problem now is to determine the orientation of the crystal from the position of the back-reflection Laue spots on the film. If we wished, we could determine the Bragg angle θ corresponding to each Laue spot from Eq. (5–2), but that would be no help in identifying the planes producing that spot, since the wavelength of the diffracted beam is unknown. We can, however, determine the orientation of the normal to the planes causing each spot, because the plane normal always bisects the angle between incident and diffracted beams. The directions of the plane normals can then be plotted on a stereographic projection, the angles between them measured, and the planes identified by comparison with a list of known interplanar angles for the crystal involved.

Our first problem, therefore, is to derive, from the measured position of each diffraction spot on the film, the position on a stereographic projection of the pole of the plane causing that spot. In doing this it is helpful to recall that all of the planes of one zone reflect beams which lie on the surface of a cone whose axis is the zone axis and whose semi-apex angle is equal to the angle ϕ at which the zone axis is inclined to the transmitted beam (Fig. 8–1). If ϕ does not exceed 45°, the cone will not intersect a film placed in the back-reflection region; if ϕ lies between 45° and 90°, the cone intersects the film in a hyperbola; and, if ϕ equals 90°, the intersection is a straight line passing through the incident beam. (If ϕ exceeds 90°, the cone shifts to a position below the transmitted beam and intersects the lower half of the film, as may be seen by viewing Fig. 8–1 upside down.) Diffraction spots on a back-reflection Laue film therefore lie on hyperbolas or straight lines, and the distance of any hyperbola from the center of the film is a measure of the inclination of the zone axis.

In Fig. 8–2 the film is viewed from the crystal. Coordinate axes are set up such that the incident beam proceeds along the z-axis in the direction Oz and the x- and y-axes lie in the plane of the film. The beam reflected by the plane shown strikes the film at S. The normal to this reflecting plane is CN and the plane itself

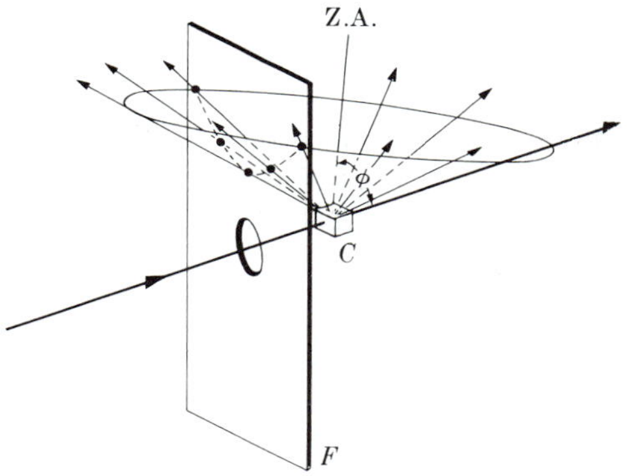

Fig. 8-1 Intersection of a conical array of diffracted beams with a film placed in the back-reflection position. C = crystal, F = film, Z.A. = zone axis.

is assumed to belong to a zone whose axis lies in the *yz*-plane. If we imagine this plane to rotate about the zone axis, it will pass through all the positions at which planes of this zone in an actual crystal might lie. During this rotation, the plane normal would cut the film in the straight line AB and the reflected beam in the hyperbola HK. AB is therefore the locus of plane normal intersections with the film and HK the locus of diffracted beam intersections. The plane which reflects a beam to S, for example, has a normal which intersects the film at N, since the incident beam, plane normal, and diffracted beam are coplanar. Since the orienta-

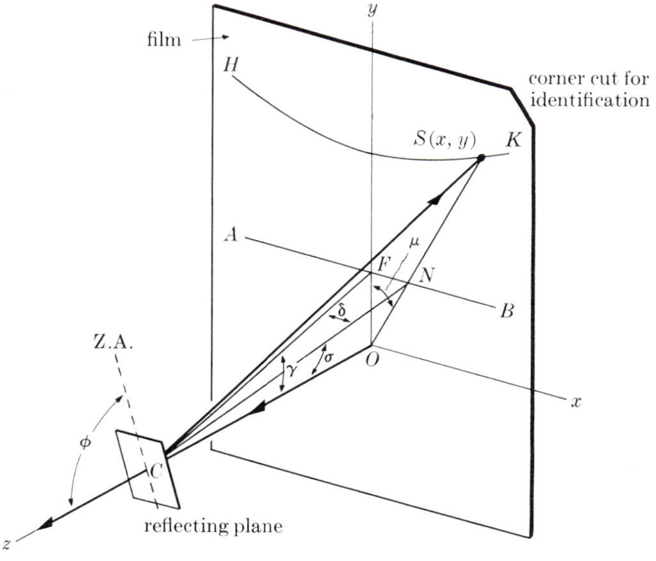

Fig. 8-2 Location of back-reflection Laue spot. Note that $\gamma = 90° - \phi$.

tion of the plane normal in space can be described by its angular coordinates γ and δ, the problem is to determine γ and δ from the measured coordinates x and y of the diffraction spot S on the film.

A graphical method of doing this was devised by Greninger [8.1] who developed a chart which, when placed on the film, gives directly the γ and δ coordinates corresponding to any diffraction spot. To plot such a chart, we note from Fig. 8–2 that

$$x = OS \sin \mu, \quad y = OS \cos \mu, \quad \text{and} \quad OS = OC \tan 2\sigma,$$

where $OC = D =$ specimen-film distance. The angles μ and σ are obtained from γ and δ as follows:

$$\tan \mu = \frac{FN}{FO} = \frac{CF \tan \delta}{CF \sin \gamma} = \frac{\tan \delta}{\sin \gamma},$$

$$\tan \sigma = \frac{ON}{OC} = \left(\frac{FN}{\sin \mu}\right)\left(\frac{1}{CF \cos \gamma}\right) = \left(\frac{CF \tan \delta}{\sin \mu}\right)\left(\frac{1}{CF \cos \gamma}\right)$$

$$= \frac{\tan \delta}{\sin \mu \cos \gamma}.$$

With these equations, the position (in terms of x and y) of any diffraction spot can be plotted for given values of γ and δ and any desired specimen-film distance D. The result is the Greninger chart, graduated at 2° intervals shown in Fig. 8–3. The hyperbolas running from left to right are curves of constant γ, and any one of these curves is the locus of diffraction spots from planes of a zone whose axis is tilted away from the plane of the film by the indicated angle γ. If points having the same value of δ are joined together, another set of hyperbolas running from top to bottom is obtained. The lower half of the chart contains a protractor whose use will be referred to later. Greninger charts should have dark lines on a transparent background and are best prepared as positive prints on photographic film [2.3].

In use, the chart is placed over the film with its center coinciding with the film center and with the edges of chart and film parallel. The γ and δ coordinates corresponding to any diffraction spot are then read directly. Note that use of the chart avoids any measurement of the actual coordinate distances x and y of the spot. The chart gives directly, not the x and y coordinates of the spot, but *the angular coordinates γ and δ of the normal to the plane causing the spot.*

Knowing the γ and δ coordinates of any plane normal, for example CN in Fig. 8–2, we can plot the pole of the plane on a stereographic projection. Imagine a reference sphere centered on the crystal in Fig. 8–2 and tangent to the film, and let the projection plane coincide with the film. The point of projection is taken as the intersection of the transmitted beam and the reference sphere. Since the plane normal CN intersects the side of the sphere nearest the x-ray source, the projection must be viewed from that side and the film "read" from that side. In order to know, after processing, the orientation the film had during the x-ray exposure, the upper right-hand corner of the film (viewed from the crystal) is cut

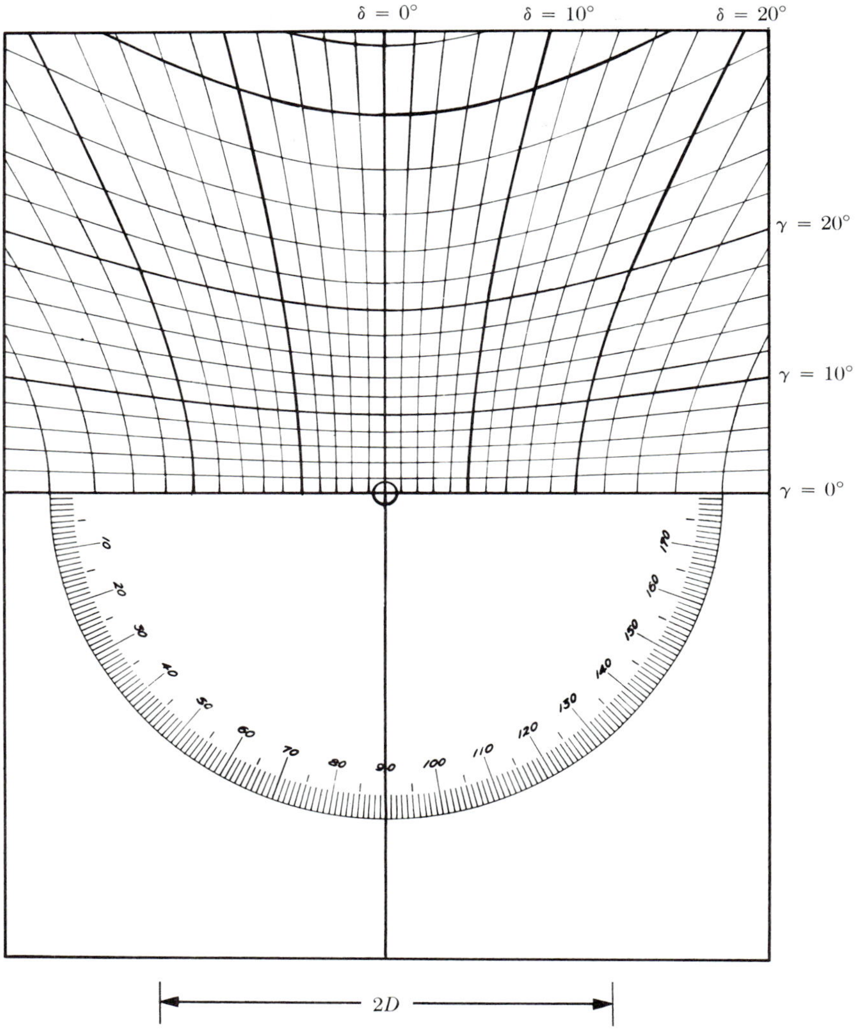

Fig. 8-3 Greninger chart for the solution of back-reflection Laue patterns, reproduced in the correct size for a specimen-to-film distance D of 3 cm.

away before it is placed in the cassette, as shown in Fig. 8-2. When the film is read, this cut corner must therefore be at the upper left, as shown in Fig. 8-4(a). The angles γ and δ, read from the chart, are then laid out on the projection as indicated in Fig. 8-4(b). Note that the underlying Wulff net must be oriented so that its meridians run from side to side, not top to bottom. The reason for this is the fact that diffraction spots which lie on curves of constant γ come from planes of a zone, and the poles of these planes must therefore lie on a great circle on the projection. The γ,δ coordinates corresponding to diffraction spots on the lower half of the film are obtained simply by reversing the Greninger chart end for end.

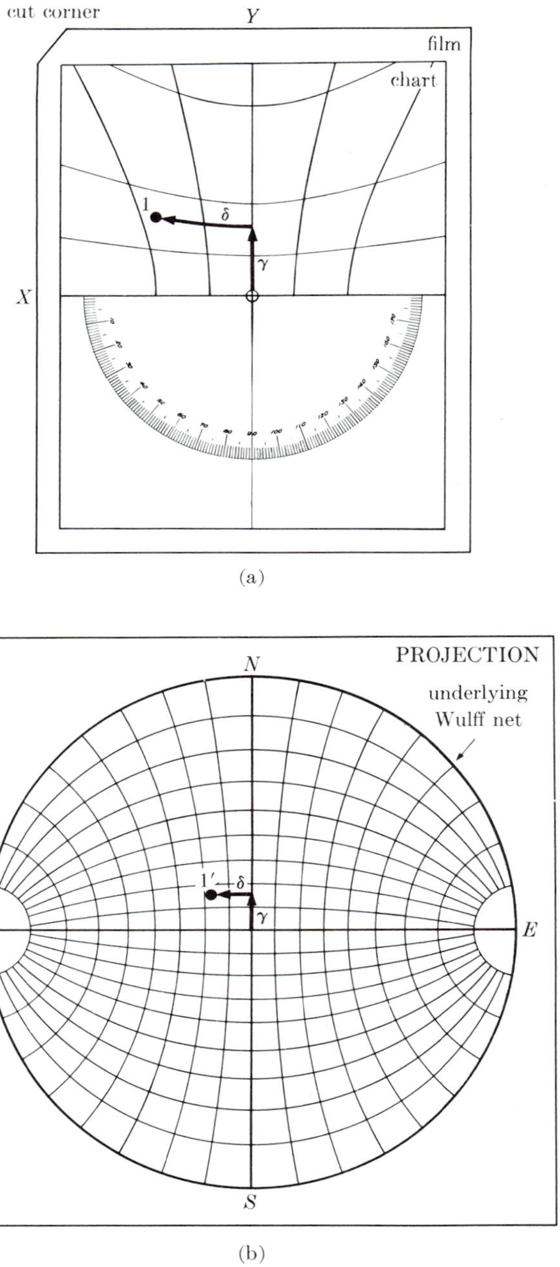

Fig. 8–4 Use of the Greninger chart to plot the pole of a reflecting plane on a stereographic projection. Pole 1′ in (b) is the pole of the plane causing diffraction spot 1 in (a).

This procedure may be illustrated by determining the orientation of the aluminum crystal whose back-reflection Laue pattern is shown in Fig. 3–6(b). Fig. 8–5 is a tracing of this photograph, showing the more important spots numbered for reference. The poles of the planes causing these numbered spots are plotted stereographically in Fig. 8–6 by the method of Fig. 8–4 and are shown as solid circles.

The problem now is to "index" these planes, i.e., to find their Miller indices, and so disclose the orientation of the crystal. With the aid of a Wulff net, great circles are drawn through the various sets of poles corresponding to the various hyperbolas of spots on the film. These great circles connect planes of a zone, and planes lying at their intersections are generally of low indices, such as {100}, {110}, {111}, and {112}. The axes of the zones themselves are also of low indices, so it is helpful to locate these axes on the projection. They are shown as open circles in Fig. 8–6, P_A being the axis of zone A, P_B the axis of zone B, etc. We then measure the angles between important poles (zone intersections and zone axes) and try to identify the poles by comparison of these measured angles with those calculated for cubic crystals (Table 2–3). The method is essentially one of trial and error. We note, for example, that the angles $P_A - P_B$, $P_A - 5'$, and $P_B - 5'$ are all 90°. This suggests that one or more of these poles might be {100} or {110}, since the angle between two {100} poles or between two {110} poles is 90°. Suppose we tentatively assume that P_A, P_B, and 5′ are all {100} poles.* Then P_E, which lies on the great circle between P_A and P_B and at an angular distance of 45° from each, must be a {110} pole. We then turn our attention to zone C and find that the distance between pole 6′ and either pole 5′ or P_E is also 45°. But reference to a standard projection, such as Fig. 2–37, shows that there is no important pole located midway on the great circle between {100}, which we have identified with 5′, and {110}, which we have identified with P_E. Our original assumption is therefore wrong. We therefore make a second assumption, which is consistent with the angles measured so far, namely that 5′ is a {100} pole, as before, but that P_A and P_B are {110} poles. P_E must then be a {100} pole and 6′ a {110} pole. We can check this assumption by measuring the angles in the triangle $a - b - 5'$. Both a and b are found to be 55° from 5′, and 71° from each other, which conclusively identifies a and b as {111} poles. We note also, from a standard projection, that a {111} pole must lie on a great circle between {100} and {110}, which agrees with the fact that a, for example, lies on the great circle between 5′, assumed to be {100}, and P_B, assumed to be {110}. Our second assumption is therefore shown to be correct.

---

* The reader may detect an apparent error in nomenclature here. Pole 5′ for example, is assumed to be a {100} pole and spot 5 on the diffraction pattern is assumed, tacitly, to be due to a 100 reflection. But aluminum is face-centered cubic and we know that there is no 100 reflection from such a lattice, since *hkl* must be unmixed for diffraction to occur. Actually, spot 5, if our assumption is correct, is due to overlapping reflections from the {200}, {400}, {600}, etc., planes. But these planes are all parallel and are represented on the stereographic projection by one pole, which is conventionally referred to as {100}. The corresponding diffraction spot is also called, conventionally but loosely, the 100 spot.

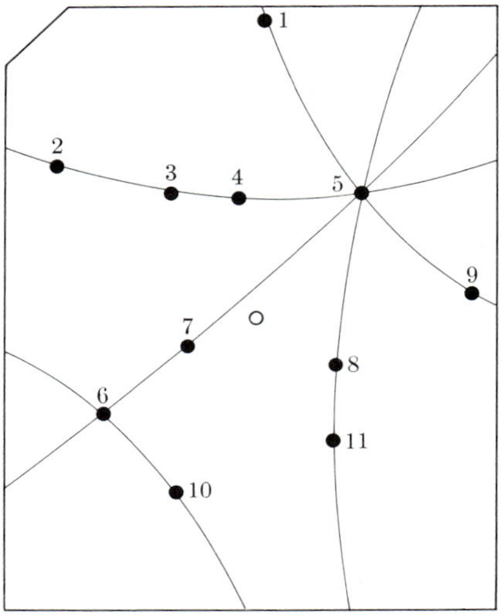

Fig. 8–5 Selected diffraction spots of back-reflection Laue pattern of an aluminum crystal, traced from Fig. 3–6(b).

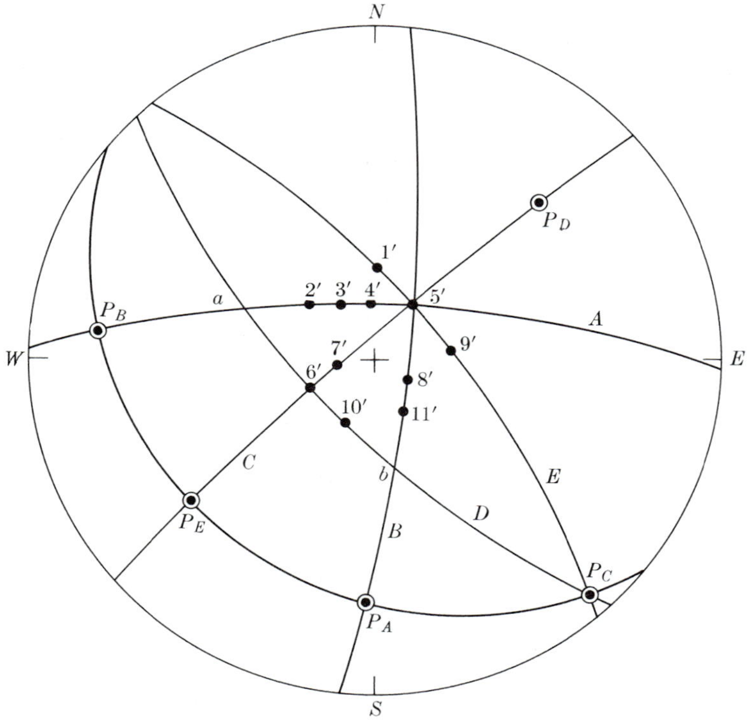

Fig. 8–6 Stereographic projection corresponding to back-reflection pattern of Fig. 8–5.

Figure 8-7 shows the stereographic projection in a more complete form, with all poles of the type {100}, {110}, and {111} located and identified. Note that it was not necessary to index all the observed diffraction spots in order to determine the crystal orientation, which is specified completely, in fact, by the locations of any two {100} poles on the projection. The information given in Fig. 8-7 is therefore all that is commonly required. Occasionally, however, we may wish to know the Miller indices of a particular diffraction spot on the film, spot 11 for example. To find these indices, we note that pole 11′ is located 35° from (001) on the great circle passing through (001) and (111). Reference to a standard projection and a table of interplanar angles shows that its indices are (112).

As mentioned above, the stereographic projection of Fig. 8-7 is a complete description of the orientation of the crystal. Other methods of description are also possible. The crystal to which Fig. 8-7 refers had the form of a square plate and was mounted with its plane parallel to the plane of the film (and the projection) and its edges parallel to the film edges, which are in turn parallel to the *NS* and *EW* axes of the projection. Since the (001) pole is near the center of the projection, which corresponds to the specimen normal, and the (010) pole near the edge of the projection and approximately midway between the *E* and *S* poles, we may very roughly describe the crystal orientation as follows: one set of cube planes is

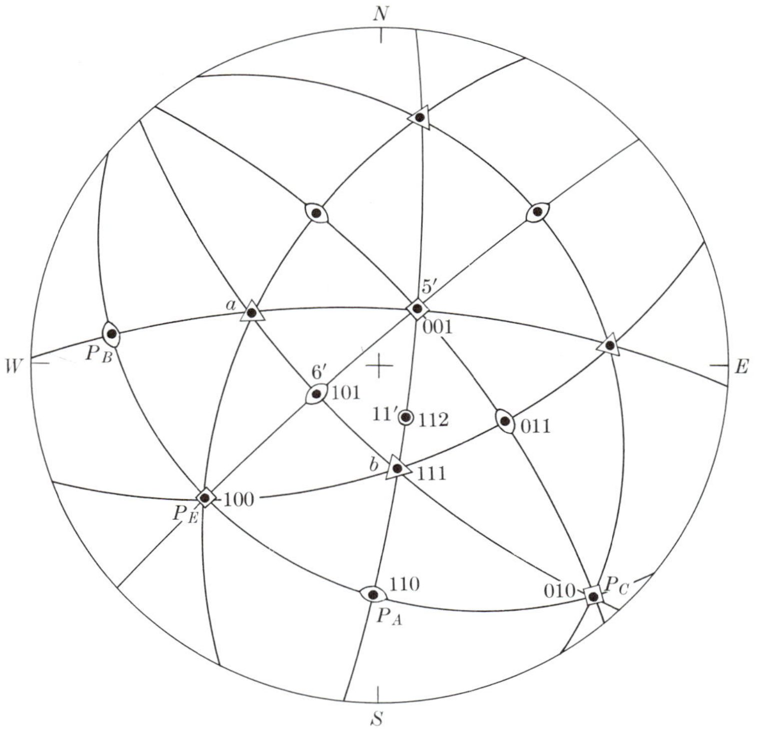

Fig. 8-7 Stereographic projection of Fig. 8-6 with poles identified.

approximately parallel to the surface of the plate while another set passes diagonally through the plate and approximately at right angles to its surface.

Another method of description may be used when only one direction in the crystal is of physical significance, such as the plate normal in the present case. For example, we may wish to make a compression test of this crystal, with the axis of compression normal to the plate surface. We are then interested in the orientation of the crystal relative to the compression axis (plate normal) or, stated inversely, in the orientation of the compression axis relative to certain directions of low indices in the crystal. Now inspection of a standard projection such as Fig. 2–36(a) shows that each half of the reference sphere is covered by 24 similar and equivalent spherical triangles, each having {100}, {110}, and {111} as its vertices. The plate normal will fall in one of these triangles and it is necessary to draw only one of them in order to describe the precise location of the normal. In Fig. 8–7, the plate normal lies in the (001)-(101)-(111) triangle which is redrawn in Fig. 8–8 in the conventional orientation, as though it formed part of a (001) standard projection. To locate the plate normal on this new drawing, we measure the angles between the center of the projection in Fig. 8–7 and the three adjacent poles. Let these angles be ρ_{001}, ρ_{101}, and ρ_{111}. These angles are then used to determine the three arcs shown in Fig. 8–8. These are circle arcs, but they are *not* centered, in general, on the corresponding poles; rather, each one is the locus of points located at an equal *angular* distance from the pole involved and their intersection therefore locates the desired point. Another method of arriving at Fig. 8–8 from Fig. 8–7 consists simply in rotating the whole projection, poles and plate normal together, from the orientation shown in Fig. 8–7 to that of a standard (001) projection.

Similarly, the orientation of a single-crystal wire or rod may be described in terms of the location of its axis in the unit stereographic triangle. Note that this method does not completely describe the orientation of the crystal, since it allows

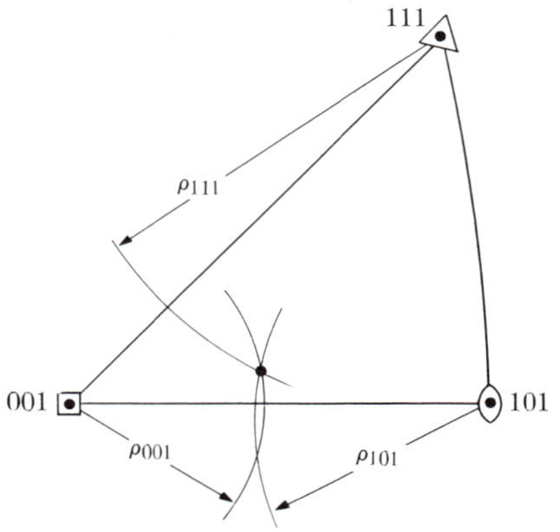

Fig. 8–8 Use of the unit stereographic triangle to describe crystal orientation. The point inside the triangle is the normal to the single crystal plate whose orientation is shown in Fig. 8–7.

one rotational degree of freedom about the specimen axis. This is of no consequence, however, when we are only interested in the value of some measured physical or mechanical property along a particular direction in the crystal.

There are other ways of manipulating both the Greninger chart and the stereographic projection, and the particular method used is purely a matter of personal preference. For example, we may ignore the individual spots on the film and focus our attention instead on the various hyperbolas on which they lie. The spots on one hyperbola are due to reflections from planes of one zone and, by means of the Greninger chart, we can plot directly the axis of this zone without plotting the poles of any of the planes belonging to it. The procedure is illustrated in Fig. 8–9. Keeping the centers of film and chart coincident, we rotate the film about this center until a particular hyperbola of spots coincides with a curve of constant γ on the chart, as in (a). The amount of rotation required is read from the intersection of a vertical pencil line, previously ruled through the center of the film and parallel to one edge, with the protractor of the Greninger chart. Suppose this angle is ε. Then the projection is rotated by the same angle ε with respect to the underlying Wulff net and the zone axis is plotted on the vertical axis of the projection at an angle γ from the *circumference*, as in (b). (Note that zone A itself is represented by a great circle located at an angle γ above the *center* of the projection. However, the plotting of the zone circle is not ordinarily necessary since the zone axis adequately represents the whole zone.)* Proceeding in this way, we plot the poles of all the important zones and, by the method of Fig. 8–4, the pole of the plane causing the most important spot or spots on the pattern. (The latter are, like spot 5 of Fig. 8–5, of high intensity, at the intersection of a number of hyperbolas, and well separated from their neighbors.) The points so obtained are always of low indices and can usually be indexed without difficulty.

An alternative method of indexing plotted poles depends on having available a set of detailed standard projections in a number of orientations, such as {100}, {110}, and {111} for cubic crystals. It is also a trial and error method and may be illustrated with reference to Fig. 8–6. First, a prominent zone is selected and an assumption is made as to its indices: for example, we might assume that zone B is a ⟨100⟩ zone. This assumption is then tested by (a) rotating the projection about its center until P_B lies on the equator of the Wulff net and the ends of the zone circle coincide with the N and S poles of the net, and (b) rotating all the important points on the projection about the NS-axis of the net until P_B lies at the center and the zone circle at the circumference. The new projection is then superimposed on a {100} standard projection and rotated about the center until all points on the projection coincide with those on the standard. If no such coincidence is obtained, another standard projection is tried. For the particular case of Fig. 8–6, a coincidence would be obtained only on a {110} standard, since P_B is actually a {110} pole. Once a match has been found, the indices of the unknown poles are given simply by the indices of the poles on the standard with which they coincide.

In the absence of a Greninger chart, the pole corresponding to any observed Laue spot may be plotted by means of an easily constructed "stereographic ruler." The construction of the ruler is based on the relations shown in Fig. 8–10. This drawing is a

* Note that, when a hyperbola of spots is lined up with a horizontal hyperbola on the chart as in Fig. 8–9(a), the vertical hyperbolas can be used to measure the difference in angle δ for any two spots and that this angle is equal to the angle between the planes causing those spots, just as the angle between two poles lying on a meridian of a Wulff net is given by their difference in latitude.

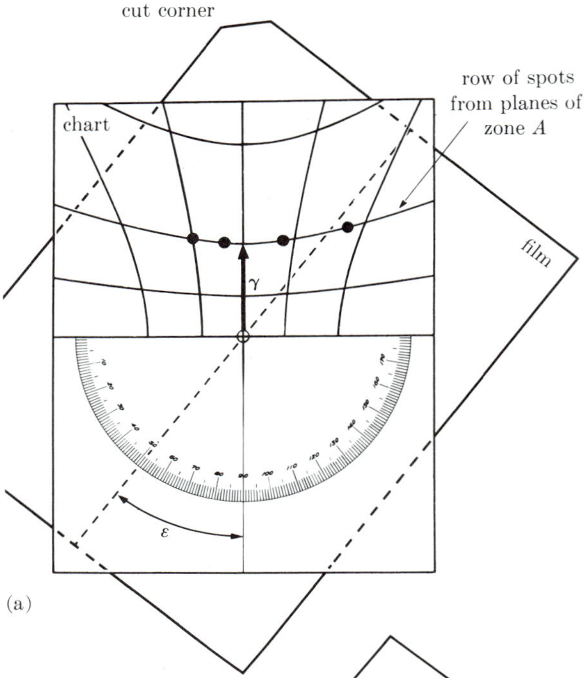

(a)

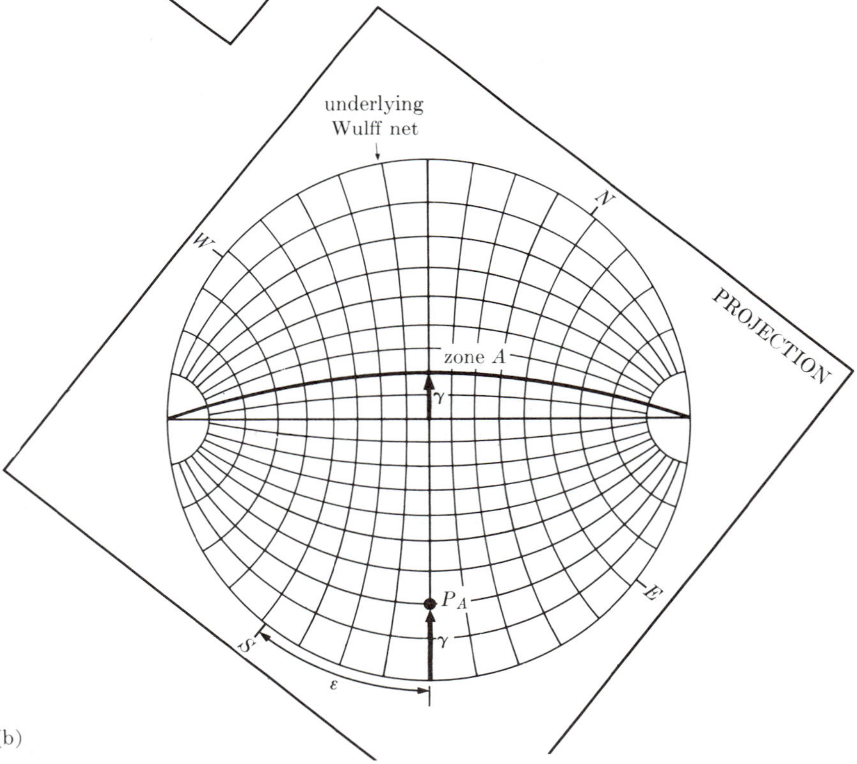

(b)

Fig. 8–9 Use of the Greninger chart to plot the axis of a zone of planes on the stereographic projection. P_A is the axis of zone A.

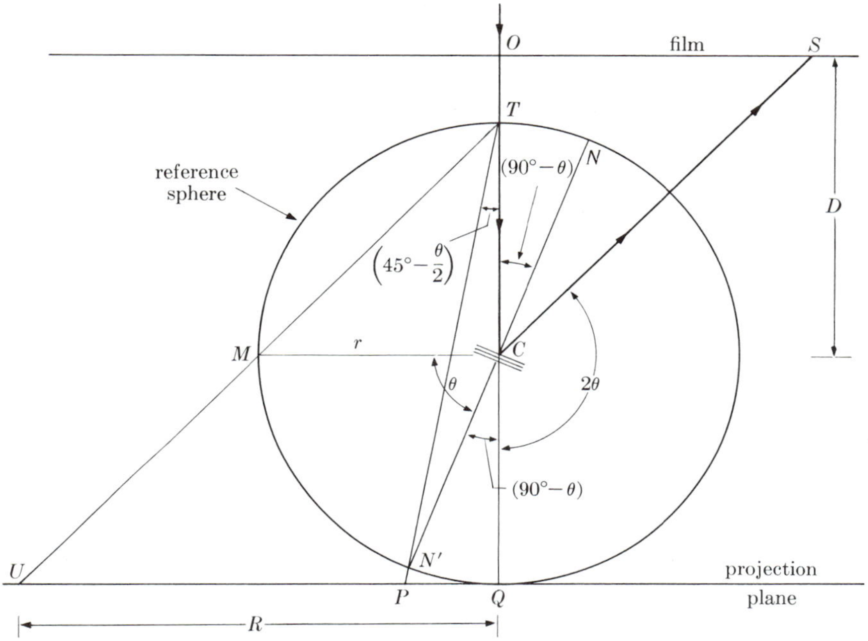

Fig. 8-10 Relation between diffraction spot S and stereographic projection P of the plane causing the spot, for back reflection.

section through the incident beam OC and any diffracted beam CS. Here it is convenient to use the plane normal CN' rather than CN and to make the projection from T, the intersection of the reference sphere with the incident beam. The projection of the pole N' is therefore at P. From the measured distance OS of the diffraction spot from the center of the film, we can find the distance PQ of the projected pole from the center of the projection, since

$$OS = OC \tan (180^\circ - 2\theta) = D \tan (180^\circ - 2\theta) \tag{8-1}$$

and

$$PQ = TQ \tan \left(45^\circ - \frac{\theta}{2}\right) = 2r \tan \left(45^\circ - \frac{\theta}{2}\right), \tag{8-2}$$

where D is the specimen-film distance and r the radius of the reference sphere. The value of r is fixed by the radius R of the Wulff net used, since the latter equals the radius of the basic circle of the projection. We note that, if the pole of the plane were in its extreme position at M, then its projection would be at U. The point U therefore lies on the basic circle of the projection, and UQ is the radius R of the basic circle. Because the triangles TUQ and TMC are similar, $R = 2r$ and

$$PQ = R \tan \left(45^\circ - \frac{\theta}{2}\right). \tag{8-3}$$

The ruler is constructed by marking off, from a central point, a scale of centimeters by which the distance OS may be measured. The distance PQ corresponding to each

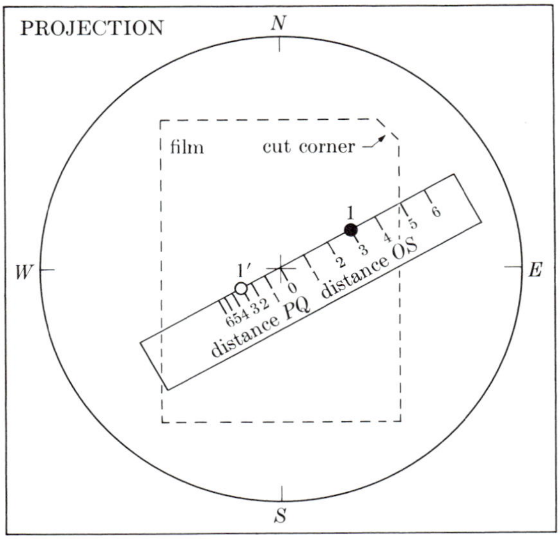

Fig. 8–11 Use of a stereographic ruler to plot the pole of a reflecting plane on a stereographic projection in the back-reflection Laue method. Pole 1′ is the pole of the plane causing diffraction spot 1.

distance *OS* is then calculated from Eqs. (8–1) and (8–3), and marked off from the center of the ruler in the opposite direction. Corresponding graduations are given the same number and the result is the ruler shown in Fig. 8–11, which also illustrates the method of using it. [Calculation of the various distances *PQ* can be avoided by use of the Wulff net itself. Figure 8–10 shows that the pole of the reflecting plane is located at an angle θ from the edge of the projection, and θ is given for each distance *OS* by Eq. (8–1). The ruler is laid along the equator of the Wulff net, its center coinciding with the net center, and the distance *PQ* corresponding to each angle θ is marked off with the help of the angular scale on the equator.]

From the choice of plane normal made in Fig. 8–10, it is apparent that the projection must be viewed from the side opposite the x-ray source. This requires that the film be read from that side also, i.e., with its cut corner in the upper right-hand position. The projection is then placed over the film, illuminated from below, as shown in Fig. 8–11. With the center of the ruler coinciding with the center of the projection, the ruler is rotated until its edge passes through a particular diffraction spot. The distance *OS* is noted and the corresponding pole plotted as shown, on the other side of center and at the corresponding distance *PQ*. This procedure is repeated for each important diffraction spot, after which the projection is transferred to a Wulff net and the poles indexed by either of the methods previously described. Note that this procedure gives a projection of the crystal from the side opposite the x-ray source, whereas the Greninger chart gives a projection of the crystal as seen from the x-ray source. A crystal orientation can, of course, be described just as well from one side as the other, and either projection can be made to coincide with the other by a 180° rotation of the projection about its *EW*-axis. Although simple to use and construct, the stereographic ruler is not as accurate as the Greninger chart in the solution of back-reflection patterns.

The methods of determining and describing crystal orientation have been presented here exclusively in terms of cubic crystals, because these are the simplest kind to consider and the most frequently encountered. These methods are quite general, however, and can be applied to a crystal of any system as long as its interplanar angles are known.

However, a noncubic crystal may have such low symmetry and/or be so oriented that the Laue pattern shows only one spot, or none at all, from a low-index plane. Plane indexing can then be difficult. Methods of coping with this problem include the following:

1. The crystal is re-oriented, in a manner suggested by the Laue pattern, and examined in a diffractometer. See Sec. 8–5.

2. A set of simulated Laue patterns, covering the unit stereographic triangle of the crystal, is generated by a computer [8.2]. The simulated pattern which most nearly matches the unknown yields tentative *(hkl)* indices for three prominent spots. Measurements on the film of the angles between these spots and the film center (incident x-ray beam) yield tentative indices for the crystal plane normal to the incident beam. A simulated pattern is then generated for this special orientation and compared with the unknown pattern to verify the plane indexing.

Specialized books on the Laue methods are those of Amoros *et al.* [G.40] and Preuss *et al.* [G.45]. The latter contains a catalog of back-reflection patterns, many generated by a computer.

8–3 TRANSMISSION LAUE METHOD

Given a specimen of sufficiently low absorption, a transmission Laue pattern can be obtained and used, in much the same way as a back-reflection Laue pattern, to reveal the orientation of the crystal.

In either Laue method, the diffraction spots on the film, due to the planes of a single zone in the crystal, always lie on a curve which is some kind of conic section. When the film is in the transmission position, this curve is a complete ellipse for sufficiently small values of ϕ, the angle between the zone axis and the transmitted beam (Fig. 8–12). For somewhat larger values of ϕ, the ellipse is incomplete because of the finite size of the film. When $\phi = 45°$, the curve becomes a parabola; when ϕ exceeds 45°, a hyperbola; and when $\phi = 90°$, a straight line. In all cases, the curve passes through the central spot formed by the transmitted beam.

The angular relationships involved in the transmission Laue method are illustrated in Fig. 8–13. Here a reference sphere is described about the crystal at *C*, the incident beam entering the sphere at *I* and the transmitted beam leaving at *O*. The film is placed tangent to the sphere at *O*, and its upper right-hand corner, viewed from the crystal, is cut off for identification of its position during the x-ray exposure. The beam reflected by the lattice plane shown strikes the film at *R*, and the normal to this plane intersects the sphere at *P*.

Suppose we consider diffraction from planes of a zone whose axis lies in the

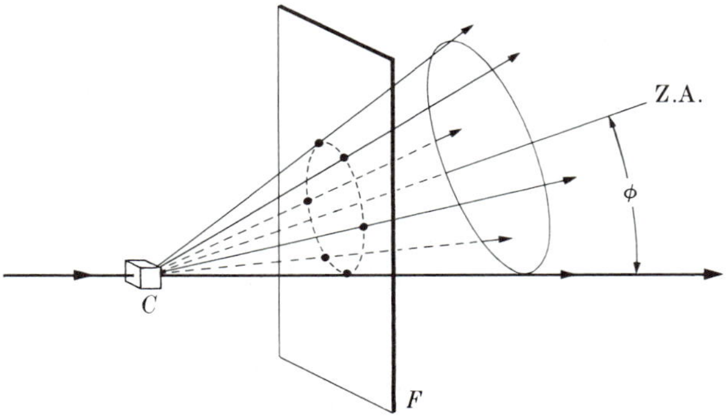

Fig. 8–12 Intersection of a conical array of diffracted beams with a film placed in the transmission position. C = crystal, F = film, Z.A. = zone axis.

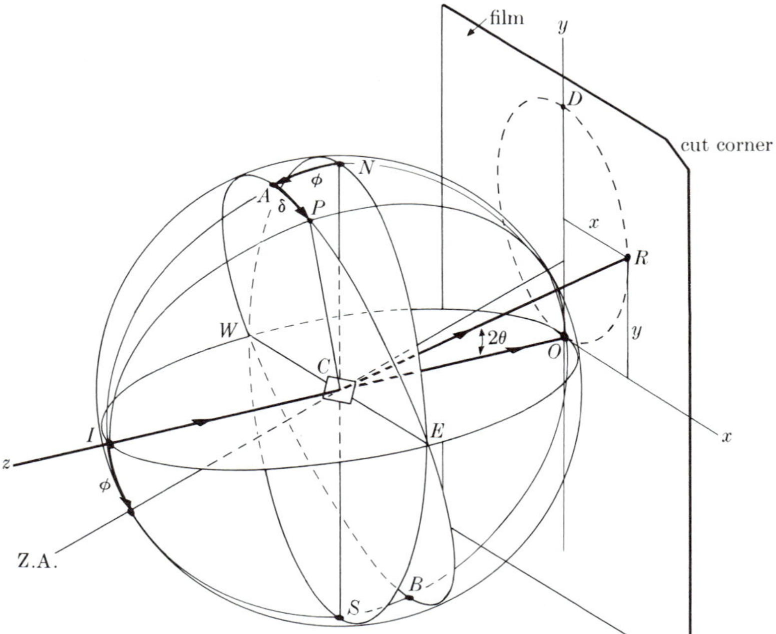

Fig. 8–13 Relation between plane normal orientation and diffraction spot position in the transmission Laue method.

yz-plane at an angle ϕ to the transmitted (or incident) beam. If a single plane of this zone is rotated so that its pole, initially at *A*, travels along the great circle *APEBWA*, then it will pass through all the orientations in which planes of this zone might occur in an actual crystal. During this rotation, the diffraction spot on the film, initially at *D*, would travel along the elliptical path *DROD* shown by the dashed line.

Any particular orientation of the plane, such as the one shown in the drawing, is characterized by particular values of ϕ and δ, the angular coordinates of its pole. These coordinates in turn, for a given crystal-film distance *D* (= *CO*), determine the *x*,*y* coordinates of the diffraction spot *R* on the film. From the spot position we can therefore determine the plane orientation, and one way of doing this is by means of the Leonhardt chart [2.3] shown in Fig. 8–14.

This chart is exactly analogous to the Greninger chart for solving back-reflection patterns and is used in precisely the same way. It consists of a grid

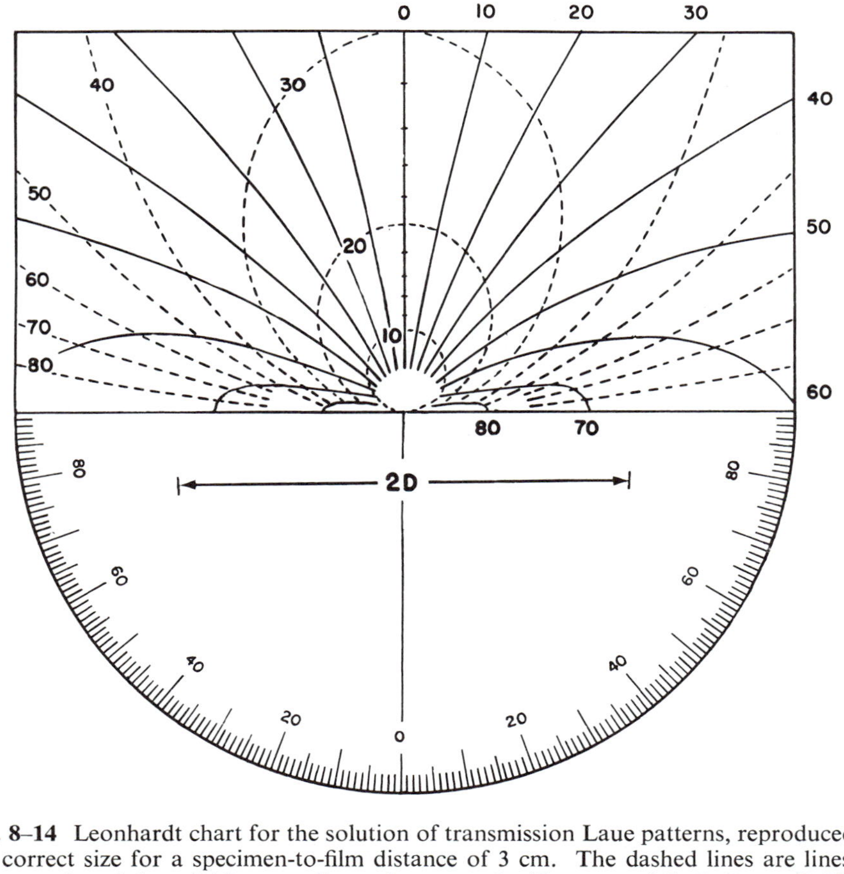

Fig. 8–14 Leonhardt chart for the solution of transmission Laue patterns, reproduced in the correct size for a specimen-to-film distance of 3 cm. The dashed lines are lines of constant ϕ, and the solid lines are lines of constant δ. (Courtesy of C. G. Dunn [8.3].)

composed of two sets of lines: the lines of one set are lines of constant ϕ and correspond to the meridians on a Wulff net, and the lines of the other are lines of constant δ and correspond to latitude lines. By means of this chart, the pole of a plane causing any particular diffraction spot may be plotted stereographically. The projection plane is tangent to the sphere at the point I of Fig. 8–13 and the projection is made from the point O. This requires that the film be read from the side facing the crystal, i.e., with the cut corner at the upper right. Figure 8–15 shows how the pole corresponding to a particular spot is plotted when the film and chart are in the parallel position. An alternate way of using the chart is to rotate it about its center until a line of constant ϕ coincides with a row of spots from planes of a single zone, as shown in Fig. 8–16; knowing ϕ and the rotation angle ε, we can then plot the axis of the zone directly.

A stereographic ruler may be constructed for the transmission method and it will give greater accuracy of plotting than the Leonhardt chart, particularly when the angle ϕ approaches 90°. Figure 8–17, which is a section through the incident beam and any diffracted beam, shows that the distance of the diffraction spot from the center of the film is given by

$$OS = D \tan 2\theta.$$

The distance of the pole of the reflecting plane from the center of the projection is given by

$$PQ = R \tan\left(45^\circ - \frac{\theta}{2}\right).$$

Figure 8–18 illustrates the use of a ruler constructed according to these equations. In this case, the projection is made on a plane located on the same side of the crystal as the film and, accordingly, the film must be read with its cut corner in the upper left-hand position.

Whether the chart or the ruler is used to plot the poles of reflecting planes, they are indexed in the same way as back-reflection patterns. The location of the projected poles is quite different for the two x-ray methods. The poles of planes responsible for observed spots on a transmission film are all located near the edge of the projection, since such planes must necessarily be inclined at small angles to the incident beam. The reverse is true of back-reflection patterns.

8-4 DIFFRACTOMETER METHOD

Still another method of determining crystal orientation involves the use of the diffractometer and a procedure radically different from that of either Laue method. With the essentially monochromatic radiation used in the diffractometer, a single crystal will produce a reflection only when its orientation is such that a certain set of reflecting planes is inclined to the incident beam at an angle θ which satisfies the Bragg law for that set of planes and the characteristic radiation employed. But when the counter, fixed in position at the corresponding angle 2θ, discloses that a

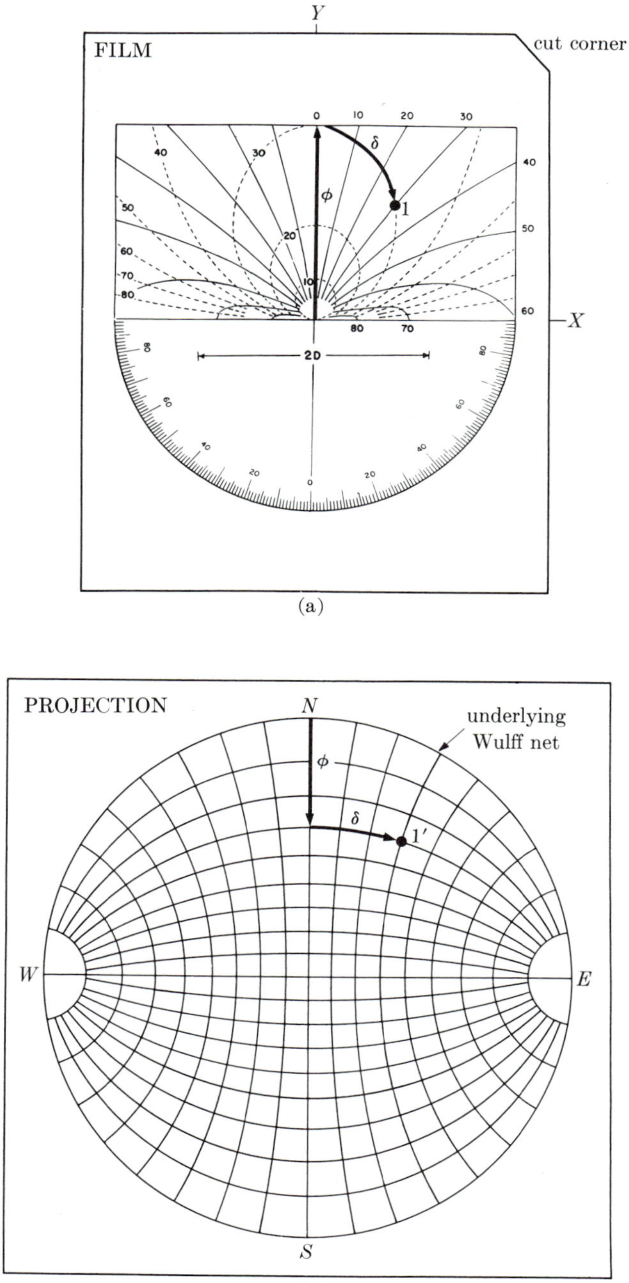

(a)

(b)

Fig. 8–15 Use of the Leonhardt chart to plot the pole of a plane on a stereographic projection. Pole 1′ in (b) is the pole of the plane causing diffraction spot 1 in (a).

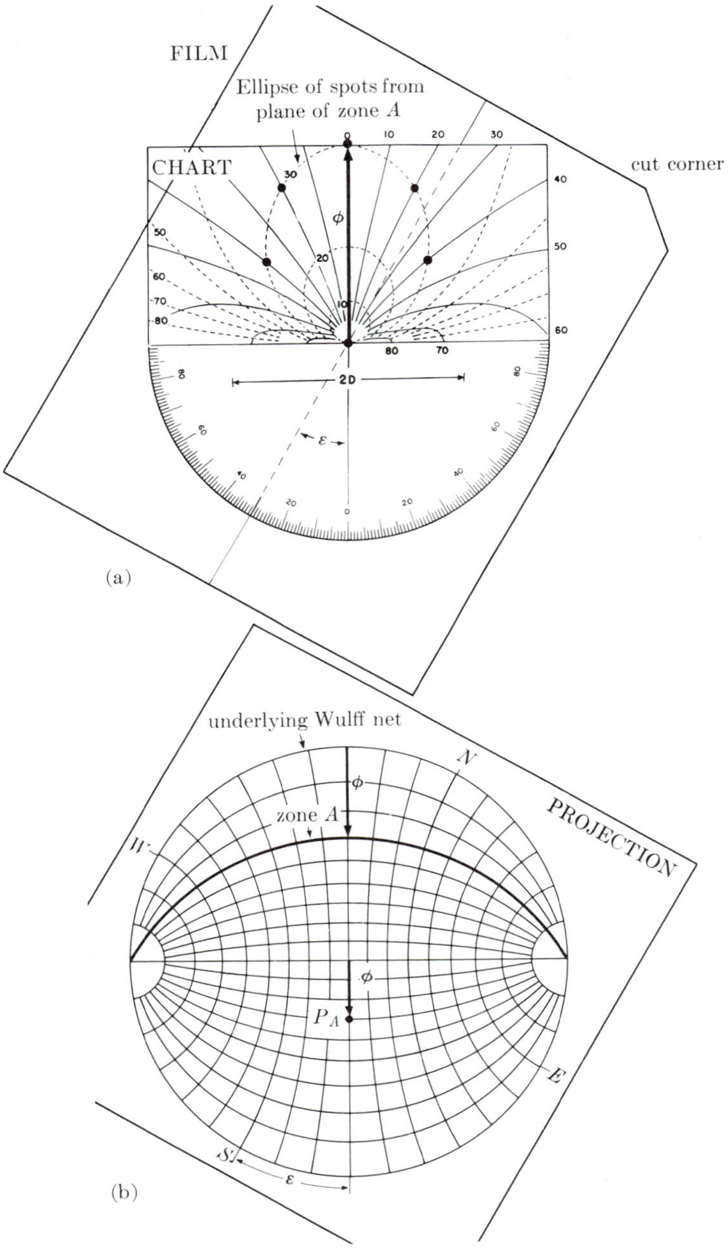

Fig. 8–16 Use of the Leonhardt chart to plot the axis of a zone of planes on the projection. P_A is the axis of zone A.

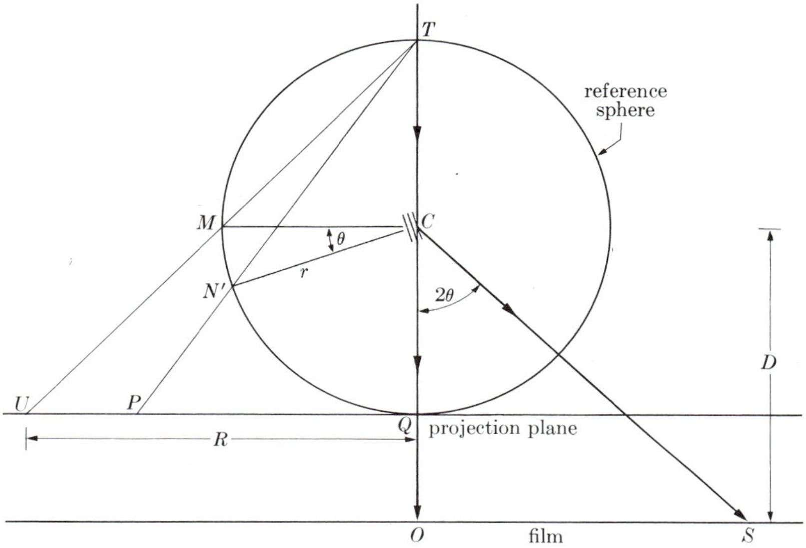

Fig. 8–17 Relation between diffraction spot S and stereographic projection P of the plane causing the spot, in transmission.

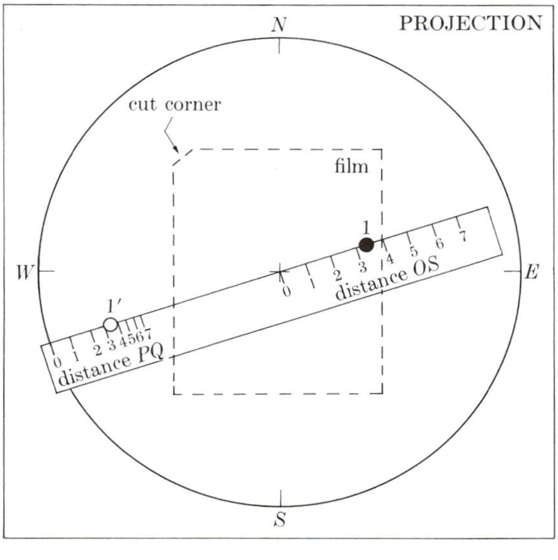

Fig. 8–18 Use of a stereographic ruler to plot the pole of a reflecting plane on a stereographic projection in the transmission Laue method. Pole 1′ is the pole of the plane causing diffraction spot 1.

reflection *is* produced, then the inclination of the reflecting planes to any chosen line or plane on the crystal surface is known from the position of the crystal. Two kinds of operation are required:

1. rotation of the crystal about various axes until a position is found for which reflection occurs,

2. location of the pole of the reflecting plane on a stereographic projection from the known angles of rotation.

The diffractometer method has many variations, depending on the particular kind of goniometer used to hold and rotate the specimen. Only one of these variations will be described here, that involving the goniometer used in the reflection method of determining preferred orientation, since that is the kind most generally available in metallurgical laboratories. This specimen holder, to be described in detail in Sec. 9–8, needs very little modification for use with single crystals, the chief one being an increase in the width of the primary beam slits in a direction parallel to the diffractometer axis in order to increase the diffracted intensity. This type of holder provides the three possible rotation axes shown in Fig. 8–19: one coincides with the diffractometer axis, the second (AA') lies in the plane of the incident beam I and diffracted beam D and tangent to the specimen surface, shown here as a flat plate, while the third (BB') is normal to the specimen surface.

Suppose the orientation of a cubic crystal is to be determined. For such crystals it is convenient to use the {111} planes as reflectors; there are four sets of these and their reflecting power is usually high. First, the 2θ value for the 111 reflection (or, if desired, the 222 reflection) is computed from the known spacing of the {111} planes and the known wavelength of the radiation used. The counter is then fixed in this 2θ position. The specimen holder is now rotated about the diffractometer axis until its surface, and the rotation axis AA', is equally inclined to the incident beam and the diffracted beam, or rather, to the line from crystal to counter with which the diffracted beam, when formed, will coincide. The specimen holder is then fixed in this position, no further rotation about the diffractometer axis being required. Then, by rotation about the axis BB', one edge of the specimen or a line drawn on it is made parallel to the diffractometer axis. This is the initial position illustrated in Fig. 8–19.

The crystal is then slowly rotated about the axes AA' and BB' until an indication of a reflection is observed on the counting-rate meter. Once a reflecting position of the crystal has been found, we know that the normal to one set of {111} planes coincides with the line CN, that is, lies in the plane of the diffractometer circle and bisects the angle between incident and diffracted beams. The pole of these diffracting planes may now be plotted stereographically, as shown in Fig. 8–20. The projection is made on a plane parallel to the specimen surface, and with the NS-axis of the projection parallel to the reference edge or line mentioned above. When the crystal is rotated β degrees about BB' from its initial position, the projection is also rotated β degrees about its center. The direction CN, which

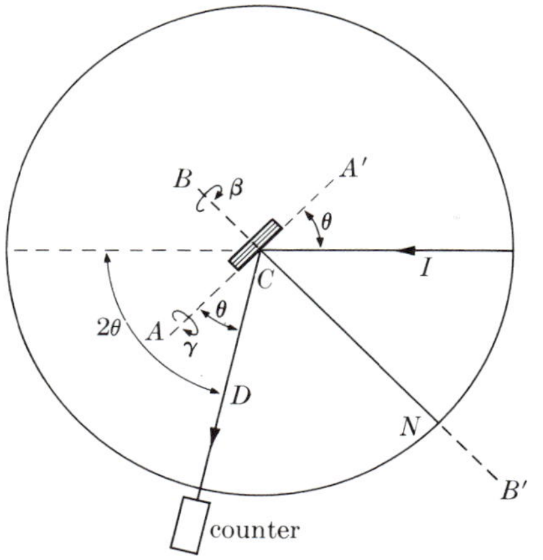

Fig. 8–19 Crystal rotation axes for the diffractometer method of determining orientation

might be called the normal to "potential" reflecting planes, is represented by the pole N', which is initially at the center of the projection but which moves γ degrees along a radius when the crystal is rotated γ degrees about AA'.

What we are trying to do, essentially, is to make N' coincide with a $\{111\}$ pole and so disclose the location of the latter on the projection. The search may be made by varying γ continuously for fixed values of β 4 or 5° apart; the projection is then covered point by point along a series of radii. It is enough to examine one quadrant in this way since there will always be at least one $\{111\}$ pole in any one quadrant. Once one pole has been located, the search for the second is aided by the knowledge that it must be 70.5° from the first. Although two $\{111\}$ poles are enough to fix the orientation of the crystal, a third should be located as a check.

Parenthetically, it should be noted that the positioning of the crystal surface and the axis AA' at equal angles to the incident and diffracted beams is done only for convenience in plotting the stereographic projection. There is no question of focusing when monochromatic radiation is reflected from an undeformed single crystal, and the ideal incident beam for the determination of crystal orientation is a parallel beam, not a divergent one.

In the hands of an experienced operator, the diffractometer method is faster than either Laue method. Furthermore, it can yield results of greater accuracy if narrow slits are used to reduce the divergence of the incident beam, although the use of extremely narrow slits will make it more difficult to locate the reflecting positions of the crystal. On the other hand, the diffractometer method furnishes no permanent record of the orientation determination, whereas Laue patterns may be filed away for future reference. But what is more important, the diffractometer method does not readily disclose the state of perfection of the crystal, whereas a Laue pattern yields this kind of information at a glance, as we will see in Sec. 8–6, and in many investigations the metallurgist is just as much interested in the relative perfection of a single crystal as he is in its orientation.

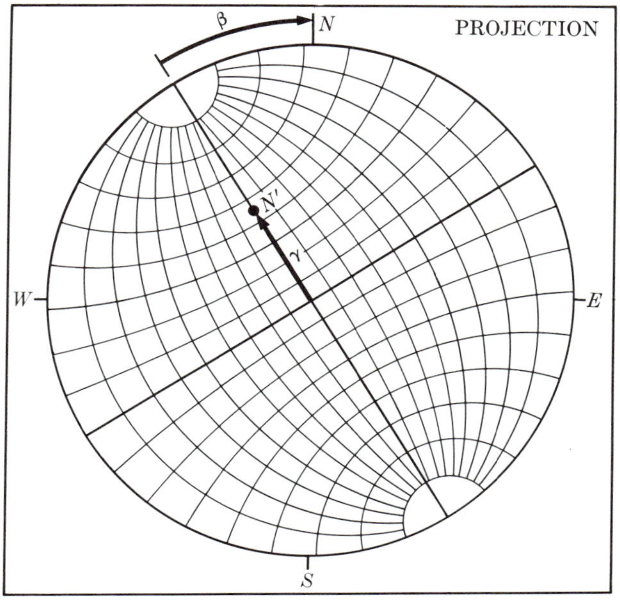

Fig. 8–20 Plotting method used when determining crystal orientation with the diffractometer. (The directions of the rotations shown here correspond to the directions of the arrows in Fig. 8–19.)

All things considered, the Laue methods are preferable when only occasional orientation determinations are required, or when there is any doubt about the quality of the crystal. When the orientations of large numbers of crystals have to be determined in a routine manner, the diffractometer method is superior. In fact, this method was developed largely for just such an application during World War II, when the orientation of large numbers of quartz crystals had to be determined. These crystals were used in radio transmitters to control, through their natural frequency of vibration, the frequency of the transmitted signal. For this purpose quartz wafers had to be cut with faces accurately parallel to certain crystallographic planes, and the diffractometer was used to determine the orientations of these planes in the crystal.

8–5 SETTING A CRYSTAL IN A REQUIRED ORIENTATION

After the orientation of a crystal is found by x-rays, it is often necessary to rotate it into some special orientation, such as one with ⟨100⟩ along the incident beam, for the purpose of either (a) subsequent x-ray examination in the special orientation, or (b) subsequent cutting along some selected plane. To obtain this orientation, the crystal is mounted in a three-circle goniometer like that shown in Fig. 5–7, whose arcs have been set at zero, and its orientation is determined by, for example, the back-reflection Laue method. A projection of the crystal is then made, and from this projection the goniometer rotations which will bring the crystal into the required orientation are determined.

For example, suppose it is required to rotate the crystal whose orientation is given by Fig. 8–7 into a position where [011] points along the incident beam and [100] points horizontally to the left, i.e., into the standard (011) orientation shown by Fig. 2–36(b) if the latter were rotated 90° about the center. The initial orientation (Position 1) is shown in Fig. 8–21 by the open symbols, referred to *NSEW*-axes. Since (011) is to be brought to the center of the projection and (100) to the left side, (010) will lie on the vertical axis of the projection when the crystal is in its final position. The first step therefore is to locate a point 90° away from (011) on the great circle joining (010) to (011), because this point must coincide with the north pole of the final projection. This is simply a construction point; in the present case it happens to coincide with the $(0\bar{1}1)$ pole, but generally it is of no crystallographic significance. The projection is then rotated 22° clockwise about the incident-beam axis to bring this point onto the vertical axis of the underlying Wulff net. (In Fig. 8–21, the latitude and longitude lines of this net have been omitted for clarity.) The crystal is now in Position 2, shown by open symbols referred to *N′S′E′W′*-axes. The next rotation is performed about the *E′W′*-axis, which requires that the underlying Wulff net be arranged with its

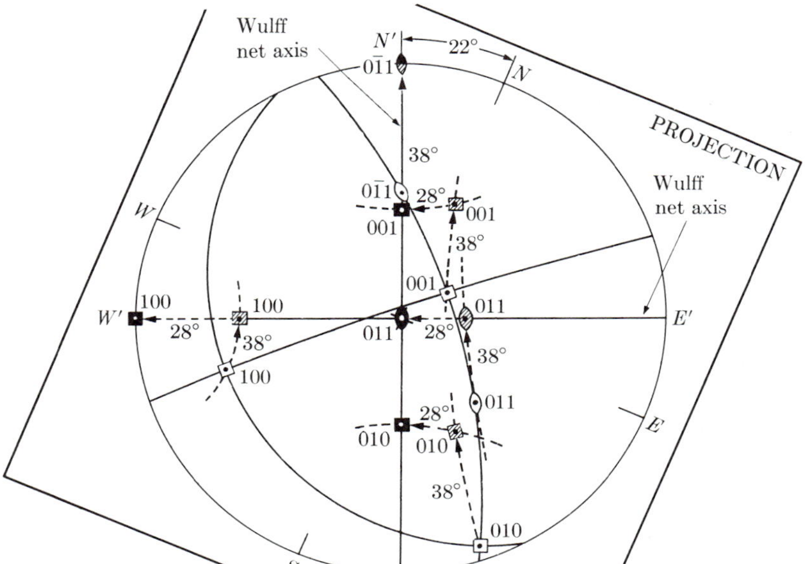

Fig. 8–21 Crystal rotation to produce specified orientation. Positions 1 and 2 are indicated by open symbols, position 3 by shaded symbols, and position 4 by solid symbols.

equator vertical so that the latitude lines will run from top to bottom. This rotation, of 38°, moves all poles along latitude lines, shown as dashed small circles, and brings $(0\bar{1}1)$ to the N'-pole, and (100) and (011) to the $E'W'$-axis of the projection, as indicated by the shaded symbols (Position 3). The final orientation is obtained by a 28° rotation about the $N'S'$-axis, with the equator of the underlying Wulff net now horizontal; the poles move to the positions shown by solid symbols (Position 4).

The necessity for selecting a construction point 90° from (011) should now be evident. If this point, which here happens to be $(0\bar{1}1)$, is brought to the N'-pole, then (011) and (100) must of necessity lie on the $E'W'$-axis; the final rotation about $N'S'$ will then move the latter to their required positions without disturbing the position of the $(0\bar{1}1)$ pole, since $[0\bar{1}1]$ coincides with the $N'S'$-axis.

The order of these three rotations is not arbitrary. The stereographic rotations correspond to physical rotations on the goniometer and must be made in such a way that one rotation does not physically alter the position of any axis about which a subsequent rotation is to be made. The goniometer used here was initially set with the axis of its uppermost arc horizontal and coincident with the primary beam, and with the axis of the next arc horizontal and at right angles to the incident beam. The first rotation about the beam axis therefore did not disturb the position of the second axis (the $E'W'$-axis), and neither of the first two rotations disturbed the position of the third axis (the vertical $N'S'$-axis). Whether or not the stereographic orientations are performed in the correct order makes a great difference in the rotation angles found, but once the right angles are determined by the correct stereographic procedure, the actual physical rotations on the goniometer may be performed in any sequence.

The back-reflection Laue pattern of an aluminum crystal rotated into the orientation described above is shown in Fig. 8–22. Note that the arrangement of spots has 2-fold rotational symmetry about the primary beam, corresponding to the 2-fold rotational symmetry of cubic crystals about their $\langle 110 \rangle$ axes. (Conversely, the observed symmetry of the Laue pattern of a crystal of unknown structure is an indication of the kind of symmetry possessed by that crystal. Thus the Laue method can be used as an aid in the determination of crystal structure.)

The crystal-setting procedure illustrated in Fig. 8–21 can be carried out whether or not the indices of the various poles are known. If the Laue pattern of a crystal is difficult to solve, any spot on it can be indexed by using a Laue camera and a diffractometer in sequence [8.4]. In addition, a goniometer is required that fits both instruments. The procedure is as follows:

1. Make a Laue pattern and a stereographic projection of the poles corresponding to a few important spots.

2. By the procedure of Fig. 8–21, rotate the pole to be indexed to the center of the projection. The corresponding rotation on the goniometer will make the unknown plane (hkl) normal to the incident beam of the Laue camera.

3. Transfer the goniometer and crystal to the diffractometer, in such a way that the (hkl) plane normal bisects the angle between the incident and (potential) diffracted beams.

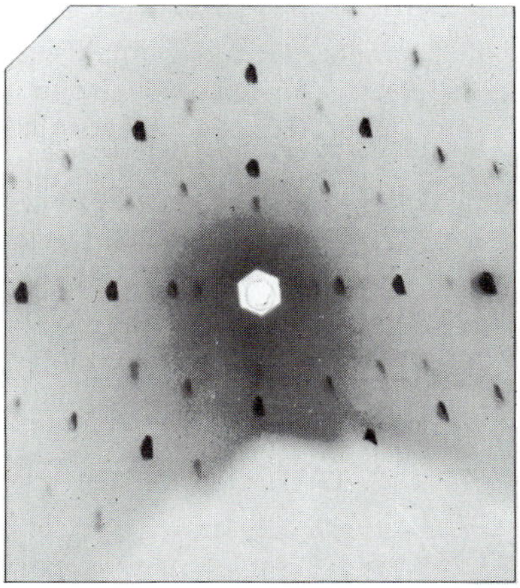

Fig. 8-22 Back-reflection Laue pattern of an aluminum crystal. The incident beam is parallel to [011], [0$\bar{1}$1] points vertically upward, and [100] points horizontally to the left. Tungsten radiation, 30 kV, 19 mA, 40 min exposure, 5 cm specimen-to-film distance. (The light shadow at the bottom is that of the goniometer which holds the specimen.)

4. Make a diffractometer scan to find the angle 2θ at which diffraction occurs from the (hkl) planes. [Higher-order reflections may also be observed, i.e., reflections from (nh, nk, nl) planes.]

5. Calculate the spacing d of the (hkl) planes from 2θ and the known value of λ.

6. From this value of d determine (hkl) by calculation from the known crystal structure or by examining a list of known d spacings for the substance involved (Sec. 14-3).

There is another method of setting a crystal in a standard orientation, which does not require either photographic registration of the diffraction pattern or stereographic manipulation of the data. It depends on the fact that the diffracted beams formed in the transmission Laue method are so intense, for a crystal of the proper thickness, that the spots they form on a fluorescent screen are visible in a dark room. The observer merely rotates the crystal about the various arcs of the goniometer until the pattern corresponding to the required orientation appears on the screen. Obviously, he must be able to recognize this pattern when it appears, but a little study of a few Laue photographs made of crystals in standard orientations will enable him to do this. The necessity for working in a darkened room may be avoided by use of a light-tight viewing box, if the job of crystal setting occurs sufficiently often to justify its construction. This box encloses the fluorescent screen which the observer views through a binocular eyepiece set in the wall of the box, either directly along the direction of the transmitted beam, or indirectly in a direction at right angles by means of a mirror or a right-angle prism. For x-ray protection, the optical system should include lead glass, and the observer's hands should be shielded during manipulation of the crystal.

More elaborate apparatus permits electronic amplification of transmission Laue spots formed on a fluorescent screen. An image of the spot pattern on the screen is projected by a lens on to the front (input) face of an image-intensifier tube. The intensified image appears on the rear (output) face of the tube and is bright enough to be photographed in 1/220 second [8.5–8.7]. This large gain in image intensity permits photography, by a motion picture or television camera, of fast changes in a Laue pattern caused, for example, by a phase change in the crystal. (The image-intensifier tube has also made it possible to obtain a transmission Laue photograph with a single 30-nanosecond pulse from a high-power pulsed x-ray tube [8.8].)

CRYSTAL QUALITY

The quality of what is nominally a "single" crystal can vary over an enormous range. At one extreme, the crystal may have undergone gross plastic deformation by bending and/or twisting, such that some portions of it are disoriented from other portions by angles as large as tens of degrees, and the dislocation density is high. At the other extreme, some carefully grown crystals are almost free of dislocations and other line or planar imperfections, and their crystal planes are flat to less than 10^{-4} degrees over distances of the order of a centimeter. In general, metal crystals tend to be more imperfect than crystals of covalent or ionic substances.

Various x-ray methods of assessing crystal quality are described below. These methods differ in sensitivity, and we will deal with the least sensitive first.

8-6 LAUE METHODS

Either Laue method, transmission or back-reflection, easily discloses gross plastic deformation. Any change in the orientation of the reflecting planes is accompanied by a corresponding change in the direction, and wavelength, of the reflected beam. In fact, Laue reflection of x-rays is often compared to the reflection of visible light by a mirror. An undistorted crystal gives sharp Laue spots. In a bent or twisted crystal, the continuous change in orientation of the reflecting planes smears the Laue spots into streaks, just as a spot of light reflected onto a screen by a flat mirror becomes elongated when the mirror is curved.

Ordinary Laue methods

If a crystal has been bent about a single axis, both the Miller indices of the bending axis and the extent of the bending can usually be determined stereographically; each Laue streak is plotted as an arc representing the range of orientation of the corresponding lattice plane, and a rotation axis that will account for the directions of these arcs on the projection is found. The angular lengths of the arcs are a measure of the amount of bending which has occurred. In measuring the amount of bending by this method, it must be remembered that the wavelengths present in the incident beam do not cover an infinite range. There is no radiation of wavelength shorter than the short-wavelength limit, and on the long-wavelength side the intensity decreases continuously as the wavelength increases. This means that, for a given degree of lattice bending, some Laue streaks may not be as long as they might be if a full range of wavelengths were available. The amount of bending estimated from the lengths of these streaks would therefore be smaller than that actually present.

Transmission and back-reflection Laue patterns made from the same deformed region usually differ markedly in appearance. The photographs in Fig. 8–23 were made, under identical conditions, of the same region of a deformed aluminum crystal having the same orientation relative to the incident beam for each photograph. Both show elongated spots, which are evidence of lattice bending, but the spots are elongated primarily in a radial direction on the transmission pattern while on the back-reflection pattern they tend to follow zone lines. The term *asterism* (from the Greek *aster* = star) was used initially to describe the starlike appearance of a transmission pattern such as Fig. 8–23(a), but it is now used to describe any form of streaking, radial or nonradial, on either kind of Laue photograph.

The striking difference between these two photographs is best understood by considering a very general case. Suppose a crystal is so deformed that the normal to a particular set of reflecting lattice planes describes a small cone of apex angle 2ε; i.e., in various parts of the crystal the normal deviates by an angle ε in all directions from its mean position. This is equivalent to rocking a flat mirror through the same angular range and, as Fig. 8–24 shows, the reflected spot S is roughly elliptical on a film placed in the transmission position. When the plane normal rocks through the angle 2ε in the plane ACN, the reflected beam moves through an angle 4ε, and the major axis of the ellipse is given approximately by

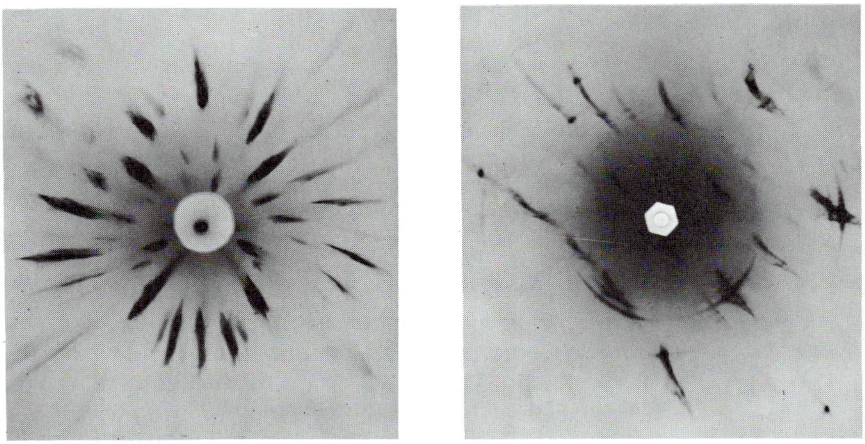

(a) Transmission (b) Back reflection

Fig. 8–23 Laue photographs of a deformed aluminum crystal. Specimen-to-film distance 3 cm, tungsten radiation, 30 kV.

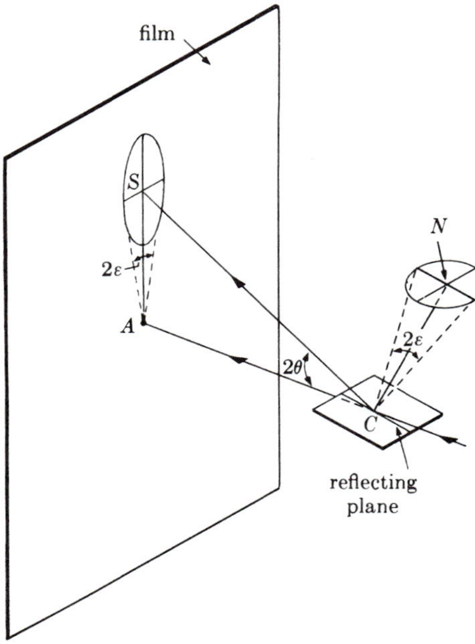

Fig. 8–24 Effect of lattice distortion on the shape of a transmission Laue spot. *CN* is the normal to the reflecting plane.

$4\varepsilon(AC)$ when 2θ is small. On the other hand, when the plane normal rocks through the angle 2ε in a direction normal to the plane of reflection *ACN*, the only effect is to rock the plane of reflection through the same angle 2ε about the incident beam. The minor axis of the elliptical spot is therefore given by $2\varepsilon(AS) \approx 2\varepsilon(AC) \tan 2\theta \approx 2\varepsilon(AC)2\theta$. The shape of the spot is characterized by the ratio

$$\frac{\text{Major axis}}{\text{Minor axis}} = \frac{4\varepsilon(AC)}{2\varepsilon(AC)2\theta} = \frac{1}{\theta}.$$

For $2\theta = 10°$, the major axis is some 12 times the length of the minor axis.

In the back-reflection region, the situation is entirely different and the spot *S* is roughly circular, as shown in Fig. 8–25. Both axes of the spot subtend an angle of approximately 4ε at the crystal. We may therefore conclude that the shape of a back-reflection spot is more directly related to the nature of the lattice distortion than is the shape of a transmission spot since, in the general case, circular motion of the end of the reflecting plane normal causes circular motion of the backward-reflected beam but elliptical motion of the forward-reflected beam. For this reason, the back-reflection method is generally preferable for studies of lattice distortion. It must not be supposed, however, that only radial streaking is possible on transmission patterns. The direction of streaking depends on the orientation of the axis about which the reflecting planes are bent and if, for example, they are bent only about an axis lying in the plane *ACN* of Fig. 8–24, then the spot will be elongated in a direction at right angles to the radius *AS*.

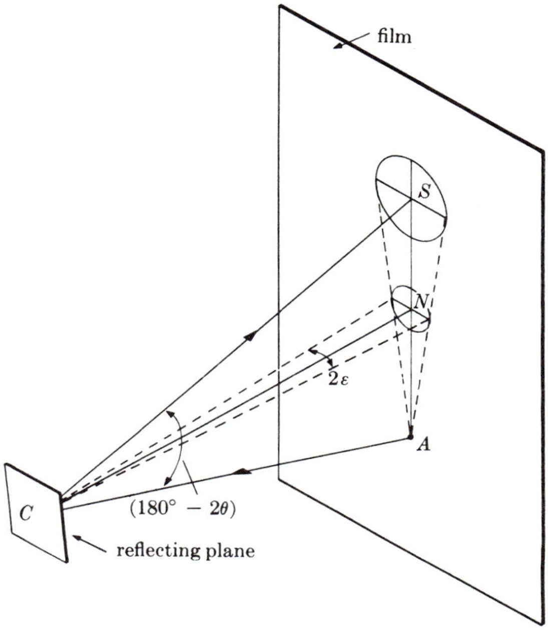

Fig. 8–25 Effect of lattice distortion on the shape of a back-reflection Laue spot. *CN* is the normal to the reflecting plane.

One feature of the back-reflection pattern of Fig. 8–23 deserves some comment, namely, the short arcs, concentric with the film center, which pass through many of the elongated Laue spots. These are portions of Debye rings, such as one might expect on a pinhole photograph made of a polycrystalline specimen with characteristic radiation (Sec. 6–9). With a polycrystalline specimen of randomly oriented grains a complete Debye ring is formed, because the normals to any particular set of planes (*hkl*) have all possible orientations in space; in a deformed single crystal, the same normals are restricted to a finite range of orientations with the result that only fragments of Debye rings appear. We may imagine a circle on the film along which a Debye ring would form if a polycrystalline specimen were used, as indicated in Fig. 8–26. If a Laue spot then becomes enlarged as a result of lattice deformation and spreads over the potential Debye ring, then a short portion of a Debye ring will form. It will be much darker than the Laue spot, since the characteristic radiation* which forms it is much more intense than the wavelengths immediately adjacent to it in the continuous spectrum. In fact, if the x-ray exposure is not sufficiently long, only the Debye arcs may be visible on the film, and the observer may be led to erroneous conclusions regarding the nature and extent of the lattice deformation.

* In Fig. 8–23(b), the characteristic radiation involved is tungsten *L* radiation. The voltage used (30 kV) is too low to excite the *K* lines of tungsten (excitation voltage = 70 kV) but high enough to excite the *L* lines (excitation voltage = 12 kV).

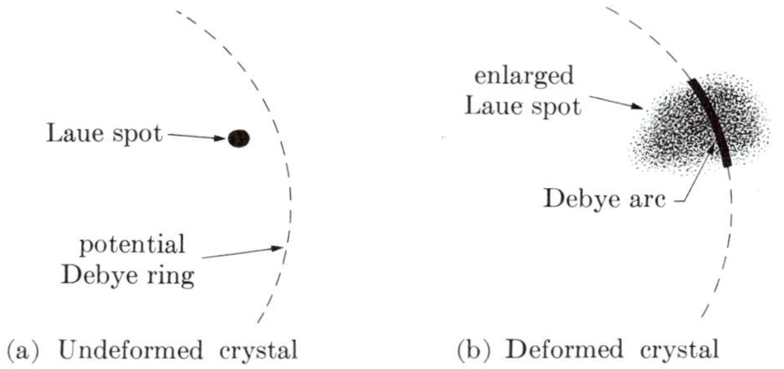

Fig. 8–26 Formation of Debye arcs on Laue patterns of deformed crystals.

With these facts in mind, re-examination of the patterns shown in Fig. 8–23 leads to the following conclusions:

1. Since the asterism on the transmission pattern is predominantly radial, lattice planes inclined at small angles to the incident beam are bent about a number of axes, in such a manner that their plane normals are confined to a small cone in space.

2. Since the asterism on the back-reflection pattern chiefly follows zone lines, the major portion of planes inclined at large angles to the incident beam are bent about a single axis. However, the existence of Debye arcs shows that there are latent Laue spots of considerable area superimposed on the visible elongated spots, and that a small portion of the planes referred to are therefore bent about a number of axes.

Laue photographs can also disclose *polygonization*, which is a process that some deformed crystals undergo when annealed at an elevated temperature. If the deformation is not too severe, plastically bent portions of the crystal break up into smaller blocks, which are relatively strain-free and disoriented by approximately the same total amount (never more than a few degrees) as the bent fragment from which they originate, as suggested by Fig. 8–27. (The term "polygonization" describes the fact that a certain crystallographic direction [*uvw*] forms part of an

Fig. 8–27 Reflection of white radiation by bent and polygonized lattices (schematic).

arc before annealing and part of a polygon afterwards.) Moreover, the mean orientation of the blocks is the same as that of the parent fragment. The effect of polygonization on a Laue pattern is therefore to replace an elongated Laue streak (from the bent lattice) with a row of small sharp spots (from the individual blocks) occupying the same position on the film, provided each block is sufficiently disoriented from its neighbor so that the beams reflected by adjoining blocks are resolved one from another. Figure 8–28 shows an example of polygonization in a crystal of silicon ferrite. A single elongated Laue spot from the deformed crystal appears in (a), while (b) and (c) show the breakup of this spot caused by annealing.

A polygonized crystal such as this may be regarded as an extreme example of the mosaic structure depicted in Fig. 3–16, extreme in the sense that the average angle between adjacent blocks (subgrains) of the polygonized crystal is much larger than normal.

The sensitivity of the ordinary Laue method in the detection of crystal disorientation may be estimated as follows. Suppose the crystal-to-film distance is 5 cm, and assume that the minimum detectible broadening of a Laue spot is 1 mm. Then the diffracted beam has diverged by about 1/50 radian or 1°. This divergence corresponds to a disorientation of the reflecting planes of about 0.5°. This disorientation applies only to the area irradiated by the incident beam, which is typically 1 mm in diameter; the irradiated area is therefore only 10^{-2} cm^2.

Guinier–Tennevin Method

This method [8.10, 8.11] is a variant of the transmission Laue method and exploits the focusing effect shown in Fig. 5–10. A large increase in sensitivity is obtained by increasing the source-to-crystal and crystal-to-film distances to the order of 50 cm. One therefore has a rather long "optical lever" that can disclose slight disorientations of the crystal.

Figure 8–29 shows the geometry of the method. This drawing is a horizontal section and shows white radiation diverging from a source S, either a point focus

(a)

(b)

(c)

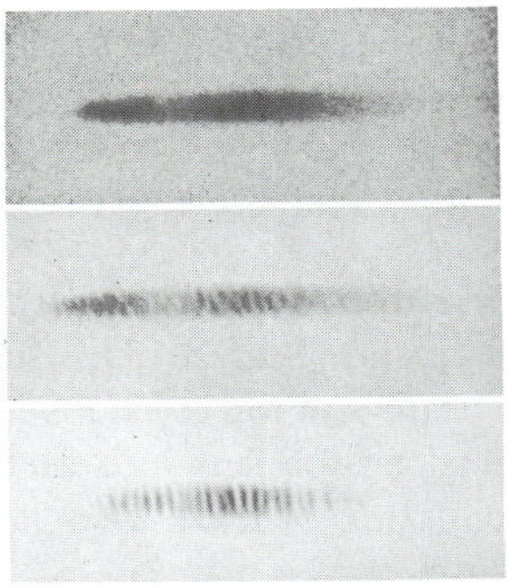

Fig. 8–28 Enlarged transmission Laue spots from a thin crystal of silicon ferrite (α-iron containing 3.3 percent silicon in solid solution): (a) as bent to a radius of 9 mm., (b) after annealing 10 min at 950°C, (c) after annealing 4 hr at 1300°C. Dunn and Daniels [8.9].

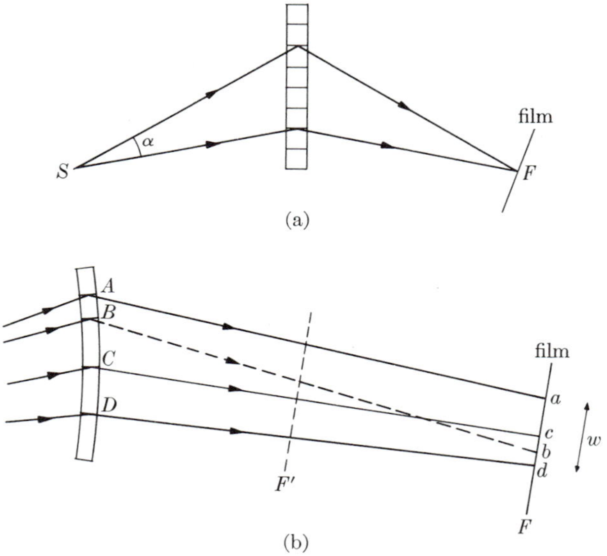

Fig. 8–29 Guinier–Tennevin method. (a) Perfect crystal. Focused diffracted beam. (b) Bent or polygonized crystal. Enlarged view of nonfocused beam.

or a fine vertical-line focus on the x-ray tube target, and incident on a thin crystal. The horizontal divergence angle α is 1°–3°. If the crystal is perfect, as in (a), the rays reflected by the transverse planes shown converge to a focus at F. The Laue spot on the film is then a thin vertical line, as in Fig. 8–30(a). Note that we are concerned with only *one* Laue spot, and not with an arrangement of spots as in an ordinary Laue pattern.

If the crystal is bent or polygonized, as in Fig. 8–29(b), the reflected beam will broaden and the Laue spot on the film will be a rectangle of width w. (The dashed lines in this drawing will be discussed later.) The disorientation of the crystal may be calculated from the magnitude of w and the other dimensions involved. Disorientations as small as 5×10^{-3} degree are detectible, and this value pertains to an irradiated crystal area of 1–2 cm^2. Fig. 8–30(b) shows a broadened reflection of this kind.

A modification of the Guinier–Tennevin method can reveal additional information about a deformed crystal [8.12, 8.13]. If a Soller slit with horizontal plates is placed in the incident beam and the crystal is twisted about a vertical axis, the original vertical-line Laue spot broadens into a striated region composed of fine inclined lines, and the inclination of these lines is a measure of the torsional strain in the crystal (Fig. 8–30(c)).

Spot (d) of Fig. 8–30 is from a plate-shaped aluminum crystal that had been plastically deformed a small amount. This spot contains considerable internal detail. The striations show that portions of the crystal are twisted, and the dark

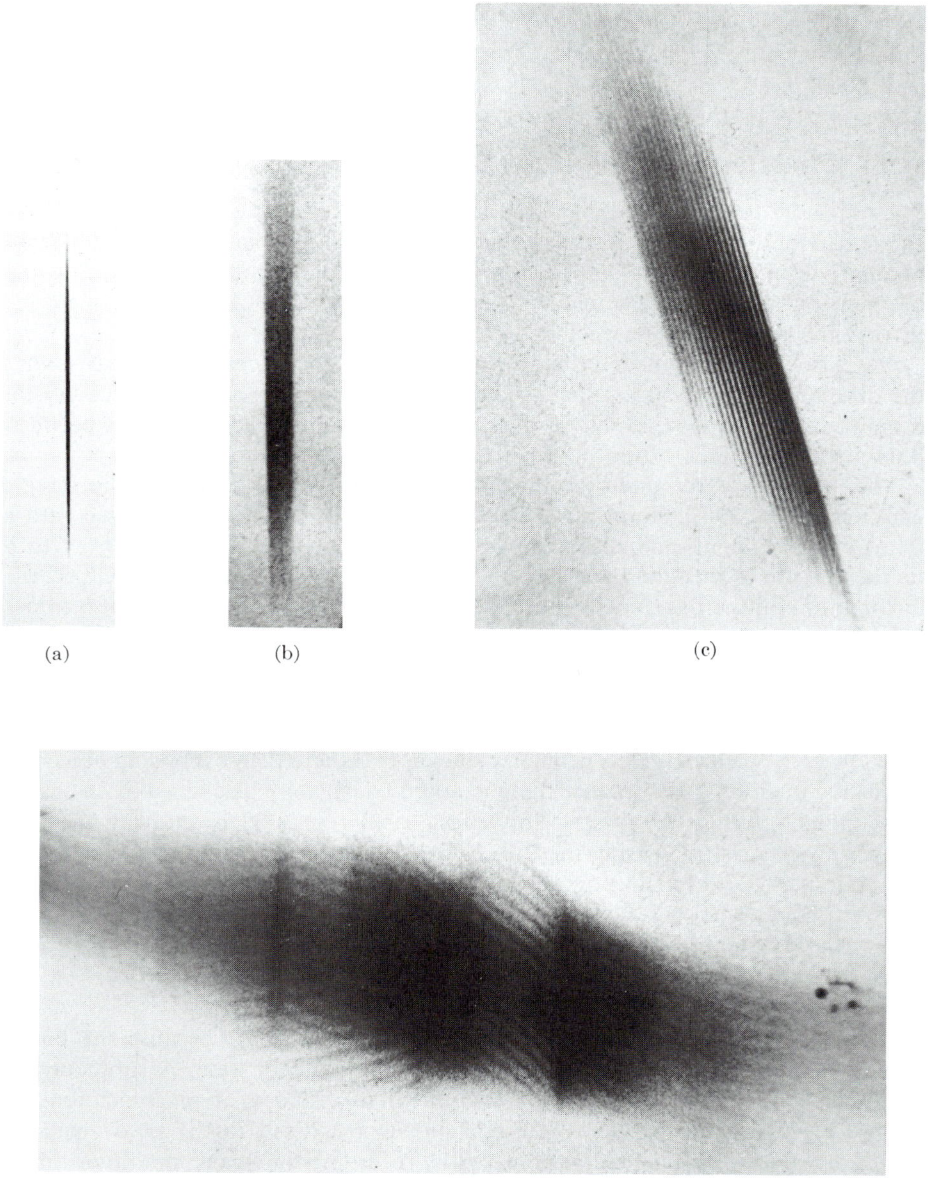

(a) (b) (c)

(d)

Fig. 8–30 Single Laue spots obtained by the Guinier–Tennevin method on a film placed at the focusing position F of Fig. 8–29(a). Spots (a), (b), and (c) are from the transverse planes of a quartz crystal plate, 37 × 13 × 0.5 mm: (a) unstrained, magnification 2X, (b) elastically bent, 2X, (c) elastically twisted, 5X. Spot (d) is from an aluminum crystal after plastic deformation, 4X. Julien *et al.* [8.12–8.14].

vertical lines, of which two are prominent, show that fragmentation (polygonization) has occurred. Each line is a focused reflection from a nearly perfect portion of the crystal.

8–7 TOPOGRAPHIC AND OTHER METHODS

We turn now to methods designed mainly for the examination of nearly perfect crystals. Here the interest is not so much in measuring the angular disorientation of one part of the crystal with respect to another, but rather in disclosing the presence of individual imperfections, such as dislocations. All of these methods involve the reflection of characteristic $K\alpha$ radiation.

Most of the methods described below are called *topographic*, in the sense that the diffraction "spot" has a fine structure that is an image or map of the crystal examined. To be topographic, a one-to-one correspondence must exist between points on the diffraction spot and points on the crystal.

To gain a clearer understanding of what constitutes a topographic image, consider Fig. 8–29(b). Suppose the crystal shown there were polygonized rather than uniformly bent, and that a subgrain at B reflected the dashed-line ray to b on the film at F. Then the Laue spot on the film would *not* be topographic, because information on the right side of the spot, viewed from the x-ray source, at b comes from the left side of the crystal, at B. But if the film were moved closer to the crystal, from F to F', the spot on the film would become a topographic image. (In fact, the only way to disclose the existence of the crossover ray shown by the dashed line is to make photographs on films moved successively closer to the crystal from position F. Note also, for the geometry shown, that there is an overriding objection to placing the film at F', because there it would be partially blackened by the beam transmitted through the crystal; this effect can be avoided by moving the film further away or by increasing the Bragg angle, or both.) We can conclude that a diffraction spot will be a topographic image if the rays of the diffracted beam are more or less parallel between crystal and film, without "crossfire." (An ordinary radiograph is topographic and derives its utility from that very fact. It is topographic because the rays forming the image proceed in straight lines from the x-ray source to the film.)

X-ray topography is sometimes called *x-ray microscopy*, because the image formed by diffracted x-rays, often only a few millimeters wide, is subsequently examined, and photographed, with an optical microscope at magnifications of 10–100 X. Such magnification is useful only if the detail in the x-ray image is sufficiently fine, and fine detail is obtained by recording the image, not on ordinary x-ray film, but on special high-resolution plates. These are glass plates coated with a thin, photographically slow, high-resolution emulsion.

Effect of Extinction

The phenomenon of extinction plays a major role in image formation in x-ray topography. As we saw in Sec. 4–12, extinction has its maximum effect in the perfect crystal. As a crystal becomes *more imperfect*, its reflecting power for x-rays *increases*. Thus, when a crystal is examined by a topographic method, the crystal's imperfect regions will reflect more strongly, and the dark regions on the topograph will reveal where the imperfections are located in the crystal.

It is not usually the imperfections themselves that are imaged on the topograph (they are much too small) but rather the strain fields around the imperfections; these strains cause the plane spacings d to vary from their equilibrium value, thereby modifying the x-ray scattering process. White [8.15] has given a vivid demonstration, on the macroscopic level, of the effect of lattice strain on reflecting power, such as occurs in x-ray topography on a microscopic scale. He reflected crystal-monochromatized Mo $K\alpha$ radiation from the surface of a thin quartz plate, elastically bent by increasing amounts, and measured the integrated intensity of the reflection (Fig. 8–31). The intensity for the unbent crystal was near the value calculated for a perfect crystal, and it increased with bending strain up to almost the value calculated for an "ideally imperfect" crystal (one in which extinction is absent). The effect was reversible and independent of whether the bending was convex or concave relative to the incident beam. The crystal plate was 0.5 mm thick, so that the surface strain was about 2×10^{-4} for the maximum bending shown in Fig. 8–31. The x-ray penetration depth (Sec. 9–5) was greater than half the thickness of the bent crystal, so that the x-rays saw strains varying from zero at the neutral axis to a maximum at the surface.

Not only does this important experiment demonstrate quite directly the relation between nonuniform lattice strain and x-ray reflecting power, but it also shows that extinction can be reversibly varied in an almost perfect crystal.

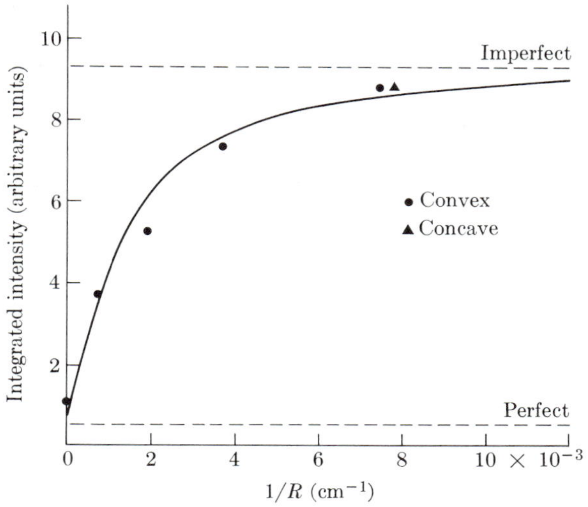

Fig. 8–31 Integrated intensity of the 202 reflection of Mo $K\alpha$ radiation from a quartz plate as a function of the amount of bending. R is the radius of curvature. The quartz plate was cut parallel to the (202) planes. After White [8.15].

Berg–Barrett Method

This method involves the reflection of $K\alpha$ radiation from the face of the specimen crystal [8.16–8.18]. There is therefore no restriction on specimen thickness.

Unfiltered radiation from the x-ray tube passes through a long collimator (Fig. 8–32), which defines a broad beam that will flood the specimen crystal. In a darkened room the crystal is adjusted so that a selected (*hkl*) plane makes the correct Bragg angle for reflection, as judged by the appearance of a bright spot on an adjacent fluorescent screen. The screen is then replaced by a photographic plate. For maximum resolution of detail the plate should be as close to the crystal as possible, preferably less than 1 mm, without touching the incident beam. The reflecting planes, as suggested in the drawing, are not normally parallel to the crystal surface; the angle 2θ should be about 90° so that the reflected rays strike the plate at right angles.

A perfect crystal would produce a uniformly dark Berg–Barrett image. Crystal imperfections affect the image mainly in two ways:

1. Local regions of the crystal with low extinction, such as those with nonuniform

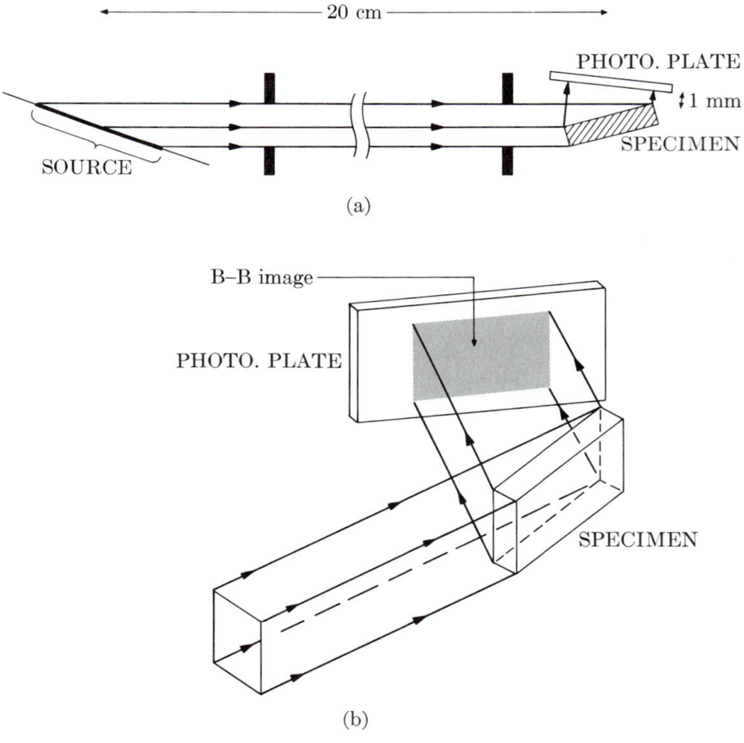

Fig. 8–32 Berg–Barrett (B–B) method. (a) Plan view. The source is the focal spot on the x-ray tube target, seen "end on" from the specimen, so that the source appears square. (b) Perspective view.

strain, will reflect more strongly and cause *darker* regions on the image, an effect called "extinction contrast."

2. Local regions that are rotated out of parallelism with the rest of the crystal will be set not quite, or not at all, at the exact Bragg angle and will reflect less strongly, causing *lighter* regions on the image.

The Berg–Barrett method has a maximum resolution of about 5 μm, and it can disclose such imperfections as subgrain boundaries, twins, and individual dislocations where they intersect or pass close to the crystal surface. The visibility of a dislocation depends on the angle between its Burgers vector **b** and the reflecting plane (hkl); maximum visibility occurs when **b** and (hkl) are at right angles; if **b** is parallel to (hkl), the dislocation will not be seen. (This condition for no contrast, which renders a dislocation invisible, is sometimes referred to as the $\mathbf{g} \cdot \mathbf{b} = 0$ criterion. The vector **g** represents the "diffraction vector," which is normal to the reflecting plane (hkl)).

The Berg–Barrett topographs in Fig. 8–33 show how the choice of reflecting plane governs dislocation visibility. The specimen is a lithium fluoride crystal, cleaved parallel to a (100) face and slightly bent. Dislocations that emerge on this face are revealed by tiny etch pits in the optical micrograph (a). Pits forming diagonal lines are due to edge dislocations; those forming horizontal lines are due to screw dislocations; both kinds of dislocations lie on {110} slip planes. Most of the screw dislocations are located to the right of the curved vertical line at the center; this line is thought to mark the place where the cleavage crack, moving from right to left, began to move so fast that dislocations had no time to form ahead of it. The x-ray topographs of the unetched surface in (b), (c), (d) reveal dislocations selectively. In (b), a 200 reflection, the emergent points of screw dislocations form dark horizontal lines, and edge dislocations are virtually invisible. On the other hand, if the image is formed by a $20\bar{2}$ reflection, as in (c), only edge dislocations are revealed, as evidenced by the dark diagonal lines. The 220 image in (d) shows both edge and screw dislocations. The thick, irregular dark lines in all three topographs are subgrain boundaries.

Lang Method

This method involves the reflection of $K\alpha$ radiation in *transmission* [8.19, 8.20]. The reflected beam will therefore be too weak if the crystal is either too thin or too thick. As will be shown in Sec. 9–8, the optimum thickness t corresponds to $\mu t \approx 1$, where μ is the linear absorption coefficient.

X-rays from the source S, a point or a line, pass through the narrow slit S_1 in Fig. 8–34 before hitting the crystal. The incident beam therefore has the shape of a very thin ribbon, intersecting the crystal along the plane AB. This beam has very little divergence in the plane of the drawing, because S_1 is a narrow slit and the source-to-slit distance D is large, several tens of centimeters. The reflecting planes, for the beam directions shown, are at right angles to the crystal surface, and these planes must make the right Bragg angle for the characteristic radiation

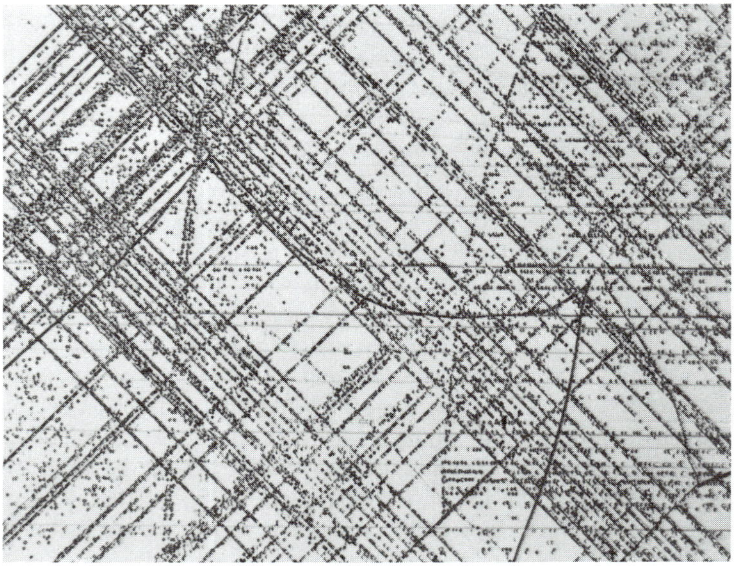

(a)

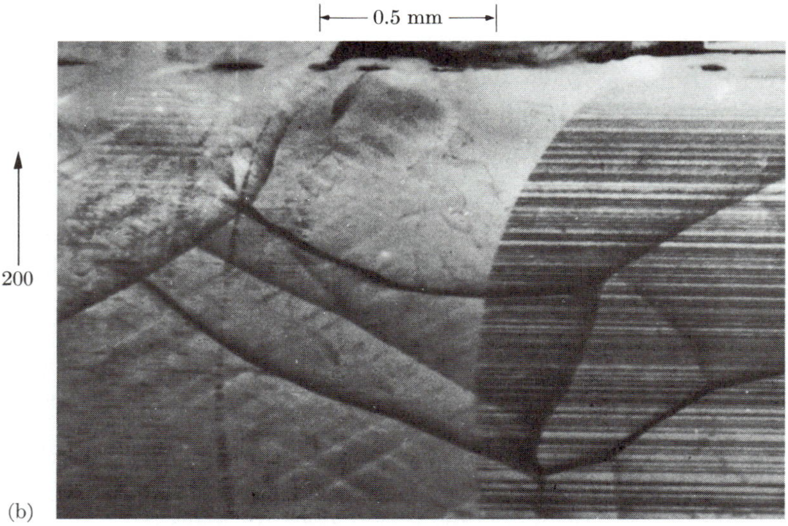

(b)

Fig. 8–33 Different aspects of the (100) face of a LiF crystal. (a) Optical micrograph of the etched surface. (b), (c), (d) Berg–Barrett x-ray topographs of the same area of the *unetched* surface, made with three different *hkl* reflections; the arrow on each topograph is the projection of the incident-beam direction on the (100) face. Newkirk [8.18].

(c)

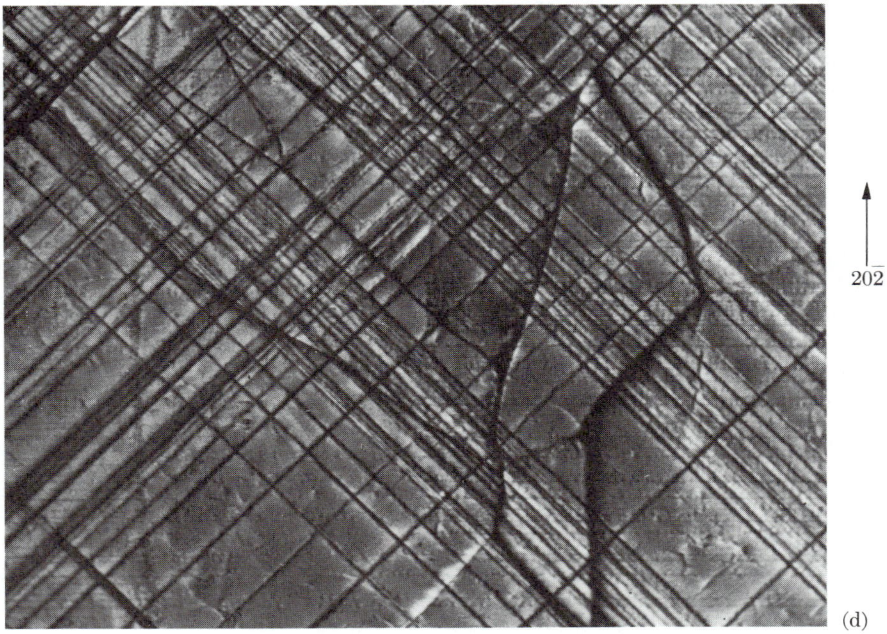

(d)

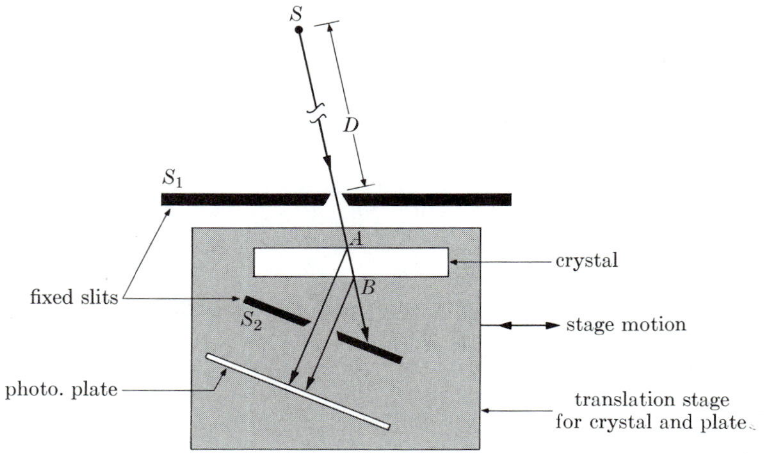

Fig. 8–34 Lang method. Slits S_1 and S_2 are fixed. The crystal and the photographic plate are mounted on a stage that can be slowly traversed in the directions shown.

involved. Slit assembly S_2 allows passage of the diffracted beam but prevents the transmitted beam from striking the plate.

If the crystal is fixed, the image on the plate is a *section topograph*, revealing the distribution of imperfections on the section AB of the crystal. The length of the crystal may be explored by making a series of section topographs, the crystal and plate being moved stepwise between exposures. Section topographs are most useful when the crystal is rather thick. The first images of individual dislocations were obtained by this method.

The crystal may be continuously explored from end to end by moving crystal and plate together, back and forth, during the exposure. The resulting image is a *projection topograph*; it is most informative when the crystal is fairly thin and the density of imperfections not too large, so that images of imperfections do not overlap. Figure 8–35 shows an example.

The Lang method can also reveal magnetic domains in ferromagnetic and ferrimagnetic materials [8.21]. Here, spontaneous magnetostrictive strains in domains and domain walls are enough to upset the regularity of the lattice and constitute crystal imperfections.

Lang cameras are commercially available and are much used in the semiconductor industry for quality control of silicon crystals. Such crystals are often centimeters wide. If a wide crystal happens to be very slightly bent, the Bragg law will not be satisfied for the highly parallel incident beam during a complete traverse of the crystal. To ensure continuous reflection Schwuttke [8.22] oscillated the crystal and plate through a very small angle about a vertical axis through A of Fig. 8–34 during the traverse, and Silver and Turner [8.23] devised an automatic control of the oscillation angle.

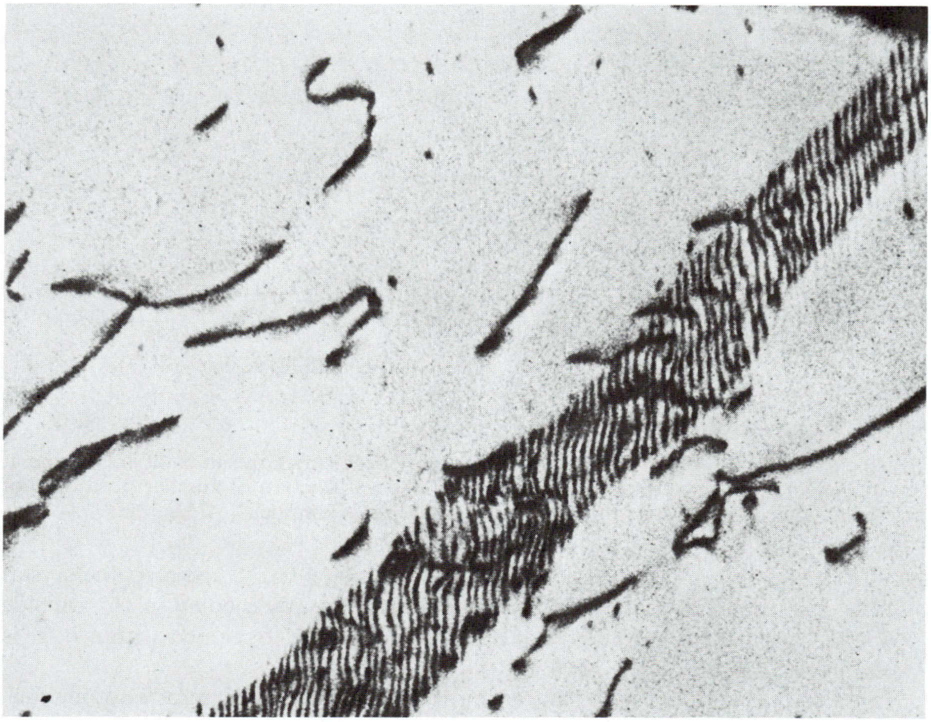

Fig. 8–35 Lang topograph of a thin crystal of lithium fluoride. Dark lines are dislocations, some randomly arranged and some nearly parallel. The latter form the band running diagonally across the photograph; these dislocations, which intersect the top and bottom surfaces of the crystal, constitute a small angle boundary between two subgrains. Mo $K\alpha_1$ radiation, 200 reflection. (Courtesy of A. R. Lang. [8.26.])

Borrmann Method

This method also involves reflection of $K\alpha$ radiation in transmission, but there are two additional restrictions: the crystal must be nearly perfect and fairly thick, such that μt has a value of the order of 10. We would then expect the transmission factor I_t/I_0 to be $e^{-10} = 5 \times 10^{-5}$, so that the transmitted and diffracted beams would be too weak to detect. Very surprisingly, Borrmann [8.24] found that both beams were fairly strong, *if* the crystal were set at the exact Bragg angle θ_B. The Borrmann effect is also called *anomalous transmission.*

The measurements of Campbell show the effect very clearly (Fig. 8–36). Crystals M and C are both calcite ($CaCO_3$); M serves as a flat crystal monochromator set to reflect $K\alpha$ radiation, and C is the crystal being studied. Two fixed counters with wide slits are arranged to receive the transmitted beam T and the beam D diffracted by the transverse planes. Crystal C is then rotated through a small angle about θ_B, and the intensities of T and D are measured as a function

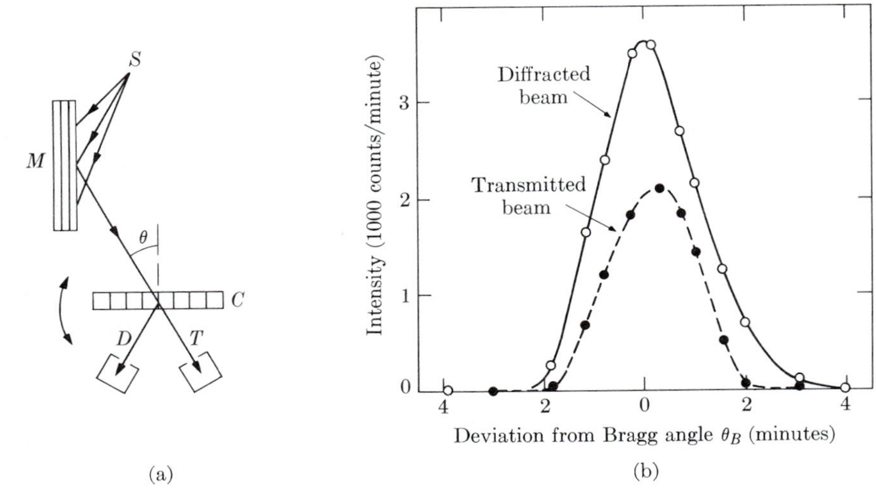

Fig. 8–36 Anomalous transmission or Borrmann effect. (a) Experimental arrangement, (b) intensities of diffracted and transmitted beams as a function of angular position θ of crystal C, which is 0.75 mm thick. Cu $K\alpha$ radiation. Campbell [8.25].

of θ (not 2θ). Within a range of ± 2 minutes of arc ($\pm 0.03°$) about θ_B the crystal both transmits and diffracts, even though μt for this crystal is equal to 14. Outside this small angular range the crystal is virtually opaque to x-rays; within it, it is transparent.

In a crystal so thick, both the transmitted *and* diffracted beams are anomalous. Figure 8–37(a) shows another unusual feature: within the crystal x-rays travel parallel to the reflecting planes, so that both the transmitted and diffracted beams emerge from the crystal at a point opposite the point of entry of the incident beam.

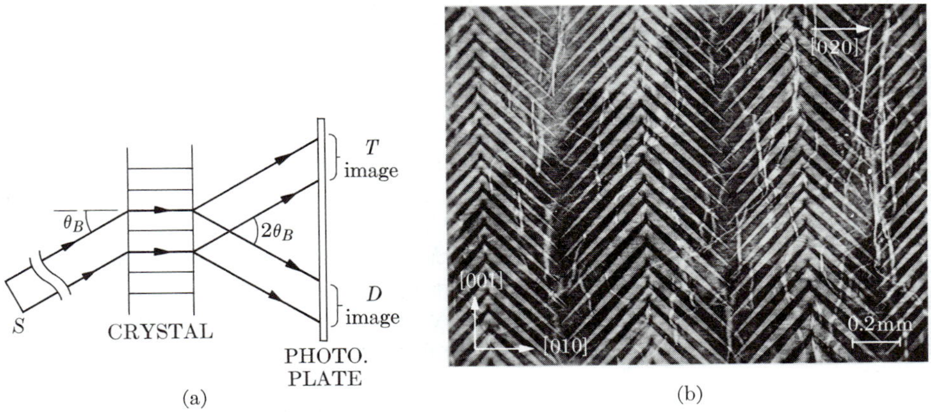

Fig. 8–37 Borrmann effect. (a) Formation of topographic images. (b) Topograph of a crystal of Fe + 3 percent Si, 0.36 mm thick, showing magnetic domains (black and white stripes) and dislocations (white lines). 020 reflection, Co $K\alpha$ radiation. (Courtesy of Carl Cm. Wu [8.26].)

Identical topographic images are formed by the T and D beams. Alternatively, the photographic plate can be placed in contact with the crystal to record a single topograph.

Imperfections in the crystal reveal themselves by a contrast mechanism opposite to that operating in the Berg–Barrett and Lang methods. Imperfections locally destroy the crystal perfection that causes the anomalously high transmission of the Borrmann effect, thereby producing less blackening of the photographic emulsion and casting a white "shadow" on the plate. Figure 8–37(b) is a topograph showing magnetic domains and dislocations in an iron alloy crystal. Because metal crystals tend to be rather imperfect, it is uncommon to find one of high enough quality to show the Borrmann effect.

The Borrmann effect is completely inexplicable by elementary theory, and an understanding of it can be gained only by a study of the complex theory of diffraction by a perfect crystal [G.30, 8.27, 8.28].

Rocking Curves

In Fig. 8–38(a), fixed crystal M, which should be as nearly perfect as possible, is set to reflect $K\alpha$ radiation in the direction of crystal C, which is to be examined. Crystal C is then rotated ("rocked") through the Bragg angle θ_B, while the beam reflected by it is measured in a fixed counter with a wide slit. The resulting curve of intensity vs. θ is called a *rocking curve* and the instrument itself a *double-crystal diffractometer*.

This instrument was much used in the early days of x-ray diffraction to compare the width and height of the rocking curve for a real crystal with the values predicted by theory for a perfect crystal [G.30]. This theory predicted a width of the order of 10 seconds (0.003°) for typical experimental conditions, and some crystals were found with rocking-curve widths approaching this value. However, most crystals exhibit widths 10 to 100 times greater.

The width of a rocking curve is a direct measure of the range of orientation present in the irradiated area of the crystal, because each block (subgrain) of a typical mosaic crystal successively comes into reflecting position as the crystal is rotated. (Note that uniform lattice strain is not revealed; strain alters d spacings, which alter 2θ values; the receiving slit of the counter is wide enough to admit all diffracted beams, whether or not small variations in 2θ exist.) Subgrain disorientation may be such that the rocking curve shows two or more subsidiary maxima, as in Fig. 8–38(c).

Figure 8–36(b) is also a rocking curve, for a calcite crystal reflecting in transmission. The surface of this crystal had been roughened by grinding it on emery paper, and the half-width of the diffraction curve is about 2 minutes (0.03°). A similar crystal that had been etched rather than ground showed a width of only 0.2 minutes (0.003°), very near the perfect-crystal value [8.25].

The double-crystal diffractometer method is not topographic. However, Weissmann and his associates [8.30] have obtained semitopographic information by photographing the beam diffracted by crystal C for various angular positions of this crystal.

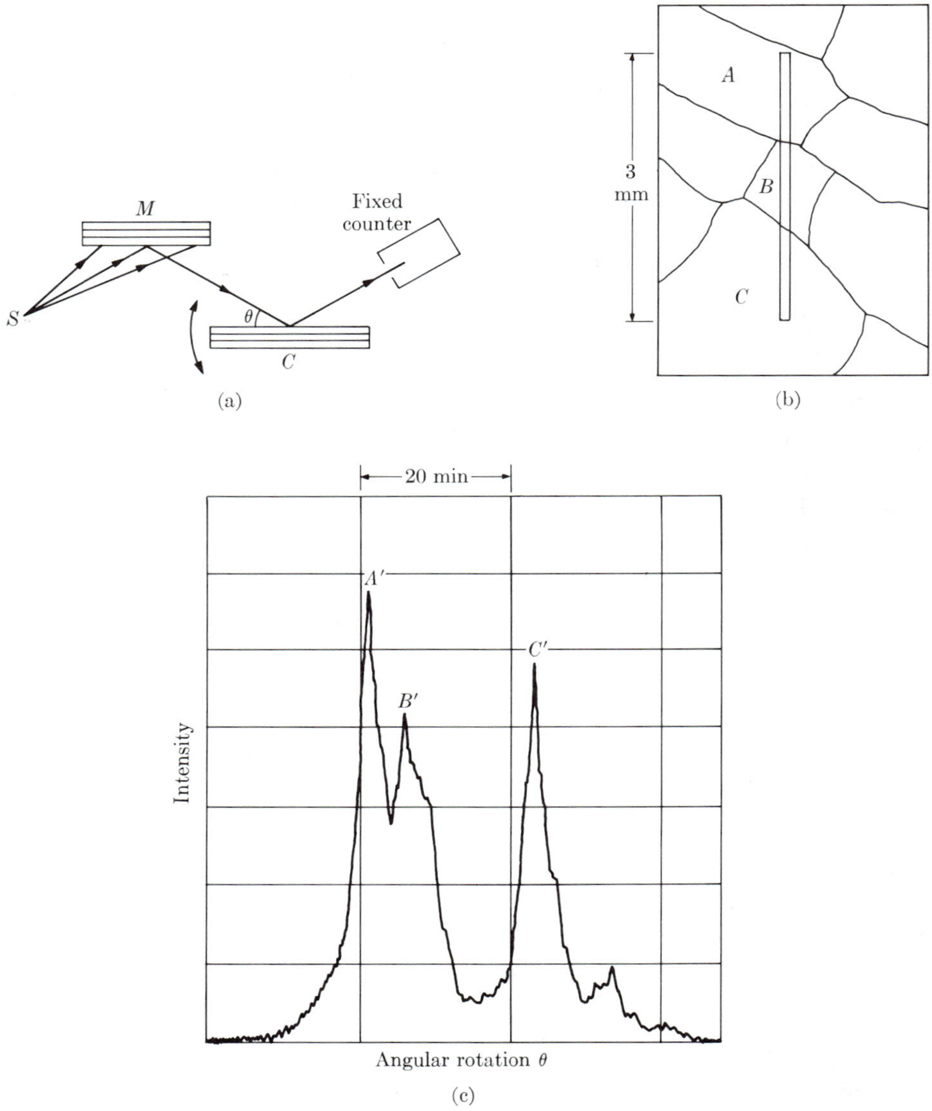

Fig. 8–38 Double-crystal diffractometer. (a) Experimental arrangement. (b) Tracing of optical photomicrograph of a polygonized aluminum crystal, etched to reveal large subgrains, produced by bending the crystal to a radius of 2 cm and annealing at 645°C for 21 hours; the superimposed rectangle shows the area irradiated (3 × 0.1 mm) by the incident x-ray beam. (c) Rocking curve (200 reflection) of the crystal in (b); the maxima A', B', C' correspond to subgrains A, B, C in (b). Intrater and Weissmann [8.29].

Table 8–1

Methods of assessing crystal quality

Method	Type*	Radiation	Area examined	Topographic?
Ordinary Laue	R,T	White	Small	No
Guinier–Tennevin	T	"	Large	No
Berg–Barrett	R	Monochromatic	Large	Yes
Lang	T	"	{Small (section) Large (projection)	Yes
Borrmann	T	"	Large	Yes
Rocking curves	R,T	"	Small	No

* R = reflection, T = transmission.

Remarks

Table 8–1 summarizes some of the features of the methods just described. Note that a crystal of any thickness can be examined in reflection, but the information received applies only to a shallow surface layer. Transmission methods, on the other hand, are restricted to rather thin crystals, of the order of 1 mm or less, but these methods reveal imperfections throughout the volume. Note also that the "area irradiated," which is related to the area of the incident beam, may differ from the "area examined" if the crystal is traversed across the beam, as in the Lang projection method.

Incidentally, the reader should note a possible confusion in terminology. When monochromatic radiation is diffracted by a single crystal in reflection (Fig. 8–38), we speak of the *Bragg case* or *Bragg reflection*. When the monochromatic beam is diffracted in transmission (Fig. 8–36), we call it the *Laue case* or *Laue reflection*, even though white radiation is not involved.

Further information about the methods listed and about other methods of assessing crystal quality may be found in various review papers [8.27, 8.30–8.33] and in the book cited in [8.27]. The production of good x-ray topographs can demand considerable experimental skill; the experimental side of the Berg–Barrett and Lang methods, including processing technique for special photographic emulsions, is described by Austerman and Newkirk [8.34].

Finally, a comparison is in order between transmission electron microscopy and x-ray topography as means of studying crystal imperfections. The electron microscope has the advantage of very much higher resolution of fine detail, but with some danger that what is seen in the thin-foil specimen does not represent the bulk crystal; dislocations may move into or out of the foil or rearrange themselves during the process of thinning the specimen. The lower resolution of x-ray topography is balanced by the fact that one topograph represents a relatively enormous area of the specimen. Roughly stated, one technique looks at the trees, the other at the forest.

PROBLEMS

***8–1** A back-reflection Laue photograph is made of an aluminum crystal with a crystal-to-film distance of 3 cm. When viewed from the x-ray source, the Laue spots have the following x, y coordinates, measured (in inches) from the center of the film:

x	y	x	y
+0.26	+0.09	−0.44	+1.24
+0.45	+0.70	−1.10	+1.80
+1.25	+1.80	−1.21	+0.40
+1.32	+0.40	−1.70	+1.19
+0.13	−1.61	−0.76	−1.41
+0.28	−1.21	−0.79	−0.95
+0.51	−0.69	−0.92	−0.26
+0.74	−0.31		

Plot these spots on a sheet of graph paper graduated in inches. By means of a Greninger chart, determine the orientation of the crystal, plot all poles of the form {100}, {110}, and {111}, and give the coordinates of the {100} poles in terms of latitude and longitude measured from the center of the projection.

8–2 A transmission Laue photograph is made of an aluminum crystal with a crystal-to-film distance of 5 cm. To an observer looking through the film toward the x-ray source, the spots have the following x, y coordinates (in inches):

x	y	x	y
+0.66	+0.88	−0.10	+0.79
+0.94	+2.44	−0.45	+2.35
+1.24	+0.64	−0.77	+1.89
+1.36	+0.05	−0.90	+1.00
+1.39	+1.10	−1.27	+0.50
+0.89	−1.62	−1.75	+1.55
+1.02	−0.95	−1.95	+0.80
+1.66	−1.10	−0.21	−0.58
		−0.59	−0.28
		−0.85	−1.31
		−1.40	−1.03
		−1.55	−0.36

Proceed as in Prob. 8–1, but use a stereographic ruler to plot the poles of reflecting planes.

***8–3** Determine the necessary angular rotations about (a) the incident beam axis, (b) the east-west axis, and (c) the north-south axis to bring the crystal of Prob. 8–2 into the "cube orientation," i.e., that shown by Fig. 2–36(a).

8–4 In the Berg–Barrett topographs of Fig. 8–33 explain why only screw dislocations are revealed in (b), only edges in (c), and screws and edges in (d). Give the indices of the operating slip systems, i.e., the indices of each slip plane and of the slip direction in that plane. (Slip in a LiF crystal occurs on {110} planes in ⟨110⟩ directions, but the experimental observations in Fig. 8–33 are inconsistent with the assumption that slip occurred on all possible slip systems. Note also that only those dislocations which intersect a crystal face will be distinctly observed by x-ray examination of that face.)

9

Structure of Polycrystalline Aggregates

9-1 INTRODUCTION

In the previous chapter we were concerned with the orientation and quality of single crystals. But the normal way in which metals and alloys are produced and used is in the form of polycrystalline aggregates, composed of a great many individual crystals usually of microscopic size. Because the properties of such aggregates are of great technological importance, they have been intensively studied in many ways. In such studies the two most useful techniques are microscopic examination and x-ray diffraction, and the wise investigator will use them both; one complements the other, and both together can provide a great deal of information about the structure of an aggregate.

The properties (mechanical, electrical, chemical, etc.) of a single-phase aggregate are determined by two factors:

1) the properties of a single crystal of the material, and

2) the way in which the single crystals are put together to form the composite mass.

In this chapter we will be concerned with the second factor, namely, the *structure* of the aggregate, using this term in its broadest sense to mean the relative size, quality, and orientation of the grains making up the aggregate. Whether these grains are large or small, strained or unstrained, oriented at random or in some particular way, frequently has very important effects on the properties of the material.

If the aggregate contains more than one phase, its properties naturally depend on the properties of each phase considered separately and on the way these phases occur in the aggregate. Such a material offers wide structural possibilities since, in general, the size, quality, and orientation of the grains of one phase may differ from those of the other phase or phases.

CRYSTAL SIZE

9-2 GRAIN SIZE

The size of the grains in a polycrystalline metal or alloy has pronounced effects on many of its properties, the best known being the increase in strength and hardness which accompanies a decrease in grain size. This dependence of properties on grain size makes the measurement of grain size a matter of some importance in the control of most metal forming operations.

The grain sizes encountered in commercial metals and alloys range from about 1000 to 1 μm. These limits are, of course, arbitrary and represent rather extreme values; typical values fall into a much narrower range, namely, about 100 to 10 μm. The most accurate method of measuring grain size in this range is by microscopic examination; the usual procedure is to determine the average number of grains per unit area of the polished section and report this in terms of an "index number" established by the American Society for Testing and Materials. The equation

$$n = 2^{N-1}$$

relates n, the number of grains per square inch when viewed at a magnification of $100\times$, and N, the ASTM "index number" or "grain-size number." Grain-size numbers of 4 and 8, for example, correspond to grain diameters of 90 and 22 μm, respectively.

Although x-ray diffraction is decidedly inferior to microscopic examination in the accurate measurement of grain size, one diffraction photograph can yield semi-quantitative information about grain size, *together with* information about crystal quality and orientation. A transmission or back-reflection pinhole photograph made with filtered radiation is best. If the back-reflection method is used, the surface of the specimen (which need not be polished) should be etched to remove any disturbed surface layer which might be present, because most of the diffracted radiation originates in a thin surface layer (see Secs. 9–4 and 9–5).

The nature of the changes produced in pinhole photographs by progressive reductions in specimen grain size is illustrated in Fig. 9–1. The governing effect here is the number of grains which take part in diffraction. This number is in turn related to the cross-sectional area of the incident beam, and its depth of penetration (in back reflection) or the specimen thickness (in transmission). When the grain size is quite coarse, as in Fig. 9–1(a), only a few crystals diffract and the photograph consists of a set of superimposed Laue patterns, one from each crystal, due to the white radiation present. A somewhat finer grain size increases the number of Laue spots, and those which lie on potential Debye rings generally are more intense than the remainder, because they are formed by the strong characteristic component of the incident radiation. Thus, the suggestion of Debye rings begins to appear, as in (b). When the grain size is further reduced, the Laue spots merge into a general background and only Debye rings are visible, as in (c). These rings are spotty, however, since not enough crystals are present in the irradiated volume of the specimen to reflect to all parts of the ring. A still finer grain size produces the smooth, continuous Debye rings shown in (d).

Several methods have been proposed for the estimation of grain size purely in terms of various geometrical factors. For example, an equation may be derived which relates the observed number of spots on a Debye ring to the grain size and other such variables as incident-beam diameter, multiplicity of the reflection, and specimen-film distance. However, many approximations are involved and the resulting equation is not very accurate. The best way to estimate grain size by

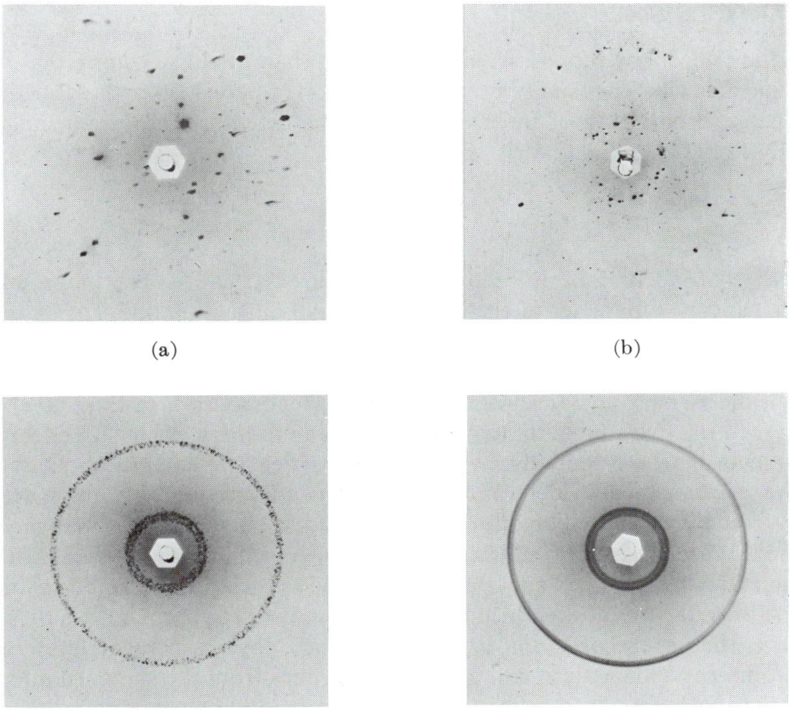

(a) (b) (c) (d)

Fig. 9-1 Back-reflection pinhole patterns of recrystallized aluminum specimens; grain size decreases in the order (a), (b), (c), (d). Filtered copper radiation.

diffraction is to obtain a set of specimens having known ASTM grain-size numbers, and to prepare from these a standard set of photographs of the kind shown in Fig. 9-1. The grain-size number of an unknown specimen of the same material is then obtained simply by matching its diffraction pattern with one of the standard photographs, *provided both are made under identical conditions.*

When the grain size reaches a value somewhere in the range 10 to 1 μm, the exact value depending on experimental conditions, the Debye rings lose their spotty character and become continuous. Between this value and 0.1 μm (1000 Å), no change occurs in the diffraction pattern. At about 0.1 μm the first signs of line broadening, due to small crystal size, begin to be detectable. There is therefore a size range, from 10 (or 1) to 0.1 μm, where x-ray diffraction is quite insensitive to variations in grain size, at least for an incident beam of normal size. With micro-beam techniques, x-ray lines remain spotty down to smaller grain sizes than are mentioned above, allowing spots from individual grains to be observed and counted [6.3].

9-3 PARTICLE SIZE

When the size of the individual crystals is less than about 0.1 μm (1000 Å), the term "particle size" is usually used. As we saw in Sec. 3-7, crystals in this size range cause broadening of the Debye rings, the extent of the broadening being given by Eq. (3-13):

$$B = \frac{0.9\lambda}{t \cos \theta}. \qquad (3\text{-}13)$$

where B = broadening of diffraction line measured at half its maximum intensity (radians) and t = diameter of crystal particle. All diffraction lines have a measurable breadth, even when the crystal size exceeds 1000 Å, due to such causes as divergence of the incident beam and size of the sample (in Debye cameras) and width of the x-ray source (in diffractometers). The breadth B in Eq. (3-13) refers, however, to the *extra* breadth, or broadening, due to the particle-size effect alone. In other words, B is essentially zero when the particle size exceeds about 1000 Å.

The chief problem in determining particle size from line breadths is to determine B from the measured breadth B_M of the diffraction line. Of the many methods proposed, Warren's is the simplest. The unknown is mixed with a standard which has a particle size greater than 1000 Å, and which produces a diffraction line near that line from the unknown which is to be used in the determination. A diffraction pattern is then made of the mixture in either a Debye camera or, preferably, a diffractometer. This pattern will contain sharp lines from the standard and broad lines from the unknown, assumed to consist of very fine particles. Let B_S be the measured breadth, at half-maximum intensity, of the line from the standard. Then B is given, not simply by the difference between B_M and B_S, but by the equation

$$B^2 = B_M^2 - B_S^2. \qquad (9\text{-}1)$$

(This equation results from the assumption that the diffraction line has the shape of an error curve [9.1].) Once B has been obtained from Eq. (9-1), it can be inserted into Eq. (3-13) to yield the particle size t. There are several other methods of finding B from B_M; compared with Warren's method, they are somewhat more accurate and considerably more intricate. These other methods involve Fourier analysis of the diffraction lines from the unknown and from the standard, and considerable computation [G.30, G.39]. Even the approach involving Eq. (9-1) can be difficult if the line from the standard is a resolved $K\alpha$ doublet; it is then simpler to use a $K\beta$ line. The Fourier methods automatically take care of the existence of a doublet.

Some investigators prefer to determine the *integral breadth* of a diffraction line rather than the breadth at half-maximum intensity. The integral breadth is given by the integrated intensity divided by the maximum intensity, i.e., it is the width of a rectangle having the same area and height as the observed line. Equation (9-1) is valid for both half-maximum and integral breadths.

The experimental difficulties involved in measuring particle size from line broadening increase with the size of the particle measured. Roughly speaking, relatively crude measurements suffice in the range 0–500 Å, but very good experimental technique is needed in the range 500–1000 Å. The maximum size measurable by line broadening was formerly placed at 1000 Å, chiefly as a result of the use of camera techniques. With the diffractometer, however, the upper limit has been pushed to almost 2000 Å. Very careful work is required and back-reflection lines are employed, since such lines exhibit the largest particle-size broadening, as shown by Eq. (3–13).

From the above discussion it might be inferred that line broadening is chiefly used to measure the particle size of loose powders rather than the size of the individual crystals in a solid aggregate. That is correct. Attempts have been made to apply Eq. (3–13) to the broadened diffraction lines from very fine-grained metal specimens and so determine the size of the individual grains. Such determinations are never very reliable, however, because the individual grains of such a material are often nonuniformly strained, and this condition, as we shall see in the next section, can also broaden the diffraction lines; an uncertainty therefore exists as to the exact cause of the observed broadening. On the other hand, the individual crystals which make up a loose powder of fine particle size can often be assumed to be strain-free, provided the material involved is a brittle (nonplastic) one, and all the observed broadening can confidently be ascribed to the particle-size effect. (But note that loose, unannealed *metal* powders, produced by filing, grinding, ball milling, etc., almost always contain nonuniform strain.) The chief applications of the line-broadening method have been in the measurement of the particle size of such materials as carbon blacks, catalysts, and industrial dusts.

Another x-ray method of measuring the size of small particles deserves some mention, although a complete description is beyond the scope of this book. This is the method of *small-angle scattering*. It is a form of diffuse scattering very near the undeviated transmitted beam, i.e., at angles 2θ ranging from 0° up to roughly 2 or 3°. From the observed variation of the scattered intensity vs. angle 2θ, the size, and to some extent the shape, of small particles can be determined, whether they are amorphous or crystalline [G.15, G.21, 9.2]. Small-angle scattering has also been used to study precipitation effects in metallic solid solutions.

CRYSTAL QUALITY

9-4 CRYSTAL QUALITY

Of the many kinds of crystal imperfection, the one we are mainly concerned with here is nonuniform strain because it is so characteristic of the *cold-worked state* of metals and alloys. When a polycrystalline piece of metal is plastically deformed, for example by rolling, slip occurs in each grain and the grain changes its shape, becoming flattened and elongated in the direction of rolling. The change in shape of any one grain is determined not only by the forces applied to the piece as a whole, but also by the fact that each grain retains contact on its boundary surfaces with all its neighbors. Because of this interaction between grains, a single grain in a polycrystalline mass is not free to deform in the same way as an isolated crystal would, if subjected to the same deformation by rolling. As a result of this restraint by its neighbors, a plastically deformed grain in a solid aggregate usually has regions of its lattice left in an elastically bent or twisted condition or, more rarely, in a state of uniform tension or compression. The metal is then said to contain *residual stress.* (Such stress is often called "internal stress" but the term is not very informative since all stresses, residual or externally imposed, are internal. The term "residual stress" emphasizes the fact that the stress remains after all external forces are removed.) Stresses of this kind are also called *microstresses* since they vary from one grain to another, or from one part of a grain to another part, on a microscopic scale. On the other hand, the stress may be quite uniform over large distances; it is then referred to as *macrostress.*

The effect of strain, both uniform and nonuniform, on the direction of x-ray reflection is illustrated in Fig. 9–2. A portion of an unstrained grain appears in (a) on the left, and the set of transverse reflecting planes shown has everywhere its equilibrium spacing d_0. The diffraction line from these planes appears on the right. If the grain is then given a uniform tensile strain at right angles to the reflecting planes, their spacing becomes larger than d_0, and the corresponding diffraction line shifts to lower angles but does not otherwise change, as shown in (b). This line shift is the basis of the x-ray method for the measurement of macrostress, as will be described in Chap. 16. In (c) the grain is bent and the strain is nonuniform; on the top (tension) side the plane spacing exceeds d_0, on the bottom (compression) side it is less than d_0, and somewhere in between it equals d_0. We may imagine this grain to be composed of a number of small regions in each of which the plane spacing is substantially constant but different from the spacing in adjoining regions. These regions cause the various sharp diffraction lines indicated on the right of (c) by the dotted curves. The sum of these sharp lines, each slightly displaced from the other, is the broadened diffraction line shown by the full curve and, of course, the broadened line is the only one experimentally observable. We can find a relation between the broadening produced and the nonuniformity of the strain by differentiating the Bragg law. We obtain

$$b = \Delta 2\theta = -2 \frac{\Delta d}{d} \tan \theta, \tag{9–2}$$

where b is the extra broadening, over and above the instrumental breadth of the line, due to a fractional variation in plane spacing $\Delta d/d$. This equation allows the variation in strain, $\Delta d/d$, to be calculated from the observed broadening. This value of $\Delta d/d$, however, includes both tensile and compressive strain and must be divided by two to obtain the maximum tensile strain alone, or maximum compressive strain alone, if these two are assumed equal. The maximum strain so found can then be multiplied by the elastic modulus E to give the maximum stress present.

When an annealed metal or alloy is cold worked, its diffraction lines become broader. This is a well-established, easily verified experimental fact, but its explanation was for many years a matter of controversy. Some investigators felt that the chief effect of cold work was to fragment the grains to a point where their

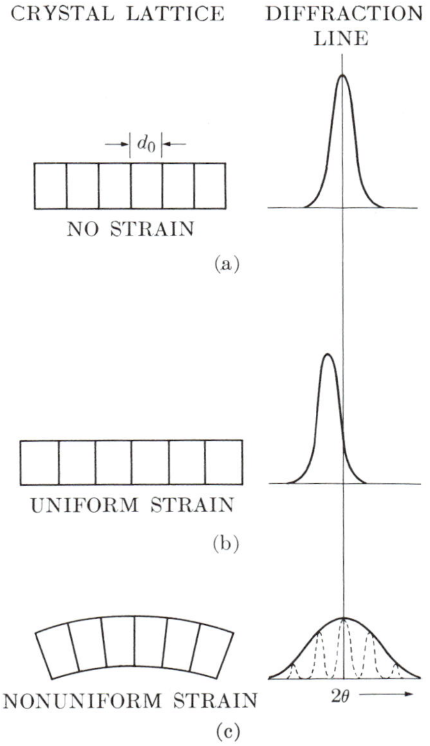

Fig. 9-2 Effect of lattice strain on Debye-line width and position.

small size alone was sufficient to account for all the observed broadening. Others concluded that the nonuniformity of strain produced by cold work was the major cause of broadening, with grain fragmentation possibly a minor contributing cause. This controversy revolved around the measurement of line widths and their interpretation in terms of either "particle-size broadening," according to Eq. (3-13), or "strain broadening," Eq. (9-2).

In 1949, however, Warren pointed out that there was important information about the state of a cold-worked metal in the *shape* of its diffraction lines, and that to base conclusions only on line *width* was to use only part of the experimental evidence. If the observed line profiles, corrected for instrumental broadening, are expressed as Fourier series, then an analysis of the Fourier coefficients discloses both particle size and strain, without the necessity for any prior assumption as to the existence of either [9.3, G.30, G.39]. Warren and Averbach [9.4] made the first measurements of this kind, on brass filings, and many similar studies followed [9.5]. Somewhat later, Paterson [9.6] showed that the Fourier coefficients of the line profile could also disclose the presence of stacking faults caused by cold work. (In FCC metals and alloys, for example, slip on {111} planes can here and there alter the normal stacking sequence *ABCABC*... of these planes to the faulted sequence *ABCBCA* ....) Thus three causes of line broadening are now recognized: small particle size, nonuniform strain, and stacking faults.

These studies of line shape showed that it was impossible to generalize about the causes of line broadening in cold-worked metals and alloys. In some materials all three causes contribute, in others only one. But there appears to be no material for which all the observed broadening can be ascribed to fine particle size. In fact, it is difficult to imagine how cold work could fragment the grains to the degree necessary to cause particle-size broadening without at the same time introducing nonuniform strains, in view of the very complex forces that must act on any one grain of an aggregate no matter how simple the forces applied to the aggregate as a whole.

The broadening of a diffraction line by cold work cannot always be observed by simple inspection of a photograph unless some standard is available for comparison. However, the separation of the $K\alpha$ doublet furnishes a very good "internal standard." In the back-reflection region, an annealed metal produces a well-resolved doublet, one component due to $K\alpha_1$ radiation and the other to $K\alpha_2$. For a given set of experimental conditions, the separation of this doublet on the film is constant and independent of the amount of cold work. But as the amount of cold work is increased, the broadening increases, until finally the two components of the doublet overlap to such an extent that they appear as one unresolved line. An unresolved $K\alpha$ doublet can therefore be taken as evidence of cold work, if the same doublet is resolved when the metal is in the annealed condition.

We are now in a position to consider some of the diffraction effects associated with the processes of *recovery*, *recrystallization*, and *grain growth*. When a cold-worked metal or alloy is annealed at a low temperature, recovery takes place; at a somewhat higher temperature, recrystallization; and at a still higher temperature, grain growth. Or at a sufficiently high constant temperature, these processes may be regarded as occurring consecutively in time. During recovery, both macro and micro residual stress are reduced in magnitude, but the strength and hardness remain high; much of this stress relief appears to be due to polygonization, which can occur in the individual grains of an aggregate just as in a single crystal. During recrystallization, new grains form, residual stress is practically eliminated, and strength and hardness decrease rather abruptly.

Because the nonuniform strain due to residual microstress is the major cause of line broadening, we usually find that the broad diffraction lines characteristic of cold-worked metal partially sharpen during recovery. When recrystallization occurs, the lines attain their maximum sharpness. During grain growth, the lines become increasingly spotty as the grain size increases.

The hardness curve and diffraction patterns of Fig. 9–3 illustrate these changes for an alpha brass, a solid solution of zinc in copper, containing 30 percent zinc by weight. The hardness remains practically constant, for an annealing period of one hour, until a temperature of 200°C is exceeded, and then decreases rapidly with increasing temperature, as shown in (a). The diffraction pattern in (b) exhibits the broad diffuse Debye lines produced by the cold-rolled, unannealed alloy. These lines become somewhat narrower for specimens annealed at 100° and 200°C, and the $K\alpha$ doublet becomes partially resolved at 250°C. At 250°, therefore, the re-

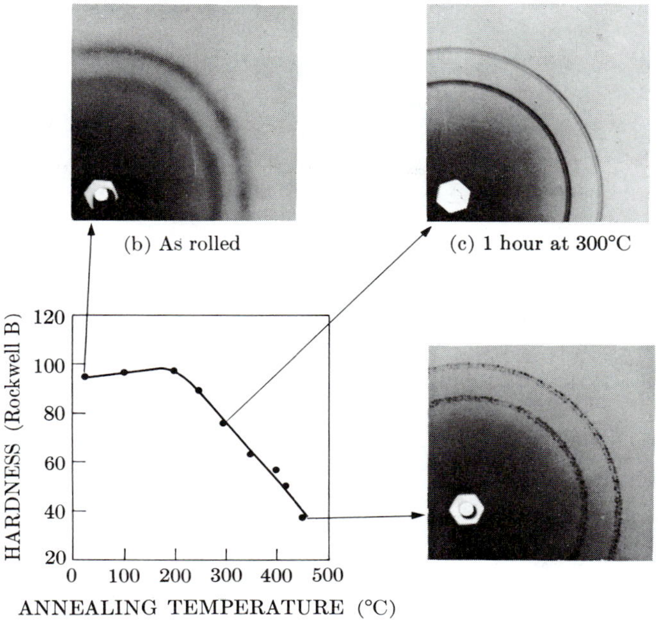

(b) As rolled (c) 1 hour at 300°C

(a) Hardness curve (d) 1 hour at 450°C

Fig. 9–3 Changes in hardness and diffraction lines of 70 Cu–30 Zn specimens, reduced in thickness by 90 percent by cold rolling, and annealed for 1 hour at the temperatures indicated in (a). (b), (c), and (d) are portions of back-reflection pinhole patterns of specimens annealed at the temperatures stated (filtered copper radiation).

covery process appears to be substantially complete in one hour and recrystallization is just beginning, as evidenced by the drop in Rockwell B hardness from 98 to 90. At 300°C the diffraction lines are quite sharp and the doublets completely resolved, as shown in (c). Annealing at temperatures above 300°C causes the lines to become increasingly spotty, indicating that the newly recrystallized grains are increasing in size. The pattern of a specimen annealed at 450°C, when the hardness had dropped to 37 Rockwell B, appears in (d).

Diffractometer measurements made on the same specimens disclose both more, and less, information. Some automatically recorded profiles of the 331 line, the outer ring of the patterns shown in Fig. 9–3, are reproduced in Fig. 9–4. It is much easier to follow changes in line shape by means of these curves than by inspection of pinhole photographs. Thus the slight sharpening of the line at 200°C is clearly evident in the diffractometer record, and so is the doublet resolution which occurs at 250°C. But note that the diffractometer cannot "see" the spotty diffraction lines caused by coarse grains. There is nothing in the diffractometer records made at 300° and 450°C which would immediately suggest that the specimen annealed at 450°C had the coarser grain size, but this fact is quite evident in the pinhole patterns shown in Figs. 9–3(c) and (d).

If an x-ray camera is not available, a piece of dental film can be placed just in

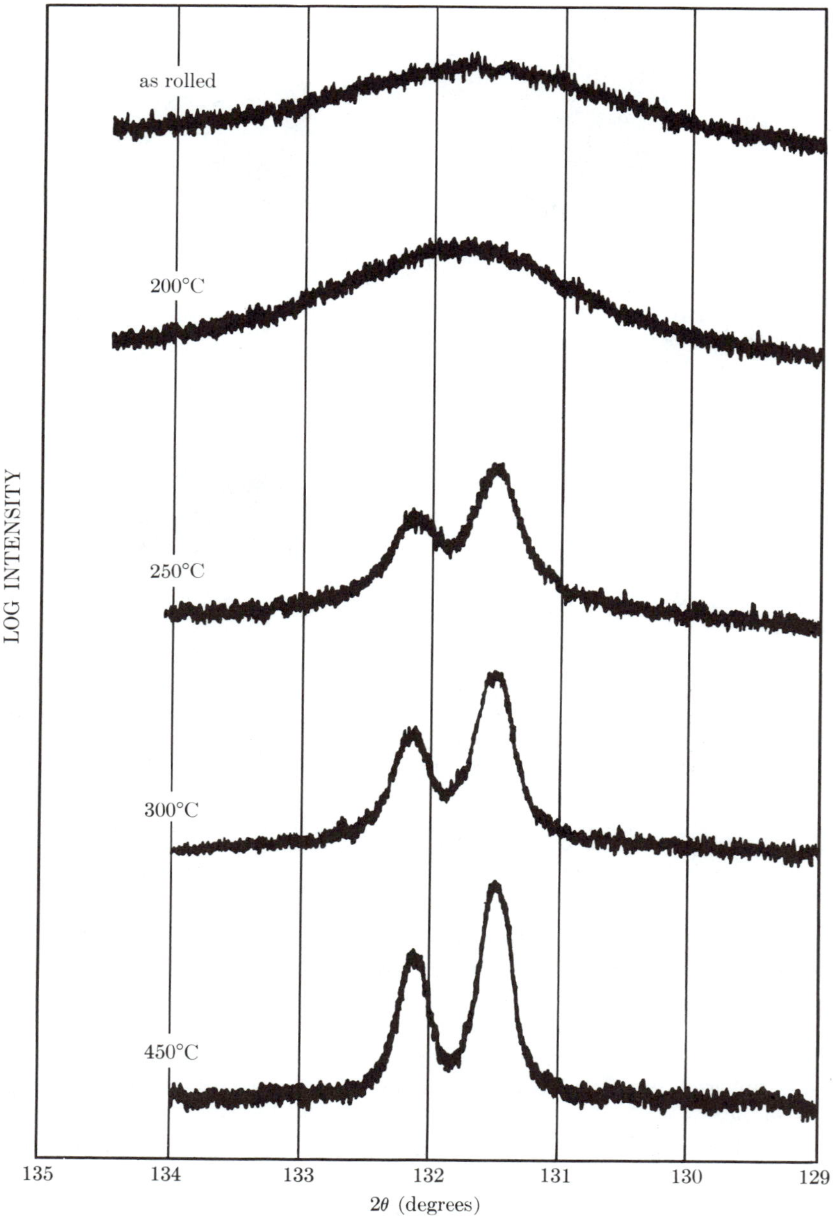

Fig. 9–4 Diffractometer traces of the 331 line of the cold-rolled and annealed 70–30 brass specimens referred to in Fig. 9–3. Filtered copper radiation. Logarithmic intensity scale. All curves displaced vertically by arbitrary amounts.

front of the counter slit of the diffractometer to disclose whether or not the diffraction lines are spotty. (Dental x-ray film, available from dental supply houses, comes in the form of a single piece of film, typically about 4 × 5 cm, enclosed in a thin envelope of light-tight plastic.) If a diffraction line is spotty on a photograph, the grain size of the specimen is too large for accurate intensity measurements with the diffractometer.

Figures 9-3 and 9-4 illustrate line sharpening by annealing. Conversely, when an annealed metal is progressively deformed, the x-ray lines progressively broaden and the hardness increases. In fact, the hardness of a particular metal or alloy can be rather accurately measured from the breadth of its diffraction lines. The relation between line breadth and hardness is not general, but must be determined experimentally for each particular material. (Very slight degrees of deformation can be detected by observation of the $K\alpha$ doublet. Rather than attempting to measure a slight increase in line width, one measures the ratio of the height of the "valley" between $K\alpha_1$ and $K\alpha_2$ to the height of $K\alpha_2$. This ratio increases rapidly as the lines broaden.)

When steel is hardened by quenching, the x-ray lines become very broad because of the microstrains due to the formation of martensite. Subsequent tempering causes progressive softening and line sharpening. Here again, a useful relation can be established experimentally between line width and hardness, applicable to a particular type of steel. (A rapid method of determining line width from the measurements used to determine residual stress is given in Sec. 16-4.)

Line-width observations are nearly always made in back reflection, whether by a photographic technique (Fig. 9-3) or by the diffractometer (Fig. 9-4). It is then necessary to remember that the observation applies only to a thin surface layer of the specimen. For example, Fig. 9-5(a) was obtained from a piece of copper and exhibits unresolved doublets in the high-angle region. The inexperienced observer might conclude that this material was highly cold worked. What the x-ray "sees" *is* cold worked, but it sees only to a limited depth. Actually, the bulk of this specimen is in the annealed condition, but the surface from which the x-ray pattern was made had had 50 μm removed by grinding on a belt sander after annealing. This treatment cold worked the surface to a considerable depth. By successive

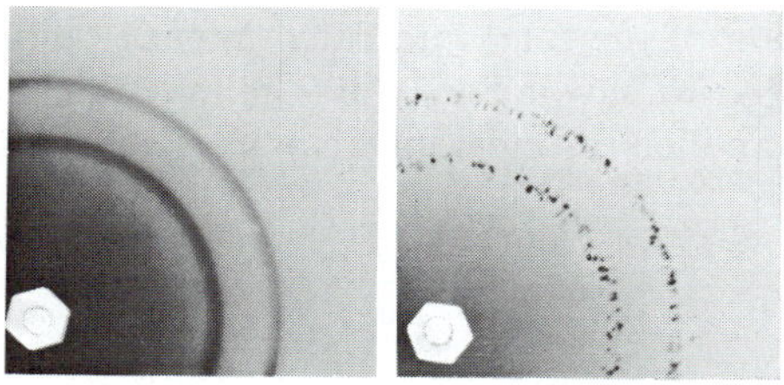

(a) (b)

Fig. 9-5 Back-reflection pinhole patterns of coarse-grained recrystallized copper. Unfiltered copper radiation: (a) from surface ground on a belt sander; (b) after removal of 0.003 in. (75 μm) from this surface by etching.

etching treatments and diffraction patterns made after each etch, the change in structure of the cold-worked layer could be followed as a function of depth below the ground surface. Not until a total of 75 μm had been removed did the diffraction pattern become characteristic of the bulk of the material; see Fig. 9–5(b), where the spotty lines indicate a coarse-grained, recrystallized structure.

9–5 DEPTH OF X-RAY PENETRATION

Observations of this kind suggest that it might be well to consider in some detail the general problem of x-ray penetration. Most metallurgical specimens strongly absorb x-rays, and the intensity of the incident beam is reduced almost to zero in a very short distance below the surface. The diffracted beams therefore originate chiefly in a thin surface layer whenever a reflection technique, as opposed to a transmission technique, is used, i.e., whenever a diffraction pattern is obtained in a back-reflection camera of any kind, a Seemann–Bohlin camera, or a diffractometer as normally used. We have just seen how a back-reflection pinhole photograph of a ground surface discloses the cold-worked condition of a thin surface layer and gives no information whatever about the bulk of the material below that layer.

These circumstances naturally pose the following question: what is the effective depth of x-ray penetration? Or, stated in a more useful manner, to what depth of the specimen does the information in such a diffraction pattern apply? This question has no precise answer because the intensity of the incident beam does not suddenly become zero at any one depth but rather decreases exponentially with distance below the surface. However, we can obtain an answer which, although not precise, is at least useful, in the following way. Equation (4–14) gives the integrated intensity diffracted by an infinitesimally thin layer located at a depth x below the surface as

$$dI_D = \frac{I_0 ab}{\sin \gamma} e^{-\mu x(1/\sin \gamma + 1/\sin \beta)} \, dx, \tag{4–14}$$

where the various symbols are defined in Sec. 4–10. This expression, integrated over any chosen depth of material, gives the total integrated intensity diffracted by that layer, but only in terms of the unknown constants I_0, a, and b. However, these constants will cancel out if we express the intensity diffracted by the layer considered as a fraction of the total integrated intensity diffracted by a specimen of infinite thickness. (As we saw in Sec. 4–10, "infinite thickness" amounts to only a few thousandths of an inch for most metals.) Call this fraction G_x. Then

$$G_x = \frac{\int_{x=0}^{x=x} dI_D}{\int_{x=0}^{x=\infty} dI_D} = [1 - e^{-\mu x(1/\sin \gamma + 1/\sin \beta)}]. \tag{9–3}$$

This expression permits us to calculate the fraction G_x of the total diffracted intensity which is contributed by a surface layer of depth x. If we arbitrarily decide that a contribution from this surface layer of 95 percent (or 99 or 99.9 percent) of the total is enough so that we can ignore the contribution from the material below that layer, then x is the effective depth of penetration. We then know that the information recorded on the diffraction pattern (or, more precisely, 95 percent of the information) refers to the layer of depth x and not to the material below it.

In the case of the diffractometer, $\gamma = \beta = \theta$, and Eq. (9-3) reduces to

$$G_x = (1 - e^{-2\mu x/\sin\theta}), \tag{9-4}$$

which shows that the effective depth of penetration decreases as θ decreases and therefore varies from one diffraction line to another. In back-reflection cameras, $\gamma = 90°$, and

$$G_x = [1 - e^{-\mu x(1 + 1/\sin\beta)}], \tag{9-5}$$

where $\beta = 2\theta - 90°$.

For example, the conditions applicable to the outer diffraction ring of Fig. 9-5 are $\mu = 473\ \text{cm}^{-1}$ and $2\theta = 136.7°$. By using Eq. (9-5), we can construct the plot of G_x as function of x which is shown in Fig. 9-6. We note that 95 percent of the information on the diffraction pattern refers to a depth of only about 25 μm. It is therefore not surprising that the pattern of Fig. 9-5(a) discloses only the presence of cold-worked metal, since we found by repeated etching treatments that the depth of the cold-worked layer was about 75 μm. Of course, the information recorded on the pattern is heavily weighted in terms of material just below the surface; thus 95 percent of the recorded information applies to a depth of 25 μm, but 50 percent of *that* information originates in the first 5 μm. (Note that an effective penetration of 25 μm means that a surface layer only one grain thick is effectively contributing to the diffraction pattern if the specimen has an ASTM grain-

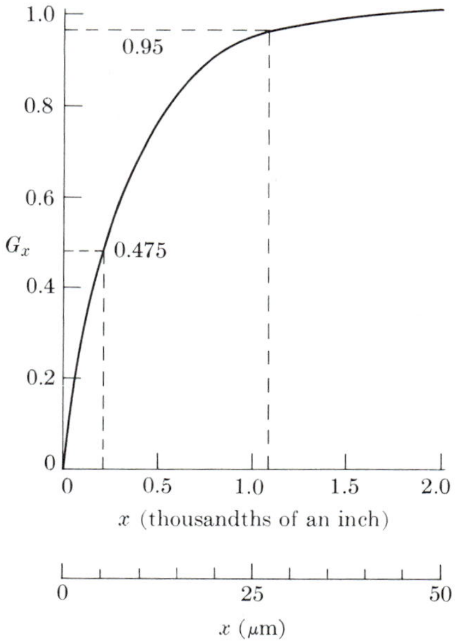

Fig. 9-6 The fraction G_x of the total diffracted intensity contributed by a surface layer of depth x, for $\mu = 473\ \text{cm}^{-1}$, $2\theta = 136.7°$, and normal incidence.

size number of 8. This layer contains some 300,000 reflecting lattice planes for the 331 diffraction line considered here.)

Equation (9–4) can be put into the following form, which is more suitable for calculation:

$$\frac{2\mu x}{\sin\theta} = \ln\left(\frac{1}{1 - G_x}\right) = K_x,$$

$$x = \frac{K_x \sin\theta}{2\mu}. \qquad \text{[Diffractometer]}$$

Similarly, we can rewrite Eq. (9–5) in the form

$$\mu x\left(1 + \frac{1}{\sin\beta}\right) = \ln\left(\frac{1}{1 - G_x}\right) = K_x,$$

$$x = \frac{K_x \sin\beta}{\mu(1 + \sin\beta)}. \qquad \text{[Back-reflection camera]}$$

Values of K_x corresponding to various assumed values of G_x are given in Table 9–1.

Table 9-1

G_x	0.50	0.75	0.90	0.95	0.99	0.999
K_x	0.69	1.39	2.30	3.00	4.61	6.91

Calculations of the effective depth of penetration can be valuable in many applications of x-ray diffraction. We may wish to make the effective depth of penetration as large as possible in some applications. Then γ and β in Eq. (9–3) must be as large as possible, indicating the use of high-angle lines, and μ as small as possible, indicating short-wavelength radiation. Other applications may demand very little penetration, as when we wish information, e.g., chemical composition or lattice parameter, from a very thin surface layer. Then we must make μ large, by using radiation which is highly absorbed, and γ and β small, by using a diffractometer at low values of 2θ.* By these means the depth of penetration can often be made surprisingly small. For instance, if a steel specimen is examined in a diffractometer with Cu $K\alpha$ radiation, 95 percent of the information afforded by the lowest angle line of ferrite (the 110 line at $2\theta = 45°$) applies to a depth of only 2 μm. There are limits, of course, to reducing the depth of x-ray penetration, and when information is required from very thin surface films, electron diffraction is a far more suitable tool (see Appendix 2).

---

* Some of these requirements may be contradictory. For example, in measuring the lattice parameter of a thin surface layer with a diffractometer, we must compromise between the low value of θ required for shallow penetration and the high value of θ required for precise parameter measurements.

Although the diffracted beam in any reflection method comes only from a very thin surface layer, it must not be supposed that the information on a diffraction pattern obtained by a transmission method is truly representative of the entire cross section of the specimen. Calculations such as those given above show that a greater proportion of the total diffracted energy originates in a layer of given thickness on the back side of the specimen (the side from which the transmitted beam leaves) than in a layer of equal thickness on the front side. If the specimen is highly absorbing, a transmission method can be almost as nonrepresentative of the entire specimen as a back-reflection method, in that most of the diffracted energy will originate in a thin surface layer. See Problem 9-5.

CRYSTAL ORIENTATION

9-6 GENERAL

Each grain in a polycrystalline aggregate normally has a crystallographic orientation different from that of its neighbors. Considered as a whole, the orientations of all the grains may be randomly distributed in relation to some selected frame of reference, or they may tend to cluster, to a greater or lesser degree, about some particular orientation or orientations. Any aggregate characterized by the latter condition is said to have a *preferred orientation*, or *texture*, which may be defined simply as a condition in which the distribution of crystal orientations is nonrandom.

Preferred orientation is a very common condition. Among metals and alloys it is most evident in wire and sheet, and the kinds of texture found in these products are dealt with below. The preferred orientation that is produced by the forming process itself (wire drawing or sheet rolling) is called a *deformation texture*. It is due to the tendency of the grains in a polycrystalline aggregate to rotate during plastic deformation; each grain undergoes slip and rotation in a complex way that is determined by the imposed forces and by the slip and rotation of adjoining grains; the result is a preferred, nonrandom orientation. When the cold-worked metal, possessed of a deformation texture, is recrystallized by annealing, the new grain structure usually has a preferred orientation too, often different from that of the cold-worked material. This is called a *recrystallization texture* or *annealing texture*. It is due to the influence which the texture of the cold-worked matrix has on the nucleation and/or growth of the new grains in that matrix.

Preferred orientation is not confined to metallurgical products. It also exists in rocks, ceramics, and in both natural and artificial polymeric fibers and sheets. In fact, preferred orientation is generally the rule, not the exception, and the preparation of an aggregate with completely random crystal orientations is a difficult matter.

The industrial importance of preferred orientation lies in the effect, often very marked, which it has on the overall, macroscopic properties of materials. Given the fact that all single crystals are anisotropic, i.e., have different properties in different directions, it follows that an aggregate having preferred orientation must also have directional properties to a greater or lesser degree. Such properties may or may not be beneficial, depending on the intended use of the material. For example, sheet steel for the cores of small electric motors should have, for magnetic reasons, all grains oriented with their {100} planes parallel to the sheet surface. But this texture would not be satisfactory if the steel were to be formed into a cup by deep drawing; here a texture with {111} planes parallel to the surface would make the steel less likely to crack during the severe deformation of deep drawing; however, if the part to be formed by deep drawing has an unsymmetrical shape, a still different texture, or none at all, might yield better results. Some control of texture is possible by the proper choice of production variables such as degree of deformation and annealing temperature, but metallurgists do not yet understand texture formation well enough to produce any desired texture in any particular metal at will.

For information on deformation and recrystallization textures observed in specific materials in wire and sheet form, the reader should consult Barrett and Massalski [G.25] and Dillamore and Roberts [9.7]. Here we are concerned only with the nature of textures and with their determination by x-ray methods.

At various places in this book, we have already noted that a pinhole photograph made of a polycrystalline specimen with characteristic radiation consists of concentric Debye rings. We have more or less tacitly assumed that these rings are always continuous and of constant intensity around their circumference, but actually such rings are not formed unless the individual crystals in the specimen have completely random orientations.* If the specimen exhibits preferred orientation, the Debye rings are of nonuniform intensity around their circumference (if the preferred orientation is slight), or actually discontinuous (if there is a high degree of preferred orientation). In the latter case, certain portions of the Debye ring are missing because the orientations which would reflect to those parts of the ring are simply not present in the specimen. Nonuniform Debye rings can therefore be taken as conclusive evidence for preferred orientation, and by analyzing the nonuniformity we can determine the kind and degree of preferred orientation present. (This nonuniformity is evident even in some Debye-Scherrer patterns; if a wire having preferred orientation is examined in a Debye-Scherrer camera, the nonuniformity of Debye rings is usually apparent even though such a camera records only a portion of the rings.)

Fiber Texture

The individual crystals in wire are so oriented that the same crystallographic direction [*uvw*] in most of the grains is parallel or nearly parallel to the wire axis. Because a similar texture occurs in natural and artificial fibers, it is called a *fiber texture* and the axis of the wire is called the *fiber axis*. Materials having a fiber texture have rotational symmetry about an axis in the sense that all crystal orientations about this axis are equally probable, like those of beads on a string. A fiber texture is therefore to be expected in any material formed by forces that have

* See the next section for one exception to this statement.

rotational symmetry about an axis, for example, in wire and rod formed by drawing, swaging, or extrusion. Less common examples of fiber texture are sometimes found in sheet formed by simple compression, in coatings formed by hot dipping, electroplating, and evaporation, and in castings among the columnar crystals next to the mold wall. The fiber axis in these is perpendicular to the plane of the sheet or coating, and parallel to the axis of the columnar crystals.

Fiber textures vary in perfection, i.e., in the scatter of the direction [*uvw*] about the fiber axis, and both single and double fiber textures have been observed. Thus, cold-drawn aluminum wire has almost a single [111] texture, but copper, also FCC, has a double [111] + [100] texture; i.e., in drawn copper wire there are two sets of grains, the fiber axis of one set being [111] and that of the other set [100].

Sheet Texture

In its simplest, most highly developed form, the texture of sheet is such that most of the grains are oriented with a certain crystallographic plane (*hkl*) roughly parallel to the sheet surface, and a certain direction [*uvw*] in that plane roughly parallel to the direction in which the sheet was rolled. Such a texture is described by the shorthand notation (*hkl*)[*uvw*]. In an ideal texture of this kind, the grain orientations in the sheet are fixed with respect to axes in the sheet; there is none of the rotational freedom of grain orientation possessed by a fiber texture.

The notation (*hkl*)[*uvw*] specifies what is called an *ideal orientation*. Some metals and alloys have sheet textures so sharp that they can be adequately described by stating the ideal orientation to which the grains of the sheet closely conform. Most sheet textures, however, have so much scatter that they can be approximated symbolically only by the sum of a number of ideal orientations or texture components, and even such a description is inadequate. Thus, the deformation texture of brass sheet (70 Cu–30 Zn) is very near the ideal orientation (110)[$\bar{1}$12]. But both the deformation and recrystallization textures of low-carbon sheet steel have so much scatter that the grain orientations present can be accurately represented only by a graphical description called a *pole figure*.

Pole Figures

A pole figure is a stereographic projection, with a specified orientation relative to the specimen, that shows the variation of pole density with pole orientation for a selected set of crystal planes. This method of describing textures was first used by the German metallurgist Wever in 1924, and its meaning can best be illustrated by the following example. Suppose we have a very coarse-grained sheet of a cubic metal containing only 10 grains, and that we determine the orientation of each of these 10 grains by one of the Laue methods. We decide to represent the orientations of all of these grains together by plotting the positions of their {100} poles on a single stereographic projection, with the projection plane parallel to the sheet surface. Since each grain has three {100} poles, there will be a total of $3 \times 10 = 30$ poles plotted on the projection. If the grains have a completely random orientation, these poles will be distributed uniformly* over the projection, as indicated in Fig. 9–7(a). But if preferred orientation is present, the poles will tend to cluster together into certain areas of the projection, leaving other areas virtually unoccupied. For example, this clustering might take the particular form shown in Fig. 9–7(b). This is called the "cube texture," because each grain is oriented with its (100) planes nearly parallel to the sheet surface and the [001] direction in these planes nearly parallel to the rolling direction. (This simple texture, which may be described by the shorthand notation (100)[001], actually forms as a recrystallization texture in many face-centered cubic metals and alloys under suitable conditions.) If we had chosen to construct a (111) pole figure, by plotting only {111} poles, the resulting pole figure would look entirely different from Fig. 9–7(b) for the same preferred orientation; in fact, it would consist of four "high-intensity" areas, one near the center of each quadrant. This illustrates the fact that the appearance of a pole figure depends on the indices of the poles plotted, and that the choice of indices depends on which aspect of the texture one wishes to show most clearly. For example, if we are interested in the plastic deformation of a certain face-centered cubic material in sheet form, we would plot a (111) pole figure, because it would show at a glance the orientation of the {111} slip planes. Similarly, if the magnetic behavior of iron sheet is of interest, a (100) pole figure would be preferred, because the directions of high magnetic permeability in iron are ⟨100⟩ directions.

The pole figure of a fiber texture necessarily has rotational symmetry about the fiber axis (Fig. 9–8). The degree of scatter of this texture is given by the angular

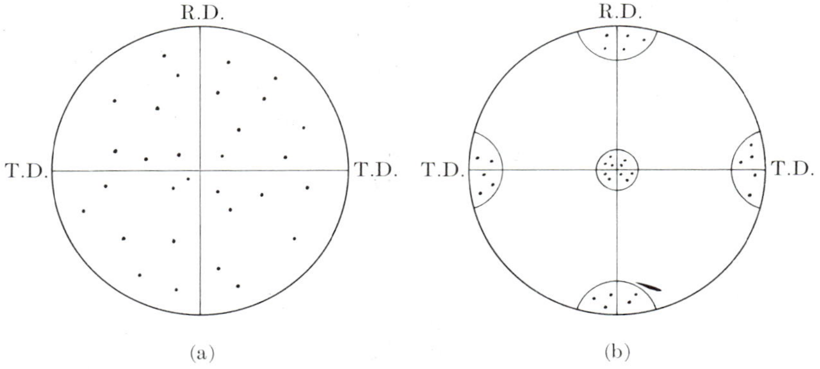

Fig. 9–7 (100) pole figures for sheet material, illustrating (a) random orientation and (b) preferred orientation. R.D. (rolling direction) and T.D. (transverse direction) are reference directions in the plane of the sheet.

---

* If the orientation is random, there will be equal numbers of poles in equal areas on the surface of a reference sphere centered on the specimen. There will not be equal numbers, however, on equal areas of the pole figure, since the stereographic projection is not area-true. This results, for randomly oriented grains, in an apparent clustering of poles at the center of the pole figure, since distances representing equal angles are much smaller in this central region than in other parts of the pole figure.

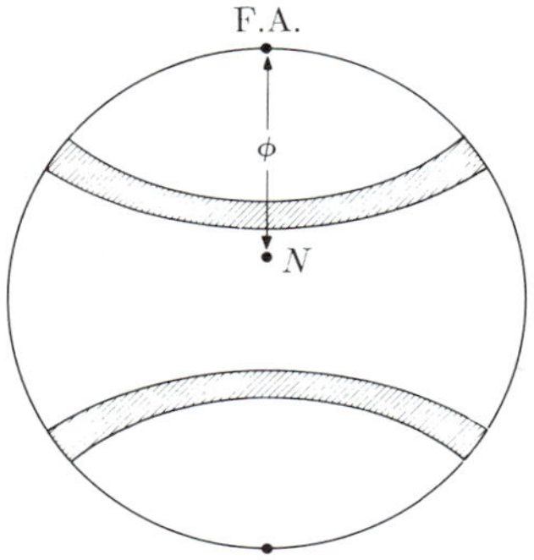

Fig. 9-8 (111) pole figure for an imperfect [100] fiber texture. F.A. = fiber axis. Cross-hatched areas are areas of high (111) pole density.

width of the bands that show where (111) poles are located. The angle ϕ is the angle between the fiber axis and the general position N of any pole being considered. For the texture shown the bands are centered on ϕ values, measured from the top and bottom of the projection, of 54.7°, because this is the angle between the [100] fiber axis and the (111) poles being displayed.

Because of its rotational symmetry a pole figure of a fiber texture displays redundant information, in the sense that the pole density along any longitude line (meridian) is the same as along any other. Thus a plot of pole density vs. angle ϕ between 0 and 90° is a simpler description of the texture; for the texture shown in Fig. 9-8, such a plot would show a single maximum at 54.7°.

When the grain size of the specimen is small, as it normally is, separate determination of the orientations of a representative number of grains, as suggested by Fig. 9-7, is out of the question. Instead, x-ray methods are used in which the diffraction effects from thousands of grains are automatically averaged. The (hkl) pole figure of a fine-grained material is constructed by analyzing the distribution of intensity around the circumference of the corresponding hkl Debye ring. Two methods of doing this are available, the photographic and the diffractometer method, and both are time consuming.

Although only a pole figure can provide a complete description of preferred orientation, some information can be obtained fairly quickly by a comparison of calculated diffraction line intensities with those observed on an ordinary diffractometer scan. As stated in Sec. 4-12, relative line intensities are given accurately by Eq. (4-21) only when the crystals of the specimen have completely random orientations. Therefore any radical disagreement between observed and calculated intensities is immediate evidence of preferred orientation in the specimen, and, from the nature of the disagreement, certain limited conclusions can usually be drawn concerning the nature of the texture. For example, if a sheet specimen is examined in the diffractometer in the usual way (the specimen making equal angles with the incident and diffracted beams), then the only grains which can contribute to the hkl reflection are those whose (hkl) planes are *parallel to the sheet surface*. If the texture is such that there are very few such grains, the intensity of the hkl reflection will be abnormally low. Or a given reflection may be of abnormally high intensity, which would indicate that the corresponding planes were preferentially oriented parallel or nearly parallel to the sheet surface. As an illustration, the 200 diffractometer reflection from a specimen having the cube texture is abnormally high, and from this fact alone it is possible to conclude that there is a preferred orientation of (100) planes parallel to the sheet surface. However, no conclusion is possible as to whether or not there is a preferred direction in the (100) plane parallel to some reference direction on the sheet surface. Such information can be obtained only by making a pole figure.

9-7 THE TEXTURE OF WIRE (PHOTOGRAPHIC METHOD)

The chief problem presented by a fiber texture is the identification of the fiber axis [*uvw*]. This can be done fairly easily with a single diffraction photograph, and the procedure is described in this section. If, in addition, we wish to determine the amount of scatter in the texture, a diffractometer method is preferable (Sec. 9-9).

The wire is examined in a transmission pinhole camera with filtered radiation and with the wire axis vertical, parallel to one edge of the flat film. The problem of finding the indices [*uvw*] of the fiber axis is best approached by considering the diffraction effects associated with an ideal case, for example, that of a wire of a cubic material having a perfect [100] fiber texture. Suppose we consider only the 111 reflection. In Fig. 9-9, the wire specimen is at *C* with its axis along *NS*, normal to the incident beam *IC*. *CP* is the normal to a set of (111) planes. Diffraction from these planes can occur only when they are inclined to the incident beam at an angle θ which satisfies the Bragg law, and this requires that the (111) pole lie somewhere

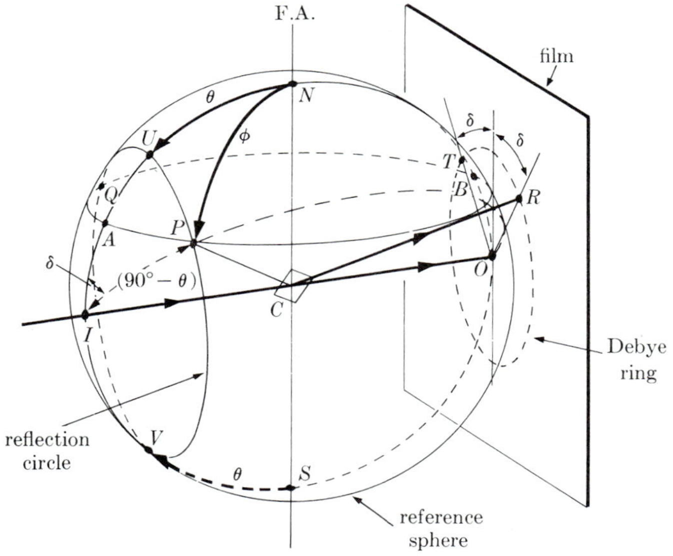

Fig. 9-9 Geometry of reflection from material having a fiber texture. F.A. = fiber axis.

on the circle $PUQV$, since then the angle PCI between the plane normal and the incident beam will always be $90° - \theta$. For this reason, $PUQV$ is called the *reflection circle*. If the grains of the wire had completely random orientations, then (111) poles would lie at all positions on the sphere surface and therefore at all positions on the reflection circle, and the 111 reflection would consist of the complete Debye ring indicated in the drawing. But if the wire has a perfect [100] fiber texture, then the diffraction pattern produced by a stationary specimen is identical with that obtained from a single crystal rotated about the axis [100], because of the rotational symmetry of the wire. During this rotation, the (111) pole is confined to the small circle $PAQB$, all points of which make a constant angle $\phi = 54.7°$ with the [100] direction N. Therefore, the only (111) planes in the specimen that are able to diffract are those with poles at the intersections of the reflection circle and the circle $PAQB$. These intersections are located at P and Q, and the corresponding diffraction spots at R and T, at an azimuthal angle δ from a vertical line through the center of the film. Two other spots, not shown, are located in symmetrical positions on the lower half of the film. If the texture is not perfect, each of these spots will broaden peripherally into an arc whose length is a function of the degree of scatter in the texture.

By solving the spherical triangle IPN, we can find the following general relation between the angles ϕ, θ, and δ:

$$\cos \phi = \cos \theta \cos \delta. \qquad (9\text{-}6)$$

These angles are shown stereographically in Fig. 9-10, projected on a plane normal to the incident beam. The (111) pole figure in (a) consists simply of two arcs which are the paths traced out by {111} poles during rotation of a single crystal about [100]. In (b), this pole figure has been superposed on a projection of the reflection circle in order to find the locations of the reflecting plane normals. Radii drawn

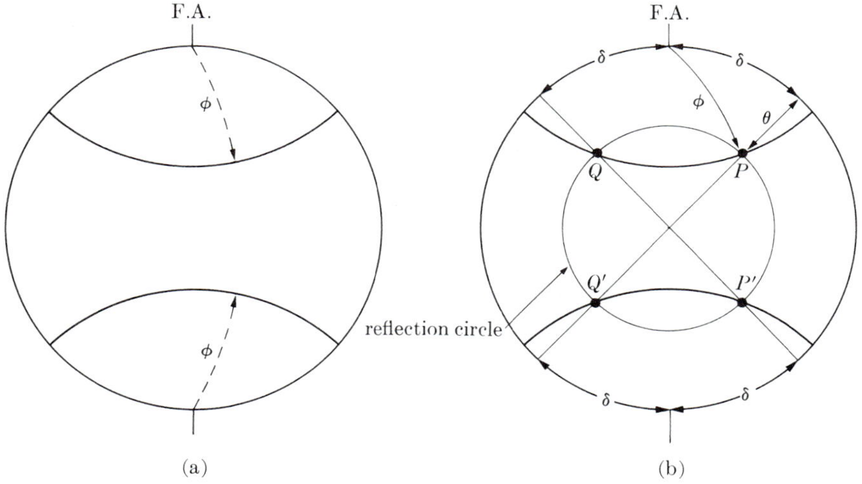

Fig. 9-10 Perfect [100] fiber texture: (a) (111) pole figure; (b) location of reflecting plane normals.

through these points (P, Q, P', and Q') then enable the angle δ to be measured and the appearance of the diffraction pattern to be predicted.

An unknown fiber axis is identified by measuring the angle δ on the film and obtaining ϕ from Eq. (9–6). When this is done for a number of different *hkl* reflections, a set of ϕ values is obtained from which the indices [*uvw*] of the fiber axis can be determined. The procedure will be illustrated with reference to the diffraction pattern of drawn aluminum wire shown in Fig. 9–11, which has the incomplete Debye rings typical of a pronounced texture. The first step is to index the Debye rings. Values of θ for each ring are calculated from measurements of ring diameter, and *hkl* indices are assigned by the use of Eq. (3–10) and Appendix 10. In this way the inner ring is identified as a 111 reflection and the outer one as 200. The angle δ is then measured from a vertical line through the center of the film to the center of each strong Debye arc. The average values of these angles are given below, together with the calculated values of ϕ:

Line	*hkl*	δ	θ	ϕ
Inner	111	69°	19.3°	70°
Outer	200	52	22.3	55

The normals to the (111) and (200) planes therefore make angles of 70° and 55°, respectively, with the fiber axis. We can determine the indices [*uvw*] of this axis either by the graphical construction shown in Fig. 8–8 or by inspection of a table of interplanar angles. In this case, inspection of Table 2–3 shows that [*uvw*] must be [111], since the angle between ⟨111⟩ and ⟨111⟩ is 70.5° and that between ⟨111⟩ and ⟨100⟩ is 54.7°, and these values agree with the values of ϕ given above within experimental error. The fiber axis of drawn aluminum wire is therefore [111].

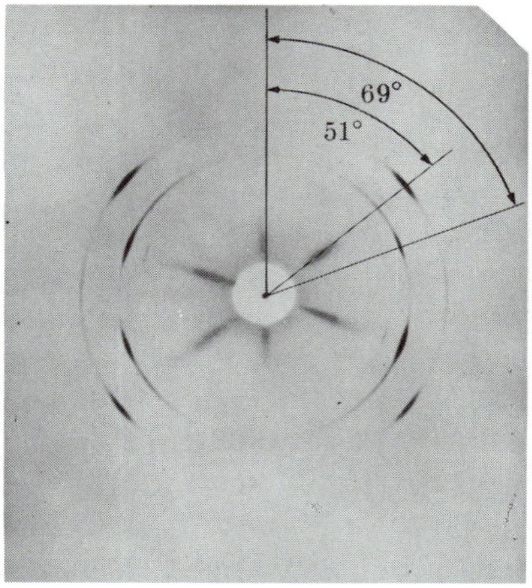

Fig. 9–11 Transmission pinhole pattern of cold-drawn aluminum wire, wire axis vertical. Filtered copper radiation. (The radial streaks near the center are formed by the white radiation in the incident beam.)

There is some scatter of the [111] direction about the wire axis, however, inasmuch as the reflections on the film are short arcs rather than sharp spots. If we wish, this can be taken into account by measuring the angular range of δ for each arc and calculating the corresponding range of ϕ. A (111) pole figure of the wire would then resemble Fig. 9-8.

Close inspection of Fig. 9-11 reveals a weak maximum on the outer ring (200 reflection) at the equator ($\delta = 90°$), indicating a weak [100] component in the texture; there is also evidence of this component on the inner 111 ring. This wire therefore has a double fiber texture: [111] strong and [100] weak. The relative amounts of these components can be measured with a diffractometer (Sec. 9-9).

If a coating, such as an electrodeposit, on a flat sheet has a fiber texture with the fiber axis [*uvw*] normal to the sheet, then the (*hkl*) planes normal to [*uvw*] will be parallel, or nearly parallel, to the sheet surface. Unusual diffraction effects can then occur:

1. *Diffractometer.* The incident and diffracted beams make the same angle with the sheet surface, so that the *hkl* reflection is abnormally strong and all others weak or absent.

2. *Back-reflection pinhole camera.* If the incident beam is normal to the sheet specimen and therefore parallel to the fiber axis, and a projection like that shown in Fig. 9-10(b) is made (projection plane parallel to sheet), then both the incident beam and the fiber axis coincide with the center of the projection. If the texture is ideally sharp, the (*hkl*) pole figure will consist of one or more concentric circles centered on the center of the projection, and the chance that one of these pole circles will coincide with the concentric reflection circle is essentially zero; no reflection will occur. But if the texture has enough scatter, one of the pole circles will broaden into a band wide enough to touch the reflection circle at all points; a Debye ring of *uniform* intensity will be formed. See Prob. 9-7. Thus a uniform Debye ring is not always evidence for randomly oriented grains.

9-8 THE TEXTURE OF SHEET (DIFFRACTOMETER METHODS)

Prior to about 1950, the texture of sheet materials was determined by a photographic method, in which reflected x-ray intensities were visually estimated from the degree of film blackening on a series of photographs. The resulting pole figures were qualitative; they were divided by contour lines into areas in which the pole density was designated by terms such as low, medium, and high.

The photographic method is now obsolete, because the diffractometer permits direct measurement of reflected intensities and yields quantitative pole figures.

The essential difference between the photographic and diffractometer methods can be understood by comparing Figs. 9-11 and 9-12. To analyze the information given in Fig. 9-11, we look successively at points around the Debye ring and estimate the degree of film blackening. With a diffractometer (Fig. 9-12) the same nonuniform Debye ring exists in space, but we cannot move the counter around the ring to explore the variation in intensity. Instead, we keep the counter fixed and *rotate the specimen.* If the sheet specimen is rotated in its own plane, for example, the Debye ring rotates with it, and the high- and low-intensity regions are

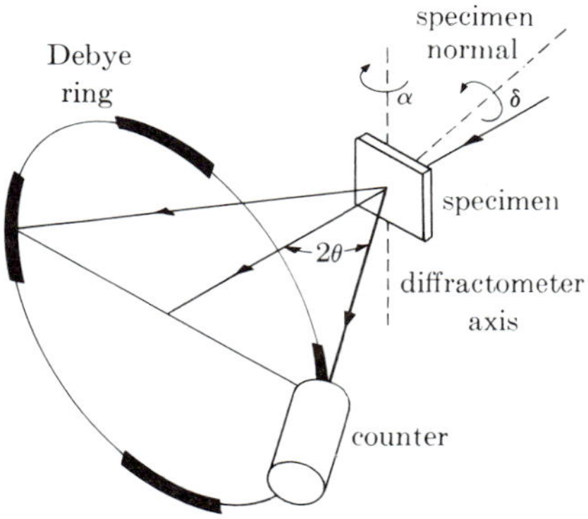

Fig. 9–12 Transmission method for pole-figure determination. After Geisler [9.8].

successively brought into a position where they can be measured by the counter. At each specimen position the intensity of the reflected beam is taken to be proportional to the volume of grains then oriented to reflect that beam, except for corrections that may be necessary; this volume is in turn proportional to pole density.

To determine an (hkl) pole figure, the counter is fixed at the proper 2θ angle to receive the hkl reflection and the sheet specimen is rotated in particular ways described below. The pole figure is a projection made on a plane parallel to the sheet surface and therefore rotates with the sheet. But, whatever the orientation of the sheet, the normal N to the reflecting planes (hkl) remains fixed in space, bisecting the angle between the incident and diffracted beams. Therefore N can be imagined as moving over a fixed projection plane. The position of N on the pole figure is derived from the specimen rotation performed, and at each position of N we can plot the pole density derived from the counter measurement of x-ray intensity.

The ideal diffractometer specimen would be a small sphere cut out of the sheet, because a sphere presents the same aspect to the incident beam whatever its orientation. Normally, however, the sheet is examined directly, and then the paths of incident and diffracted beams within the sheet change with sheet orientation. It is then necessary to correct the measured intensities for these geometrical effects or to design the x-ray optics so that corrections are not required.

There is not one, but several, diffractometer methods for measuring sheet texture. They fall into two groups, transmission and reflection, both being normally necessary for complete coverage of the pole figure.

Transmission Methods

The two methods described below both require a specimen for which μt is of the order of 1, where μ is the linear absorption coefficient and t the thickness. This means a thickness of the order of 35 μm for iron examined with Mo $K\alpha$ radiation or 75 μm for aluminum with Cu $K\alpha$. Thicker sheet has to be thinned by grinding and etching.

The method of *Decker, Asp, and Harker* [9.9] was the first application of the diffractometer to texture measurements. The sheet specimen, in a special holder, is positioned initially with the rolling direction vertical and coincident with the diffractometer axis,* and with the plane of the specimen bisecting the angle between incident and diffracted beams (Figs. 9-12 and 9-13). The specimen holder allows rotation of the sheet in its own plane and about the diffractometer axis.

Figure 9-13 shows how to plot the pole N of the reflecting plane. The angle α measures the amount of rotation about the diffractometer axis, positive when clockwise; α is zero when the sheet bisects the angle between incident and diffracted beams. The angle δ measures the amount by which the sheet is rotated in its own plane and is zero when the transverse direction is horizontal. (Note that other writers often have different symbols and definitions for these angles.) The pole N of the reflecting plane coincides initially, when α and δ are both zero, with the left transverse direction. A rotation of the specimen by δ degrees in its own plane then moves the pole of the reflecting plane δ degrees around the circumference of the pole figure, and a rotation of α degrees about the diffractometer axis then moves it α degrees from the circumference along a radius. To explore the pole figure, it is convenient to make intensity readings at intervals of 5° or 10° of α for a fixed value of δ: the pole figure is thus mapped out along a series of radii.† By this procedure

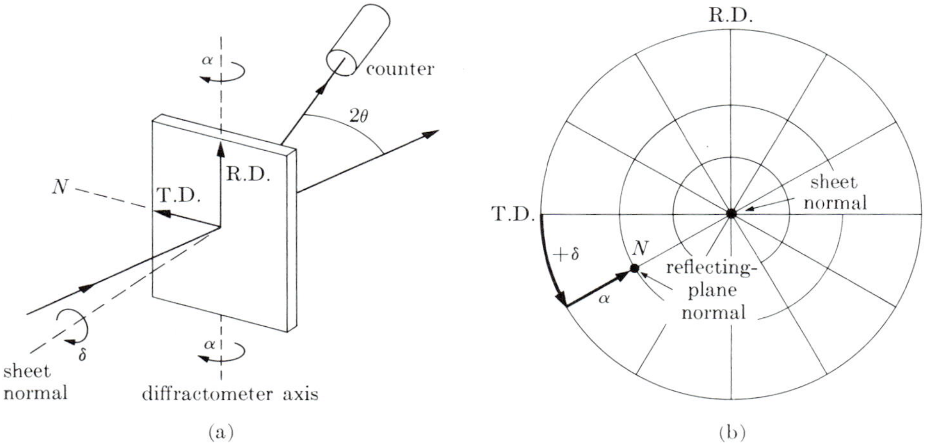

Fig. 9-13 Angular relations in the transmission method, (a) in space, with $\alpha = \delta = 0$, and (b) on the projection, with the reflecting plane normal N at $\alpha = \delta = 30°$. R.D. = rolling direction, T.D. = transverse direction.

* For simplicity, the method is described here only in terms of a vertical-axis diffractometer.

† The chart shown in skeleton form in Fig. 9-13(b) is useful for this purpose. It is called a polar stereographic net, because it shows the latitude lines (circles) and longitude lines (radii) of a ruled globe projected on a plane normal to the polar NS-axis. In the absence of such a net, the equator or central meridian of a Wulff net can be used to measure the angle α.

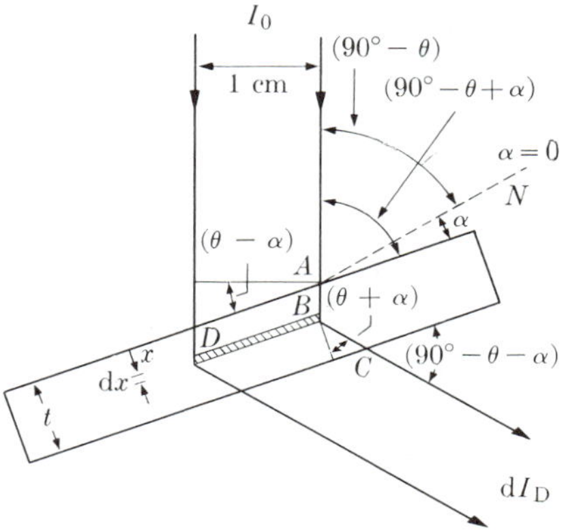

Fig. 9–14 Path length and irradiated volume in the transmission method.

the entire pole figure can be determined except for a region at the center extending from about $\alpha = 50°$ to $\alpha = 90°$; in this region not only does the absorption correction become inaccurate but the frame of the specimen holder obstructs the diffracted x-ray beam; the practical maximum value of α is thus reached before the theoretical, which is $(90° - \theta)$.

An absorption correction is necessary in this method because variations in α cause variations in both the volume of diffracting material and the path length of the x-rays within the specimen. Variations in δ have no effect. We can determine the angular dependence of the absorption factor by a method similar to that used for the reflection case considered in Sec. 4–10. The incident beam in Fig. 9–14 has intensity I_0 (ergs/cm^2/sec) and is 1 cm square in cross section. It is incident on a sheet specimen of thickness t and linear absorption coefficient μ, and the individual grains of this specimen are assumed to have a completely random orientation. Let a be the volume fraction of the specimen containing grains correctly oriented for reflection of the incident beam, and b the fraction of the incident energy diffracted by unit volume. Then the total energy per second in the diffracted beam outside the specimen, originating in a layer of thickness dx located at a depth x, is given by

$$dI_D = ab(DB)I_0 e^{-\mu(AB+BC)}\,dx \quad \text{(ergs/sec)},$$

where

$$DB = \frac{1}{\cos(\theta - \alpha)}, \qquad AB = \frac{x}{\cos(\theta - \alpha)}, \qquad \text{and} \qquad BC = \frac{t - x}{\cos(\theta + \alpha)}.$$

By substitution, we obtain

$$dI_D = \frac{abI_0}{\cos(\theta - \alpha)} e^{-\mu t/\cos(\theta+\alpha)} e^{-\mu x[1/\cos(\theta-\alpha) - 1/\cos(\theta+\alpha)]}\,dx. \qquad (9\text{–}7)$$

If we put $\alpha = 0$ in Eq. (9–7) and integrate from $x = 0$ to $x = t$, we obtain the total diffracted energy per second, the integrated intensity, for this position of the specimen:*

$$I_D\,(\alpha = 0) = \frac{abtI_0}{\cos\theta}\, e^{-\mu t/\cos\theta}. \tag{9-8}$$

When α is not zero, the same integration gives

$$I_D\,(\alpha = \alpha) = \frac{abI_0[e^{-\mu t/\cos(\theta-\alpha)} - e^{-\mu t/\cos(\theta+\alpha)}]}{\mu\{[\cos(\theta-\alpha)/\cos(\theta+\alpha)] - 1\}}. \tag{9-9}$$

We are interested only in the ratio of these two integrated intensities, namely,

$$R = \frac{I_D\,(\alpha = \alpha)}{I_D\,(\alpha = 0)} = \frac{\cos\theta[e^{-\mu t/\cos(\theta-\alpha)} - e^{-\mu t/\cos(\theta+\alpha)}]}{\mu t e^{-\mu t/\cos\theta}\{[\cos(\theta-\alpha)/\cos(\theta+\alpha)] - 1\}}. \tag{9-10}$$

A plot of R vs. α is given in Fig. 9-15 for typical values involved in the 111 reflection from aluminum with Cu $K\alpha$ radiation, namely, $\mu t = 1.0$ and $\theta = 19.25°$. (Values of $100/R$ are tabulated in [G.11, vol. 2, p. 307] and values of $1/R$ by Taylor [G.19].) Figure 9-15 shows that the integrated intensity of the reflection decreases as α increases in the clockwise direction from zero, even for a specimen containing randomly oriented grains. In the measurement of preferred orientation, it is therefore necessary to *divide* each measured intensity by the appropriate value of the correction factor R in order to arrive at a figure proportional to the pole density. From the way in which the correction factor R was derived, it follows that we must measure the *integrated intensity* of the diffracted beam. To do this with a fixed

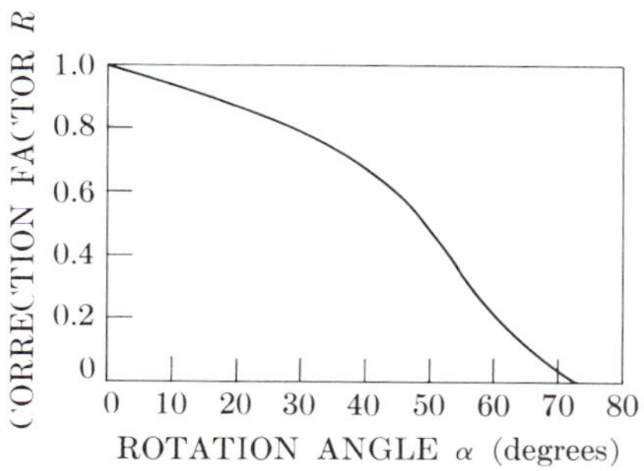

Fig. 9-15 Variation of the correction factor R with α for clockwise rotation from the zero position. $\mu t = 1.0$, $\theta = 19.25°$.

---

* In Sec. 6-9 mention was made of the fact that the diffracted beams in any transmission method were of maximum intensity when the thickness of the specimen was made equal to $1/\mu$. This result follows from Eq. (9-8). If we put $\theta = \alpha = 0$, then the primary beam will be incident on the specimen at right angles (see Fig. 9-14), as in the usual transmission pinhole method, and our result will apply approximately to diffracted beams formed at small angles 2θ. The intensity of such a beam is given by

$$I_D = abtI_0e^{-\mu t}.$$

By differentiating this expression with respect to t and setting the result equal to zero, we can find that I_D is a maximum when $t = 1/\mu$.

counter, the counter slits must be as wide as the diffracted beam for all values of α so that the whole width of the beam can enter the counter. The ideal incident beam for this method is a parallel one. However, a divergent beam may be used without too much error, provided the divergence is not too great. There is no question of focusing here: if the incident beam is divergent, the diffracted beam will diverge also and very wide counter slits will be required to admit its entire width. Clockwise rotation of the specimen about the diffractometer axis makes the diffracted beam narrower and is therefore preferred to counterclockwise rotation.

The value of μt used in Eq. (9–10) must be obtained by direct measurement, since it is not sufficiently accurate to use a tabulated value of μ together with the measured thickness t of the specimen. To determine μt we use a strong diffracted beam from any convenient material and measure its intensity when the sheet specimen is inserted in the diffracted beam and again when it is not. The value of μt is then obtained from the absorption law, $I_t = I_0 e^{-\mu t}$, where I_0 and I_t are the intensities incident on and transmitted by the sheet specimen, respectively.

The *Schulz transmission method* [9.10] involves the use of a divergent, rather than a parallel, incident beam and a counter slit narrow enough to intercept only the central portion of the diffracted beam. For these conditions Schulz derived an equation, analogous to Eq. (9–9), relating the intensity diffracted by a random specimen (for brevity "random specimen" will be written for "specimen with randomly oriented grains") to the angular setting α. He also showed that the intensity diffracted by a random specimen was constant within a few percent up to an α value of about 30° for a sufficiently thin specimen (μt = 0.4 to 0.7) and small values of θ (less than about 20°); under these conditions the correction equation is not needed.

In tests of the correction equations by means of random specimens, Newkirk and Bruce [9.11] found good agreement with Eq. (9–10) of Decker, Asp, and Harker. On the other hand, the data of Aoki *et al.* [9.12] show better agreement between experiment and theory for the Schulz method than that of Decker *et al.* However, in judging the validity of correction equations one must always keep in mind the necessity not only of a truly random specimen, but also of a close match between the actual x-ray optics (nature of incident beam and width of counter slit) and those assumed in the derivation of the equations.

Reflection Methods

The central region of the pole figure is inaccessible to any transmission method and can be explored only by a reflection technique. The specimen must be of effectively infinite thickness or extra corrections will be required.

The most popular is the *Schulz reflection method* [9.13]. It requires a special specimen holder which allows rotation of the specimen in its own plane about an axis normal to its surface and about a horizontal axis; these axes are shown as BB' and AA' in Fig. 9–16. The horizontal axis AA' lies in the specimen surface and is initially adjusted, by rotation about the diffractometer axis, to make equal angles with the incident and diffracted beams. After this is done, no further rotation about the diffractometer axis is made. Since the axis AA' remains in a fixed posi-

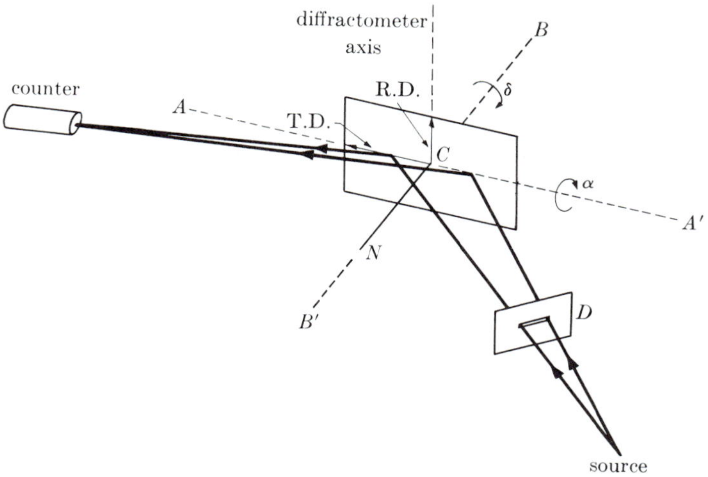

Fig. 9-16 Schulz reflection method.

tion during the other rotations of the specimen, the irradiated surface of the specimen is always tangent to a focusing circle passing through the x-ray source and counter slit. A divergent beam may therefore be used since the diffracted beam will converge to a focus at the counter.

When the specimen is rotated about the axis AA', the axis BB' normal to the specimen surface rotates in a vertical plane, but CN, the reflecting plane normal, remains fixed in a horizontal position normal to AA'. The rotation angles α and δ are shown in Fig. 9-16. The angle α is zero when the sheet is horizontal and has a value of 90° when the sheet is in the vertical position shown in the drawing. In this position of the specimen, the reflecting plane normal is at the center of the projection. The angle δ measures the amount by which the rolling direction is rotated away from the left end of the axis AA' and has a value of $+90°$ for the position illustrated. With these conventions the angles α and δ may be plotted on the pole figure in the same way as in the transmission method [Fig. 9-13(b)].

The great virtue of the Schulz method is that no absorption correction is required for values of α between 90° and about 40°, i.e., up to about 50° from the center of the pole figure. In other words, a random specimen can be rotated over this range of α values without any change in the measured intensity of the diffracted beam, provided the specimen has effectively infinite thickness. Under these circumstances, the intensity of the diffracted beam is directly proportional to the pole density in the specimen, without any correction. The constancy of the absorption factor is due essentially to the narrow horizontal slit placed in the primary beam at D (Fig. 9-16), close to the specimen. The vertical opening in this slit is only about 0.5 mm, which means that the specimen is irradiated only over a long narrow rectangle centered on the fixed axis AA'. It can be shown [9.13] that a change in absorption does occur, as the specimen is rotated about AA', but it is exactly canceled by a change in the volume of diffracting material, the net result being a constant diffracted intensity for a random specimen when α lies between 90° and about 40°. To achieve this condition, the reflecting surface of the specimen must be adjusted to accurately coincide with the axis AA' for all values of α and δ. This adjustment is extremely important.

When the specimen is rotated out of the vertical position ($\alpha < 90°$) in the sense shown in Fig. 9–16, the top part moves behind, and the bottom in front of, the focusing circle. The diffracted beam therefore widens at the counter slit, and the measured diffracted intensity from a random specimen may decrease as α departs from 90°. This effect is called the *defocusing error*. It may be minimized by slit adjustment (widening the counter slit and decreasing the vertical opening in slit D) or corrected by calculation [9.14, 9.15].

Figure 9–17 shows a specimen holder suitable for either transmission method and for the Schulz reflection method.

The *Field and Merchant reflection method* [9.16] is designed for a parallel incident beam, shown simply as a single line in Fig. 9–18. The specimen is placed initially with the rolling direction vertical, coincident with the diffractometer axis, the transverse direction horizontal, and the plane of the sheet equally inclined to the

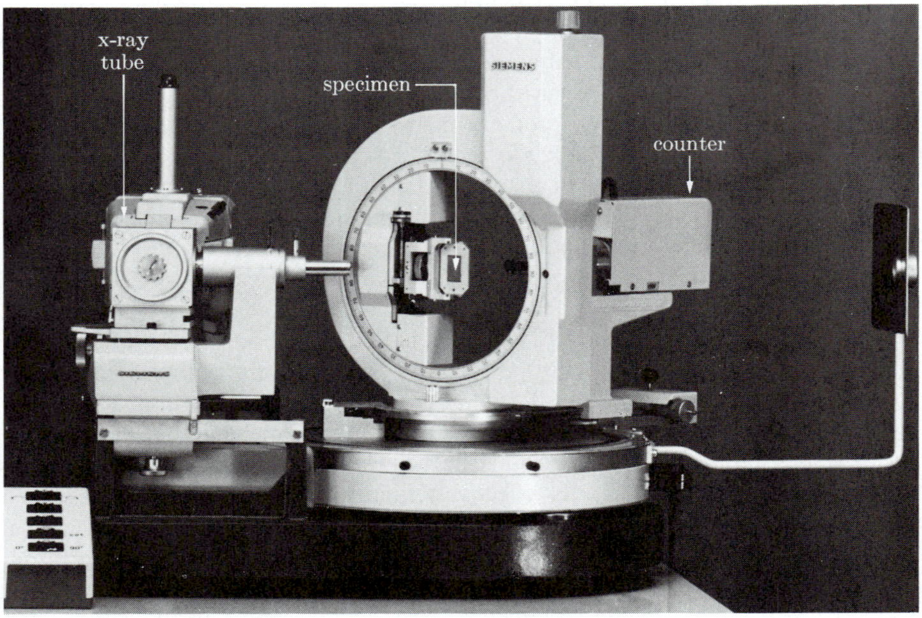

Fig. 9–17 Pole-figure goniometer. The specimen shown is positioned for measurements by the transmission method, and a simple change in the orientation of the specimen holder allows measurements by the Schulz reflection method. The x-ray tube is seen here end-on. This instrument is designed for automatic operation. (Courtesy of Siemens Corporation.)

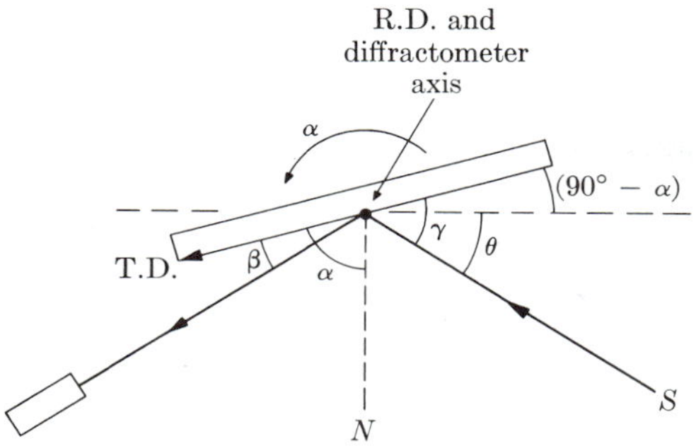

Fig. 9–18 Field and Merchant reflection method. N = diffracting plane normal.

incident and diffracted beams; the pole N of the reflecting plane is then at the center of the projection ($\alpha = 90°$). Counterclockwise rotation of the specimen about the diffractometer axis, which incidentally will narrow the diffracted beam, moves N to the left along the equator ($\delta = 0$) of the pole figure. The angle δ is changed by rotating the specimen in its own plane.

Reflected intensities must be corrected for change in absorption due to change in α. The angles γ and β of incidence and reflection are the same as shown in Fig. 4–19, and Eq. (4–14) applies:

$$dI_D = \frac{I_0 ab}{\sin \gamma} e^{-\mu x(1/\sin \gamma + 1/\sin \beta)} \, dx. \tag{4–14}$$

For counterclockwise rotation $\gamma = \theta + (90° - \alpha)$ and $\beta = \alpha - (90° - \theta)$. Making these substitutions into Eq. (4–14) and integrating from $x = 0$ to ∞, we find the integrated intensity:

$$I_D = \frac{I_0 ab}{\mu\{1 - [\cos(\alpha - \theta)/\cos(\alpha + \theta)]\}}. \tag{9–11}$$

We are interested only in the ratio of this quantity to the same quantity for $\alpha = 90°$. This ratio is

$$S = \frac{I_D(\alpha = \alpha)}{I_D(\alpha = 90°)} = 1 - \cot \alpha \cot \theta. \tag{9–12}$$

Diffracted intensities must be divided by S, which is independent of μ, to give values proportional to pole density. The correction is less severe (S closer to 1), the larger the value of θ; it is therefore advantageous to measure a higher order of the *hkl* reflection measured in transmission. The specimen holder can be identical with that used in the transmission method.

Plotting the Pole Figure

A transmission method yields pole densities covering the outer part of the pole figure, from $\alpha = 0$ to about 50°. A reflection method covers the inner part, from α = about 40° to 90°. The pole densities are in arbitrary units, either directly measured diffracted intensities or corrected intensities, depending on the method used. Along those radii of the pole figure where substantial pole density exists in the region of overlap of the two methods ($\alpha = 40°$ to $50°$), a normalizing factor is found which will make the pole densities from the transmission method agree with those from the reflection method. The match in the overlap region is rarely perfect, but a substantial disagreement between normalizing factors for different radii points to experimental or computational errors.

Once one set of data is normalized to match the other set, numbers proportional to pole density can be written on the pole figure at each point at which a measurement was made. Contour lines are then drawn at selected levels to connect points of the same pole density, and the result is a pole figure such as Fig. 9–19. Many, but not all, textures are symmetrical with respect to reflection planes normal to the rolling and transverse directions, and many published pole figures have been determined from measurements made only in one quadrant, with the other quadrants found by assuming symmetry, without supporting data.

The deformation texture of brass sheet (Fig. 9–19) is fairly sharp, and it is then of interest to know whether or not it can be approximated by an "ideal orientation." To find this orientation we successively lay several standard projections over the pole figure, looking for a match between (111) poles and high-density regions. The solid triangles in Fig. 9–19 show such a match; they represent the (111) poles of a single crystal oriented so that its (110) plane is parallel to the sheet and the $[\bar{1}12]$ direction parallel to the rolling direction. Reflection of these poles in the

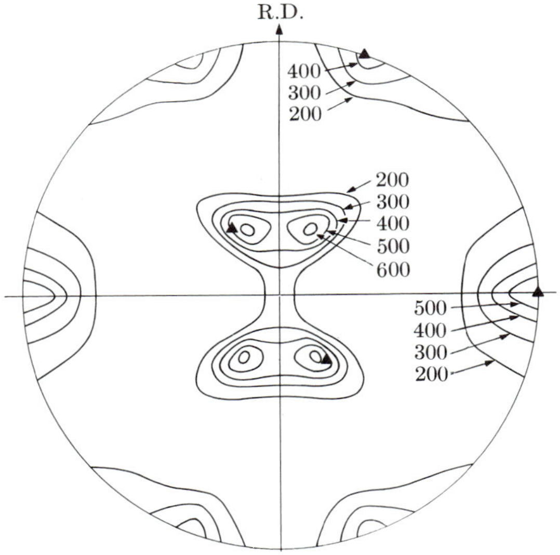

Fig. 9–19 (111) pole figure of alpha brass sheet (70 Cu–30 Zn), cold rolled to a reduction in thickness of 95 percent. Pole densities in arbitrary units. The outer parts of all four quadrants were determined experimentally; the inner parts of the upper right and lower left quadrants were measured, and the other two constructed by reflection. The solid triangles show the (110) $[\bar{1}12]$ orientation. Hu, Sperry, and Beck [9.17].

symmetry planes of the texture, which is equivalent to adding orientations like $(110)[1\bar{1}2]$, will approximately account for all the high-density regions of the pole figure. This texture can therefore be represented by the ideal orientation $\{110\}$ $\langle\bar{1}12\rangle$. But the pole figure itself must be regarded as a far better description of the texture than any bare statement of an ideal orientation, which says nothing about the scatter. A quantitative pole figure has about the same relation to an ideal orientation as an accurate contour map of a hill has to a statement of the height, width, and length of the hill.

Pole densities in arbitrary units are not as informative as those expressed in multiples of the pole density of a random specimen, so called "times random" units. The contour lines in Fig. 9-20 are marked with these units, and one can see at a glance those regions of the pole figure that have a higher, or lower, pole density than random; this pole figure, incidentally, was measured in all four quadrants. The texture represented there, of considerable industrial interest, is messy and cannot be well characterized by ideal orientations. It is approximately a $\langle 111\rangle$ fiber texture, with the fiber axis normal to the plane of the sheet, containing $\{111\}$ $\langle 110\rangle$, called the "cube-on-corner" texture, as its strongest single component [9.18, 9.19].

Diffracted intensities, proportional to pole densities, may be put on a times-random basis by comparing them with intensities diffracted by a random specimen [9.20]. The random specimen should be of the same material as the textured specimen and, for a transmission method, it should have the same value of μt; if not, a correction has to be made that will depend on the transmission method involved. The random specimen itself is usually made by compressing and sintering a powder [9.11, 9.12]. The randomness of grain orientation in this specimen must be checked by determining its diffraction pattern with a diffractometer in the usual way; the measured integrated intensities of all lines should agree with those calculated by Eq. (4-21).

General

The conditions for optimum specimen thickness in transmission and infinite thickness in reflection are such that the same specimen can serve for both methods. The penalty for exceeding the optimum thickness is not severe: a thickness of double the optimum value for transmission at $\alpha = 0$ reduces the diffracted intensity by only 26 percent (Problem 9-8).

It may be difficult to make a thin specimen, particularly of a heavy metal, which has the required low, and *uniform*, value of μt throughout. Some investigators have therefore avoided a transmission method altogether by determining only the central portion of the pole figure by reflection; such partial pole figures are useful for some purposes. Others have obtained a complete pole figure by reflecting x-rays from a surface or surfaces inclined to the sheet surface. Several pieces of the sheet are stacked with rolling directions parallel and a composite specimen made by clamping or cementing the stack. If this specimen is cut to expose a surface whose normal makes the same angle, of 54.7°, with the sheet normal and rolling and transverse directions, then measurements on this surface

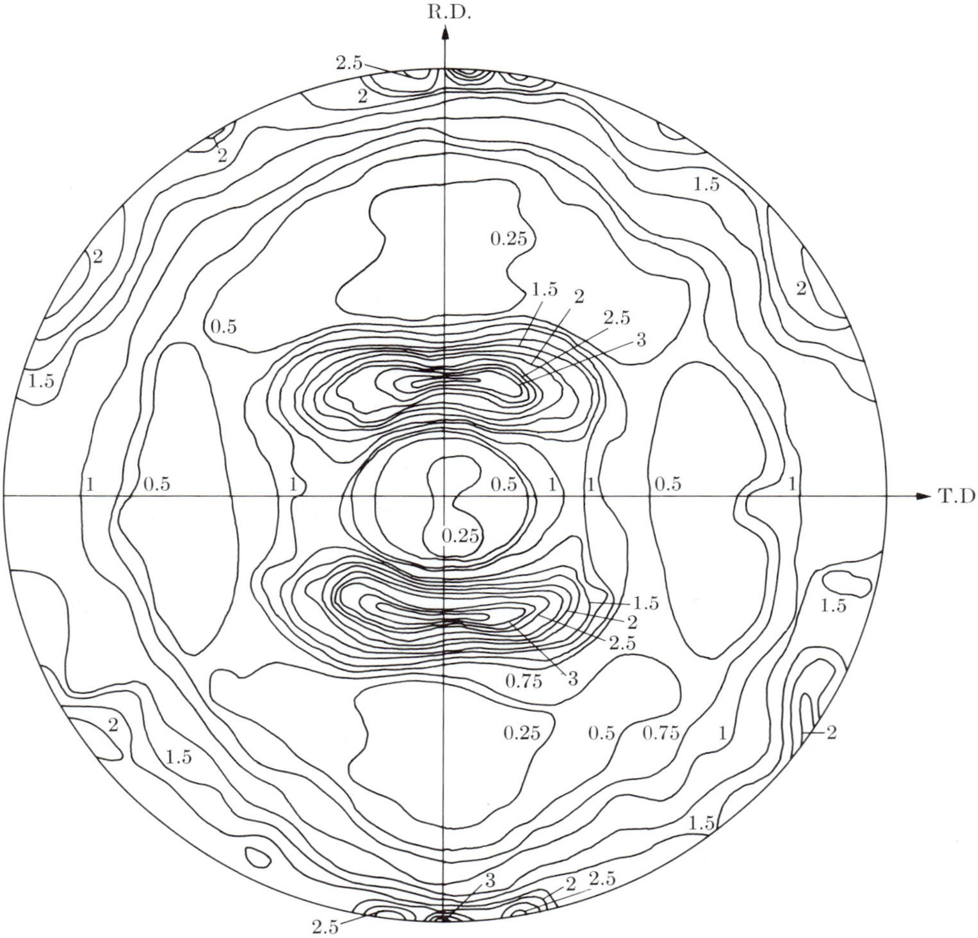

Fig. 9–20 (110) pole figure of recrystallized commercial low-carbon (0.04 percent) sheet steel, aluminum killed, 0.9 mm thick. Pole densities in "times random" units. Determined by a reflection method from composite specimens; see text under "General." Bunge and Roberts [9.18].

by a reflection method will cover one quadrant of the pole figure [9.21]. The pole figure of Fig. 9–20 was derived from reflection measurements on the sheet plane and on sections normal to the rolling and transverse directions.

Some errors that can affect texture measurements are the following:

1. If the grain size is too large, as in some recrystallized specimens, the incident beam will not sample enough grains for good statistical accuracy. The specimen holder should then include a mechanical device that will translate the specimen back and forth in its own plane in order to expose more grains to the incident beam.

2. The texture at the surface of a sheet can differ from that in the midplane [9.17].

When such a texture gradient exists, it must be remembered that the transmission and reflection methods preferentially sample different layers of the same specimen and that the thicknesses of the layers so sampled vary with α in a systematic way.

3. The x-ray optics, particularly slit sizes, must conform to those required by the x-ray method involved. Too narrow a counter slit will exclude a wanted part of the diffraction line; too wide a slit will include some of the background. Too often the background is simply ignored, with the result that measured intensities include both line and background; it would be highly beneficial to eliminate or reduce the background by means of a crystal monochromator, balanced filters, or a pulse-height analyzer. Errors from these sources can be large. They can be almost entirely eliminated by comparing intensities from the textured specimen with intensities measured under identical conditions from a truly random specimen, provided the two specimens differ *only* in grain orientation. Such a random specimen can be very hard to make. (Note, for example, that an annealed random specimen will have narrower diffraction lines than a cold-worked textured specimen.)

Because the manual determination of preferred orientation is rather tedious, the process has been automated by laboratories that need large amounts of this kind of data [9.22–9.27]. Under computer control, the angular settings α and δ of the specimen are varied in a prescribed sequence by motor drive, and the diffracted intensity at each setting is recorded, on paper tape, for example. These output data can also be used to control an automatic pole-figure plotter. Apparatus of this kind is commercially available, such as that shown in Fig. 9-17.

The *analysis* of preferred orientation has now gone beyond pole figures, in the direction of a more complete description of the texture. When we establish experimentally that a certain crystal in a sheet has a (100) pole, for example, located at a point on the pole figure specified by the angles α and δ, we have not fully specified the orientation of the crystal. These two angles merely describe the orientation of the direction [100], and the crystal might have any rotational position about this axis. A third angle is needed to fix the orientation of the crystal. A full description of the texture would require the specification of three angles (called ψ, θ, ϕ) for each crystal in the sheet. This information is contained in the *crystal orientation distribution*, and methods of calculating this distribution from pole figures have been developed by Roe, Bunge, and Williams [9.29–9.32]. The mathematics involved are complex and the calculations extensive, but this approach is powerful and of great generality.

The crystal orientation distribution for a cubic material in sheet form can be calculated from any two experimental pole figures, for example, the (100) and (110). Once the distribution is known, any other desired pole figure can be calculated, for example, the (111); it need not be measured. It is even possible to calculate the orientation distribution from a set of partial pole figures, determined by a reflection method out to 60° from the center of the pole figure ($\alpha = 30°$) [9.28]. The crystal orientation distribution itself is usually presented in the form of crystal density plots, in which the density is shown as a contour map on, for example, ψ, θ axes with ϕ held constant; a series of such constant-ϕ maps is needed to present all the information. Applications of the orientation distribution approach may be found in [9.18, 9.19, 9.28, 9.33, 9.34].

Finally, we must not forget that the practical reason for investigating preferred orientation is to understand the properties of the aggregate. The old problem still remains: how do we calculate the physical and mechanical properties of an aggregate from the corresponding properties of the single crystal and the measured texture? The crystal orientation distribution affords the most rational basis so far available for this calculation, and considerable progress has been made in this direction [9.18, for example].

9-9 THE TEXTURE OF WIRE (DIFFRACTOMETER METHOD)

As mentioned in Sec. 9–6, if a wire or rod has a true fiber texture, its pole figure will have rotational symmetry about the fiber axis and will resemble Fig. 9–8. We therefore have to measure pole density only along a single radius. The angle between the pole N and the fiber axis F.A. is usually called ϕ, rather than α, when dealing with fiber textures.

The Field and Merchant method may be used to measure pole density, and two specimens are required to cover the entire 90° range of ϕ:

1) *Low-ϕ region.* X-rays are reflected from the cross section of the wire, as in Fig. 9–21(a). The specimen is a bundle of wires, packed together and cemented into a rectangular hole cut in a thick plastic disc; the wire ends are then ground, polished, and etched. This cross section is made initially parallel to the diffractometer axis and equally inclined to the incident and diffracted beams. The angle ϕ measures the counterclockwise rotation of the specimen about the diffractometer axis from this initial position, and we define a new angle ρ as the acute angle between the specimen surface and the reflecting-plane normal N. The angle α of Eq. (9–12) and Fig. 9–18 becomes $\rho = \rho_L = 90° - \phi$, so that

$$W = \frac{I_D(\rho = \rho)}{I_D(\rho = 90°)} = 1 - \cot\rho \cot\theta. \tag{9–13}$$

Diffracted intensities are to be divided by W to obtain numbers proportional to pole density.

2) *High-ϕ region.* X-rays are reflected from the side of the wire, as in Fig. 9–21(b). The specimen is a set of wires glued to a grooved plate. Equation (9–13) still applies, but now $\alpha = \rho = \rho_H = \phi$.

When the diffracted intensities I_D given by each method have been divided by W and normalized in the region of overlap, we have a set of numbers I proportional to pole density. Figure 9–22 shows an I, ϕ curve obtained in this way for the inside texture of cold-drawn aluminum wire. The peaks at $\phi = 0$ and 70° are due to the strong [111] component of the texture and the peak at 55° to a weak [100] component.

By analysis of an I, ϕ pole density curve we can (a) put pole densities on a times-random basis and (b) determine the relative amounts of the components in a double fiber texture [9.36–9.38].

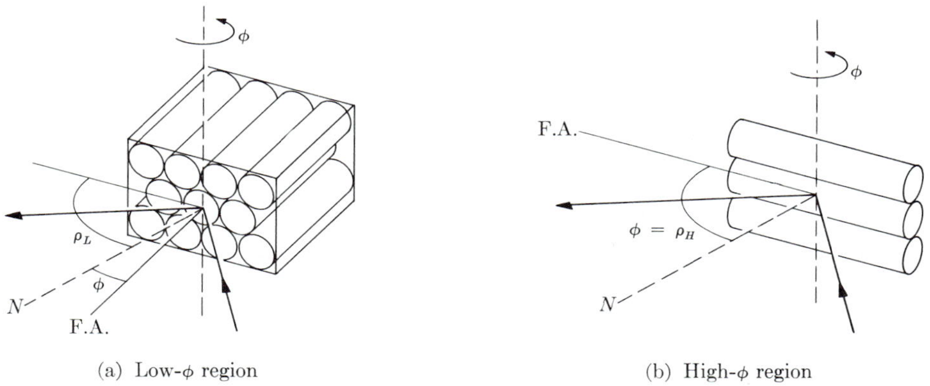

Fig. 9–21 Reflection from composite wire specimens. ϕ is angle between fiber axis F.A. and reflecting-plane normal N. ρ is angle between N and specimen surface.

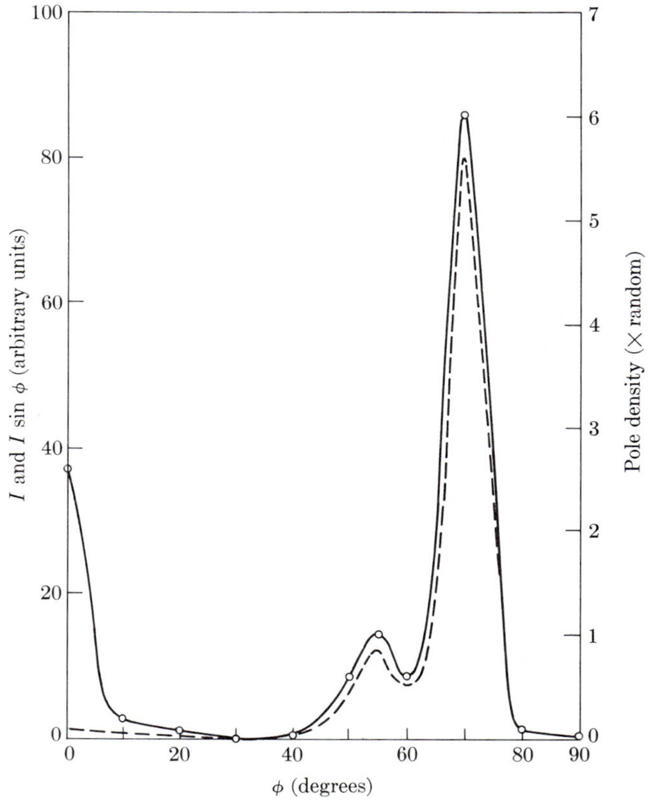

Fig. 9–22 (111) pole density I (full curve) and $I \sin \phi$ (dashed) as a function of ϕ for a cold-drawn aluminum wire, reduced in area 95 percent by drawing, and etched to 80 percent of the as-drawn diameter. Final specimen diameter 1.3 mm, Cr $K\alpha$ radiation, 222 reflection. Freda *et al.* [9.35].

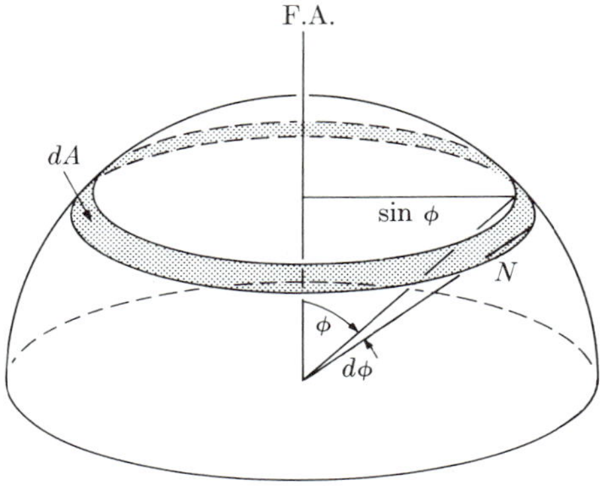

Fig. 9–23 Element of area on reference sphere.

Imagine a reference sphere of unit radius centered on the wire (Fig. 9–23). The element of area on the surface is $\mathrm{d}A = 2\pi \sin \phi \, \mathrm{d}\phi$. If I is the pole density, the number of (hkl) poles in this area is $\mathrm{d}n = I \, \mathrm{d}A = 2\pi I \sin \phi \, \mathrm{d}\phi$, and the total number of poles on the surface of a hemisphere is

$$n = 2\pi \int_0^{\pi/2} I \sin \phi \, \mathrm{d}\phi. \qquad (9\text{–}14)$$

If I_r is the pole density of a random specimen, then $n = I_r(2\pi)$. Therefore

$$I_r = \int_0^{\pi/2} I \sin \phi \, \mathrm{d}\phi. \qquad (9\text{–}15)$$

This integral is simply the area under a curve of $I \sin \phi$ vs. ϕ, and this area in turn is equal to the average ordinate of the curve $\langle I \sin \phi \rangle_{\text{av}}$ times $\pi/2$. Therefore,

$$I_r = \frac{\pi}{2} \langle I \sin \phi \rangle_{\text{av}}. \qquad (9\text{–}16)$$

This relation is valid whether the n poles are distributed randomly on the sphere or in some preferred manner, and it enables us to find I_r from measurements on a textured specimen. From experimental I, ϕ data we construct a curve of $I \sin \phi$ vs. ϕ, shown dashed in Fig. 9–22, determine its average ordinate, and find I_r from Eq. (9–16). Once I_r is known (14.3 units for this wire), the I, ϕ pole density curve can be put on a times-random basis (right-hand ordinate). (Because the angular aperture of the counter slit is not small relative to ϕ when ϕ is small, the true pole density I near $\phi = 0$ can only be approximated [9.37]. We therefore extrapolate the $I \sin \phi$ curve near $\phi = 0$ rather than extend it to zero, as is mathematically indicated.)

The amount of each component in a double texture is proportional to the area under the peak(s) due to that component on a curve of $I \sin \phi$ vs. ϕ; knowledge of the multiplicities involved can reduce the number of peaks that need be considered. The multiplicity of {111} poles is 8 for a sphere and 4 for a hemisphere. In Fig. 9–22 the [100] component puts four {111} poles per grain under the peak at 55°; the [111] component, on the other hand, has one {111} pole at 0° and three at 70°. Because of the uncertainty in the area of the peak at 0°, the area ascribed to the [111] component can be taken as

$\frac{4}{3}$ times the area of the peak at 70°. Therefore,

$$\frac{\text{Volume fraction of [111] component}}{\text{Volume fraction of [100] component}} = \frac{(\frac{4}{3})(\text{area of } 70° \text{ peak})}{(\text{area of } 55° \text{ peak})}.$$

For the aluminum wire, this calculation led to volume fractions of 0.85 for the [111] component and 0.15 for the [100].

Note that this result was obtained without making any use of the measurements made at ϕ values less than 40°. Thus a complete pole density curve is not necessary for the evaluation of a texture, provided the texture is sharp enough to produce well resolved peaks in the high-ϕ region. This is a fortunate circumstance, because high-ϕ measurements require little or no specimen preparation.

Preferred orientation in wire does not always take the form of a pure fiber texture. For example, the deformation texture of iron wire is usually considered to be a [110] fiber texture, but Leber [9.39] showed that a *cylindrical texture* was also present. Such a texture may be regarded as a sheet texture, (100) [011] for iron, wrapped around the wire axis. Thus at any point on the wire surface, a (100) plane is tangent to the surface and a [011] direction parallel to the wire axis. The presence of a cylindrical component in a fiber texture is disclosed by anomalies in the $I \sin \phi$ curve: the areas under the peaks ascribed to the fiber-texture component will not be in the ratio to be expected from the multiplicities [9.40].

9-10 INVERSE POLE FIGURES

A pole figure shows the distribution of a selected crystallographic direction relative to certain directions in the specimen. Texture data may also be presented in the form of an *inverse pole figure*, which shows the distribution of a selected direction *in the specimen* relative to the crystal axes. The projection plane for an inverse pole figure is therefore a standard projection of the crystal, of which only the unit stereographic triangle need be shown. Both wire and sheet textures may be represented.

Figure 9-24(a) is an inverse pole figure for the inside texture of an extruded aluminum rod, showing the density distribution of the rod axis on a times-random basis. It was derived by a trial-and-error method [9.36] from pole density curves, as in Fig. 9-22, for the (001), (111), and (113) poles. We note concentrations of the rod axis at [001] and [111], indicating a double fiber texture; the volume fractions of the [001] and [111] components were estimated as 0.53 and 0.47, respectively. Note that an inverse pole figure shows immediately the crystallographic "direction" of the scatter. In this double texture, there is a larger scatter of each component toward one another than toward [011].

Sheet textures may also be represented by inverse pole figures. Here three separate projections are needed to show the distribution of the sheet normal, rolling direction, and transverse direction. Figure 9-24(b) is such a projection for the normal direction of the steel sheet whose (110) pole figure was given in Fig. 9-20; it was calculated from the crystal orientation distribution mentioned in Sec. 9-8. The distribution of the normal direction is also shown in (c), for the same material. This distribution was measured directly in the following way. A powder pattern is made of the sheet in a diffractometer by the usual method, with the sheet equally

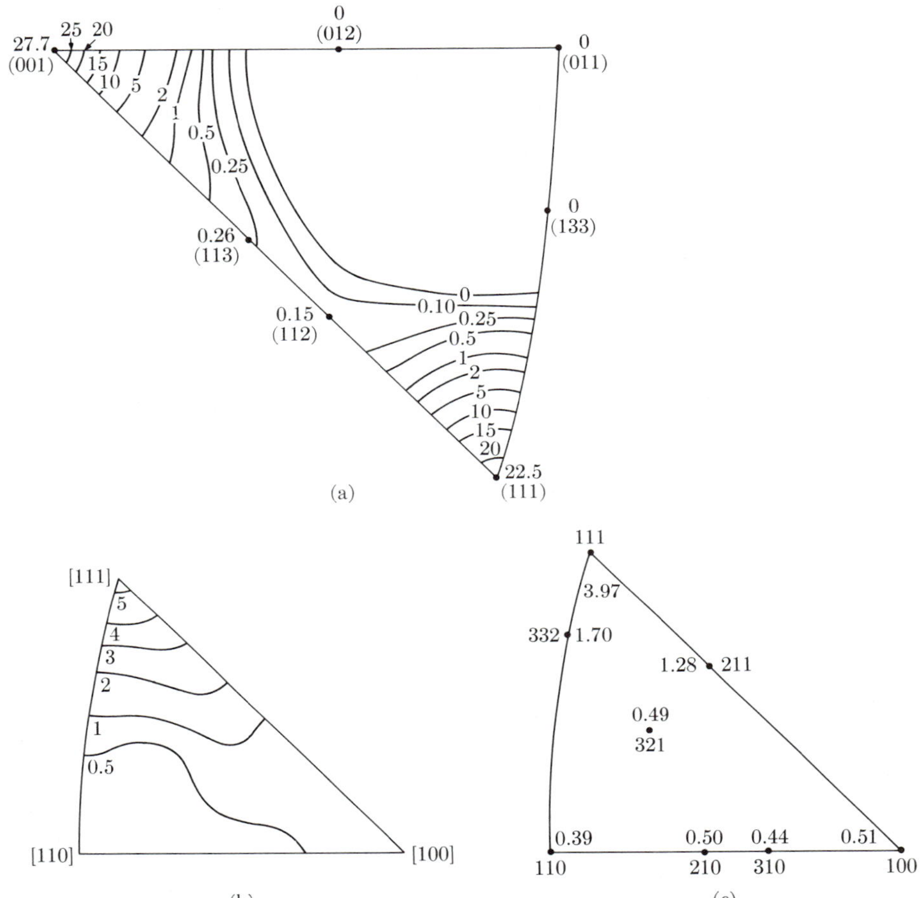

Fig. 9–24 Inverse pole figures. (a) Distribution of axis of aluminum rod, extruded at 450°F to a reduction in area of 92 percent and a final diameter of 23 mm. Jetter, McHargue, and Williams [9.36]. (b) and (c) show the distribution of the sheet normal for the steel sheet of Fig. 9–20. Bunge and Roberts [9.18].

inclined to the incident and diffracted beams. The intensity of any *hkl* reflection, relative to that from a random specimen, is then proportional to the volume fraction of grains having their (*hkl*) planes parallel to the sheet surface, or to the volume fraction of grains having the sheet normal parallel to the (*hkl*) normal. Stated another way, and with specific reference to Fig. 9–24(c), the probability of the sheet normal in this steel being parallel to [111] is 3.97 times normal. Similar data for the rolling direction, for example, are obtained by reflecting x-rays from a surface normal to the rolling direction, a surface exposed by sectioning a stack of sheets. This method produces only as many data points in the stereographic triangle as there are lines on the powder pattern and is therefore better suited to materials of low crystal symmetry than to cubic materials.

The inverse pole figures of Figs. 9–24(b) and (c) both show a high density of (111) poles parallel to the sheet normal and are therefore consistent with the [111] quasi-fiber texture mentioned in Sec. 9–8.

The inverse pole figure is the best way to represent a fiber texture, but it offers no advantage over a direct pole figure in the description of a sheet texture. Inverse or direct, a pole figure is a two-dimensional plot that fixes, at a point, only a direction in space, be it crystal space or specimen space. Only the three-dimensional "plot" afforded by the crystal orientation distribution (Sec. 9–8) can completely describe the orientations present, and this approach, being quite general, is just as applicable to fiber textures as it is to sheet.

AMORPHOUS SOLIDS

9–11 AMORPHOUS AND SEMI-AMORPHOUS SOLIDS

This chapter has concentrated on crystalline materials. X-ray diffraction can also furnish structural information about amorphous and semi-amorphous solids, even though the "structure" is much more diffuse.

Most glasses are amorphous. They yield a pinhole pattern consisting only of a broad, diffuse halo around the central spot, corresponding to the single maximum in the scattering curve of Fig. 3–18. Analysis of this curve yields such information as average interatomic distances and average number of atoms around a given atom. The theory of diffraction by amorphous substances is treated in the books of Guinier [G.21] and Warren [G.30].

Many plastics are partly crystalline. These polymeric substances are composed of very long molecules, generally in a state of great disarray but here and there organized into ordered regions usually called "crystallites." These regions, typically very small and highly strained, produce very broad diffraction lines. By comparing the integrated intensity of these lines with that of the broad halo due to the amorphous regions, the volume fraction of crystallites, called the "degree of crystallinity" of the polymer, can be estimated. X-ray diffraction can also disclose crystallite size, usually by means of the Scherrer equation for line broadening (Eq. 3–13), and preferred orientation. The latter condition is quite common in fibers and sheets and is studied by one or more of the methods described in this chapter. In fact, the alignment of crystallites in natural fibers like cotton and silk has long been known and is the origin of the term "fiber texture" to describe the preferred orientation found in metal wires.

The particular applications of x-ray methods to the study of polymers is the subject of a book by Alexander [9.41].

9–12 SUMMARY

In this chapter we have considered various aspects of the structure of polycrystalline aggregates and the quantitative effects of variations in crystal size, quality, and orientation on the diffraction pattern. Although a complete investigation of the structure of an aggregate requires a considerable amount of time and rather complex apparatus, the very great utility of the simple pinhole photograph should not be overlooked. It is surprising how much information an experienced observer can obtain simply by inspection of a pinhole photograph, without making any measurements on the film and without any knowledge of the specimen, i.e., without knowing its chemical identity, crystal structure, or even whether it is amorphous or crystalline. The latter point can be settled at a glance, since Debye rings indicate crystallinity and broad haloes an amorphous condition. If the specimen is crystalline, the conclusions that can be drawn from the appearance of the lines are summarized in Table 9–2.

Table 9–2

Appearance of diffraction lines	Condition of specimen
Continuous	Fine-grained (or coarse-grained and cold-worked)
Spotty	Coarse-grained
Narrow (1)	Strain-free
Broad (1)	Residual stress and possibly small particle size (if specimen is a solid aggregate) Small particle size (if specimen is a brittle powder)
Uniform intensity	Random orientation (2)
Nonuniform intensity	Preferred orientation

(1) Best judged by noting whether or not the $K\alpha$ doublet is resolved in back reflection.
(2) Or possibly presence of a fiber texture, if the incident beam is parallel to the fiber axis.

PROBLEMS

9–1 Prove the statement made in Sec. 4–10 that the effective irradiated volume of a flat plate specimen in a diffractometer is constant and independent of θ.

***9–2** For given values of θ and μ, which results in a greater effective depth of x-ray penetration, a back-reflection pinhole camera or a diffractometer?

9–3 Assume that the effective depth of penetration of an x-ray beam is that thickness of material which contributes 99 percent of the total energy diffracted by an infinitely thick specimen. Calculate the penetration depth in μm for a low-carbon steel specimen under the following conditions:

a) Diffractometer; lowest-angle reflection; Cu $K\alpha$ radiation.
b) Diffractometer; highest-angle reflection; Cu $K\alpha$ radiation.
c) Diffractometer; highest-angle reflection; Cr $K\alpha$ radiation.
d) Back-reflection pinhole camera; highest-angle reflection; Cr $K\alpha$ radiation.

9–4 If the same *hkl* reflection from a given material is examined in a diffractometer with successively different wavelengths, how does the penetration depth x vary with λ? (Assume the wavelengths used lie on the same branch of the absorption curve of the material.)

***9–5** (a) A transmission pinhole photograph is made of a sheet specimen of thickness t and linear absorption coefficient μ. Show that the fraction of the total diffracted energy in any one reflection contributed by a layer of thickness w is given by

$$T = \frac{e^{-\mu[x+(t-x)/\cos 2\theta]}[e^{-\mu w(1-1/\cos 2\theta)} - 1]}{e^{-\mu t} - e^{-\mu t/\cos 2\theta}},$$

where x is the distance to the side of the layer involved, measured from the side of the specimen on which the primary beam is incident.

b) A transmission pinhole photograph is made of a sheet of aluminum 0.5 mm thick with Cu $K\alpha$ radiation. Consider only the 111 reflection which occurs at $2\theta = 38.4°$. Imagine the sheet to be divided into four layers, the thickness of each being equal to one-fourth of the total thickness. Calculate T for each layer.

9–6 A transmission pinhole pattern is made with Co $K\alpha$ radiation of an iron wire having an almost perfect [110] fiber texture. The wire axis is vertical. How many high-intensity maxima will appear on the lowest-angle 110 Debye ring and what are their azimuthal angles on the film?

9–7 An electroplated layer of copper on sheet steel is examined in a back-reflection pinhole camera with Cu $K\alpha$ radiation incident at right angles to the sheet surface. Assume the copper has a fiber texture with the fiber axis [uvw] scattered by an angle β in every direction about the sheet normal. How large must β be for the 420 Debye ring (see Table 4–2) to appear on the film if the fiber axis [uvw] is (a) [110], (b) [100]?

***9–8** Consider the diffraction geometry for $\alpha = 0$ in the transmission method for determining preferred orientation and for $\alpha = 90°$ in the reflection method. Let t_{inf} be the infinite thickness required in the reflection method, and assume t_{inf} is that thickness which would diffract 99 percent of the intensity diffracted by a specimen of truly infinite thickness. Let t_{opt} be the optimum thickness for the transmission method.

a) Show that $t_{inf}/t_{opt} = 2.30 \tan \theta$.

b) If the thickness t of a transmission specimen is $2t_{opt}$, by how much is the diffracted intensity decreased?

9–9 On a stereographic projection parallel to the surface of a rolled sheet, show (a) the positions of the (110) poles, represented by small ellipses, for the ideal orientation {111} ⟨110⟩, including the positions due to reflection symmetry, and (b) the lines showing the positions of the (110) poles for a ⟨111⟩ fiber texture, with the fiber axis normal to the plane of the sheet.

Compare your result with the experimental (110) pole figure for low-carbon sheet steel in Fig. 9–20.

10

Determination of Crystal Structure

10-1 INTRODUCTION

Since 1913, when W. L. Bragg solved the structure of NaCl, the structures of many thousands of crystals, organic and inorganic, have been determined. This vast body of knowledge is of fundamental importance in such fields as crystal chemistry, solid-state physics, and the biological sciences because, to a large extent, structure determines properties and the properties of a substance are never fully understood until its structure is known. In metallurgy, a knowledge of crystal structure is a necessary prerequisite to any understanding of such phenomena as plastic deformation, alloy formation, or phase transformations.

The work of structure determination goes on continuously since there is no dearth of unsolved structures. New substances are constantly being synthesized, and the structures of many old ones are still unknown. In themselves crystal structures vary widely in complexity: the simplest can be solved in a few hours, while the more complex may require months or even years for their complete solution. (Proteins form a notable example of the latter kind; some protein structures are now known, but others still defy solution.) Complex structures require complex methods of solution, and structure determination in its entirety is more properly the subject of a book than of a single chapter. All we can do here is to consider some of the principles involved and how they can be applied to the solution of fairly simple structures. Moreover, we will confine our attention to the methods of determining structure from powder patterns alone, because such patterns are the kind most often encountered by the metallurgist.

The basic principles involved in structure determination have already been introduced in Chaps. 3 and 4. We saw there that the crystal structure of a substance determines the diffraction pattern of that substance or, more specifically, that the shape and size of the unit cell determines the angular positions of the diffraction lines, and the arrangement of the atoms within the unit cell determines the relative intensities of the lines. It may be worthwhile to state this again in tabular form:

Crystal structure		Diffraction pattern
Unit cell	↔	Line positions
Atom positions	↔	Line intensities

Since structure determines the diffraction pattern, it should be possible to go in the other direction and deduce the structure from the pattern. It is possible, *but not in any direct manner*. Given a structure, we can calculate its diffraction pattern in a very straightforward fashion, and examples of such calculations were given in Sec. 4-13; but the reverse problem, that of directly calculating the structure from the observed pattern, has not yet been solved for the general case (Sec. 10-8). The procedure adopted is essentially one of trial and error. On the basis of an educated guess, a structure is assumed, its diffraction pattern calculated, and the calculated pattern compared with the observed one. If the two agree in all detail, the assumed structure is correct; if not, the process is repeated as often as is necessary to find the correct solution. The problem is not unlike that of deciphering a code, and requires of the crystallographer the same qualities possessed by a good crypt-analyst, namely, knowledge, perseverance, and not a little intuition.

The determination of an unknown structure proceeds in three major steps:

1. The shape and size of the unit cell are deduced from the angular positions of the diffraction lines. An assumption is first made as to which of the seven crystal systems the unknown structure belongs to, and then, on the basis of this assumption, the correct Miller indices are assigned to each reflection. This step is called "indexing the pattern" and is possible only when the correct choice of crystal system has been made. Once this is done, the shape of the unit cell is known (from the crystal system), and its size is calculable from the positions and Miller indices of the diffraction lines.

2. The number of atoms per unit cell is then computed from the shape and size of the unit cell, the chemical composition of the specimen, and its measured density.

3. Finally, the positions of the atoms within the unit cell are deduced from the relative intensities of the diffraction lines.

Only when these three steps have been accomplished is the structure determination complete. The third step is generally the most difficult, and there are many structures which are known only incompletely, in the sense that this final step has not yet been made. Nevertheless, a knowledge of the shape and size of the unit cell, without any knowledge of atom positions, is in itself of very great value in many applications.

The average metallurgist is rarely, if ever, called upon to determine an unknown crystal structure. If the structure is at all complex, its determination is a job for a specialist in x-ray crystallography, who can bring special techniques, both experimental and mathematical, to bear on the problem. The metallurgist should, however, know enough about structure determination to unravel any simple structures he may encounter and, what is more important, he must be able to index the powder patterns of substances of *known* structure, as this is a routine problem in almost all diffraction work. The procedures given below for indexing patterns are applicable whether the structure is known or not, but they are of course very much easier to apply if the structure is known beforehand.

10-2 PRELIMINARY TREATMENT OF DATA

The powder pattern of the unknown is obtained with a Debye–Scherrer camera or a diffractometer, the object being to cover as wide an angular range of 2θ as possible. A camera such as the Seemann–Bohlin, which records diffraction lines over only a limited angular range, is of very little use in structure analysis. The specimen preparation must ensure random orientation of the individual particles of powder, if the observed relative intensities of the diffraction lines are to have any meaning in terms of crystal structure. After the pattern is obtained, the value of $\sin^2\theta$ is calculated for each diffraction line; this set of $\sin^2\theta$ values is the raw material for the determination of cell size and shape. Or one can calculate the d value of each line and work from this set of numbers.

Since the problem of structure determination is one of finding a structure which will account for all the lines on the pattern, in both position and intensity, the investigator must make sure at the outset that the observed pattern does not contain any extraneous lines. The ideal pattern contains lines formed by x-rays of a single wavelength, diffracted only by the substance whose structure is to be determined. There are therefore two sources of extraneous lines:

1. *Diffraction of x-rays having wavelengths different from that of the principal component of the radiation.* If filtered radiation is used, then $K\alpha$ radiation is the principal component, and characteristic x-rays of any other wavelength may produce extraneous lines. The chief offender is $K\beta$ radiation, which is never entirely removed by a filter and may be a source of extraneous lines when diffracted by lattice planes of high reflecting power. The presence of $K\beta$ lines on a pattern can usually be revealed by calculation, since if a certain set of planes reflect $K\beta$ radiation at an angle θ_β, they must also reflect K_α radiation at an angle θ_α (unless θ_α exceeds 90°), and one angle may be calculated from the other. It follows from the Bragg law that

$$\left(\frac{\lambda_{K\alpha}^2}{\lambda_{K\beta}^2}\right)\sin^2\theta_\beta = \sin^2\theta_\alpha, \tag{10-1}$$

where $\lambda_{K\alpha}^2/\lambda_{K\beta}^2$ has a value near 1.2 for most radiations. If it is suspected that a particular line is due to $K\beta$ radiation, multiplication of its $\sin^2\theta$ value by $\lambda_{K\alpha}^2/\lambda_{K\beta}^2$ will give a value equal, or nearly equal, to the value of $\sin^2\theta$ for some $K\alpha$ line on the pattern, unless the product exceeds unity. The $K\beta$ line corresponding to a given $K\alpha$ line is always located at a smaller angle 2θ and has lower intensity. However, since $K\alpha$ and $K\beta$ lines (from different planes) may overlap on the pattern, Eq. (10-1) alone can establish only the possibility that a given line is due to $K\beta$ radiation, but it can never prove that it is. Another possible source of extraneous lines is L characteristic radiation from tungsten contamination on the target of the x-ray tube, particularly if the tube is old. If such contamination is suspected, equations such as (10-1) can be set up to test the possibility that certain lines are due to tungsten radiation.

2. *Diffraction by substances other than the unknown.* Such substances are usually impurities in the specimen but may also include the specimen mount or badly

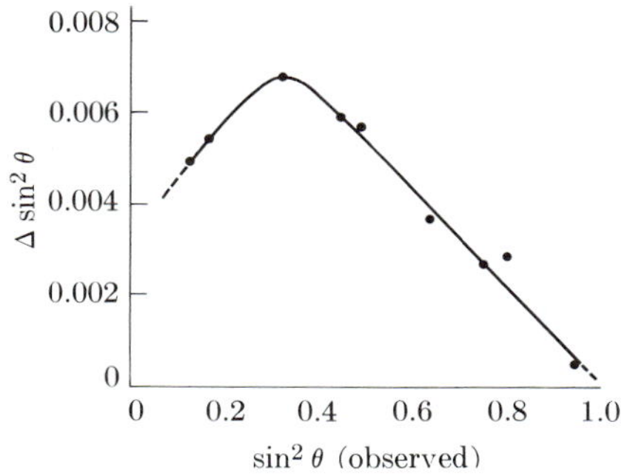

Fig. 10-1 An example of a correction curve for $\sin^2\theta$ values.

aligned slits. Careful specimen preparation and good experimental technique will eliminate extraneous lines due to these causes.

For reasons to be discussed in Chap. 11, the observed values of $\sin^2\theta$ always contain small systematic errors. These errors are not large enough to cause any difficulty in indexing patterns of cubic crystals, but they can seriously interfere with the determination of some noncubic structures. The best method of removing such errors from the data is to calibrate the camera or diffractometer with a substance of known lattice parameter, mixed with the unknown. The difference between the observed and calculated values of $\sin^2\theta$ for the standard substance gives the error in $\sin^2\theta$, and this error can be plotted as a function of the observed values of $\sin^2\theta$. Figure 10-1 shows a correction curve of this kind, obtained with a particular specimen and a particular Debye-Scherrer camera.* The errors represented by the ordinates of such a curve can then be applied to each of the observed values of $\sin^2\theta$ for the diffraction lines of the unknown substance. For the particular determination represented by Fig. 10-1, the errors shown are to be subtracted from the observed values.

10-3 INDEXING PATTERNS OF CUBIC CRYSTALS

A cubic crystal gives diffraction lines whose $\sin^2\theta$ values satisfy the following equation, obtained by combining the Bragg law with the plane-spacing equation for the cubic system, as in Eq. (3-10):

$$\frac{\sin^2\theta}{(h^2+k^2+l^2)} = \frac{\sin^2\theta}{s} = \frac{\lambda^2}{4a^2}. \tag{10-2}$$

Since the sum $s = (h^2 + k^2 + l^2)$ is always integral and $\lambda^2/4a^2$ is a constant for any one pattern, the problem of indexing the pattern of a cubic substance is one of finding a set of integers s which will yield a constant quotient when divided one by one into the observed $\sin^2\theta$ values. (Certain integers, such as 7, 15, 23, 28, 31,

* For the shape of this curve, see Prob. 11-5.

etc., are impossible because they cannot be formed by the sum of three squared integers.) Once the proper integers s are found, the indices hkl of each line can be written down by inspection or from the tabulation in Appendix 10.

The proper set of integers s is not hard to find because there are only a few possible sets. Each of the four common cubic lattice types has a characteristic sequence of diffraction lines, described by their sequential s values:

Simple cubic: 1, 2, 3, 4, 5, 6, 8, 9, 10, 11, 12, 13, 14, 16, . . .
Body-centered cubic: 2, 4, 6, 8, 10, 12, 14, 16, . . .
Face-centered cubic: 3, 4, 8, 11, 12, 16, . . .
Diamond cubic: 3, 8, 11, 16, . . .

Each set can be tried in turn. Longer lists can be prepared from Appendix 10. If a set of integers satisfying Eq. (10–2) cannot be found, then the substance involved does not belong to the cubic system, and other possibilities (tetragonal, hexagonal, etc.) must be explored.

The following example will illustrate the steps involved in indexing the pattern of a cubic substance and finding its lattice parameter. In this particular example, Cu $K\alpha$ radiation was used and eight diffraction lines were observed. Their $\sin^2\theta$ values are listed in the second column of Table 10–1. After a few trials, the integers s listed in the third column were found to produce the reasonably constant quotients listed in the fourth column, when divided into the observed $\sin^2\theta$ values. The fifth column lists the lattice parameter calculated from each line position, and the sixth column gives the Miller indices of each line. The systematic error in $\sin^2\theta$ shows up as a gradual decrease in the value of $\lambda^2/4a^2$, and a gradual increase in the value of a, as θ increases. We shall find in Chap. 11 that the systematic error in a decreases as θ increases; therefore we can select the value of a for the highest-angle line, namely, 3.62 Å, as being the most accurate of those listed. Our analysis of line positions therefore leads to the conclusion that the substance involved, copper in this case, is cubic in structure with a lattice parameter of 3.62 Å.

We can also determine the Bravais lattice of the specimen by observing which lines are present and which absent. Examination of the sixth column of Table 10–1

Table 10–1

1	2	3	4	5	6
Line	$\sin^2\theta$	$s = (h^2 + k^2 + l^2)$	$\dfrac{\lambda^2}{4a^2}$	a(Å)	hkl
1	0.140	3	0.0467	3.57	111
2	0.185	4	0.0463	3.59	200
3	0.369	8	0.0461	3.59	220
4	0.503	11	0.0457	3.61	311
5	0.548	12	0.0457	3.61	222
6	0.726	16	0.0454	3.62	400
7	0.861	19	0.0453	3.62	331
8	0.905	20	0.0453	3.62	420

shows that all lines which have mixed odd and even indices, such as 100, 110, etc., are absent from the pattern. Reference to the rules relating Bravais lattices to observed and absent reflections, given in Table 4-1, shows that the Bravais lattice of this specimen is face-centered. We now have certain information about the arrangement of atoms within the unit cell, and it should be noted that we have had to make use of observed line intensities in order to obtain this information. In this particular case, the observation consisted simply in noting which lines had zero intensity.

The characteristic line sequences for cubic lattices are shown graphically in Fig. 10-2, in the form of calculated diffraction patterns. The calculations are made for Cu $K\alpha$ radiation and a lattice parameter a of 3.50 Å. The positions of all the diffraction lines which would be formed under these conditions are indicated as they would appear on a film or chart of the length shown. (For comparative purposes, the pattern of a hexagonal close-packed structure is also illustrated, since this structure is frequently encountered among metals and alloys. The line positions are calculated for Cu $K\alpha$ radiation, $a = 2.50$ Å, and $c/a = 1.633$, which corresponds to the close packing of spheres.)

Powder patterns of cubic substances can usually be distinguished at a glance from those of noncubic substances, since the latter patterns normally contain many more lines. In addition, the Bravais lattice can usually be identified by inspection: there is an almost regular sequence of lines in simple cubic and body-centered cubic patterns, but the former contains almost twice as many lines, while a face-centered cubic pattern is characterized by a pair of lines, followed by a single line, followed by a pair, another single line, etc.

The problem of indexing a cubic pattern is of course very much simplified if the substance involved is *known* to be cubic and if the lattice parameter is also known. The simplest procedure then is to calculate the value of $\lambda^2/4a^2$ and divide this value into the observed $\sin^2 \theta$ values to obtain the value of s for each line.

There is one difficulty that may arise in the interpretation of cubic powder patterns, and that is due to a possible ambiguity between simple cubic and body-centered cubic patterns. There is a regular sequence of lines in both patterns up to the sixth line; the sequence then continues regularly in body-centered cubic patterns, but is interrupted in simple cubic patterns since $s = 7$ is impossible. Therefore, if λ is so large, or a so small, that six lines or less appear on the pattern, the two Bravais lattices are indistinguishable. For example, suppose that the substance involved is actually body-centered cubic but the investigator mistakenly indexes it as simple cubic, assigning the value $s = 1$ to the first line, $s = 2$ to the second line, etc. He thus obtains a value of $\lambda^2/4a^2$ twice as large as the true one, and a value of a which is $1/\sqrt{2}$ times the true one. This mistake becomes apparent when the number of atoms per unit cell is calculated from the measured density of the specimen (Sec. 10-7); the wrong cell size will give a nonintegral value for the number of atoms per cell, and such a value is impossible. The ambiguity in the diffraction pattern itself can be avoided by choosing a wavelength short enough to produce at least seven lines on the pattern.

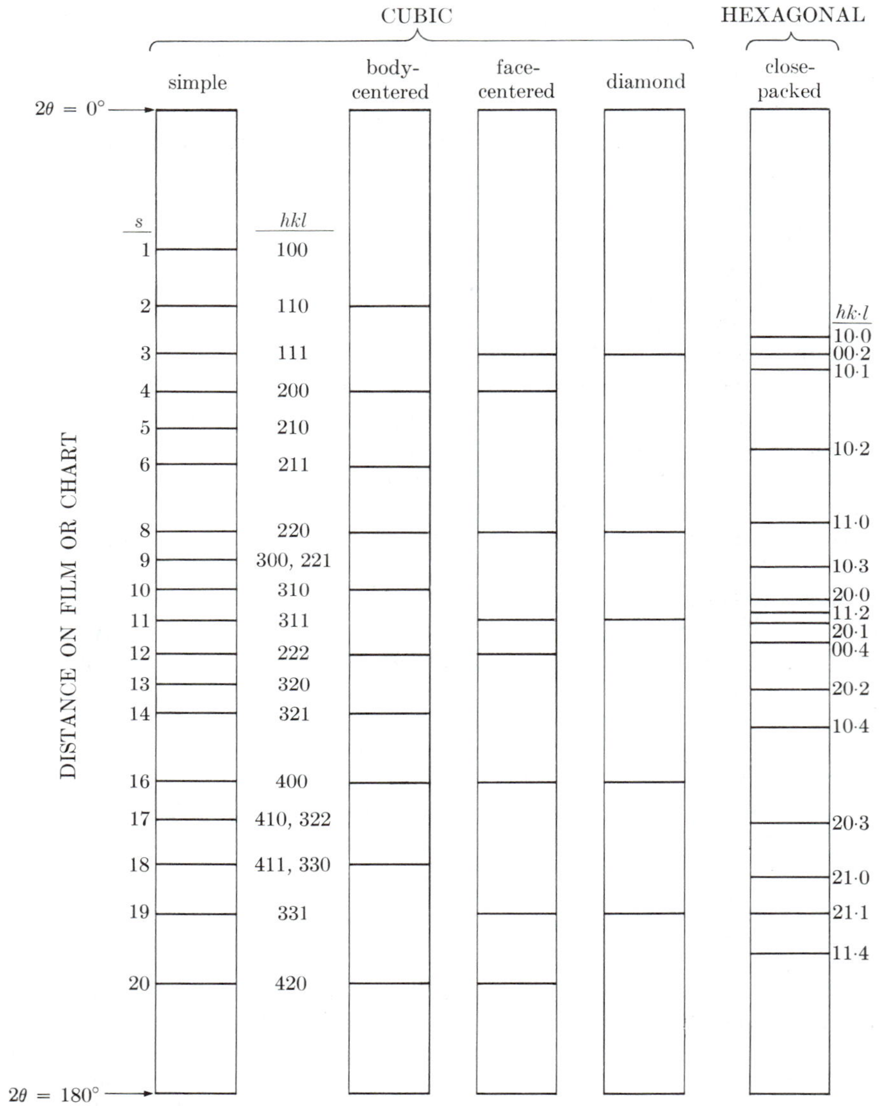

Fig. 10–2 Calculated diffraction patterns for various lattices. $s = h^2 + k^2 + l^2$.

10–4 INDEXING PATTERNS OF NONCUBIC CRYSTALS (GRAPHICAL METHODS)

The problem of indexing powder patterns becomes more difficult as the number of unknown parameters increases. There is only one unknown parameter for cubic crystals, the cell edge a, but noncubic crystals have two or more, and special graphical and analytical techniques have had to be devised in order to index the patterns of such crystals.

Tetragonal System

The plane-spacing equation for this system involves two unknown parameters, a and c:

$$\frac{1}{d^2} = \frac{h^2 + k^2}{a^2} + \frac{l^2}{c^2}. \tag{10-3}$$

This may be rewritten in the form

$$\frac{1}{d^2} = \frac{1}{a^2}\left[(h^2 + k^2) + \frac{l^2}{(c/a)^2}\right],$$

or

$$2 \log d = 2 \log a - \log\left[(h^2 + k^2) + \frac{l^2}{(c/a)^2}\right]. \tag{10-4}$$

Suppose we now write Eq. (10-4) for any two planes of a tetragonal crystal, distinguishing the two planes by subscripts 1 and 2, and then subtract the two equations. We obtain

$$2 \log d_1 - 2 \log d_2 = -\log\left[(h_1^2 + k_1^2) + \frac{l_1^2}{(c/a)^2}\right] + \log\left[(h_2^2 + k_2^2) + \frac{l_2^2}{(c/a)^2}\right]. \tag{10-5}$$

This equation shows that the difference between the 2 log d values for any two planes is independent of a and depends only on the axial ratio c/a and the indices hkl of each plane. This fact was used by Hull and Davey as the basis for a graphical method of indexing the powder patterns of tetragonal crystals [10.1].

The construction of a Hull–Davey chart is illustrated in Fig. 10-3. First, the variation of the quantity $[(h^2 + k^2) + l^2/(c/a)^2]$ with c/a is plotted on two-range semilog paper for particular values of hkl. Each set of indices hkl, as long as they correspond to planes of different spacing, produces a different curve, and when $l = 0$ the curve is a straight line parallel to the c/a axis. Planes of different indices but the same spacing, such as (100) and (010), are represented by the same curve on the chart, which is then marked with the indices of either one of them, in this case (100). [The chart shown is for a simple tetragonal lattice; one for a body-centered tetragonal lattice is made simply by omitting all curves for which $(h + k + l)$ is an odd number.] A single-range logarithmic d scale is then constructed; it extends over two ranges of the $[(h^2 + k^2) + l^2/(c/a)^2]$ scale and runs in the opposite direction, since the coefficient of log d in Eq. (10-4) is -2 times the coefficient of log $[(h^2 + k^2) + l^2/(c/a)^2]$. This means that the d values of two planes, for a given c/a ratio, are separated by the same distance on the scale as the horizontal separation, at the same c/a ratio, of the two corresponding curves on the chart.

The chart and scale are used for indexing in the following manner. The spacing d of the reflecting planes corresponding to each line on the diffraction pattern is calculated. Suppose that the first seven of these values for a particular pattern are 6.00, 4.00, 3.33, 3.00, 2.83, 2.55, and 2.40 Å. A strip of paper is first laid alongside the d scale in position I of Fig. 10–3, and the observed d values are marked off on its edge with a pencil. The paper strip is then placed on the chart and moved about, both vertically and horizontally, until a position is found where each mark on the strip coincides with a line on the chart. Vertical and horizontal movements correspond to trying various c/a and a values, respectively, and the only restriction on these movements is that the edge of the strip must always be horizontal. When a correct fit has been obtained, as shown by position II of Fig. 10–3, the indices of each line are simply read from the corresponding curves, and the approximate value of c/a from the vertical position of the paper strip. In the present example, the c/a ratio is 1.5 and the first line on the pattern (formed by planes of spacing 6.00 Å) is a 001 line, the second a 100 line, the third a 101 line, etc. After all the lines have been indexed in this way, the d values of the two highest-angle lines are used to set up two equations of the form of Eq. (10–3), and these are solved simultaneously to yield the values of a and c. From these values, the axial ratio c/a may then be calculated with more precision than it can be found graphically.

Figure 10–3 is only a partial Hull–Davey chart. A complete one, showing curves of higher indices, is reproduced on a small scale in Fig. 10–4, which applies to body-centered tetragonal lattices. Note that the curves of high indices are often so crowded that it is difficult to assign the proper indices to the observed lines. It may then be necessary to calculate the indices of these high-angle lines on the basis of a and c values derived from the already indexed low-angle lines.

Some Hull–Davey charts, like the one shown in Fig. 10–4, are designed for use with $\sin^2\theta$ values rather than d values. No change in the chart itself is involved, only a change in the accompanying scale. This is possible because an equation similar to Eq. (10–4) can be set up in terms of $\sin^2\theta$ rather than d, by combining Eq. (10–3) with the Bragg law. This equation is

$$\log \sin^2\theta = \log \frac{\lambda^2}{4a^2} + \log\left[(h^2 + k^2) + \frac{l^2}{(c/a)^2}\right]. \qquad (10\text{–}6)$$

The $\sin^2\theta$ scale is therefore a two-range logarithmic one (from 0.01 to 1.0), equal in length to the two-range $[(h^2 + k^2) + l^2/(c/a)^2]$ scale on the chart and running in the same direction. A scale of this kind appears at the top of Fig. 10–3.

When the c/a ratio becomes equal to unity, a tetragonal cell becomes cubic. It follows that a cubic pattern can be indexed on a tetragonal Hull–Davey chart by keeping the paper strip always on the horizontal line corresponding to $c/a = 1$. (In fact, one can make one-dimensional Hull–Davey charts for the rapid indexing of cubic patterns.) It is instructive to consider a tetragonal cell as a departure from a cubic one and to examine a Hull–Davey chart in that light, since the chart shows at a glance how the powder pattern changes for any given change in the c/a ratio. It shows, for example, how certain lines split into two as soon as the c/a ratio departs from unity, and how even the order of the lines on the pattern can change with changes in c/a.

Another graphical method of indexing tetragonal patterns has been devised by Bunn [10.2]. Like the Hull–Davey chart, a Bunn chart consists of a network of curves, one for each value of hkl, but the curves are based on somewhat different functions of hkl and c/a than those used by Hull and Davey, with the result that

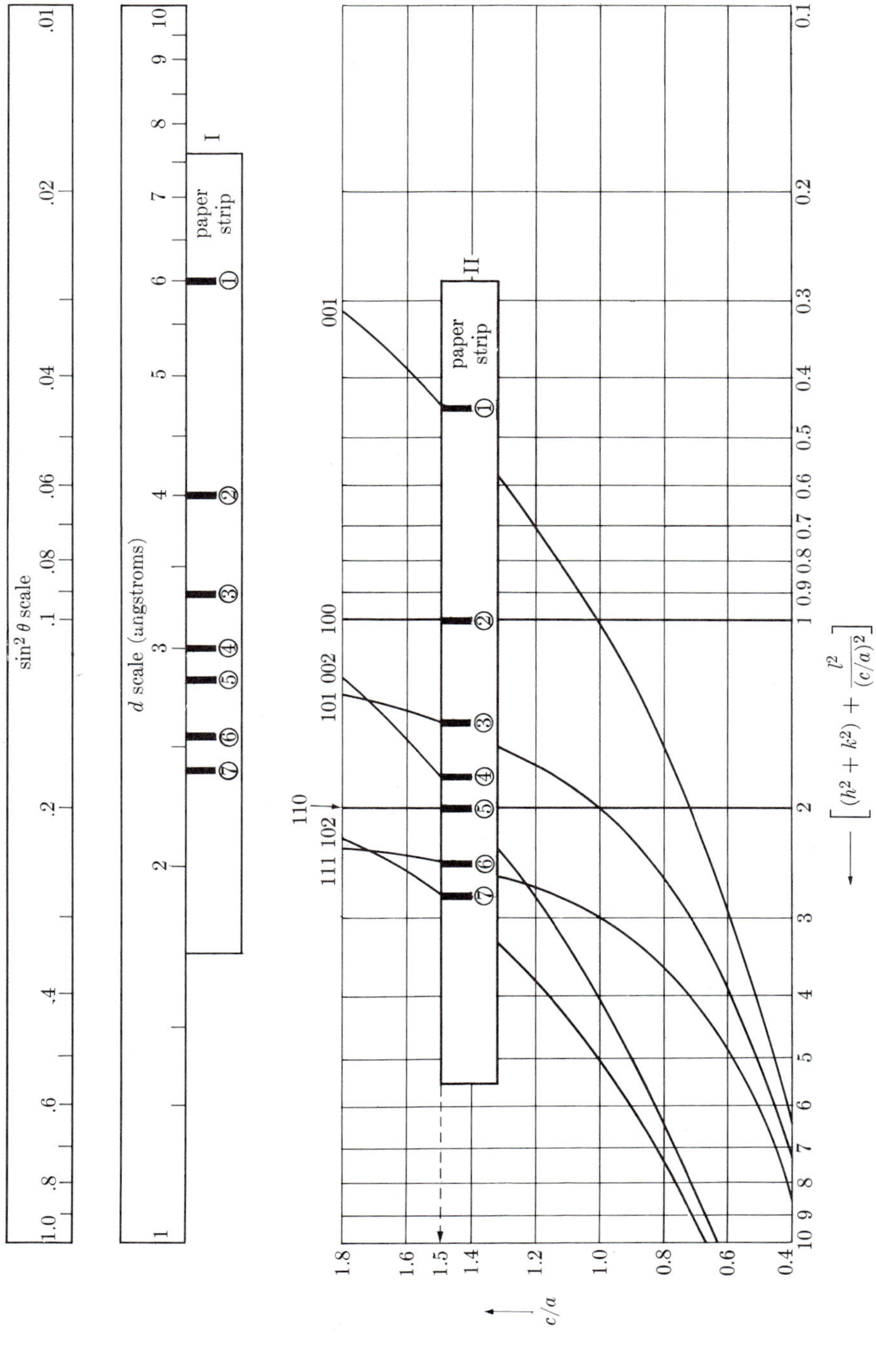

Fig. 10–3 Partial Hull–Davey chart for simple tetragonal lattices.

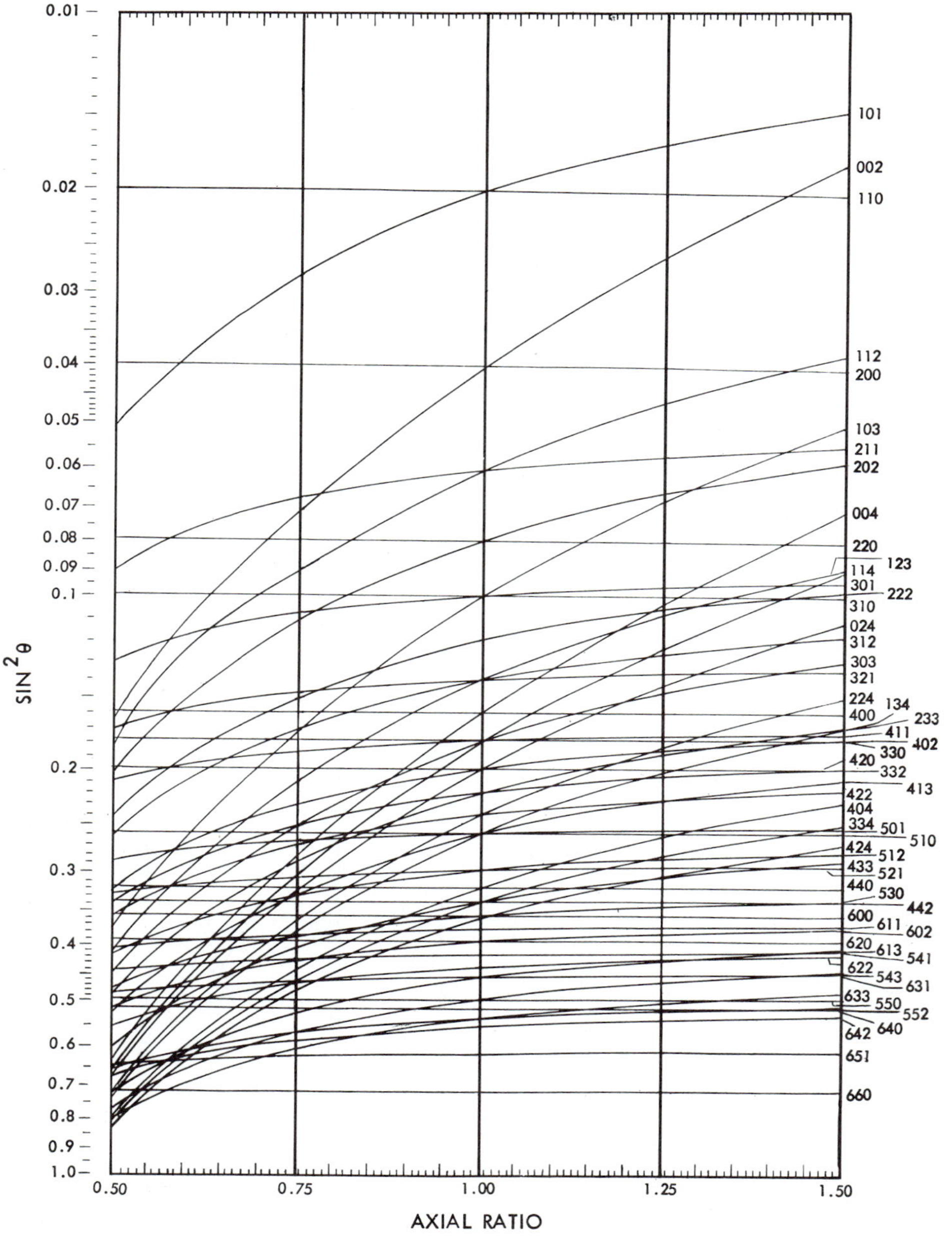

Fig. 10–4 Complete Hull–Davey chart for body-centered tetragonal lattices.

the curves are less crowded in certain regions of the chart. The Bunn chart is accompanied by a logarithmic scale of d values, and the combination of chart and scale is used in exactly the same way as a Hull–Davey chart and scale. Bunn charts may be purchased [10.3] or constructed from numerical data in [G.11, vol. 2, p. 207].

Hexagonal System

Patterns of hexagonal crystals can also be indexed by graphical methods, since the hexagonal unit cell, like the tetragonal, is characterized by two variable parameters, a and c. The plane-spacing equation is

$$\frac{1}{d^2} = \frac{4}{3} \cdot \frac{h^2 + hk + k^2}{a^2} + \frac{l^2}{c^2}.$$

After some manipulation, this becomes

$$2 \log d = 2 \log a - \log \left[\frac{4}{3} (h^2 + hk + k^2) + \frac{l^2}{(c/a)^2} \right],$$

which is of exactly the same form as Eq. (10–4) for the tetragonal system. A Hull–Davey chart for the hexagonal system can therefore be constructed by plotting the variation of $\log \left[\frac{4}{3}(h^2 + hk + k^2) + l^2/(c/a)^2\right]$ with c/a. A Bunn chart may also be constructed for this system. Special charts for hexagonal close-packed lattices may also be prepared by omitting all curves for which $(h + 2k)$ is an integral multiple of 3 and l is odd.

The powder pattern of zinc made with Cu $K\alpha$ radiation (Fig. 3–13) will serve to illustrate how the pattern of a hexagonal substance is indexed. Thirteen lines were observed on this pattern; their $\sin^2 \theta$ values and relative intensities are listed in Table 10–2. A fit was obtained on a Hull–Davey chart for hexagonal close-packed lattices at an approximate c/a ratio of 1.87. The chart lines disclosed the indices listed in the fourth column of the table. In the case of line 5, two chart lines (10 · 3 and 11 · 0) almost intersect at $c/a = 1.87$, so the observed line is evidently the sum of two lines, almost overlapping, one from the (10 · 3) planes and the other from (11 · 0) planes. The same is true of line 11. Four lines on the chart, namely, 20 · 0, 10 · 4, 21 · 0, and 20 · 4, do not appear on the pattern, and it must be inferred that these are too weak to be observed. On the other hand, all the observed lines are accounted for, so we may conclude that the lattice of zinc is actually hexagonal close-packed. The next step is to calculate the lattice parameters. Combination of the Bragg law and the plane-spacing equation gives

$$\sin^2 \theta = \frac{\lambda^2}{4} \left[\frac{4}{3} \cdot \frac{(h^2 + hk + k^2)}{a^2} + \frac{l^2}{c^2} \right],$$

where $\lambda^2/4$ has a value of 0.594 Å^2 for Cu $K\alpha$ radiation. Writing this equation out for the two highest-angle lines, namely, 12 and 13, we obtain:

$$0.806 = 0.594 \left(\frac{4}{3} \cdot \frac{7}{a^2} + \frac{1}{c^2} \right),$$

$$0.879 = 0.594 \left(\frac{4}{3} \cdot \frac{7}{a^2} + \frac{4}{c^2} \right).$$

Table 10–2

Line	Intensity	$\sin^2\theta$	$hk{\cdot}l$
1	s	0.097	00·2
2	s	0.112	10·0
3	vs	0.136	10·1
4	m	0.209	10·2
5	s	0.332	10·3, 11·0
6	vw	0.390	00·4
7	m	0.434	11·2
8	m	0.472	20·1
9	vw	0.547	20·2
10	w	0.668	20·3
11	m	0.722	11·4, 10·5
12	m	0.806	21·1
13	w	0.879	21·2

Simultaneous solution of these two equations gives $a = 2.66$ Å, $c = 4.94$ Å, and $c/a = 1.86$.

Rhombohedral System

Rhombohedral crystals are also characterized by unit cells having two parameters, in this case a and α. No new chart is needed, however, to index the patterns of rhombohedral substances, because, as mentioned in Sec. 2–4, any rhombohedral crystal may be referred to hexagonal axes. A hexagonal Hull–Davey or Bunn chart may therefore be used to index the pattern of a rhombohedral crystal. The indices so found will, of course, refer to a hexagonal cell, and the method of converting them to rhombohedral indices is described in Appendix 4.

We can conclude that the pattern of any two-parameter crystal (tetragonal, hexagonal, or rhombohedral) can be indexed on the appropriate Hull–Davey or Bunn chart. If the structure is known, the procedure is quite straightforward. The best method is to calculate the c/a ratio from the known parameters, lay a straightedge on the chart to discover the proper line sequence for this value of c/a, calculate the value of $\sin^2\theta$ for each line from the indices found on the chart, and then determine the indices of the observed lines by a comparison of calculated and observed $\sin^2\theta$ values.

If the structure is unknown, the problem of indexing is not always so easy as it seems in theory. The most common source of trouble is the presence of extraneous lines, as defined in Sec. 10–2, in the observed pattern. Such lines can be very confusing and, if any difficulty in indexing is encountered, every effort should be made to eliminate them from the pattern, either experimentally or by calculation. In addition, the observed $\sin^2\theta$ values usually contain systematic errors which make a simultaneous fit of all the pencil marks on the paper strip to curves on the chart impossible, even when the paper strip is at the correct c/a position. Because of these errors, the strip has to be shifted slightly from line to line in order to make successive pencil marks coincide with curves on the chart. Two important rules must always be kept in mind when using Hull–Davey or Bunn charts:

1. Every mark on the paper strip must coincide with a curve on the chart, except for extraneous lines. A structure which accounts for only a portion of the observed lines is not correct: *all* the lines in the pattern must be accounted for, either as due to the structure of the substance involved or as extraneous lines.

2. There need not be a mark on the paper strip for every curve on the chart, because some lines may have zero intensity or be too weak to be observed.

Orthorhombic, Monoclinic, and Triclinic Systems

Substances with these low-symmetry structures yield powder patterns which are almost impossible to index by graphical methods, although the patterns of some orthorhombic crystals have been indexed by a combination of graphical and analytical methods. The essential difficulty is the large number of variable parameters involved. In the orthorhombic system there are three such parameters (a, b, c), in the monoclinic four (a, b, c, β), and in the triclinic six (a, b, c, α, β, γ). If the structure is known, patterns of substances in these crystal systems can be indexed by comparison of the observed $\sin^2\theta$ values with those calculated for all possible values of *hkl*.

10-5 INDEXING PATTERNS OF NONCUBIC CRYSTALS (ANALYTICAL METHODS)

Analytical methods of indexing involve arithmetical manipulation of the observed $\sin^2\theta$ values in an attempt to find certain relationships among them. Since each crystal system is characterized by particular relationships between $\sin^2\theta$ values, recognition of these relationships identifies the crystal system and leads to a solution of the line indices. These analytical methods are due mainly to Hesse and Lipson [10.4, 10.5, G.8, G.32, G.17].

Tetragonal System

Here the $\sin^2\theta$ values must obey the relation:

$$\sin^2\theta = A(h^2 + k^2) + Cl^2, \tag{10-7}$$

where A $(= \lambda^2/4a^2)$ and C $(= \lambda^2/4c^2)$ are constants for any one pattern. The problem is to find these constants, since, once found, they will disclose the cell parameters a and c and enable the line indices to be calculated. The value of A is obtained from the *hk*0 lines. When $l = 0$, Eq. (10-7) becomes

$$\sin^2\theta = A(h^2 + k^2).$$

The permissible values of $(h^2 + k^2)$ are 1, 2, 4, 5, 8, etc. Therefore the *hk*0 lines must have $\sin^2\theta$ values in the ratio of these integers, and A will be some number which is 1, $\frac{1}{2}$, $\frac{1}{4}$, $\frac{1}{5}$, $\frac{1}{8}$, etc., times the $\sin^2\theta$ values of these lines. C is obtained from the other lines on the pattern and the use of Eq. (10-7) in the form

$$\sin^2\theta - A(h^2 + k^2) = Cl^2.$$

Differences represented by the left-hand side of the equation are set up, for various assumed values of h and k, in an attempt to find a consistent set of Cl^2 values, which must be in the ratio 1, 4, 9, 16, etc. Once these values are found, C can be calculated.

Hexagonal System

For hexagonal crystals, an exactly similar procedure is used. In this case, $\sin^2 \theta$ values are given by

$$\sin^2 \theta = A(h^2 + hk + k^2) + Cl^2,$$

where $A = \lambda^2/3a^2$ and $C = \lambda^2/4c^2$. Permissible values of $(h^2 + hk + k^2)$ are tabulated in Appendix 10; they are 1, 3, 4, 7, 9, etc. The indexing procedure is best illustrated by means of a specific example, namely, the powder pattern of zinc, whose observed $\sin^2 \theta$ values are listed in Table 10–2. We first divide the $\sin^2 \theta$ values by the integers 1, 3, 4, etc., and tabulate the results, as shown by Table 10–3, which applies to the first six lines of the pattern. We then examine these numbers, looking for quotients which are equal to one another or equal to one of the observed $\sin^2 \theta$ values. In this case, the two starred entries, 0.112 and 0.111, are the most nearly equal, so we assume that lines 2 and 5 are $hk0$ lines. We then tentatively put $A = 0.112$ which is equivalent to saying that line 2 is 100. Since the $\sin^2 \theta$ value of line 5 is very nearly 3 times that of line 2, line 5 should be 110. To find the value of C, we must use the equation

$$\sin^2 \theta - A(h^2 + hk + k^2) = Cl^2.$$

We now subtract from each $\sin^2 \theta$ value the values of A (= 0.112), $3A$ (= 0.336), $4A$ (= 0.448), etc., and look for remainders (Cl^2) which are in the ratio of 1, 4, 9, 16, etc. These figures are given in Table 10–4. Here the five starred entries are of interest, because these numbers (0.024, 0.097, 0.221, and 0.390) are very nearly in the ratio 1, 4, 9, and 16. We therefore put $0.024 = C(1)^2$, $0.097 = C(2)^2$, $0.221 = C(3)^2$, and $0.390 = C(4)^2$. This gives $C = 0.024$ and immediately identifies line 1 as 002 and line 6 as 004. Since line 3 has a $\sin^2 \theta$ value equal to the sum of A and C, its indices must be 101. Similarly, the indices of lines 4 and 5 are found to be 102 and 103, respectively. In this way, indices are assigned to all the lines on the pattern, and a final check on their correctness is made in the usual manner, by a comparison of observed and calculated $\sin^2 \theta$ values.

Table 10–3

Line	$\sin^2\theta$	$\frac{\sin^2\theta}{3}$	$\frac{\sin^2\theta}{4}$	$\frac{\sin^2\theta}{7}$	hkl
1	0.097	0.032	0.024	0.014	
2	0.112*	0.037	0.028	0.016	100
3	0.136	0.045	0.034	0.019	
4	0.209	0.070	0.052	0.030	
5	0.332	0.111*	0.083	0.047	110
6	0.390	0.130	0.098	0.056	

Table 10-4

Line	$\sin^2\theta$	$\sin^2\theta - A$	$\sin^2\theta - 3A$	hkl
1	0.097*			002
2	0.112	0.000		100
3	0.136	0.024*		101
4	0.209	0.097*		102
5	0.332	0.221*		110, 103
6	0.390*	0.278	0.054	004

Orthorhombic System

The basic equation governing the $\sin^2 \theta$ values is now

$$\sin^2 \theta = Ah^2 + Bk^2 + Cl^2.$$

The indexing problem is considerably more difficult here, in that three unknown constants, A, B, and C, have to be determined. The general procedure, which is too lengthy to illustrate here, is to search for significant differences between various pairs of $\sin^2 \theta$ values. For example, consider any two lines having indices $hk0$ and $hk1$, with hk the same for each, such as 120 and 121; the difference between their $\sin^2 \theta$ values is C. Similarly, the difference between the $\sin^2 \theta$ values of two lines such as 310 and 312 is $4C$, and so on. If the structure is such that there are many lines missing from the pattern, because of a zero structure factor for the corresponding planes, then the difficulties of indexing are considerably increased, inasmuch as the missing lines may be the very ones which would supply the most easily recognized clues if they were present. Despite such difficulties, this analytical method has been applied successfully to a number of orthorhombic patterns. One requisite for its success is fairly high accuracy in the $\sin^2 \theta$ values (at least ± 0.0005), and the investigator should therefore correct his observations for systematic errors before attempting to index the pattern.

Monoclinic and Triclinic Systems

These crystal systems involve four and six independent constants, respectively. The corresponding powder patterns are of great complexity and may contain more than a hundred lines. Such patterns are seldom solved without the aid of the computer.

General

Analytical methods of indexing are search procedures designed to reveal certain numerical relationships among the observed $\sin^2 \theta$ values. The digital computer is therefore a natural tool to use, and many computer programs have been written for the indexing of powder patterns. Klug and Alexander [G.39] have surveyed the general problems involved and give references to specific programs.

Computer indexing is not always successful. The computer may yield not one but many sets of indices that approximately conform to the input data; it is then up to the investigator's experience and judgment to select the correct set. Extraneous diffraction lines and inaccurate $\sin^2 \theta$ values can mislead a computer as well as a human searcher.

The powder patterns of low-symmetry substances are so difficult to solve that the crystal structures of such substances are almost always determined by examining a single crystal, by either the rotating-crystal method or one of its variations. With these methods the x-ray crystallographer can, without much difficulty, determine the shape and size of an unknown unit cell, no matter how low its symmetry. Many substances are very difficult to prepare in single-crystal form, but, on the other hand, if the substance involved is one of low symmetry, the time spent in trying to obtain a single crystal is usually more fruitful than the time spent in trying to solve the powder pattern. The single-crystal specimen need not be large: a crystal as small as 0.1 mm in any dimension can be successfully handled and will give a satisfactory diffraction pattern.

10-6 THE EFFECT OF CELL DISTORTION ON THE POWDER PATTERN

At this point we might digress slightly from the main subject of this chapter, and examine some of the changes produced in a powder pattern when the unit cell of the substance involved is distorted in various ways. As we have already seen, there are many more lines on the pattern of a substance of low symmetry, such as triclinic, than on the pattern of a substance of high symmetry, such as cubic, and we may take it as a general rule that any distortion of the unit cell which decreases its symmetry, in the sense of introducing additional variable parameters, will increase the number of lines on the powder pattern.

Figure 10-5 graphically illustrates this point. On the left is the calculated diffraction pattern of the body-centered cubic substance whose unit cell is shown at the top. The line positions are computed for $a = 4.00$ Å and Cr $K\alpha$ radiation. If this cell is expanded or contracted uniformly but still remains cubic, the diffraction lines merely shift their positions but do not increase in number, since no change in cell symmetry is involved. However, if the cubic cell is distorted along only one axis, then it becomes tetragonal, its symmetry decreases, and more diffraction lines are formed. The center pattern shows the effect of stretching the cubic cell by 4 percent along its [001] axis, so that c is now 4.16 Å. Some lines are unchanged in position, some are shifted, and new lines have appeared. If the tetragonal cell is now stretched by 8 percent along its [010] axis, it becomes orthorhombic, with $a = 4.00$ Å, $b = 4.32$ Å, and $c = 4.16$ Å, as shown on the right. The result of this last distortion is to add still more lines to the pattern. The increase in the number of lines is due essentially to the introduction of new plane spacings, caused by nonuniform distortion. Thus, in the cubic cell, the (200), (020), and (002) planes all have the same spacing and only one line is formed, called the 200 line, but this line splits into two when the cell becomes tetragonal, since now the (002) plane spacing differs from the other two. When the cell becomes orthorhombic, all three spacings are different and three lines are formed.

Changes of this nature are not uncommon among phase transformations and ordering reactions. For example, the powder pattern of slowly cooled plain carbon steel shows lines due to ferrite (body-centered cubic) and cementite (Fe_3C,

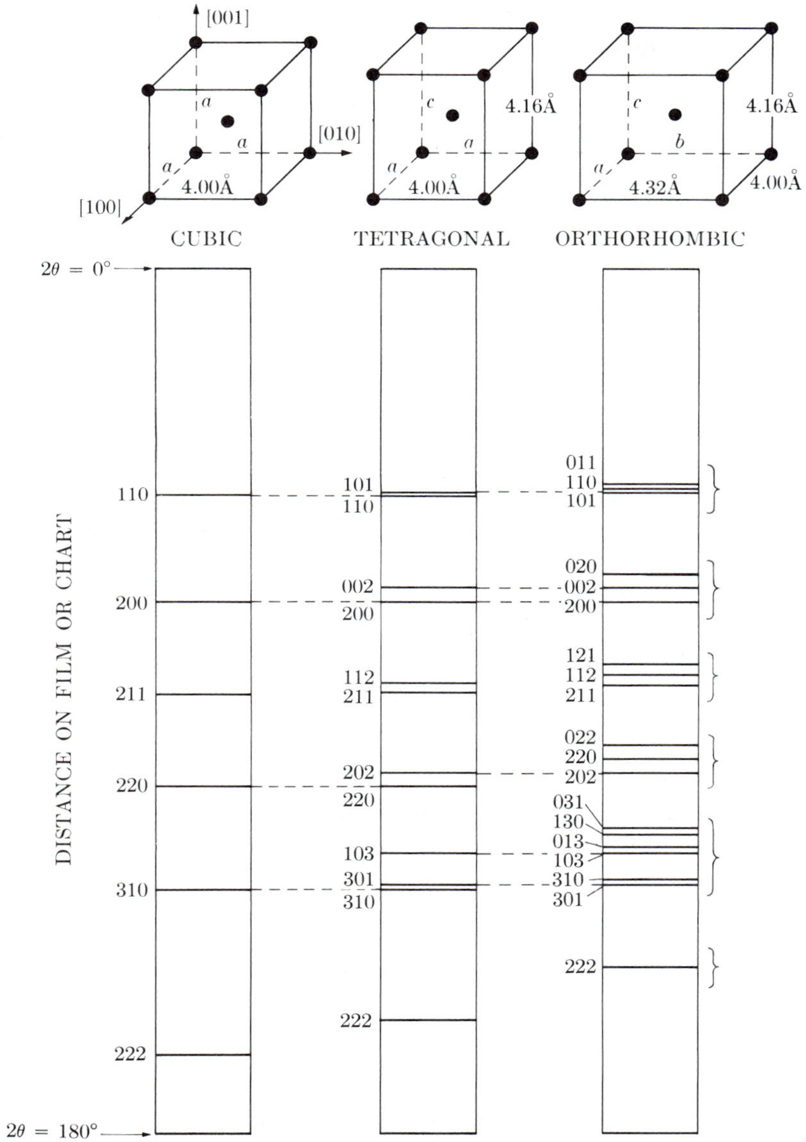

Fig. 10–5 Effects of cell distortion on powder patterns. Lines unchanged in position are connected by dashed lines.

orthorhombic). When the same steel is quenched from the austenite region, the phases present are martensite (body-centered tetragonal) and, possibly, some untransformed austenite (face-centered cubic). The a and c parameters of the martensite cell do not differ greatly from the a parameter of the ferrite cell (see Fig. 12–5). The result is that the diffraction pattern of a quenched steel shows pairs of martensite lines occurring at about the same 2θ positions as the individual lines of ferrite in the previous pattern. (These line pairs, however, are seldom resolved; martensite contains so much microstrain that each line of a pair is so broad that it merges with its neighbor.) If the quenched steel is now tempered, the martensite will ultimately decompose into ferrite and cementite, and each pair of martensite lines will coalesce into a single ferrite line. Somewhat similar effects can be produced in a copper-gold alloy having the composition represented by the formula AuCu. This alloy is cubic in the disordered state but becomes either tetragonal or orthorhombic when ordered, depending on the ordering temperature (see Sec. 13–3).

The changes produced in a powder pattern by cell distortion depend, in degree, on the amount of distortion. If the latter is small, the pattern retains the main features of the pattern of the original undistorted cell. Thus, in Fig. 10–5, the nineteen lines of the orthorhombic pattern fall into the six bracketed groups shown, each group corresponding to one of the single lines on the cubic pattern. In fact, an experienced crystallographer, if confronted with this orthorhombic pattern, might recognize this grouping and guess that the unit cell of the substance involved was not far from cubic in shape, and that the Bravais lattice was either simple or body-centered, since the groups of lines are spaced in a fairly regular manner. But if the distortion of the cubic cell had been much larger, each line of the original pattern would split into such widely separated lines that no features of the original pattern would remain.

10–7 DETERMINATION OF THE NUMBER OF ATOMS IN A UNIT CELL

To return to the subject of structure determination, the next step after establishing the shape and size of the unit cell is to find the number of atoms in that cell, because the number of atoms must be known before their positions can be determined. To find this number we use the fact that the volume of the unit cell, calculated from the lattice parameters by means of the equations given in Appendix 3, multiplied by the measured density of the substance equals the weight of all the atoms in the cell. From Eq. (3–7), we have

$$\sum A = \frac{\rho V'}{1.66042},$$

where $\sum A$ is the sum of the atomic weights of the atoms in the unit cell, ρ is the density (gm/cm^3), and V' is the volume of the unit cell (Å^3). If the substance is an element of atomic weight A, then

$$\sum A = n_1 A,$$

where n_1 is the number of atoms per unit cell. If the substance is a chemical compound, or an intermediate phase whose composition can be represented by a simple chemical formula, then

$$\sum A = n_2 M,$$

where n_2 is the number of "molecules" per unit cell and M the molecular weight. The number of atoms per cell can then be calculated from n_2 and the composition of the phase.

When determined in this way, the number of atoms per cell is always an integer, within experimental error, except for a very few substances which have "defect structures." In these substances, atoms are simply missing from a certain fraction of those lattice sites which they would be expected to occupy, and the result is a nonintegral number of atoms per cell. FeO and the β phase in the Ni-Al system are examples.

10–8 DETERMINATION OF ATOM POSITIONS

We now have to find the positions of a known number of atoms in a unit cell of known shape and size. To solve this problem, we must make use of the observed relative *intensities* of the diffracted beams, since these intensities are determined by atom positions. In finding the atom positions, however, we must again proceed by trial and error, because there is no known general method of directly calculating atom positions from observed intensities.

To see why this is so, we must consider the two basic equations involved, namely,

$$I = |F|^2 p \left(\frac{1 + \cos^2 2\theta}{\sin^2 \theta \cos \theta} \right), \tag{4–19}$$

which gives the relative intensities of the reflected beams, and

$$F = \sum_1^N f_n e^{2\pi i(hu_n + kv_n + lw_n)}, \tag{4–11}$$

which gives the value of the structure factor F for the hkl reflection in terms of the atom positions uvw. Since the relative intensity I, the multiplicity factor p, and the Bragg angle θ are known for each line on the pattern, we can find the value of $|F|$ for each reflection from Eq. (4–19). But $|F|$ measures only the relative amplitude of each reflection, whereas, in order to use Eq. (4–11) for calculating atom positions, we must know the value of F, which measures both the amplitude *and* phase of one reflection relative to another. This is the crux of the problem. The intensities of two reflected beams are proportional to the squares of their amplitudes but independent of their relative phase. Since all we can measure is intensity, we can determine amplitude but not phase, which means that we cannot, in general, compute the structure factor but only its absolute value. This "phase problem," which had baffled crystallographers for years, has now been partially solved, in the sense that direct methods of structure determination, applicable to some structures, now exist. These methods have a success rate of 80 to 90 percent when applied to crystals with a hundred atoms or less per unit cell. Protein crystals, however, contain thousands of atoms per cell. No direct method has yet been found powerful enough to solve the structure of *any* crystal.

Atom positions, therefore, can be determined only by trial and error. A set of atom positions is assumed, the intensities corresponding to these positions are calculated, and the calculated intensities are compared with the observed ones, the process being repeated until satisfactory agreement is reached. The problem of selecting a structure for trial is not as hopelessly broad as it sounds, since the investigator has many aids to guide him. Foremost among these is the accumulated knowledge of previously solved structures. From these known structures he may be able to select a few likely candidates, and then proceed on the assumption that his unknown structure is the same as, or very similar to, one of these known ones. A great many known structures may be classified into groups according to the kind of bonding (ionic, covalent, metallic, or mixtures of these) which holds the atoms together, and a selection among these groups is aided by a knowledge of the probable kind of atomic bonding in the unknown phase, as judged from the positions of its constituent elements in the periodic table. For example, suppose the phase of unknown structure has the chemical formula AB, where A is strongly electropositive and B strongly electronegative, and that its powder pattern is characteristic of a simple cubic lattice. Then the bonding is likely to be ionic, and the CsCl structure is strongly suggested. But the FeSi structure shown in Fig. 2–20 is also a possibility. In this particular case, one or the other can be excluded by a density measurement, since the CsCl cell contains one "molecule" and the FeSi cell four. If this were not possible, diffracted intensities would have to be calculated on the basis of each cell and compared with the observed ones. It is this simple kind of structure determination, illustrated by an example in the next section, that the metallurgist should be able to carry out unaided.

Needless to say, many structures are too complex to be solved by this simple approach and the crystallographer must turn to more powerful methods. Chief among these are space-group theory and Fourier series. Although any complete description of these subjects is beyond the scope of this book, a few general remarks may serve to show their utility in structure determination. The *theory of space groups*, one of the triumphs of mathematical crystallography, relates crystal symmetry, on the atomic scale, to the possible atomic arrangements which possess that symmetry. For example, if a given substance is known to be hexagonal and to have n atoms in its unit cell, then space-group theory lists all possible arrangements of n atoms which will have hexagonal symmetry. This listing of possible arrangements aids tremendously in the selection of trial structures. A further reduction in the number of possibilities can then be made by noting the indices of the reflections absent from the diffraction pattern. By such means alone, i.e., before any detailed consideration is given to relative diffracted intensities, space-group theory can often exclude all but two or three possible atomic arrangements. There are 230 different space groups, and the possible atomic arrangements in each group are listed in [G.11, Vol. 1].

A *Fourier series* is a type of infinite trigonometric series by which any kind of periodic function may be expressed. Now the one essential property of a crystal is that its atoms are arranged in space in a periodic fashion. But this means that the density of electrons is also a periodic function of position in the crystal, rising to a maximum at the point where an atom is located and dropping to a low value in the region between atoms. To regard a crystal in this manner, as a positional variation of electron density rather than as an arrangement of atoms, is particularly appropriate where diffraction is involved, in that x-rays are scattered by electrons and not by atoms as such. Since the electron density is a periodic function of position, a crystal may be described analytically by means of Fourier series. This method of description is very useful in structure determination because it can be shown that the coefficients of the various terms in the series are related to the F values of the various x-ray reflections. But such a series is not of immediate use, since the structure factors are not usually known both in magnitude and phase. However, another kind of series has been devised, called the Patterson function, whose coefficients are related to the experimentally observable $|F|$ values and which gives, not electron density, but information regarding the various interatomic vectors in the unit cell. This information is frequently enough to determine the phase of the various structure factors; then the first kind of series can be used to map out the actual electron density throughout the cell and thus disclose the atom positions.

10-9 EXAMPLE OF STRUCTURE DETERMINATION

As a simple example, we will consider an intermediate phase which occurs in the cadmium-tellurium system. Chemical analysis of the specimen, which appeared essentially one phase under the microscope, showed it to contain 46.6 weight percent Cd and 53.4 weight percent Te. This is equivalent to 49.8 atomic percent Cd and can be represented by the formula CdTe. The specimen was reduced to powder and a diffraction pattern obtained with a Debye-Scherrer camera and Cu $K\alpha$ radiation.

The observed values of $\sin^2\theta$ for the first 16 lines are listed in Table 10-5, together with the visually estimated relative line intensities. This pattern can be indexed on the basis of a cubic unit cell, and the indices of the observed lines are given in the table. The lattice parameter, calculated from the $\sin^2\theta$ value for the highest-angle line, is 6.46 Å.

The density of the specimen, as determined by weighing a quantity of the powder in a pyknometer bottle, was 5.82 gm/cm³. We then find, from Eq. (3-7), that

$$\sum A = \frac{(5.82)(6.46)^3}{1.66042} = 945.$$

Since the molecular weight of CdTe is 240.02, the number of "molecules" per unit cell is 945/240.02 = 3.94, or 4, within experimental error.

At this point, we know that the unit cell of CdTe is cubic and that it contains 4 "molecules" of CdTe, i.e., 4 atoms of cadmium and 4 atoms of tellurium. We must now consider possible arrangements of these atoms in the unit cell. First we examine the indices listed in Table 10-5 for evidence of the Bravais lattice. Since the indices of the observed lines are all unmixed, the Bravais lattice must be face-centered. (Not all possible sets of unmixed indices are present, however: 200, 420,

Table 10–5

Line	Intensity	$\sin^2\theta$	*hkl*
1	s	0.0462	111
2	vs	0.1198	220
3	vs	0.1615	311
4	vw	0.1790	222
5	m	0.234	400
6	m	0.275	331
7	s	0.346	422
8	m	0.391	511, 333
9	w	0.461	440
10	m	0.504	531
11	m	0.575	620
12	w	0.616	533
13	w	0.688	444
14	m	0.729	711, 551
15	vs	0.799	642
16	s	0.840	731, 553

600, 442, 622, and 640 are missing from the pattern. But these reflections may be too weak to be observed, and the fact that they are missing does not invalidate our conclusion that the lattice is face-centered.) Now there are two common face-centered cubic structures of the AB type, i.e., containing two different atoms in equal proportions, and both contain four "molecules" per unit cell: these are the NaCl structure [Fig. 2–18(b)] and the zinc-blende form of ZnS [Fig. 2–19(b)]. Both of these are logical possibilities even though the bonding in NaCl is ionic and in ZnS covalent, since both kinds of bonding have been observed in telluride structures.

The next step is to calculate relative diffracted intensities for each structure and compare them with experiment, in order to determine whether or not one of these structures is the correct one. If CdTe has the NaCl structure, then its structure factor for unmixed indices [see Example (e) of Sec. 4–6] is given by

$$
\begin{aligned}
F^2 &= 16(f_{\mathrm{Cd}} + f_{\mathrm{Te}})^2, && \text{if } (h + k + l) \text{ is even,} \\
F^2 &= 16(f_{\mathrm{Cd}} - f_{\mathrm{Te}})^2, && \text{if } (h + k + l) \text{ is odd.}
\end{aligned}
\qquad (10\text{–}8)
$$

On the other hand, if the ZnS structure is correct, then the structure factor for unmixed indices (see Sec. 4–13) is given by

$$
\begin{aligned}
|F|^2 &= 16(f_{\mathrm{Cd}}^2 + f_{\mathrm{Te}}^2), && \text{if } (h + k + l) \text{ is odd,} \\
|F|^2 &= 16(f_{\mathrm{Cd}} - f_{\mathrm{Te}})^2, && \text{if } (h + k + l) \text{ is an odd multiple of 2,} \\
|F|^2 &= 16(f_{\mathrm{Cd}} + f_{\mathrm{Te}})^2, && \text{if } (h + k + l) \text{ is an even multiple of 2.}
\end{aligned}
\qquad (10\text{–}9)
$$

Even before making a detailed calculation of relative diffracted intensities by means of Eq. (4–19), we can almost rule out the NaCl structure as a possibility simply by inspection of Eqs. (10–8). The atomic numbers of cadmium and tellurium are 48 and 52, respectively, so the value of $(f_{Cd} + f_{Te})^2$ is several hundred times greater than the value of $(f_{Cd} - f_{Te})^2$, for all values of $\sin \theta/\lambda$. Then, if CdTe has the NaCl structure, the 111 reflection should be very weak and the 200 reflection very strong. Actually, 111 is strong and 200 is not observed. Further evidence that the NaCl structure is incorrect is given in the fourth column of Table 10-6, where the calculated intensities of the first eight possible lines are listed: there is no agreement whatever between these values and the observed intensities.

On the other hand, if the ZnS structure is assumed, intensity calculations lead to the values listed in the fifth column. The agreement between these values and the observed intensities is excellent, except for a few minor inconsistencies among the low-angle reflections, and these are due to neglect of the absorption factor. In particular, we note that the ZnS structure satisfactorily accounts for all the missing reflections (200, 420, etc.), since the calculated intensities of these reflections are all extremely low. We can therefore conclude that CdTe has the structure of the zinc-blende form of ZnS.

Table 10-6

1	2	3	4	5
Line	*hkl*	Observed intensity	Calculated intensity	
			NaCl structure	ZnS structure
1	111	s	0.05	12.4
	200	nil	13.2	0.03
2	220	vs	10.0 ←→	10.0
3	311	vs	0.02	6.2
4	222	vw	3.5	0.007
5	400	m	1.7	1.7
6	331	m	0.01	2.5
	420	nil	4.6	0.01
7	422	s	......	3.4
8	511, 333	m	......	1.8
9	440	w	......	1.1
10	531	m	......	2.0
	600, 442	nil	......	0.005
11	620	m	......	1.8
12	533	w	......	0.9
	622	nil	......	0.004
13	444	w	......	0.6
14	711, 551	m	......	1.8
	640	nil	......	0.005
15	642	vs	......	4.0
16	731, 553	s	......	3.3

(N.B. Calculated intensities have been adjusted so that the 220 line has an intensity of 10.0 for both structures.)

After a given structure has been shown to be in accord with the diffraction data, it is advisable to calculate the interatomic distances involved in that structure. This calculation not only is of interest in itself, but serves to disclose any gross errors that may have been made, since there is obviously something wrong with a proposed structure if it brings certain atoms impossibly close together. In the present structure, the nearest neighbor to the Cd atom at 0 0 0 is the Te atom at $\frac{1}{4}\,\frac{1}{4}\,\frac{1}{4}$. The Cd-Te interatomic distance is therefore $\sqrt{3}\,a/4 = 2.80$ Å. For comparison, we can calculate a "theoretical" Cd-Te interatomic distance simply by averaging the distances of closest approach in the pure elements. In doing this, we regard the atoms as rigid spheres in contact, and ignore the effects of coordination number and type of bonding on atom size. These distances of closest approach are 2.98 Å in pure cadmium and 2.86 Å in pure tellurium, the average being 2.92 Å. The observed Cd-Te interatomic distance is 2.80 Å, or some 4.1 percent smaller than the calculated value; this difference is not unreasonable and can be largely ascribed to the covalent bonding which characterizes this structure. In fact, it is a general rule that the A-B interatomic distance in an intermediate phase A_xB_y is always somewhat smaller than the average distance of closest approach in pure A and pure B, because the mere existence of the phase shows that the attractive forces between unlike atoms is greater than that between like atoms. If this were not true, the phase would not form.

PROBLEMS

***10–1** The powder pattern of aluminum, made with Cu $K\alpha$ radiation, contains ten lines, whose $\sin^2\theta$ values are 0.1118, 0.1487, 0.294, 0.403, 0.439, 0.583, 0.691, 0.727, 0.872, and 0.981. Index these lines and calculate the lattice parameter.

10–2 A pattern is made of a cubic substance with unfiltered chromium radiation. The observed $\sin^2\theta$ values and intensities are 0.265(m), 0.321(vs), 0.528(w), 0.638(s), 0.793(s), and 0.958(vs). Index these lines and state which are due to $K\alpha$ and which to $K\beta$ radiation. Determine the Bravais lattice and lattice parameter. Identify the substance by reference to Appendix 5.

10–3 Construct a Hull–Davey chart, and accompanying $\sin^2\theta$ scale, for hexagonal close-packed lattices. Use two-range semilog graph paper, $8\frac{1}{2} \times 11$ in. Cover a c/a range of 0.5 to 2.0, and plot only the curves $00\cdot2$, $10\cdot0$, $10\cdot1$, $10\cdot2$, and $11\cdot0$.

***10–4** Use the chart constructed in Prob. 10–3 to index the first five lines on the powder pattern of α-titanium. With Cu $K\alpha$ radiation, these lines have the following $\sin^2\theta$ values: 0.091, 0.106, 0.117, 0.200, and 0.268.

In each of the following problems the powder pattern of an element is represented by the observed $\sin^2\theta$ *values of the first seven or eight lines on the pattern, made with Cu $K\alpha$ radiation. In each case, index the lines, find the crystal system, Bravais lattice, and approx-*

imate lattice parameter (or parameters), and identify the element from the tabulation given in Appendix 5.

10–5	***10–6**	**10–7**	***10–8**
0.0806	0.0603	0.1202	0.0768
0.0975	0.1610	0.238	0.0876
0.1122	0.221	0.357	0.0913
0.210	0.322	0.475	0.1645
0.226	0.383	0.593	0.231
0.274	0.484	0.711	0.274
0.305	0.545	0.830	0.308
0.321	0.645		0.319

11

Precise Parameter Measurements

11-1 INTRODUCTION

Many applications of x-ray diffraction require precise knowledge of the lattice parameter (or parameters) of the material under study. In the main, these applications involve solid solutions; since the lattice parameter of a solid solution varies with the concentration of the solute, the composition of a given solution can be determined from a measurement of its lattice parameter. Thermal expansion coefficients can also be determined, without a dilatometer, by measurements of lattice parameter as a function of temperature in a high-temperature camera. Since, in general, a change in solute concentration or temperature produces only a small change in lattice parameter, rather precise parameter measurements must be made in order to measure these quantities with any accuracy. In this chapter we shall consider the methods that are used to obtain high precision, leaving the various applications to be discussed at a later time. Cubic substances will be dealt with first, because they are the simplest, but our general conclusions will also be valid for noncubic materials, which will be discussed in detail later.

The process of measuring a lattice parameter is a very indirect one, and is fortunately of such a nature that high precision is fairly easily obtainable. The parameter a of a cubic substance is directly proportional to the spacing d of any particular set of lattice planes. If we measure the Bragg angle θ for this set of planes, we can use the Bragg law to determine d and, knowing d, we can calculate a. But it is $\sin \theta$, not θ, which appears in the Bragg law. Precision in d, or a, therefore depends on precision in $\sin \theta$, a derived quantity, and not on precision in θ, the measured quantity. This is fortunate because the value of $\sin \theta$ changes very slowly with θ in the neighborhood of 90°, as inspection of Fig. 11-1 or a table of sines will show. For this reason, a very accurate value of $\sin \theta$ can be obtained from a measurement of θ which is itself not particularly precise, *provided that θ is near 90°*. For example, an error in θ of 1° leads to an error in $\sin \theta$ of 1.7 percent at $\theta = 45°$ but only 0.15 percent at $\theta = 85°$. Stated in another way, the angular position of a diffracted beam is much more sensitive to a given change in plane spacing when θ is large than when it is small.

We can obtain the same result directly by differentiating the Bragg law with respect to θ. We obtain

$$\frac{\Delta d}{d} = -\cot \theta \Delta \theta. \tag{11-1}$$

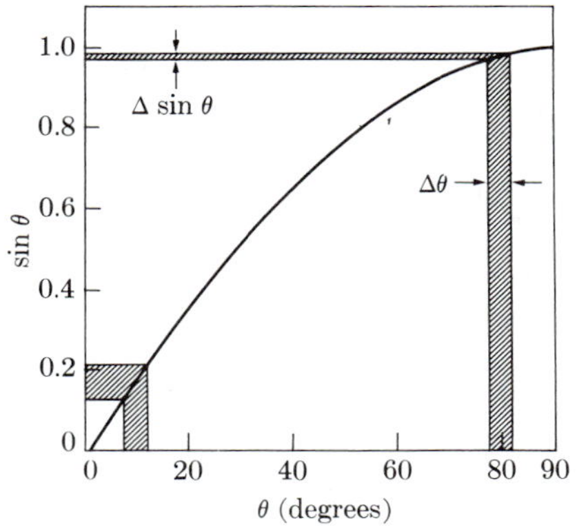

Fig. 11-1 The variation of sin θ with θ. The error in sin θ caused by a given error in θ decreases as θ increases ($\Delta\theta$ exaggerated).

In the cubic system,

$$a = d\sqrt{h^2 + k^2 + l^2}.$$

Therefore

$$\frac{\Delta a}{a} = \frac{\Delta d}{d} = -\cot\theta\Delta\theta. \tag{11-2}$$

Since cot θ approaches zero as θ approaches 90°, $\Delta a/a$, the fractional error in a caused by a given error in θ, also approaches zero as θ approaches 90°, or as 2θ approaches 180°. The key to precision in parameter measurements therefore lies in the use of backward-reflected beams having 2θ values as near to 180° as possible.

Although the parameter error disappears as 2θ approaches 180°, we cannot observe a reflected beam at this angle. But since the values of a calculated for the various lines on the pattern approach the true value more closely as 2θ increases, we should be able to find the true value of a simply by plotting the measured values against 2θ and extrapolating to $2\theta = 180°$. Unfortunately, this curve is not linear and the extrapolation of a nonlinear curve is not accurate. However, it may be shown that if the measured values of a are plotted against certain functions of θ, rather than against θ or 2θ directly, the resulting curve is a straight line which may be extrapolated with confidence. The bulk of this chapter is devoted to showing how these functions can be derived and used. Because the exact form of the function depends on the kind of camera employed, we shall have to consider successively the various cameras that are normally used for parameter measurements.

But first we might ask: what sort of precision is possible with such methods? Without any extrapolation or any particular attention to good experimental technique, simply by selection of the parameter calculated for the highest-angle line on the pattern, we can usually obtain a precision of 0.01 Å. Since the lattice parameters of most substances of metallurgical interest are in the neighborhood of 3 to 4 Å, this represents a precision of about 0.3 percent. With good experimental technique and the use of the proper extrapolation function, this precision can be increased to 0.001 Å, or 0.03 percent, without much difficulty. Finally, about the best precision that can be expected is 0.0001 Å, or 0.003 percent, but this can be obtained only by the expenditure of considerable effort, both experimental and computational.

In work of high precision it is imperative that the units in which the measured parameter is expressed, kX or Å, be correctly stated. In order to avoid confusion on this point, the reader is advised to review the discussion of these units given in Sec. 3–4. The actual *numerical value* of the wavelength or wavelengths used in the determination of the parameter should be explicitly stated.

Methods of determining lattice parameters with high precision are reviewed by Barrett and Massalski [G.25], Klug and Alexander [G.39], Parrish and Wilson [11.1], Azaroff and Buerger [G.17], and Lipson and Steeple [G.32].

11–2 DEBYE–SCHERRER CAMERAS

The general approach in finding an extrapolation function is to consider the various effects which can lead to errors in the measured values of θ, and to find out how these errors in θ vary with the angle θ itself. For a Debye–Scherrer camera, the chief sources of error in θ are the following:

1. Film shrinkage.
2. Incorrect camera radius.
3. Off-centering of specimen.
4. Absorption in specimen.

Since only the back-reflection region is suitable for precise measurements, we shall consider these various errors in terms of the quantities S' and ϕ, defined in Fig. 11–2. S' is the distance on the film between two corresponding back-reflection lines; 2ϕ is the supplement of 2θ, i.e., $\phi = 90^\circ - \theta$. These quantities are related to the camera radius R by the equation

$$\phi = \frac{S'}{4R}. \tag{11–3}$$

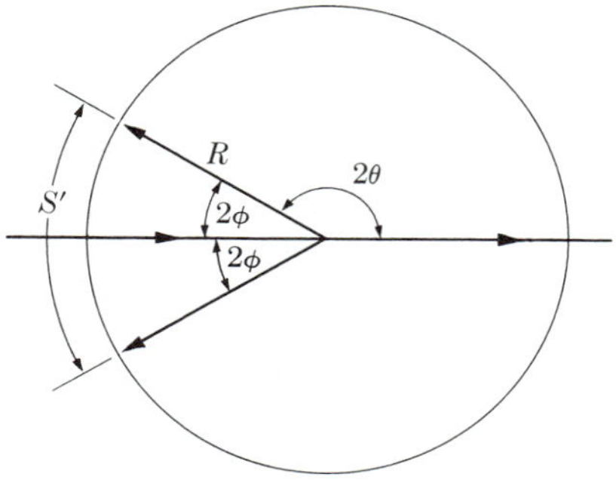

Fig. 11–2

Shrinkage of the film, caused by processing and drying, causes an error $\Delta S'$ in the quantity S'. The camera *radius* may also be in error by an amount ΔR. The effects of these two errors on the value of ϕ may be found by writing Eq. (11–3) in logarithmic form:

$$\ln \phi = \ln S' - \ln 4 - \ln R.$$

Differentiation then gives

$$\frac{\Delta\phi}{\phi} = \frac{\Delta S'}{S'} - \frac{\Delta R}{R}. \tag{11–4}$$

The error in ϕ due to shrinkage and the radius error is therefore given by

$$\Delta\phi_{S',R} = \left(\frac{\Delta S'}{S'} - \frac{\Delta R}{R}\right)\phi. \tag{11–5}$$

The shrinkage error can be minimized by loading the film so that the incident beam enters through a hole in the film, since corresponding back-reflection lines are then only a short distance apart on the film, and their separation S' is little affected by film shrinkage. The method of film loading shown in Fig. 6–5(a) is not at all suitable for precise measurements. Instead, methods (b) or (c) of Fig. 6–5 should be used. Method (c), the unsymmetrical or Staumanis method of film loading, is particularly recommended since no knowledge of the camera radius is required.

An *off-center specimen* also leads to an error in ϕ. Whatever the displacement of the specimen from the camera center, this displacement can always be broken up into two components, one (Δx) parallel to the incident beam and the other (Δy) at right angles to the incident beam. The effect of the parallel displacement is illustrated in Fig. 11–3(a). Instead of being at the camera center C', the specimen is displaced a distance Δx to the point O. The diffraction lines are registered at D and C instead of at A and B, the line positions for a properly centered specimen.

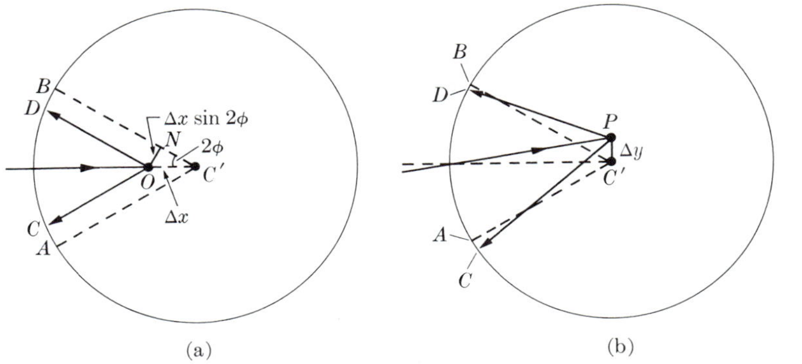

Fig. 11–3 Effect of specimen displacement on line positions.

The error in S' is then $(AC + DB) = 2DB$, which is approximately equal to $2ON$, or

$$\Delta S' = 2ON = 2\Delta x \sin 2\phi. \tag{11-6}$$

The effect of a specimen displacement at right angles to the incident beam [Fig. 11–3(b)] is to shift the lines from A to C and from B to D. When Δy is small, AC is very nearly equal to BD and so, to a good approximation, no error in S' is introduced by a right-angle displacement.

The total error in S' due to specimen displacement in some direction inclined to the incident beam is therefore given by Eq. (11–6). This error in S' causes an error in the computed value of ϕ. Inasmuch as we are considering the various errors one at a time, we can now put the radius error ΔR equal to zero, so that Eq. (11–4) becomes

$$\frac{\Delta\phi}{\phi} = \frac{\Delta S'}{S'}, \tag{11-7}$$

which shows how an error in S' alone affects the value of ϕ. By combining Eqs. (11–3), (11–6), and (11–7), we find that the error in ϕ due to the fact that the specimen is off center is given by

$$\Delta\phi_C = \frac{\phi\Delta S'}{S'} = \frac{\phi(2\Delta x \sin 2\phi)}{4R\phi} = \frac{\Delta x}{R}\sin\phi\cos\phi. \tag{11-8}$$

It should not be assumed that the centering error is removed when the specimen is so adjusted, relative to the rotating shaft of the camera, that no perceptible wobble can be detected when the shaft is rotated. This sort of adjustment is taken for granted in this discussion. The off-center error refers to the possibility that the axis of rotation of the *shaft* is not located at the center of the camera, due to improper construction of the camera.

Absorption in the specimen also causes an error in ϕ. This effect, often the largest single cause of error in parameter measurements, is unfortunately very difficult to calculate with any accuracy. But we have seen, in Fig. 4–18(b), that back-reflected rays come almost entirely from that side of the specimen which faces the collimator. Therefore, to a rough approximation, the effect of a centered, highly absorbing specimen is the same as that of a nonabsorbing specimen displaced from the camera center in the manner shown in Fig. 11–3(a). Consequently we can assume that the error in ϕ due to absorption, $\Delta\phi_A$, is included in the centering error given by Eq. (11–8).

Thus, the overall error in ϕ due to film shrinkage, radius error, centering error, and absorption, is given by the sum of Eqs. (11–5) and (11–8):

$$\Delta\phi_{S',R,C,A} = \left(\frac{\Delta S'}{S'} - \frac{\Delta R}{R}\right)\phi + \frac{\Delta x}{R}\sin\phi\cos\phi. \tag{11-9}$$

But

$$\phi = 90^\circ - \theta, \quad \Delta\phi = -\Delta\theta, \quad \sin\phi = \cos\theta, \quad \text{and} \quad \cos\phi = \sin\theta.$$

Therefore Eq. (11–2) becomes

$$\frac{\Delta d}{d} = -\frac{\cos\theta}{\sin\theta}\Delta\theta = \frac{\sin\phi}{\cos\phi}\Delta\phi$$

and

$$\frac{\Delta d}{d} = \frac{\sin\phi}{\cos\phi}\left[\left(\frac{\Delta S'}{S'} - \frac{\Delta R}{R}\right)\phi + \frac{\Delta x}{R}\sin\phi\cos\phi\right]. \qquad (11\text{-}10)$$

In the back-reflection region, ϕ is small and may be replaced, in the first term of Eq. (11-10), by $\sin\phi\cos\phi$, since $\sin\phi \approx \phi$ and $\cos\phi \approx 1$, for small values of ϕ. We then have

$$\frac{\Delta d}{d} = \left(\frac{\Delta S'}{S'} - \frac{\Delta R}{R} + \frac{\Delta x}{R}\right)\sin^2\phi.$$

The bracketed terms are constant for any one film, so that

$$\frac{\Delta d}{d} = K\sin^2\phi = K\cos^2\theta, \qquad (11\text{-}11)$$

where K is a constant. Accordingly, we have the important result that the fractional errors in d are directly proportional to $\cos^2\theta$, and therefore approach zero as $\cos^2\theta$ approaches zero or as θ approaches 90°. In the cubic system,

$$\frac{\Delta d}{d} = \frac{\Delta a}{a} = \frac{a - a_0}{a_0} = K\cos^2\theta$$

$$a = a_0 + a_0 K\cos^2\theta \qquad (11\text{-}12)$$

Hence, for cubic substances, if the value of a computed for each line on the pattern is plotted against $\cos^2\theta$, a straight line should result, and a_0, the true value of a, can be found by extrapolating this line to $\cos^2\theta = 0$. (Or, since $\sin^2\theta = 1 - \cos^2\theta$, the various values of a may be plotted against $\sin^2\theta$, and the line extrapolated to $\sin^2\theta = 1$. Values of $\sin^2\theta$ are given in Appendix 9.)

From the various approximations involved in the derivation of Eq. (11-12), it is clear that this equation is true only for large values of θ (small values of ϕ). Therefore, only lines having θ values greater than about 60° should be used in the extrapolation, and the more lines there are with θ greater than 80°, the more precise is the value of a_0 obtained. To increase the number of lines in the back-reflection region, it is common practice to employ unfiltered radiation so that $K\beta$ as well as $K\alpha$ can be reflected. If the x-ray tube is demountable, special alloy targets can also be used to increase the number of lines; or two exposures can be made on the same film with different characteristic radiations. In any case, it must never be assumed that the process of extrapolation can automatically produce a precise value of a_0 from careless measurements made on a film of poor quality. For high precision, the lines must be sharp and the $K\alpha$ doublets well resolved at high angles, which means in turn that the individual particles of the specimen must be strain-free and not too fine. The line positions must be determined carefully, and it is best to measure each one two or three times and average the results. In computing a for each line, the proper wavelength must be assigned to each component of the $K\alpha$ doublet when that line is resolved, and when it is not resolved, the weighted mean wavelength should be used.

To illustrate this extrapolation method, we shall consider a powder pattern of tungsten made in a Debye-Scherrer camera 5.73 cm in diameter with unfiltered

Table 11–1

Line	hkl	Radiation	θ	$\sin^2\theta$	a(Å)
6	400	$K\beta$	61.71°	0.7754	3.162
5	321	$K\alpha$	65.91	0.8334	3.160
4	411, 330	$K\beta$	69.05	0.8722	3.162
3	400	$K\alpha_1$	76.73	0.9473	3.166
2	400	$K\alpha_2$	77.48	0.9530	3.164
1	420	$K\beta$	79.67	0.9678	3.164

copper radiation. The data for all lines having θ values greater than 60° are given in Table 11–1. The drift in the computed a values is obvious: in general they increase with θ and tend to approach the true value a_0 at high angles. In Fig. 11–4, these values of a are plotted against $\sin^2\theta$, and a_0 is found by extrapolation to be 3.165 Å.

Other functions of θ, besides $\sin^2\theta$ or $\cos^2\theta$, may be used as a basis for extrapolation. For example, if we replace $\sin\phi\cos\phi$ in Eq. (11–10) by ϕ, instead of replacing ϕ by $\sin\phi\cos\phi$, we obtain

$$\frac{\Delta d}{d} = K\phi\tan\phi.$$

Therefore, a plot of a against $\phi\tan\phi$ will also be linear and will extrapolate to a_0 at $\phi\tan\phi = 0$. In practice, there is not much difference between an extrapolation against $\phi\tan\phi$ and one against $\cos^2\theta$ (or $\sin^2\theta$), and either will give satisfactory results. (Values of $\phi\tan\phi$ are given in [G.11, Vol. 2, p. 232].) Nelson and Riley [11.2] and Taylor and Sinclair [11.3] analyzed the various sources of error, particularly absorption, more rigorously than we have done and showed that the relation

$$\frac{\Delta d}{d} = K\left(\frac{\cos^2\theta}{\sin\theta} + \frac{\cos^2\theta}{\theta}\right)$$

holds quite accurately down to very low values of θ and not just at high angles. The bracketed terms are sometimes called the Nelson–Riley function. The value

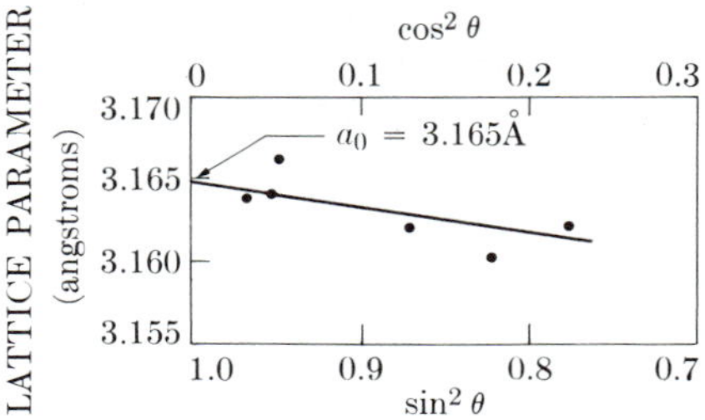

Fig. 11–4 Extrapolation of measured lattice parameters against $\sin^2\theta$ (or $\cos^2\theta$).

of a_0 can be found by plotting a against the N-R function, which approaches zero as θ approaches 90°. Although it is doubtful whether any advantage results from using the N-R function instead of $\cos^2\theta$ in the back-reflection region, the greater range of linearity of the N-R function is an advantage when there are only a few lines in the high-angle region. Values of the N-R function are given in [G.8, G.13, G.17, G.19, G.32, and Vol. 2 of G.11].

Noncubic crystals present additional difficulties, regardless of the particular extrapolation function chosen. Consider, for example, hexagonal and tetragonal crystals. The difficulty is simply this: the position of a line which has indices *hkl* is determined by two parameters, a and c, and it is impossible to calculate both of them from the observed $\sin^2\theta$ value of that line alone. One way of avoiding this difficulty is to ignore the *hkl* lines and divide the remainder into two groups, those with indices *hk*0 and those with indices 00*l*. A value of a is calculated for each *hk*0 line and a value of c from each 00*l* line; two separate extrapolations are then made to find a_0 and c_0. Since there are usually very few *hk*0 and 00*l* lines in the back-reflection region, some low-angle lines have to be included, which means that the extrapolations must be made against $(\cos^2\theta/\sin\theta + \cos^2\theta/\theta)$ and not against $\cos^2\theta$. And if there are *no* lines of the type *hk*0 and 00*l* with θ greater than 80°, even the former function will not assure an accurate extrapolation. Because of these difficulties with graphical extrapolation, it is better to use Cohen's analytical method (Sec. 11-7) for noncubic substances.

To conclude this section, a few general remarks on the nature of errors may not be amiss. In the measurement of a lattice parameter, as in many other physical observations, two kinds of error are involved, *systematic* and *random*. A systematic error is one which varies in a regular manner with some particular parameter. Thus the fractional errors in a due to the various effects considered above (film shrinkage, incorrect radius, off-center specimen, absorption) are all systematic errors because they vary in a regular way with θ, decreasing as θ increases. Further, a systematic error is always of the same sign: for example, the effect of absorption in a Debye-Scherrer camera is always to make the computed value of a less than the true value. Random errors, on the other hand, are the ordinary chance errors involved in any direct observation. For example, the errors involved in measuring the positions of the various lines on a film are random errors; they may be positive or negative and do not vary in any regular manner with the position of the line on the film.

As we have already seen, the systematic errors in a approach zero as θ approaches 90°, and may be eliminated by use of the proper extrapolation function. The magnitude of these errors is proportional to the slope of the extrapolation line and, if these errors are small, the line will be quite flat. In fact, if we purposely increase the systematic errors, say, by using a slightly incorrect value of the camera radius in our calculations, the slope of the line will increase but the extrapolated value of a_0 will remain the same. The random errors involved in measuring line positions show up as random errors in a, and are responsible for the deviation of the various points from the extrapolation line. The random errors in a also decrease in magnitude as θ increases, due essentially to the slow variation of $\sin\theta$ with θ at large angles.

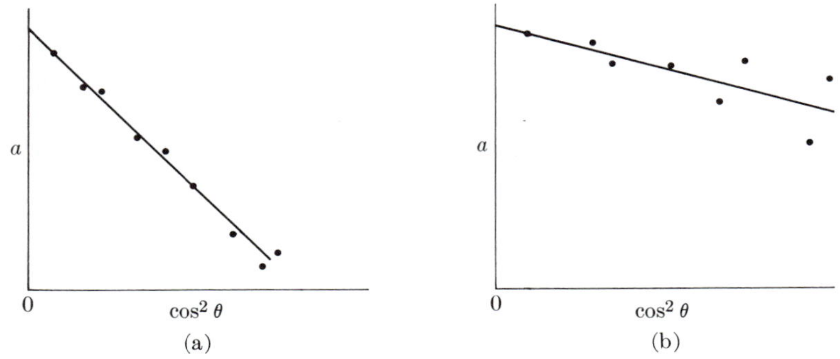

Fig. 11–5 Extreme forms of extrapolation curves (schematic): (a) large systematic errors, small random errors; (b) small systematic errors, large random errors.

These various effects are summarized graphically in Fig. 11–5. In (a) the calculated points conform quite closely to the line, indicating small random errors, but the line itself is quite steep because of large systematic errors. The opposite situation is shown in (b): here the systematic error is small, but the wide scatter of the points shows that large random errors have been made. Inasmuch as the difficulty of drawing the line increases with the degree of scatter, it is obvious that every possible effort should be made to minimize random errors at the start.

11–3 BACK-REFLECTION FOCUSING CAMERAS

A camera of this kind is preferred for work of the highest precision, since the position of a diffraction line on the film is twice as sensitive to small changes in plane spacing with this camera as it is with a Debye–Scherrer camera of the same diameter. It is, of course, not free from sources of systematic error. The most important of these are the following:

1. Film shrinkage.

2. Incorrect camera radius.

3. Displacement of specimen from camera circumference.

4. Absorption in specimen. (If the specimen has very *low* absorption, many of the diffracted rays will originate at points outside the camera circumference even though the specimen surface coincides with the circumference.)

A detailed analysis of these various sources of error shows that they produce fractional errors in d which are very closely proportional to $\phi \tan \phi$, where ϕ is again equal to $(90° - \theta)$. This function is therefore the one to use in extrapolating lattice parameters measured with this camera.

11–4 PINHOLE CAMERAS

The pinhole camera, used in back reflection, is not really an instrument of high precision in the measurement of lattice parameters, but it is mentioned here because of its very great utility in metallurgical work. Since both the film and the specimen surface are flat, no focusing of the diffracted rays occurs, and the result is that the diffraction lines are much broader than is normally desirable for precise measurement of their positions. The chief sources of systematic error are the following:

1. Film shrinkage.
2. Incorrect specimen-to-film distance.
3. Absorption in the specimen.

In this case it may be shown that the fractional error in d is proportional to $\sin 4\phi \tan \phi$, or to the equivalent expression $\cos^2 \theta (2 \cos^2 \theta - 1)$, where $\phi = (90° - \theta)$. With either of these extrapolation functions a fairly precise value of the lattice parameter can be obtained; in addition, the back-reflection pinhole camera has the particular advantage that mounted metallographic specimens may be examined directly. This means that a parameter determination can be made on the same part of a specimen as that examined under the microscope. A dual examination of this kind is quite valuable in many problems, especially in the determination of phase diagrams.

11-5 DIFFRACTOMETERS

The diffractometer is a more complex apparatus than a powder camera and therefore more subject to misalignment of its component parts. A further difficulty with most commercial diffractometers is the impossibility of observing the same back-reflected cone of radiation on both sides of the incident beam. Thus, the experimenter has no automatic check on the accuracy of the angular scale of the instrument or the precision of its alignment.

When a diffractometer is used to measure plane spacings, the more important sources of systematic error in d are the following:

1. Misalignment of the instrument. In particular, the center of the incident beam must intersect the diffractometer axis and the 0° position of the counter slit.
2. Use of a flat specimen instead of a specimen curved to conform to the focusing circle. This error is minimized, with loss of intensity, by decreasing the irradiated width of the specimen by means of an incident beam of small horizontal divergence.
3. Absorption in the specimen. Specimens of low absorption should be made as thin as possible.
4. Displacement of the specimen from the diffractometer axis. This is usually the largest single source of error. It causes an error in d given by

$$\frac{\Delta d}{d} = -\frac{D}{R}\frac{\cos^2 \theta}{\sin \theta} \tag{11-13}$$

where D is the specimen displacement parallel to the reflecting-plane normal (positive when the displacement is in front of the axis) and R is the diffractometer radius (Problem 11-6).

5. Vertical divergence of the incident beam. This error is minimized, with loss of intensity, by decreasing the vertical opening of the counter slit.

No single extrapolation function can be completely satisfactory, because

$\Delta d/d$ varies as $\cos^2 \theta$ for errors (2) and (3) but as $\cos^2 \theta/\sin \theta$ for error (4). Presumably the function that gives the better straight line will disclose what error is predominant.

The suggested procedure is therefore:

a) Carefully align the component parts of the instrument in accordance with the manufacturer's instructions.

b) Adjust the specimen surface to coincide as closely as possible with the diffractometer axis.

c) Extrapolate the calculated parameters against $\cos^2 \theta/\sin \theta$ or $\cos^2 \theta$.

Every effort should, of course, be made to measure line positions precisely. If one aims at a precision of 3 parts in 100,000, equivalent to ± 0.0001 Å in the lattice parameter, then Eq. (11–1) shows that the 2θ position of a line at $2\theta = 160°$ must be measured to within 0.02° and lower angle lines even more closely. It is better to determine the line profile by step-counting with a scaler than by chart-recording with a ratemeter. In parameter measurements it is usual to take the 2θ value of maximum intensity as the line position; strangely, curve-fitting techniques for establishing line position, which are standard in the field of stress measurement (Sec. 16–4), are practically never used.

11–6 METHOD OF LEAST SQUARES

All the previously described methods of accurately measuring lattice parameters depend in part on graphical extrapolation. Their accuracy therefore depends on the accuracy with which a straight line can be drawn through a set of experimental points, each of which is subject to random errors. However, different persons will in general draw slightly different lines through the same set of points, so that it is desirable to have an objective, analytical method of finding the line which best fits the data. This can be done by the method of least squares. Since this method can be used in a variety of problems, it will be described here in a quite general way; in the next section, its application to parameter measurements will be taken up in detail.

If a number of measurements are made of the same physical quantity and if these measurements are subject only to random errors, then the theory of least squares states that the most probable value of the measured quantity is that which makes the sum of the squares of the errors a minimum.

This theorem is applied as follows to the problem of finding the straight line which best fits a set of experimentally determined points. If there are only two points, there is no problem, because the two constants which define a straight line can be unequivocally determined from these two points. But, in general, there will be more points available than constants to be determined. Suppose that the various points have coordinates x_1y_1, x_2y_2, x_3y_3, . . . and that it is known that x and y are related by an equation of the form

$$y = a + bx. \tag{11–14}$$

Our problem is to find the values of the constants a and b, since these define the straight line. In general, the line will not pass exactly through any of the points since each is subject to a random error. Therefore each point is in error by an amount given by its deviation from the straight line. For example, Eq. (11–14) states that the value of y corresponding to $x = x_1$ is $(a + bx_1)$. Yet the first experimental point has a value of $y = y_1$. Therefore e_1, the error in the first point, is given by

$$e_1 = (a + bx_1) - y_1.$$

We can calculate the errors in the other points in similar fashion, and then write down the expression for the sum of the squares of these errors:

$$\sum (e^2) = (a + bx_1 - y_1)^2 + (a + bx_2 - y_2)^2 + \cdots. \tag{11-15}$$

According to the theory of least squares, the "best" straight line is that which makes the sum of the squared errors a minimum. Therefore, the best value of a is found by differentiating Eq. (11–15) with respect to a and equating the result to zero:

$$\frac{\partial \sum (e^2)}{\partial a} = 2(a + bx_1 - y_1) + 2(a + bx_2 - y_2) + \cdots = 0,$$

or

$$\sum a + b \sum x - \sum y = 0. \tag{11-16}$$

The best value of b is found in a similar way:

$$\frac{\partial \sum (e^2)}{\partial b} = 2x_1(a + bx_1 - y_1) + 2x_2(a + bx_2 - y_2) + \cdots = 0,$$

or

$$a \sum x + b \sum x^2 - \sum xy = 0. \tag{11-17}$$

Equations (11–16) and (11–17) are the *normal equations*. Simultaneous solution of these two equations yields the best values of a and b, which can then be substituted into Eq. (11–14) to give the equation of the line.

The normal equations as written above can be rearranged as follows:

$$\sum y = \sum a + b \sum x$$

and

$$\sum xy = a \sum x + b \sum x^2. \tag{11-18}$$

A comparison of these equations and Eq. (11–14) shows that the following rules can be laid down for the formation of the normal equations:

a) Substitute the experimental values of x and y into Eq. (11–14). If there are n experimental points, n equations in a and b will result.

b) To obtain the first normal equation, multiply each of these n equations by the coefficient of a in each equation, and add.

c) To obtain the second normal equation, multiply each equation by the coefficient of b, and add.

As an illustration, suppose that we determine the best straight line through the following four points:

x	10	18	30	42
y	15	11	11	8

The normal equations are obtained in three steps:

a) Substitution of the given values:

$$\begin{aligned} 15 &= a + 10b \\ 11 &= a + 18b \\ 11 &= a + 30b \\ 8 &= a + 42b \end{aligned}$$

b) Multiplication by the coefficient of a:

$$\begin{aligned} 15 &= a + 10b \\ 11 &= a + 18b \\ 11 &= a + 30b \\ 8 &= a + 42b \\ \hline 45 &= 4a + 100b \quad \text{(first normal equation)} \end{aligned}$$

c) Multiplication by the coefficient of b:

$$\begin{aligned} 150 &= 10a + 100b \\ 198 &= 18a + 324b \\ 330 &= 30a + 900b \\ 336 &= 42a + 1764b \\ \hline 1014 &= 100a + 3088b \quad \text{(second normal equation)} \end{aligned}$$

Simultaneous solution of the two normal equations gives $a = 16.0$ and $b = -0.189$. The required straight line is therefore

$$y = 16.0 - 0.189x.$$

This line is shown in Fig. 11–6, together with the four given points.

The least-squares method is not confined to finding the constants of a straight line; it can be applied to any kind of curve. Suppose, for example, that x and y are known to be related by a parabolic equation

$$y = a + bx + cx^2.$$

Since there are three unknown constants here, we need three normal equations. These are

$$\begin{aligned} \sum y &= \sum a + b \sum x + c \sum x^2, \\ \sum xy &= a \sum x + b \sum x^2 + c \sum x^3, \\ \sum x^2y &= a \sum x^2 + b \sum x^3 + c \sum x^4. \end{aligned} \tag{11-19}$$

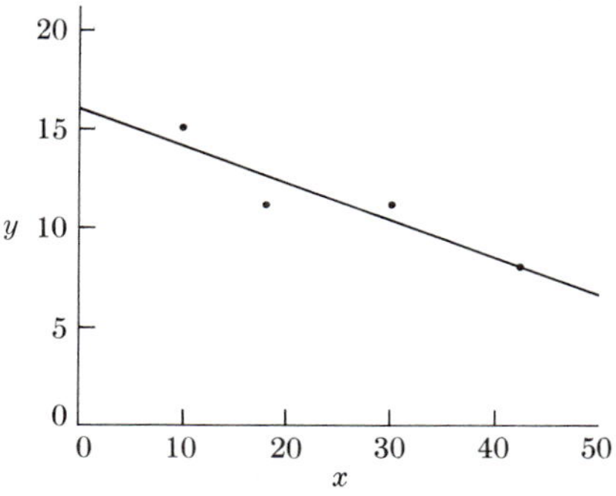

Fig. 11-6 Best straight line, determined by least-squares method.

These normal equations can be found by the same methods as were used for the straight-line case, i.e., successive multiplication of the n observational equations by the coefficients of a, b, and c, followed by addition of the equations in each set.

It should be noted that the least-squares method is not a way of finding the best curve to fit a given set of observations. The investigator must know at the outset, from his understanding of the phenomenon involved, the kind of relation (linear, parabolic, exponential, etc.) the two quantities x and y are supposed to obey. All the least-squares method can do is give him the best values of the constants in the equation he selects, but it does this in a quite objective and unbiased manner.

11-7 COHEN'S METHOD

In preceding sections we have seen that the most accurate value of the lattice parameter of a cubic substance is found by plotting the value of a calculated for each reflection against a particular function, which depends on the apparatus used, and extrapolating to a value a_0 at $\theta = 90°$. Two different things are accomplished by this procedure: (a) systematic errors are eliminated by selection of the proper extrapolation function, and (b) random errors are reduced in proportion to the skill of the investigator in drawing the best straight line through the experimental points. Cohen [11.4] proposed, in effect, that the least-squares method be used to find the best straight line so that the random errors would be minimized in a reproducible and objective manner.

Cubic system

Suppose a cubic substance is being examined in a Debye-Scherrer camera. Then Eq. (11-11), namely,

$$\frac{\Delta d}{d} = \frac{\Delta a}{a} = K \cos^2 \theta, \tag{11-11}$$

defines the extrapolation function. But instead of using the least-squares method to find the best straight line on a plot of a against $\cos^2 \theta$, Cohen applied the method to the observed $\sin^2\theta$ values directly. By squaring the Bragg law and taking logarithms of each side, we obtain

$$\ln \sin^2\theta = \ln\left(\frac{\lambda^2}{4}\right) - 2\ln d.$$

Differentiation then gives

$$\frac{\Delta \sin^2\theta}{\sin^2\theta} = -\frac{2\Delta d}{d}. \tag{11-20}$$

By substituting this into Eq. (11–11) we find how the error in $\sin^2\theta$ varies with θ:

$$\Delta \sin^2\theta = -2K \sin^2\theta \cos^2\theta = D \sin^2 2\theta, \tag{11-21}$$

where D is a new constant. [This equation is valid only when the $\cos^2\theta$ extrapolation function is valid. If some other extrapolation function is used, Eq. (11–21) must be modified accordingly.] Now the true value of $\sin^2\theta$ for any diffraction line is given by

$$\sin^2\theta\ (\text{true}) = \frac{\lambda^2}{a4_0^2}(h^2 + k^2 + l^2),$$

where a_0, the true value of the lattice parameter, is the quantity we are seeking. But

$$\sin^2\theta\ (\text{observed}) - \sin^2\theta\ (\text{true}) = \Delta \sin^2\theta,$$

$$\sin^2\theta - \frac{\lambda^2}{4a_0^2}(h^2 + k^2 + l^2) = D\sin^2 2\theta,$$

$$\sin^2\theta = C\alpha + A\delta, \tag{11-22}$$

where

$$C = \frac{\lambda^2}{4a_0^2}, \quad \alpha = (h^2 + k^2 + l^2), \quad A = \frac{D}{10}, \quad \text{and} \quad \delta = 10\sin^2 2\theta.$$

(The factor 10 is introduced into the definitions of the quantities A and δ solely to make the coefficients of the various terms in the normal equations of the same order of magnitude.)

The experimental values of $\sin^2\theta$, α, and δ are now substituted into Eq. (11–22) for each of the n back-reflection lines used in the determination. This gives n equations in the unknown constants C and A, and these equations can be solved for the most probable values of C and A by the method of least squares. Once C is found, a_0 can be calculated directly from the relation given above; the constant A is related to the amount of systematic error involved and is constant for any one film, but varies slightly from one film to another. The two normal equations we need to find C and A are found from Eq. (11–22) and the rules previously given. They are

$$\sum \alpha \sin^2\theta = C\sum \alpha^2 + A\sum \alpha\delta,$$

$$\sum \delta \sin^2\theta = C\sum \alpha\delta + A\sum \delta^2.$$

To illustrate the way in which such calculations are carried out, we will apply Cohen's method to a determination of the lattice parameter of tungsten from measurements made on the pattern shown in Fig. 6-10. Since this pattern was made with a symmetrical back-reflection focusing camera, the correct extrapolation function is

$$\frac{\Delta d}{d} = K\phi \tan \phi.$$

Substituting this into Eq. (11-20), we have

$$\begin{aligned}\Delta \sin^2 \theta &= -2K\phi \sin^2 \theta \tan \phi \\ &= -2K\phi \cos^2 \phi \tan \phi \\ &= D\phi \sin 2\phi,\end{aligned}$$

where D is a new constant. We can therefore write, for each line on the pattern,

$$\sin^2 \theta = \cos^2 \phi = \frac{\lambda^2}{4a_0^2}(h^2 + k^2 + l^2) + D\phi \sin 2\phi, \qquad (11\text{-}23)$$

$$\cos^2 \phi = C\alpha + A\delta, \qquad (11\text{-}24)$$

where

$$C = \frac{\lambda^2}{4a_0^2}, \quad \alpha = (h^2 + k^2 + l^2), \quad A = \frac{D}{10}, \quad \text{and} \quad \delta = 10\phi \sin 2\phi.$$

Equation 11-23 cannot be applied directly because lines due to three different wavelengths (Cu $K\alpha_1$, Cu $K\alpha_2$, and Cu $K\beta$) are present on the pattern, which means that λ varies from line to line, whereas in Eq. (11-23) it is treated as a constant. But the data can be "normalized" to any one wavelength by use of the proper multiplying factor. For example, suppose we decide to normalize all lines to the $K\beta$ wavelength. Then for a particular line formed by $K\alpha_1$ radiation, for instance, we have

$$\cos^2 \phi_{K\alpha_1} = \frac{\lambda_{K\alpha_1}^2}{4a_0^2}\alpha + A\delta_{K\alpha_1},$$

$$\left(\frac{\lambda_{K\beta}^2}{\lambda_{K\alpha_1}^2}\right)\cos^2 \phi_{K\alpha_1} = \left(\frac{\lambda_{K\beta}^2}{\lambda_{K\alpha_1}^2}\right)\left(\frac{\lambda_{K\alpha_1}^2}{4a_0^2}\right)\alpha + \left(\frac{\lambda_{K\beta}^2}{\lambda_{K\alpha_1}^2}\right)A\delta_{K\alpha_1}.$$

From the Bragg law,

$$\left(\frac{\lambda_{K\beta}^2}{\lambda_{K\alpha_1}^2}\right)\cos^2 \phi_{K\alpha_1} = \cos^2 \phi_{K\beta}.$$

Therefore,

$$\cos^2 \phi_{K\beta} = \left(\frac{\lambda_{K\beta}^2}{4a_0^2}\right)\alpha + A\left(\frac{\lambda_{K\beta}^2}{\lambda_{K\alpha_1}^2}\right)\delta_{K\alpha_1}, \qquad (11\text{-}25)$$

where $(\lambda_{K\beta}^2/\lambda_{K\alpha_1}^2)\delta_{K\alpha_1}$ is a normalized δ. Equation (11-25) now refers only to the $K\beta$ wavelength, i.e., it gives the position, in terms of $\cos^2 \phi$, that a $K\alpha_1$ line would have if it were formed by $K\beta$ radiation. Lines due to $K\alpha_2$ radiation can be normalized in a similar manner. When this has been done for all lines, the quantity C in Eq. (11-24) is then a true constant, equal to $\lambda_{K\beta}^2/4a_0^2$. The values of the two normalizing factors, for copper radiation, are

$$\frac{\lambda_{K\beta}^2}{\lambda_{K\alpha_1}^2} = 0.816688 \qquad \text{and} \qquad \frac{\lambda_{K\beta}^2}{\lambda_{K\alpha_2}^2} = 0.812644.$$

Table 11–2 shows the observed and normalized values of $\cos^2 \phi$ and δ for each line on the tungsten pattern. The values of δ need not be calculated to more than two significant figures, since δ occurs in Eq. (11–24) in only the last term which is very small compared to the other two. From the data in Table 11–2, we obtain

$$\sum \alpha^2 = 1628, \quad \sum \delta^2 = 21.5, \quad \sum \alpha\delta = 157.4,$$

$$\sum \alpha \cos^2 \phi = 78.6779, \quad \sum \delta \cos^2 \phi = 7.6044.$$

The normal equations are

$$78.6779 = 1628C + 157.4A,$$

$$7.6044 = 157.4C + 21.5A.$$

Solving these, we find

$$C = \frac{\lambda_{K\beta}^2}{4a_0^2} = 0.0483651 \qquad \text{and} \qquad a_0 = 3.1653\ \text{Å}^*,$$

$$A = -0.000385.$$

The constant A, called the drift constant, is a measure of the total systematic error involved in the determination. The value of a_0 is based on the following wavelengths: 1.392218 Å* (Cu $K\beta$), 1.540562 Å* (Cu $K\alpha_1$), 1.544390 Å* (Cu $K\alpha_2$).

Noncubic systems

Cohen's method of determining lattice parameters is even more valuable when applied to noncubic substances, since, as we saw in Sec. 11–2, straightforward graphical extrapolation cannot be used when there is more than one lattice parameter involved. Cohen's method, however, provides a direct means of determining these parameters, although the equations are naturally more complex than those needed for cubic substances. For example, suppose that the substance involved is hexagonal. Then

$$\sin^2 \theta \text{ (true)} = \frac{\lambda^2}{4} \cdot \frac{4}{3} \cdot \frac{h^2 + hk + k^2}{a_0^2} + \frac{\lambda^2}{4} \cdot \frac{l^2}{c_0^2}$$

Table 11–2

Line	hkl	λ	α	ϕ	Observed		Normalized to $K\beta$	
					$\cos^2 \phi$	δ	$\cos^2 \phi$	δ
1	321	$K\alpha_1$	14	24.518°	0.82779	3.2	0.67605	2.6
2	321	$K\alpha_2$	14	24.193	0.83205	3.2	0.67616	2.6
3	411, 330	$K\beta$	18	21.167	0.86962	2.5	0.86962	2.5
4	400	$K\alpha_1$	16	13.302	0.94706	1.0	0.77345	0.8
5	400	$K\alpha_2$	16	12.667	0.95191	1.0	0.77356	0.8
6	420	$K\beta$	20	10.454	0.96708	0.7	0.96708	0.7

and

$$\sin^2\theta - \frac{\lambda^2}{3a_0^2}(h^2 + hk + k^2) - \frac{\lambda^2}{4c_0^2}(l^2) = D\sin^2 2\theta,$$

if the pattern is made in a Debye–Scherrer camera. By rearranging this equation and introducing new symbols, we obtain

$$\sin^2\theta = C\alpha + B\gamma + A\delta, \tag{11-26}$$

where

$$C = \frac{\lambda^2}{3a_0^2}, \quad \alpha = (h^2 + hk + k^2), \quad B = \frac{\lambda^2}{4c_0^2}, \quad \gamma = l^2,$$

$$A = \frac{D}{10}, \quad \text{and} \quad \delta = 10\sin^2 2\theta.$$

The values of C, B, and A, of which only the first two are really needed, are found from the three normal equations:

$$\begin{aligned}\sum \alpha \sin^2\theta &= C\sum \alpha^2 + B\sum \alpha\gamma + A\sum \alpha\delta,\\ \sum \gamma \sin^2\theta &= C\sum \alpha\gamma + B\sum \gamma^2 + A\sum \gamma\delta,\\ \sum \delta \sin^2\theta &= C\sum \alpha\delta + B\sum \delta\gamma + A\sum \delta^2.\end{aligned}$$

11-8 GENERAL

This chapter has dealt with powder methods exclusively. But there is also interest in measuring the lattice parameters of single crystals. Bond has devised a method capable of very high precision, provided the crystal specimen is of sufficiently high quality [11.5, 11.6]. With this method Baker *et al.* [11.7] were able to measure the thermal expansion coefficient of an MgO crystal over a temperature interval of only 1.0°C, and these investigators have attained a precision of one part in ten million in parameter measurements. Their measurements are made automatically with a computer-controlled goniometer of special design.

In work of the highest precision a small correction for refraction is sometimes applied (see, for example, [11.1]). The index of refraction for x-rays in matter differs very slightly from unity, so that an x-ray has slightly different wavelengths in air and in a crystal. The correction to a lattice parameter is less than 0.0001 Å.

Again, in work of the highest precision, it is necessary to control and report the temperature of the specimen during the parameter determination (Problem 11-1).

Investigators who wish to check their measurement techniques against those of a standardizing laboratory can buy a sample of silicon powder from the U.S. Bureau of Standards. These samples, known as Standard Reference Material 640, were made available in 1974 [11.8]. The Bureau states the weighted average of the lattice parameter of this material to be 5.43088 Å, with an estimated standard error of 3.5×10^{-5} Å.

In the measurement of any physical quantity one should always be aware of the distinction between precision and accuracy:

a) *Precision* is reproducibility,

b) *Accuracy* is the approach to the "true" value.

It is therefore quite possible to make highly precise but inaccurate measurements through the use, for example, of improperly calibrated instruments or because of inadequate correction for systematic errors. Even careful investigators usually overestimate the accuracy of their measurements, probably misled by their high precision.

For example, in the late 1950s the International Union of Crystallography distributed samples of a single lot of silicon powder to fifteen laboratories interested in precision parameter measurements [11.9]. The mean of the reported parameter values was 5.43054 Å at 25°C, referred to the same wavelength scale and corrected for refraction. The average precision claimed by the participating laboratories was ±0.00011 Å (2 parts per 100,000), but the spread in reported parameters (highest minus lowest) was 0.00067 Å (12 parts per 100,000). If the mean value of the parameter is taken as the true value for this lot of material, then some laboratories are about three times as inaccurate as their precision might suggest.

This result is not an isolated example. Similar interlaboratory comparisons have been made with respect to other physical measurements, sometimes on the same specimen, with similar results. We all think we are better than we are.

PROBLEMS

***11–1** The lattice parameter of copper is to be determined to a precision of ±0.0001 Å at 20°C. Within what limits must the temperature of the specimen be controlled if errors due to thermal expansion are to be avoided? The linear coefficient of thermal expansion of copper is 16.6×10^{-6} per °C.

11–2 The following data were obtained from a Debye–Scherrer pattern of a simple cubic substance, made with copper radiation. The given $\sin^2\theta$ values are for the $K\alpha_1$ lines only.

$h^2 + k^2 + l^2$	$\sin^2\theta$
38	0.9114
40	0.9563
41	0.9761
42	0.9980

Determine the lattice parameter a_0, accurate to four significant figures, by graphical extrapolation of a against $\cos^2\theta$.

***11–3** From the data given in Prob. 11–2, determine the lattice parameter to four significant figures by Cohen's method.

11–4 From the data given in Table 11–2, determine the lattice parameter of tungsten to five significant figures by graphical extrapolation of a against $\phi \tan \phi$.

***11–5** If the fractional error in the plane spacing d is accurately proportional to the function $(\cos^2\theta/\sin\theta + \cos^2\theta/\theta)$ over the whole range of θ, show that a plot of $\Delta \sin^2\theta$ against $\sin^2\theta$ has a maximum, as illustrated for a particular case by Fig. 10–1. At approximately what value of θ does the maximum occur?

11–6 Derive Eq. (11–13).

12
Phase-diagram Determination

12-1 INTRODUCTION

An alloy is a combination of two or more metals, or of metals and nonmetals. It may consist of a single phase or of a mixture of phases, and these phases may be of different types, depending only on the composition of the alloy and the temperature,* provided the alloy is at equilibrium. The changes in the constitution of the alloy produced by given changes in composition or temperature may be conveniently shown by means of a *phase diagram*, also called an equilibrium diagram or constitution diagram. It is a plot of temperature vs. composition, divided into areas wherein a particular phase or mixture of phases is stable. As such it forms a sort of map of the alloy system involved. Phase diagrams are therefore of great importance in metallurgy, and much time and effort have been devoted to their determination. In this chapter we will consider how x-ray methods can be used in the study of phase diagrams of binary systems.

X-ray methods are, of course, not the only ones which can be used in investigations of this kind. The two classical methods are thermal analysis and microscopic examination, and many diagrams have been determined by these means alone. X-ray diffraction, however, supplements these older techniques in many useful ways and provides, in addition, the only means of determining the crystal structures of the various phases involved. Most phase diagrams today are therefore determined by a combination of all three methods. In addition, measurements of other physical properties may be used to advantage in some alloy systems: the most important of these subsidiary techniques are measurements of the change in length and of the change in electric resistance as a function of temperature.

In general, the various experimental techniques differ in sensitivity, and therefore in usefulness, from one portion of the phase diagram to another. Thus, thermal analysis is the best method for determining the liquidus and solidus, including eutectic and peritectic horizontals, but it may fail to reveal the existence of eutectoid and peritectoid horizontals because of the sluggishness of some solid-state reactions or the small heat effects involved. Such features of the diagram are best determined by microscopic examination or x-ray diffraction, and the same applies to the determination of solvus (solid solubility) curves.

The general principles of phase-diagram determination are described by Taylor [G.19], with special emphasis on x-ray methods.

* The pressure on the alloy is another effective variable, but it is usually held constant at that of the atmosphere and may be neglected.

12-2 GENERAL PRINCIPLES

The key to the interpretation of the powder patterns of alloys is the fact that each phase produces its own pattern independently of the presence or absence of any other phase. Thus a single-phase alloy produces a single pattern while the pattern of a two-phase alloy consists of two superimposed patterns, one due to each phase.

Assume, for example, that two metals A and B are *completely soluble* in the solid state, as illustrated by the phase diagram of Fig. 12-1. The solid phase α, called a *continuous solid solution*, is of the substitutional type; it varies in composition, but not in crystal structure, from pure A to pure B, which must necessarily have the same structure. The lattice parameter of α also varies continuously from that of pure A to that of pure B. Since all alloys in a system of this kind consist of the same single phase, their powder patterns appear quite similar, the only effect of a change in composition being to shift the diffraction-line positions in accordance with the change in lattice parameter.

More commonly, the two metals A and B are only *partially soluble* in the solid state. The first additions of B to A go into solid solution in the A lattice, which may expand or contract as a result, depending on the relative sizes of the A and B atoms and the type of solid solution formed (substitutional or interstitial). Ultimately the solubility limit of B in A is reached, and further additions of B cause the precipitation of a second phase. This second phase may be a B-rich solid solution with the same structure as B, as in the alloy system illustrated by Fig. 12-2(a). Here the solid solutions α and β are called *primary solid solutions* or *terminal solid solutions*. Or the second phase which appears may have no connection with the B-rich solid solution, as in the system shown in Fig. 12-2(b). Here the effect of supersaturating α with metal B is to precipitate the phase designated γ. This phase is called an *intermediate solid solution* or *intermediate phase*. It usually has a crystal structure entirely different from that of either α or β, and it is separated from each of these terminal solid solutions, on the phase diagram, by at least one two-phase region.

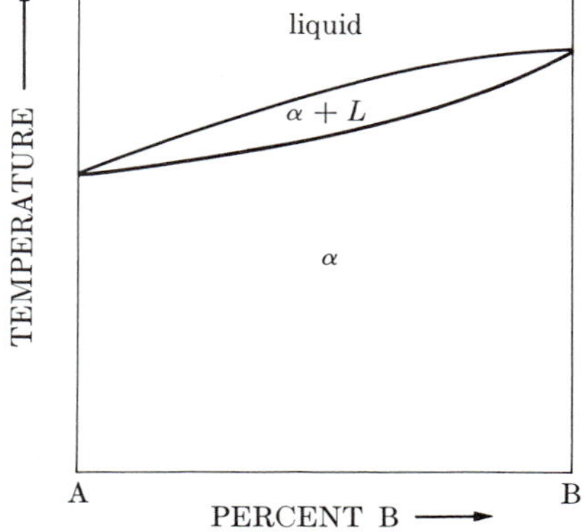

Fig. 12-1 Phase diagram of two metals, showing complete solid solubility.

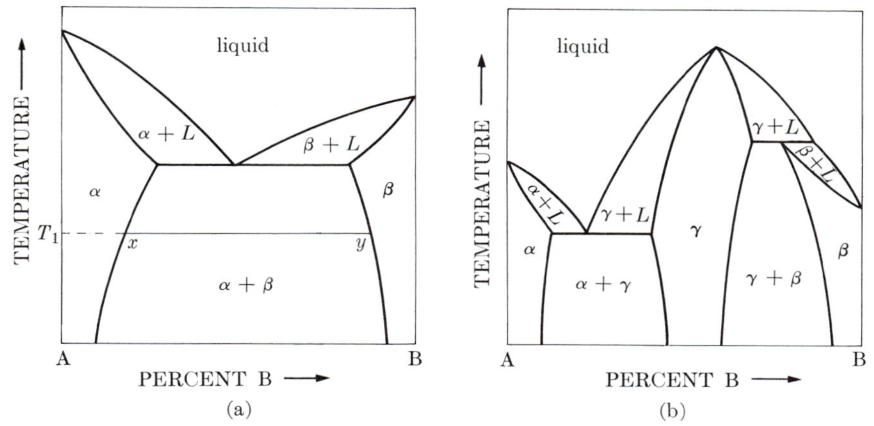

Fig. 12–2 Phase diagrams showing (a) partial solid solubility, and (b) partial solid solubility together with the formation of an intermediate phase.

Phase diagrams much more complex than those just mentioned are often encountered in practice, but they are always reducible to a combination of fairly simple types. When an unknown phase diagram is being investigated, it is best to make a preliminary survey of the whole system by preparing a series of alloys at definite composition intervals, say 5 or 10 atomic percent, from pure A to pure B. The powder pattern of each alloy and each pure metal is then prepared. These patterns may appear quite complex but, no matter what the complexities, the patterns may be unraveled and the proper sequence of phases across the diagram may be established, if proper attention is paid to the following principles:

1. *Equilibrium.* Each alloy must be at equilibrium at the temperature where the phase relations are being studied.

2. *Phase sequence.* A horizontal (constant temperature) line drawn across the diagram must pass through single-phase and two-phase regions alternately.

3. *Single-phase regions.* In a single-phase region, a change in composition generally produces a change in lattice parameter and therefore a shift in the positions of the diffraction lines of that phase.

4. *Two-phase regions.* In a two-phase region, a change in composition of the alloy produces a change in the relative amounts of the two phases but no change in their compositions. These compositions are fixed at the intersections of a horizontal "tie line" with the boundaries of the two-phase field. Thus, in the system illustrated in Fig. 12–2(a), the tie line drawn at temperature T_1 shows that the compositions of α and β at equilibrium at this temperature are x and y respectively. The powder pattern of a two-phase alloy brought to equilibrium at temperature T_1 will therefore consist of the superimposed patterns of α of composition x and β of composition y. The patterns of a series of alloys in the xy range will all contain the same diffraction lines at the same positions, but the intensity of the lines of the α phase relative to the intensity of the lines of the β phase will decrease in a regular manner as the concentration of B in the alloy changes from x to y, since this change in total composition decreases the amount of α relative to the amount of β.

These principles are illustrated with reference to the hypothetical alloy system shown in Fig. 12–3. This system contains two substitutional terminal solid solutions α and β, both assumed to be face-centered cubic, and an intermediate phase γ, which is body-centered cubic. The solubility of either A or B in γ is assumed to be negligibly small: the lattice parameter of γ is therefore constant in all alloys in which this phase appears. On the other hand, the parameters of α and β vary with composition in the manner shown by the lower part of Fig. 12–3. Since the B atom is assumed to be larger than the A atom, the addition of B expands the A lattice, and the parameter of α increases from a_1 for pure A to a_3 for a solution of composition x, which represents the limit of solubility of B in A at room temperature. In two-phase ($\alpha + \gamma$) alloys containing more than x percent B, the parameter of α remains constant at its saturated value a_3. Similarly, the addition of A to B causes the parameter of β to decrease from a_2 to a_4 at the solubility limit, and then remain constant in the two-phase ($\gamma + \beta$) field.

Calculated powder patterns are shown in Fig. 12–4 for the eight alloys designated by number in the phase diagram of Fig. 12–3. It is assumed that the alloys have been brought to equilibrium at room temperature by slow cooling. Examination of these patterns reveals the following:

1. Pattern of pure A (face-centered cubic).

2. Pattern of α almost saturated with B. The expansion of the lattice causes the lines to shift to smaller angles 2θ.

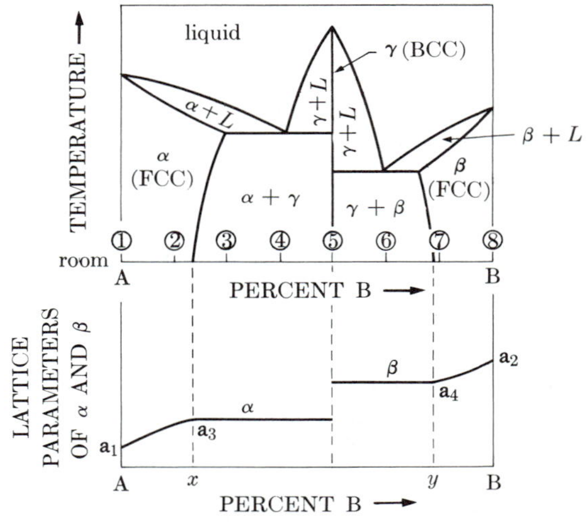

Fig. 12–3 Phase diagram and lattice constants of a hypothetical alloy system.

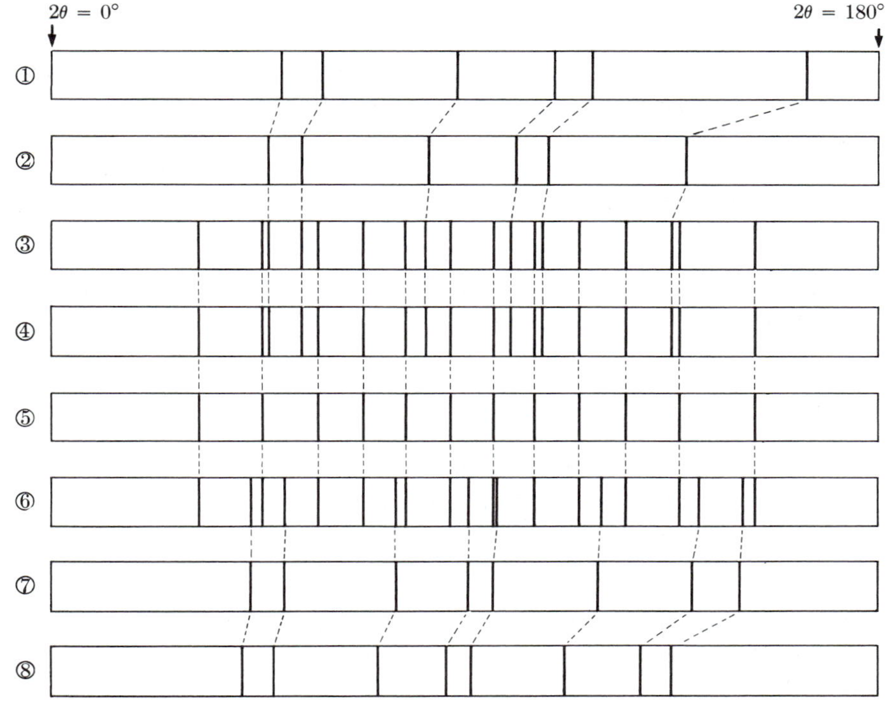

Fig. 12-4 Calculated powder patterns of alloys 1 to 8 in the alloy system shown in Fig. 12-3.

3. Superimposed patterns of α and γ. The α phase is now saturated and has its maximum parameter a_3.

4. Same as pattern 3, except for a change in the relative intensities of the two patterns which is not indicated on the drawing.

5. Pattern of pure γ (body-centered cubic).

6. Superimposed patterns of γ and of saturated β with a parameter of a_4.

7. Pattern of pure β with a parameter somewhat greater than a_4.

8. Pattern of pure B (face-centered cubic).

When an *unknown* phase diagram is being determined, the investigator must, of course, work in the reverse direction and deduce the sequence of phases across the diagram from the observed powder patterns. This is done by visual comparison of patterns prepared from alloys ranging in composition from pure A to pure B, and the previous example illustrates the nature of the changes which can be expected from one pattern to another. Corresponding lines in different patterns are identified by placing the films side by side as in Fig. 12-4 and noting which lines are common to the two patterns.* This may be difficult in some alloy systems where the phases involved have complex diffraction patterns, or where it is suspected that lines due to $K\beta$ radiation may be present in some patterns and not in others. It is important to remember that a diffraction pattern of a given phase is characterized not only by line positions but also by line intensities. This means that the presence of phase X in a mixture of phases cannot be proved merely by coincidence of the lines of phase X with a set of lines in the pattern of the mixture; the lines in the pattern of the mixture which coincide with the lines of phase X must also have the *same relative intensities* as the lines of phase X. The addition of one or more phases to a particular phase weakens the diffraction lines of that phase, simply by dilution, but it cannot change the intensities of those lines relative to one another. Finally, it should be noted that the crystal structure of a phase need not be known for the presence of that phase to be detected in a mixture: it is enough to know the positions and intensities of the diffraction lines of that phase.

Phase diagram determination by x-ray methods usually begins with a determination of the room-temperature equilibria. The first step is to prepare a series of alloys by melting and casting, or by melting and solidification in the melting crucible. The resulting ingots are homogenized at a temperature just below the solidus to remove segregation, and very slowly cooled to room temperature.† Powder specimens are then prepared by grinding or filing, depending on whether the alloy is brittle or not. If the alloy is brittle enough to be ground into powder, the resulting powder is usually sufficiently stress-free to give sharp diffraction lines. Filed powders, however, must be re-annealed to remove the stresses produced by plastic deformation during filing before they are ready for x-ray examination. Only relatively low temperatures are needed to relieve stresses, but the filings should again be slowly cooled, after the stress-relief anneal, to ensure equilibrium at room temperature. Screening is usually necessary to obtain fine enough particles for x-ray examination, and when two-phase alloys are being screened, the precautions mentioned in Sec. 6–3 should be observed.

After the room-temperature equilibria are known, a determination of the phases present at high temperatures can be undertaken. Powder specimens are sealed in small evacuated silica tubes, heated to the desired temperature long enough for equilibrium to be attained, and rapidly quenched. Diffraction patterns of the quenched powders are then made at room temperature. This method works very well in many alloy systems, in that the quenched powder retains the structure it had at the elevated temperature. In some alloys, however, phases stable at high temperature will decompose on cooling to room temperature, no matter how rapid

* Superposition of the two films is generally confusing and may make some of the weaker lines almost invisible. A better method of comparison consists of slitting each Debye–Scherrer film lengthwise down its center and placing the center of one film adjacent to the center of another. The curvature of the diffraction lines then does not interfere with the comparison of line positions.

† Slow cooling alone may not suffice to produce room-temperature equilibrium, which is often very difficult to achieve. It may be promoted by cold working and recrystallizing the cast alloy, in order to decrease its grain size and thus accelerate diffusion, prior to homogenizing and slow cooling.

the quench, and such phases can be studied only by means of a high-temperature camera or diffractometer.

The latter instrument is of particular value in work of this kind because it allows continuous observation of a diffraction line. For example, the temperature below which a high-temperature phase is unstable, such as a eutectoid temperature, can be determined by setting the diffractometer counter to receive a prominent diffracted beam of the high-temperature phase, and then measuring the intensity of this beam as a function of temperature as the specimen is slowly cooled. The temperature at which the intensity falls to that of the general background is the temperature required, and any hysteresis in the transformation can be detected by a similar measurement on heating.

12-3 SOLID SOLUTIONS

Inasmuch as solid solubility, to a greater or lesser extent, is so common between metals, we might digress a little at this point to consider how the various kinds of solid solutions may be distinguished experimentally. Irrespective of its extent or its position on the phase diagram, any solid solution may be classified as one of the following types, solely on the basis of its crystallography:

1. Interstitial.

2. Substitutional.
   a) Random.
   b) Ordered. (Because of its special interest, this type is described separately in Chap. 13.)
   c) Defect. (A very rare type.)

Information on specific solid solutions, particularly on the variation of lattice parameter with composition, is given by Pearson [G.16].

An *interstitial solid solution* of B in A is to be expected only when the B atom is so small compared to the A atom that it can enter the interstices of the A lattice without causing much distortion. As a consequence, about the only interstitial solid solutions of any importance in metallurgy are those formed between a metal and one of the elements, carbon, nitrogen, hydrogen, and boron, all of which have atoms less than 2 Å in diameter. The interstitial addition of B to A is always accompanied by an increase in the volume of the unit cell. If A is cubic, then the single lattice parameter a must increase. If A is not cubic, then one parameter may increase and the other decrease, as long as these changes result in an increase in cell volume. Thus, in austenite, which is an interstitial solid solution of carbon in face-centered cubic γ-iron, the addition of carbon increases the cell edge a. But in martensite, a metastable interstitial solid solution of carbon in α-iron, the c parameter of the body-centered tetragonal cell increases while the a parameter decreases, when carbon is added. These effects are illustrated in Fig. 12-5.

The density of an interstitial solid solution is given by the basic density equation

$$\rho = \frac{1.66042 \sum A}{V'}, \tag{3-7}$$

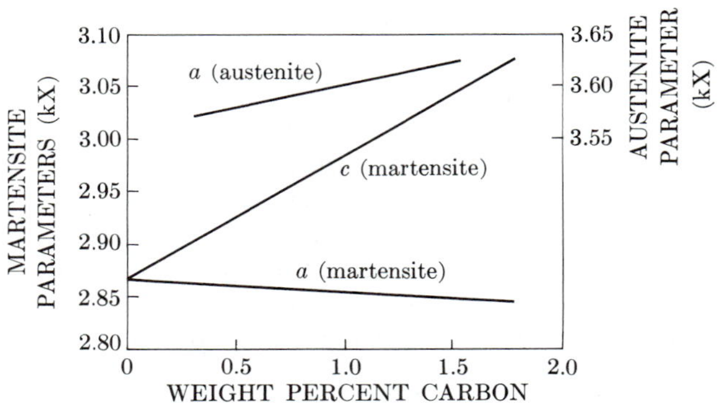

Fig. 12–5 Variation of martensite and austenite lattice parameters with carbon content. After Roberts [12.1].

where

$$\sum A = n_s A_s + n_i A_i; \tag{12-1}$$

n_s and n_i are numbers of solvent and interstitial atoms, respectively, per unit cell; and A_s and A_i are atomic weights of solvent and interstitial atoms, respectively. Note that the value of n_s is an integer and independent of the concentration of the interstitial element, and that n_i is normally a small fraction of unity.

The formation of a *random substitutional solid solution* of B and A may be accompanied by either an increase or a decrease in cell volume, depending on whether the B atom is larger or smaller than the A atom. In continuous solid solutions of ionic salts, such as KCl-KBr, the lattice parameter of the solution is directly proportional to the atomic percent solute present. This relationship, known as Vegard's law, is not strictly obeyed by metallic solid solutions and, in fact, there is no reason why it should be. However, it is often used as a sort of yardstick by which one solution may be compared with another. Figure 12–6 shows both positive and negative deviations from Vegard's law among solutions of face-centered cubic metals, and even larger deviations have been found in hexagonal close-packed solutions. In terminal and intermediate solid solutions, the lattice parameter may or may not vary linearly with the atomic percent solute and, when the variation *is* linear, the parameter found by extrapolating to 100 percent solute does not usually correspond to the atom size deduced from the parameter of the pure solute, even when allowance is made for a possible change in coordination number.

The density of a random substitutional solid solution is found from Eq. (3–7) with the $\sum A$ factor being given by

$$\sum A = n_{\text{solvent}} A_{\text{solvent}} + n_{\text{solute}} A_{\text{solute}}, \tag{12-2}$$

where n again refers to the number of atoms per cell and A to the atomic weight. But here $(n_{\text{solvent}} + n_{\text{solute}})$ is a constant integer, equal to the total number of atoms per cell. Whether a given solution is interstitial or substitutional may be decided by determining whether the x-ray density calculated according to Eq. (12–1) or that calculated according to Eq. (12–2) agrees with the directly measured density.

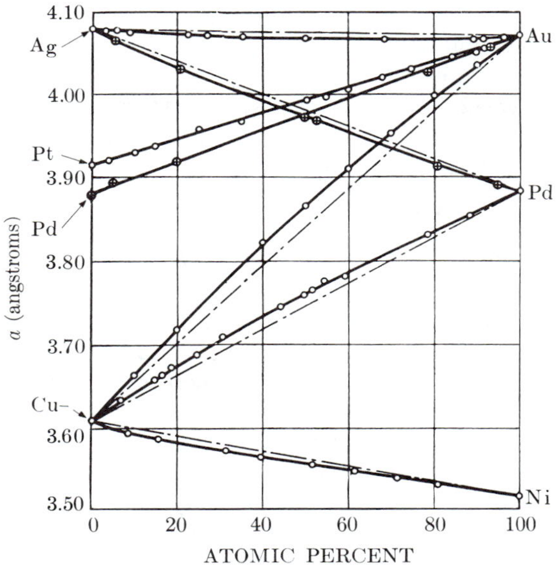

Fig. 12-6 Lattice parameters of some continuous solid solutions. Dot-dash lines indicate Vegard's law. Barrett [1.7].

Defect substitutional solid solutions are ones in which some lattice sites, normally occupied by atoms at certain compositions, are simply vacant at other compositions. Solutions of this type are rare among metals; the best-known example is the intermediate β solution in the nickel-aluminum system [p. 387, G.19]. A defect solution is disclosed by anomalies in the curves of density and lattice parameter vs. composition. Suppose, for example, that the solid solution of B in A is perfectly normal up to x percent B, but beyond that point a defect lattice is formed; i.e., further increases in B content are obtained, not by further substitution of B for A, but by dropping A atoms from the lattice to leave vacant sites. Under these circumstances, the density and parameter curves will show sudden changes in slope, or even maxima or minima, at the composition x. Furthermore, the x-ray density calculated according to Eq. (12-2) will no longer agree with the direct density simply because Eq. (12-2), as usually used, applies only to normal solutions where all lattice sites are occupied. The actual structure of a defect solid solution, including the proportion of vacant lattice sites at any given composition, can be determined by a comparison of the direct density with the x-ray density, calculated according to Eq. (12-2), and an analysis of the diffracted intensities.

12-4 DETERMINATION OF SOLVUS CURVES (DISAPPEARING-PHASE METHOD)

To return to the main subject of this chapter, we might now consider the methods used for determining the position of a solvus curve (solid solubility curve) on a phase diagram. Such a curve forms the boundary between a single-phase solid region and a two-phase solid region, and the single-phase solid may be a primary or intermediate solid solution.

One method of locating such curves is based on the "lever law." This law, with reference to Fig. 12–7 for example, states that the relative proportions of α and β in an alloy of composition z in equilibrium at temperature T_1 is given by the relative lengths of the lines zy and zx, or that

$$W_\alpha(z - x) = W_\beta(y - z), \qquad (12\text{–}3)$$

where W_α and W_β denote the relative weights of α and β if x, y, and z are expressed in weight percent. It follows from Eq. (12–3) that the weight fraction of β in the alloy varies linearly with composition from 0 at point x to 1 at point y. The intensity of any diffraction line from the β phase also varies from zero at x to a maximum at y, but the variation with weight percent B is not generally linear.* Nevertheless, this variation may be used to locate the point x. A series of alloys in the two-phase region is brought to equilibrium at temperature T_1 and quenched. From diffraction patterns made at room temperature, the ratio of the intensity I_β of a prominent line of the β phase to the intensity I_α of a prominent line of the α phase is plotted as a function of weight percent B. The composition at which the ratio I_β/I_α extrapolates to zero is taken as the point x. (Use of the ratio I_β/I_α rather than I_β alone eliminates the effect of any change which may occur in the intensity of the incident beam from one diffraction pattern to another. However, this ratio also varies nonlinearly with weight percent B.) Other points on the solvus curve are located by similar experiments on alloys quenched from other temperatures. This method is known, for obvious reasons, as the disappearing-phase method.

Since the curve of I_β/I_α vs. weight percent B is not linear, high accuracy in the extrapolation depends on having several experimental points close to the phase boundary which is being determined. The accuracy of the disappearing-phase method is therefore governed by the sensitivity of the x-ray method in detecting small amounts of a second phase in a mixture, and this sensitivity varies widely

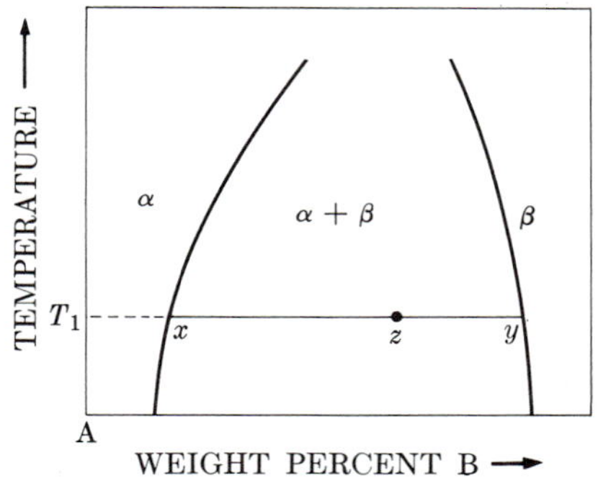

Fig. 12–7 Lever-law construction for finding the relative amounts of two phases in a two-phase field.

---

* The reasons for nonlinearity are discussed in Sec. 14–10.

from one alloy system to another. The intensity of a diffraction line depends on, among other things, the atomic scattering factor f, which in turn is almost directly proportional to the atomic number Z. Therefore, if A and B have nearly the same atomic number, the α and β phases will consist of atoms having almost the same scattering powers, and the intensities of the α and β diffraction patterns will also be roughly equal when the two phases are present in equal amounts. Under favorable circumstances such as these, an x-ray pattern can reveal the presence of less than 1 percent of a second phase. On the other hand, if the atomic number of B is considerably less than that of A, the intensity of the β pattern may be so much lower than that of the α pattern that a relatively large amount of β in a two-phase mixture will go completely undetected. This amount may exceed 50 percent in extreme cases, where the atomic numbers of A and B differ by some 70 or 80 units. Under such circumstances, the disappearing-phase x-ray method is practically worthless. On the whole, the microscope is superior to x-rays when the disappearing-phase method is used, inasmuch as the sensitivity of the microscope in detecting the presence of a second phase is generally very high and independent of the atomic numbers of the elements involved. However, this sensitivity does depend on the particle size of the second phase, and if this is very small, as it often is at low temperatures, the second phase may not be detectable under the microscope. Hence the method of microscopic examination is not particularly accurate for the determination of solvus curves at low temperatures.

Whichever technique is used to detect the second phase, the accuracy of the disappearing-phase method increases as the width of the two-phase region decreases. If the ($\alpha + \beta$) region is only a few percent wide, then the relative amounts of α and β will vary rapidly with slight changes in the total composition of the alloy, and this rapid variation of W_α/W_β will enable the phase boundary to be fixed quite precisely. This is true, for the x-ray method, even if the atomic numbers of A and B are widely different, because, if the ($\alpha + \beta$) region is narrow, the compositions of α and β do not differ very much and neither do their x-ray scattering powers.

12-5 DETERMINATION OF SOLVUS CURVES (PARAMETRIC METHOD)

As we have just seen, the disappearing-phase method of locating the boundary of the α field is based on a determination of the composition at which the β phase just disappears from a series of ($\alpha + \beta$) alloys. The parametric method, on the other hand, is based on observations of the α solid solution itself. This method depends on the fact, previously mentioned, that the lattice parameter of a solid solution generally changes with composition up to the saturation limit, and then remains constant beyond that point.

Suppose the exact location of the solvus curve shown in Fig. 12-8(a) is to be determined. A series of alloys, 1 to 7, is brought to equilibrium at temperature T_1, where the α field is thought to have almost its maximum width, and quenched to room temperature. The lattice parameter of α is measured for each alloy and plotted against alloy composition, resulting in a curve such as that shown in Fig. 12-8(b).

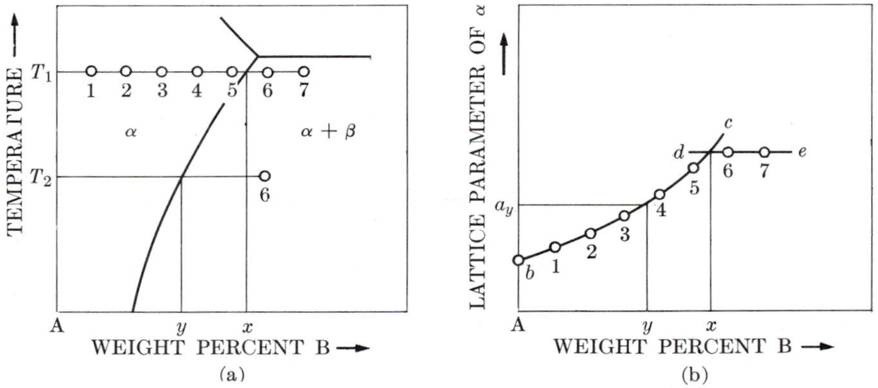

Fig. 12–8 Parametric method for determining a solvus curve.

This curve has two branches: an inclined branch *bc*, which shows how the parameter of α varies with the composition of α, and a horizontal branch *de*, which shows that the α phase in alloys 6 and 7 is saturated, because its lattice parameter does not change with change in alloy composition. In fact, alloys 6 and 7 are in a two-phase region at temperature T_1, and the only difference between them is in the amounts of saturated α they contain. The limit of the α field at temperature T_1 is therefore given by the intersection of the two branches of the parameter curve. In this way, we have located one point on the solvus curve, namely x percent B at T_1.

Other points could be found in a similar manner. For example, if the same series of alloys were equilibrated at temperature T_2, a parameter curve similar to Fig. 12–8(b) would be obtained, but its inclined branch would be shorter and its horizontal branch lower. But heat treatments and parameter measurements on all these alloys are unnecessary, once the parameter-composition curve of the solid solution has been established. Only one two-phase alloy is needed to determine the rest of the solvus. Thus, if alloy 6 is equilibrated at T_2 and then quenched, it will contain α saturated at that temperature. Suppose the measured parameter of α in this alloy is a_y. Then, from the parameter-composition curve, we find that α of parameter a_y contains y percent B. This fixes a point on the solvus at temperature T_2. Points on the solvus at other temperatures may be found by equilibrating the same alloy, alloy 6, at various temperatures, quenching, and measuring the lattice parameter of the contained α.

The parameter-composition curve, branch *bc* of Fig. 12–8(b), thus serves as a sort of master curve for the determination of the whole solvus. For a given precision of lattice parameter measurement, the accuracy with which the solvus can be located depends markedly on the slope of the parameter-composition curve. If this curve is nearly flat, i.e., if changes in the composition of the solid solution produce very small changes in parameter, then the composition, as determined from the parameter, will be subject to considerable error and so will the location of the solvus. However, if the curve is steep, just the opposite is true, and relatively crude parameter measurements may suffice to fix the location of the solvus quite

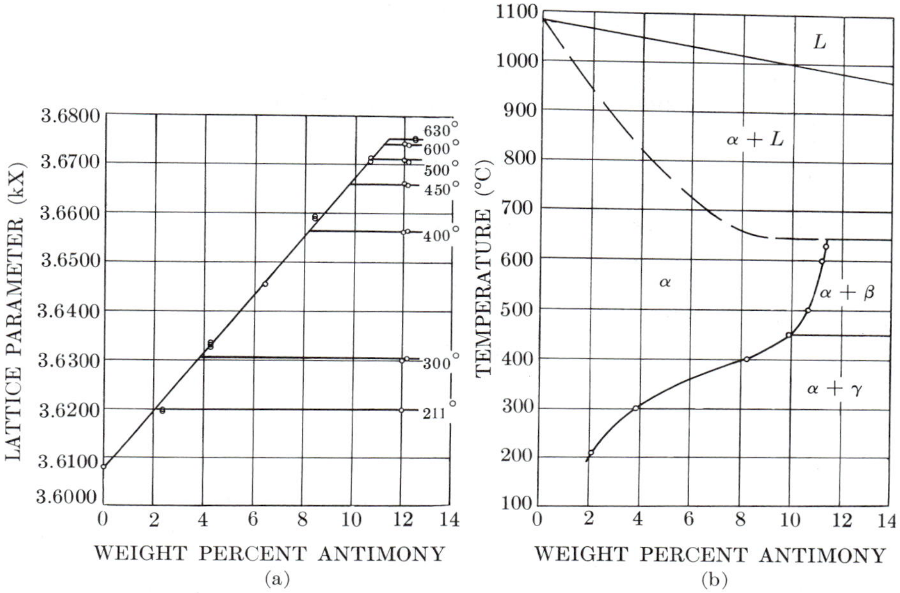

Fig. 12-9 Solvus curve determination in the copper-antimony system by the parametric method: (a) parameter versus composition curve; (b) solubility versus temperature curve. Mertz and Mathewson [12.2].

accurately. In the parametric method, precision in parameter measurement is more important than accuracy.

Figure 12-9 illustrates the use of the parametric method in determining the solid solubility of antimony in copper as a function of temperature. The sloping curve in (a) was found from parameter measurements made on a series of alloys, containing from 0 to about 12 weight percent Sb, equilibrated at 630°C. The horizontal lines represent the parameters of two-phase alloys, containing about 12 weight percent Sb, equilibrated at the temperatures indicated. The solvus curve constructed from these data is given in (b), together with adjoining portions of the phase diagram.

In most cases, the parametric method is more accurate than the disappearing-phase method, whether based on x-ray measurements or microscopic examination, in the determination of solvus curves at low temperatures. As mentioned earlier, both x-ray diffraction and microscopic examination may fail to disclose the presence of small amounts of a second phase, although for different reasons. When this occurs, the disappearing-phase method always results in a measured extent of solubility *higher* than the actual extent. But the parametric method, since it is based on measurements made on the phase whose range of solubility is being determined (the α phase), is not influenced by any property of the second phase (the β phase). The β phase may have an x-ray scattering power much higher or lower than that of the α phase, and the β phase may precipitate in the form of large particles or small ones, without affecting the parameter measurements made on the α phase.

Note that the parametric method is not confined to determining the extent of primary solid solutions, as in the examples given above. It may also be used to determine the solvus curves which bound an intermediate solid solution on the phase diagram. Note also that the parametric method may be employed even when the crystal structure of the α phase is so complex that its diffraction lines cannot be indexed. In this case, the plane spacing d corresponding to some high-angle line, or, even more directly, the 2θ value of the line, is plotted against composition and the resulting curve used in exactly the same way as a parameter-composition curve. In fact, the "parametric" method could be based on the measurement of any property of the solid solution which changes with the composition of the solid solution, e.g., its electric resistivity.

PROBLEMS

***12–1** Metals A and B form a terminal solid solution α, cubic in structure. The variation of the lattice parameter of α with composition, determined by quenching single-phase alloys from an elevated temperature, is found to be linear, the parameter varying from 3.6060 Å for pure A to 3.6140 Å in α containing 4.0 weight percent B. The solvus curve is to be determined by quenching a two-phase alloy containing 5.0 weight percent B from a series of temperatures and measuring the parameter of the contained α. How precisely must the parameter be measured if the solvus curve is to be located within ± 0.1 weight percent B at any temperature?

12–2 The two-phase alloy mentioned in Prob. 12–1, after being quenched from a series of temperatures, contains α having the following measured parameters:

Temperature	Parameter
100°C	3.6082 Å
200	3.6086
300	3.6091
400	3.6098
500	3.6106
600	3.6118

Plot the solvus curve over this temperature range. What is the solubility of B in A at 440°C?

13

Order-disorder Transformations

13-1 INTRODUCTION

In most substitutional solid solutions, the two kinds of atoms A and B are arranged more or less at random on the atomic sites of the lattice. In solutions of this kind the only major effect of a change in temperature is to increase or decrease the amplitude of thermal vibration. But, as noted in Sec. 2-7, there are some solutions which have this random structure only at elevated temperatures. When these solutions are cooled below a certain critical temperature T_c, the A atoms arrange themselves in an orderly, periodic manner on one set of atomic sites, and the B atoms do likewise on another set. The solution is then said to be *ordered* or to possess a *superlattice*. When this periodic arrangement of A and B atoms persists over very large distances in the crystal, it is known as *long-range order*. If the ordered solution is heated above T_c, the atomic arrangement becomes more or less random again and the solution is said to be *disordered*.

The change in atom arrangement which occurs on ordering produces changes in a large number of physical and chemical properties, and the existence of ordering may be inferred from some of these changes. However, the only conclusive evidence for a disorder-order transformation is a particular kind of change in the x-ray diffraction pattern of the substance. Evidence of this kind was first obtained by the American metallurgist Bain in 1923, for a gold-copper solid solution having the composition $AuCu_3$ [13.1]. Since that time, the same phenomenon has been discovered in many other alloy systems.

Data on the ordered structures found in particular alloys are given by Barrett and Massalski [G.25] and Pearson [G.16]. The theory of the diffraction phenomena involved is treated by Warren [G.30] and Guinier [G.21].

13-2 LONG-RANGE ORDER IN $AuCu_3$

The gold and copper atoms of $AuCu_3$, above a critical temperature of about 390°C, are arranged more or less at random on the atomic sites of a face-centered cubic lattice, as illustrated in Fig. 13-1(a). If the disorder is complete, the probability that a particular site is occupied by a gold atom is simply $\frac{1}{4}$, the atomic fraction of gold in the alloy, and the probability that it is occupied by a copper atom is $\frac{3}{4}$, the atomic fraction of copper. These probabilities are the same for every site and, considering the structure as a whole, we can regard each site as being occupied by a statistically "average" gold-copper atom. Below the critical tem-

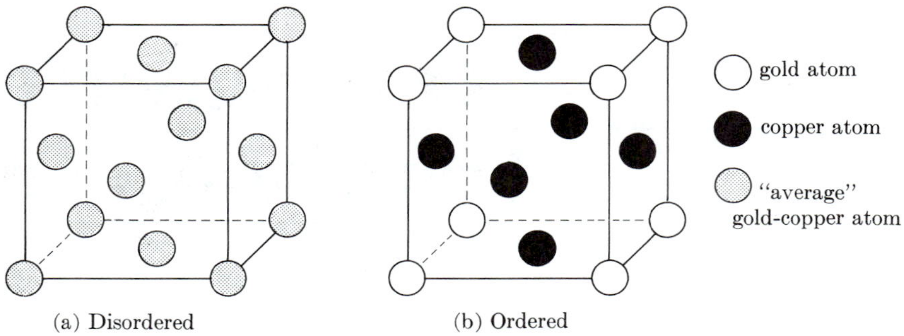

Fig. 13–1 Unit cells of the disordered and ordered forms of $AuCu_3$.

perature, the gold atoms in a perfectly ordered alloy occupy only the corner positions of the unit cube and the copper atoms the face-centered positions, as illustrated in Fig. 13–1(b). Both structures are cubic and have practically the same lattice parameters. Figure 13–2 shows how the two atomic arrangements differ on a particular lattice plane. The same kind of ordering has been observed in $PtCu_3$, $FeNi_3$, $MnNi_3$, and $(MnFe)Ni_3$.

What differences will exist between the diffraction patterns of ordered and disordered $AuCu_3$? Since there is only a very slight change in the size of the unit cell on ordering, and none in its shape, there will be practically no change in the positions of the diffraction lines. But the change in the positions of the atoms must necessarily cause a change in line intensities. We can determine the nature of these changes by calculating the structure factor F for each atom arrangement:

a) *Complete disorder*. The atomic scattering factor of the "average" gold-copper atom is given by

$$f_{av} = (\text{atomic fraction Au})\, f_{Au} + (\text{atomic fraction Cu})\, f_{Cu},$$

$$f_{av} = \tfrac{1}{4} f_{Au} + \tfrac{3}{4} f_{Cu}.$$

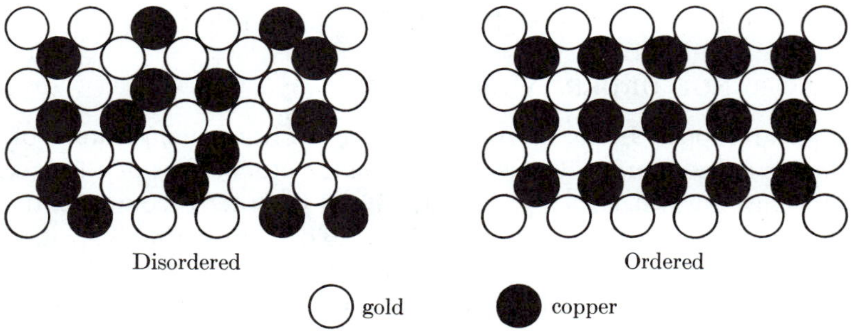

Fig. 13–2 Atom arrangements on a {100} plane, disordered and ordered $AuCu_3$.

There are four "average" atoms per unit cell, at 0 0 0, $\frac{1}{2}\,\frac{1}{2}$ 0, $\frac{1}{2}$ 0 $\frac{1}{2}$, and 0 $\frac{1}{2}\,\frac{1}{2}$. Therefore the structure factor is given by

$$F = \sum f e^{2\pi i(hu+kv+lw)},$$

$$F = f_{av}[1 + e^{\pi i(h+k)} + e^{\pi i(h+l)} + e^{\pi i(k+l)}].$$

By example (d) of Sec. 4-6, this becomes

$$F = 4f_{av} = (f_{Au} + 3f_{Cu}), \qquad \text{for } hkl \text{ unmixed,}$$

$$F = 0, \qquad \text{for } hkl \text{ mixed.}$$

We therefore find, as might be expected, that the disordered alloy produces a diffraction pattern similar to that of any face-centered cubic metal, say pure gold or pure copper. No reflections of mixed indices are present.

b) *Complete order*. Each unit cell now contains one gold atom, at 0 0 0, and three copper atoms, at $\frac{1}{2}\,\frac{1}{2}$ 0, $\frac{1}{2}$ 0 $\frac{1}{2}$, and 0 $\frac{1}{2}\,\frac{1}{2}$.

$$F = f_{Au} + f_{Cu}[e^{\pi i(h+k)} + e^{\pi i(h+l)} + e^{\pi i(k+l)}];$$

$$\left.\begin{aligned} F &= (f_{Au} + 3f_{Cu}), \qquad \text{for } hkl \text{ unmixed,} \\ F &= (f_{Au} - f_{Cu}), \qquad \text{for } hkl \text{ mixed.} \end{aligned}\right\} \tag{13-1}$$

The ordered alloy thus produces diffraction lines for all values of *hkl*, and its diffraction pattern therefore resembles that of a simple cubic substance. In other words, there has been a change of Bravais lattice on ordering; the Bravais lattice of the disordered alloy is face-centered cubic and that of the ordered alloy simple cubic.

The diffraction lines from planes of unmixed indices are called *fundamental lines*, since they occur at the same positions and with the same intensities in the patterns of both ordered and disordered alloys. The extra lines which appear in the pattern of an ordered alloy, arising from planes of mixed indices, are called *superlattice lines*, and their presence is direct evidence that ordering has taken place. The physical reason for the formation of superlattice lines may be deduced from an examination of Fig. 13-1. Consider reflection from the (100) planes of the disordered structure, and let an incident beam of wavelength λ make such an angle of incidence θ that the path difference between rays scattered by adjacent (100) planes is one whole wavelength. But there is another plane halfway between these two, containing, on the average, exactly the same distribution of gold and copper atoms. This plane scatters a wave which is therefore $\lambda/2$ out of phase with the wave scattered by either adjacent (100) plane and of exactly the same amplitude. Complete cancellation results and there is no 100 reflection. In the ordered alloy, on the other hand, adjacent (100) planes contain both gold and copper atoms, but the plane halfway between contains only copper atoms. The rays scattered by the (100) planes and those scattered by the midplanes are still exactly out of phase, but they now differ in amplitude because of the difference in scattering power of the gold and copper atoms. The ordered structure therefore produces a weak 100 reflection. And, as Eqs. (13-1) show, all the superlattice lines are much weaker than the fundamental lines, since their structure factors involve the difference, rather than the sum, of the atomic scattering factors of each atom. This effect is shown quite clearly in Fig. 13-3, where f and s are used to designate the fundamental and superlattice lines, respectively.

At low temperatures, the long-range order in $AuCu_3$ is virtually perfect but, as T_c is approached, some randomness sets in. This departure from perfect order can be described by means of the *long-range order parameter S*, defined as follows:

$$S = \frac{r_A - F_A}{1 - F_A}, \tag{13-2}$$

where r_A = fraction of A sites occupied by the "right" atoms, i.e., A atoms, and F_A = fraction of A atoms in the alloy. When the long-range order is perfect, $r_A = 1$ by definition, and therefore $S = 1$. When the atomic arrangement is completely random, $r_A = F_A$ and $S = 0$. For example, consider 100 atoms of $AuCu_3$, i.e., 25 gold atoms and 75 copper atoms. Suppose the ordering is not perfect and only 22 of these gold atoms are on "gold sites," i.e., cube corner positions, the other 3 being on "copper sites." Then, considering the gold atom as the A atom in Eq. (13-2), we find that $r_A = \frac{22}{25} = 0.88$ and $F_A = \frac{25}{100} = 0.25$. Therefore,

$$S = \frac{0.88 - 0.25}{1.00 - 0.25} = 0.84$$

describes the degree of long-range order present. The same result is obtained if we consider the distribution of copper atoms.

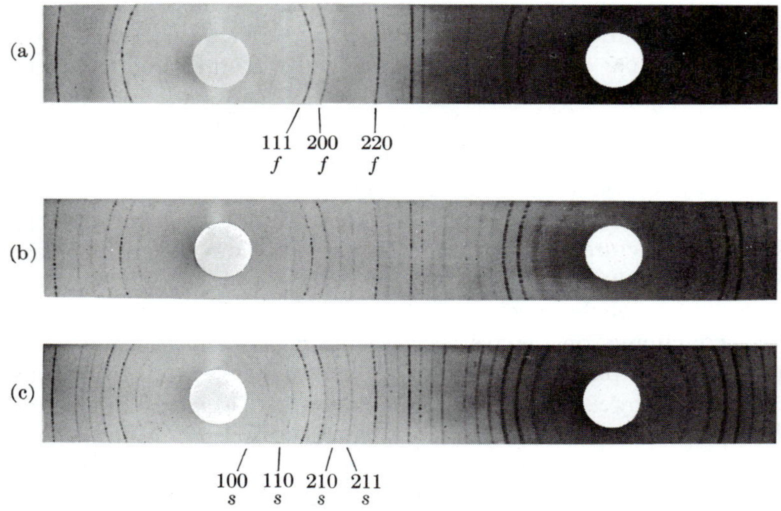

Fig. 13-3 Powder patterns of $AuCu_3$ (very coarse-grained) made with filtered copper radiation: (a) quenched from 440°C (disordered); (b) held 30 min at 360°C and quenched (partially ordered); (c) slowly cooled from 360°C to room temperature (completely ordered).

Any departure from perfect long-range order in a superlattice causes the superlattice lines to become weaker. It may be shown [G.30] that the structure factors of partially ordered $AuCu_3$ are given by

$$F = (f_{Au} + 3f_{Cu}), \qquad \text{for } hkl \text{ unmixed,}$$
$$F = S(f_{Au} - f_{Cu}), \qquad \text{for } hkl \text{ mixed.} \tag{13-3}$$

Comparing these equations with Eqs. (13-1), we note that only the superlattice lines are affected. But the effect is a strong one, because the intensity of a superlattice line is proportional to $|F|^2$ and therefore to S^2. For example, a decrease in order from $S = 1.00$ to $S = 0.84$ decreases the intensity of a superlattice line by about 30 percent. The weakening of superlattice lines by partial disorder is illustrated in Fig. 13-3. By comparing the integrated intensity ratio of a superlattice and fundamental line, we can determine S experimentally.

Values of S obtained in this way are shown in Fig. 13-4 as a function of the absolute temperature T, expressed as a fraction of the critical temperature T_c. For $AuCu_3$ the value of S decreases gradually, with increasing temperature, to about 0.8 at T_c and then drops abruptly to zero. Above T_c the atomic distribution is random and there are no superlattice lines. Recalling the approximate law of conservation of diffracted energy, already alluded to in Sec. 4-11, we might expect that the energy lost from the superlattice lines should appear in some form in the pattern of a completely disordered alloy. As a matter of fact it does, in the form of a weak diffuse background extending over the whole range of 2θ. This diffuse scattering is due to randomness, and is another illustration of the general law that any departure from perfect periodicity of atom arrangement results in some diffuse scattering at non-Bragg angles.

Von Laue showed that if two kinds of atoms A and B are distributed completely at random in a solid solution, then the intensity of the diffuse scattering produced is given by

$$I_D = k(f_A - f_B)^2, \tag{13-4}$$

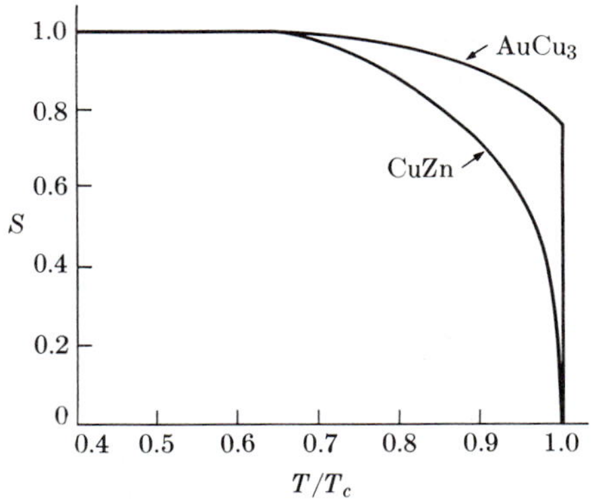

Fig. 13-4 Variation of the long-range order parameter S with temperature, for $AuCu_3$ and CuZn. ($AuCu_3$ data from Keating and Warren [13.2]; CuZn data from Chipman and Warren [13.3]).

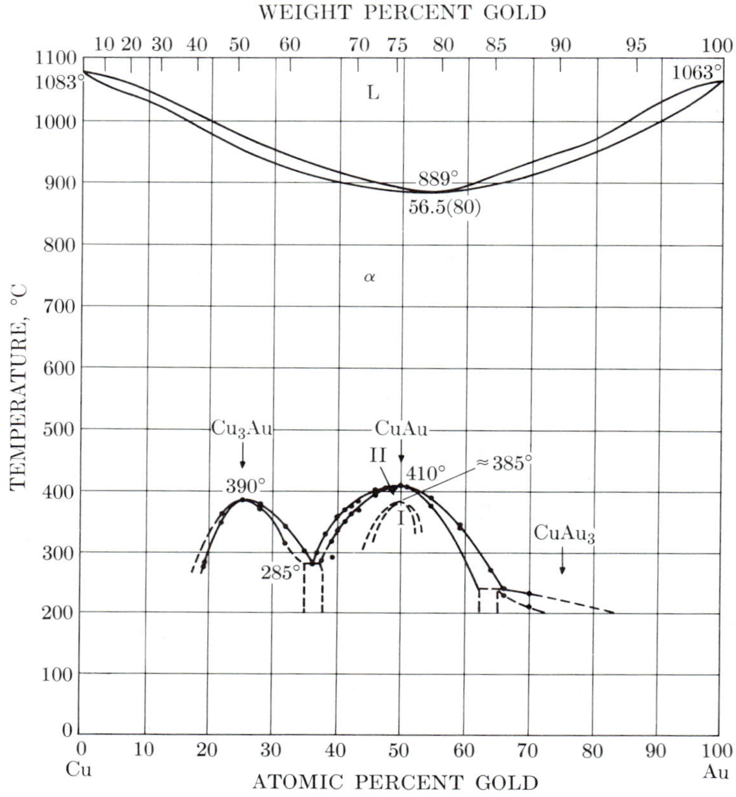

Fig. 13–5 Phase diagram of the copper-gold system. Hansen and Anderko [13.4].

where k is a constant for any one composition, and f_A and f_B are atomic scattering factors [G.30]. Both f_A and f_B decrease as $(\sin\theta)/\lambda$ increases, and so does their difference; therefore I_D is a maximum at $2\theta = 0$ and decreases as 2θ increases. This diffuse scattering is very difficult to measure experimentally. It is weak to begin with and is superimposed on other forms of diffuse scattering that may also be present, namely, Compton modified scattering, temperature-diffuse scattering, etc. It is worth noting, however, that Eq. (13–4) is quite general and applies to any random solid solution, whether or not it is capable of undergoing ordering at low temperatures. We will return to this point in Sec. 13–5.

Another aspect of long-range order that requires some mention is the effect of change in composition. Since the ratio of corner sites to face-centered sites in the $AuCu_3$ lattice is 1:3, it follows that perfect order can be attained only when the ratio of gold to copper atoms is also exactly 1:3. But ordering can also take place in alloys containing somewhat more, or somewhat less, than 25 atomic percent gold, as shown by the phase diagram of Fig. 13–5. In an ordered alloy containing somewhat more than 25 atomic percent gold, all the corner sites are occupied by gold atoms, and the remainder of the gold atoms occupy some of the face-centered sites normally occupied by copper atoms. Just the reverse is true for an alloy containing less than 25 atomic percent gold. But, as the phase diagram shows, there are limits to the variation in composition which the ordered lattice will accept without becoming unstable. In fact, if the gold content is increased to about 50 atomic percent, an entirely different ordered alloy, AuCu, can be formed.

13-3 OTHER EXAMPLES OF LONG-RANGE ORDER

Before considering the ordering transformation in AuCu, which is rather complex, we might examine the behavior of β-brass. This alloy is stable at room temperature over a composition range of about 46 to almost 50 atomic percent zinc, and so may be represented fairly closely by the formula CuZn. At high temperatures its structure is, statistically, body-centered cubic, with the copper and zinc atoms distributed at random. Below a critical temperature of about 460°C, ordering occurs; the cell corners are then occupied only by copper atoms and the cell centers only by zinc atoms, as indicated in Fig. 13-6. The ordered alloy therefore has the CsCl structure and its Bravais lattice is simple cubic. Other alloys which have the same ordered structure are CuBe, CuPd, and FeCo.

By calculations similar to those made in the previous section, the structure factors of β-brass, for the ideal composition CuZn, can be shown to be

$$\begin{aligned} F &= (f_{Cu} + f_{Zn}), &\quad \text{for } (h + k + l) \text{ even}, \\ F &= S(f_{Cu} - f_{Zn}), &\quad \text{for } (h + k + l) \text{ odd}. \end{aligned} \tag{13-5}$$

In other words, there are fundamental lines, those for which $(h + k + l)$ is even, which are unchanged in intensity whether the alloy is ordered or not. And there are superlattice lines, those for which $(h + k + l)$ is odd, which are present only in the pattern of an alloy exhibiting some degree of order, and then with an intensity which depends on the degree of order present.

Figure 13-4 indicates how the degree of long-range order in CuZn varies with the temperature. The order parameter for CuZn decreases continuously to zero as T approaches T_c, whereas for $AuCu_3$ it remains fairly high right up to T_c and

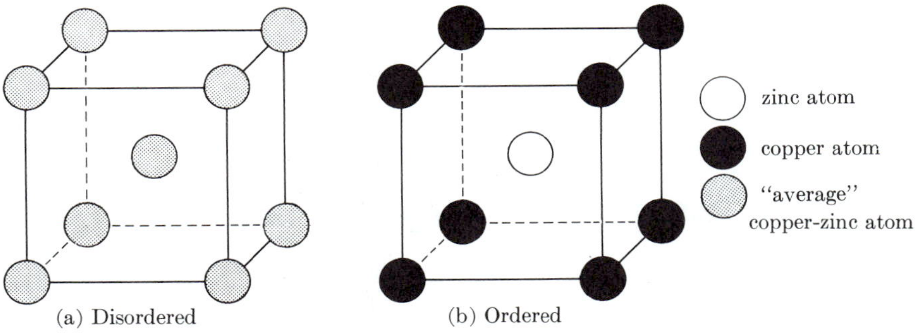

Fig. 13-6 Unit cells of the disordered and ordered forms of CuZn.

then drops abruptly to zero. There is also a notable difference in the velocity of the disorder-order transformation in these two alloys. The transformation in $AuCu_3$ is relatively so sluggish that the structure of this alloy at any temperature can be retained by quenching to room temperature, as evidenced by the diffraction patterns in Fig. 13–3, which were made at room temperature. In CuZn, on the other hand, ordering is so rapid that disorder existing at an elevated temperature cannot be retained at room temperature, no matter how rapid the quench. Therefore, any specimen of CuZn at room temperature can be presumed to be almost completely ordered. (The S vs. T/T_c curve for CuZn, shown in Fig. 13–4, was necessarily based on measurements made *at* temperature with a high-temperature diffractometer.)

Not all order-disorder transformations are as simple, crystallographically speaking, as those occurring in $AuCu_3$ and CuZn. Complexities are encountered, for example, in gold-copper alloys at or near the composition AuCu; these alloys become ordered below a critical temperature of about 410°C or lower, depending on the composition (see Fig. 13–5). Whereas the ratio of gold to copper atoms in $AuCu_3$ is 1:3, this ratio is 1:1 for AuCu, and the structure of ordered AuCu must therefore be such that the ratio of gold sites to copper sites is also 1:1. Two ordered forms are produced, depending on the ordering temperature, and these have different crystal structures:

a) Tetragonal AuCu, designated I, formed by slow cooling from high temperatures or by isothermal ordering below about 385°C. The unit cell is shown in Fig. 13–7(a). It is almost cubic in shape, since c/a equals about 0.93, and the gold and copper atoms occupy alternate (002) planes.

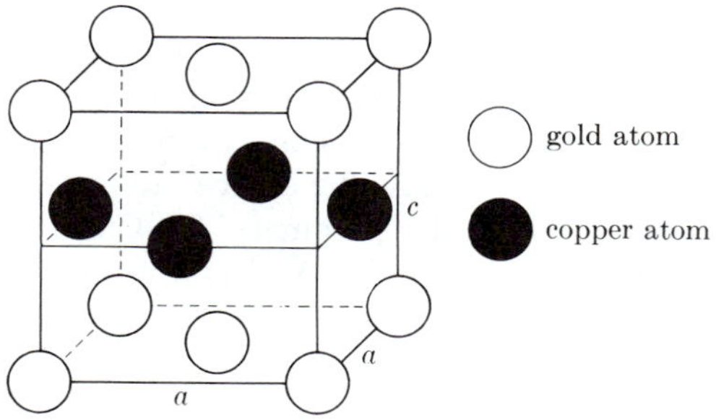

(a) I-tetragonal

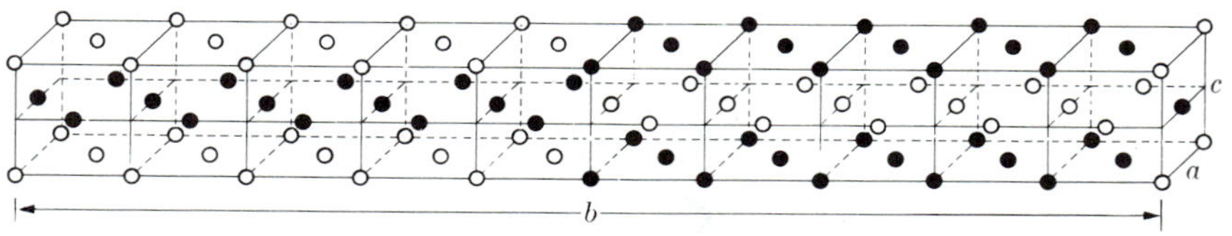

(b) II-orthorhombic

Fig. 13–7 Unit cells of the two ordered forms of AuCu.

b) Orthorhombic AuCu, designated II, formed by isothermal ordering between about 410° and 385°C. Its very unusual unit cell, shown in Fig. 13-7(b), is formed by placing ten tetragonal cells like that of I side by side and then translating five of them by the vectors **c/2** and **a/2** with respect to the other five. (Some distortion occurs, with the result that each of the ten component cells, which together make up the true unit cell, is not tetragonal but orthorhombic; i.e., b is not exactly ten times a, but equal to about $10.02a$. The c/a ratio is about 0.92.) The result is a structure in which the atoms in any one (002) plane are wholly gold for a distance of $b/2$, then wholly copper for a distance of $b/2$, and so on. It is called a long-period superlattice.

From a crystallographic viewpoint, there is a fundamental difference between the kind of ordering which occurs in $AuCu_3$ or CuZn, on the one hand, and that which occurs in AuCu, on the other. In $AuCu_3$ there is a change in Bravais lattice, but no change in crystal system, accompanying the disorder-order transformation: both the disordered and ordered forms are cubic. In AuCu, the ordering process changes both the Bravais lattice and the crystal system, the latter from cubic to tetragonal, AuCu(I), or orthorhombic, AuCu(II). These changes are due to changes in the symmetry of atom arrangement, because the crystal system to which a given structure belongs depends ultimately on the symmetry of that structure (see Sec. 2-4). In the gold-copper system, the disordered phase α is cubic, because the arrangement of gold and copper atoms on a face-centered lattice has cubic symmetry, in a statistical sense, at any composition. In $AuCu_3$, the ordering process puts the gold and copper atoms in definite positions in each cell (Fig. 13-1), but this arrangement still has cubic symmetry so the cell remains cubic. In ordered AuCu, on the other hand, to consider only the tetragonal modification, the atom arrangement is such that there is no longer three-fold rotational symmetry about directions of the form $\langle 111 \rangle$. Inasmuch as this is the minimum symmetry requirement for the cubic system, this cell [Fig. 13-7(a)] is not cubic. There is, however, four-fold rotational symmetry about [001], but not about [010] or [100]. The ordered form is accordingly tetragonal. The segregation of gold and copper atoms on alternate (002) planes causes c to differ from a, in this case in the direction of a small contraction of c relative to a, because of the difference in size between the gold and copper atoms. But even if c were equal to a, the cell shown in Fig. 13-7(a) would still be classified as tetragonal on the basis of its symmetry.

13-4 DETECTION OF SUPERLATTICE LINES

We have already seen that the intensity of a superlattice line from an ordered solid solution is much lower than that of a fundamental line. Will it ever be so low that the line cannot be detected? We can make an approximate estimate by ignoring the variation in multiplicity factor and Lorentz-polarization factor from line to line, and assuming that the relative integrated intensities of a superlattice and fundamental line are given by their relative $|F|^2$ values. For fully ordered $AuCu_3$, for example, we find from Eqs. (13-1) that

$$\frac{\text{Intensity (superlattice line)}}{\text{Intensity (fundamental line)}} \approx \frac{|F|_s^2}{|F|_f^2} = \frac{(f_{Au} - f_{Cu})^2}{(f_{Au} + 3f_{Cu})^2}. \tag{13-6}$$

At $(\sin\theta)/\lambda = 0$ we can put $f = Z$ and, since the atomic numbers of gold and copper are 79 and 29, respectively, Eq. (13–6) becomes, for small scattering angles,

$$\frac{I_s}{I_f} \approx \frac{(79 - 29)^2}{[79 + 3(29)]^2} \approx 0.09.$$

Superlattice lines are therefore only about one-tenth as strong as fundamental lines, but they can still be detected without any difficulty, as shown by Fig. 13–3.

But in CuZn, even when fully ordered, the situation is much worse. The atomic numbers of copper and zinc are 29 and 30, respectively, and, making the same assumptions as before, we find that

$$\frac{I_s}{I_f} \approx \frac{(f_{\text{Cu}} - f_{\text{Zn}})^2}{(f_{\text{Cu}} + f_{\text{Zn}})^2} \approx \frac{(29 - 30)^2}{(29 + 30)^2} \approx 0.0003.$$

This ratio is so low that the superlattice lines of ordered CuZn can be detected by x-ray diffraction only under very special circumstances. (The powder pattern of this alloy, ordered or disordered, ordinarily appears to be that of a body-centered cubic substance.) The same is true of any superlattice of elements A and B which differ in atomic number by only one or two units, because the superlattice-line intensity is generally proportional to $(f_A - f_B)^2$.

There is one way, however, of increasing the intensity of a superlattice line relative to that of a fundamental line, when the two atoms involved have almost the same atomic numbers, and that is by the proper choice of the incident wavelength. The atomic scattering factor f of any element is usually considered to be independent of the wavelength of the scattered radiation, as long as the quantity $(\sin\theta)/\lambda$ is constant. This is not quite true. When the incident wavelength λ is nearly equal to the wavelength λ_K of the K absorption edge of the scattering element, then the atomic scattering factor of that element may be several units lower than it is when λ is very much shorter than λ_K. This change in f is the *anomalous dispersion* mentioned in Sec. 4–3. Physically, it can be regarded as a resonance effect in which the oscillations of the K electrons, which are the cause of the scattering, are disturbed when the radiation being scattered has a frequency near that at which K electrons can be actually ejected from the atom. If we put f_0 = atomic scattering factor for $\lambda \ll \lambda_K$ (this is the usual value as tabulated, for example, in Appendix 12) and Δf = change in f_0 when λ is near λ_K, then the quantity $f = f_0 + \Delta f$ gives the value of the atomic scattering factor when λ is near λ_K. Figure 13–8 shows approximately how Δf varies with λ/λ_K, and this curve may be used to estimate the correction Δf which must be applied for any particular combination of wavelength and scattering element.

Strictly speaking, Δf depends also on the atomic number of the scattering element, which means that a different correction curve is required for every element. But the variation of Δf with Z is not very large, and Fig. 13–8, which is computed for an element of medium atomic number (about 50), can be used with fairly good accuracy as a master correction curve for any element. Figure 13–8 was calculated from data in James [G.7, p. 608]. Anomalous dispersion is also discussed by Guinier [G.21]. Values of Δf as a function of Z at constant λ, for five characteristic wavelengths, can be calculated from data in [G.11, Vol. 4].

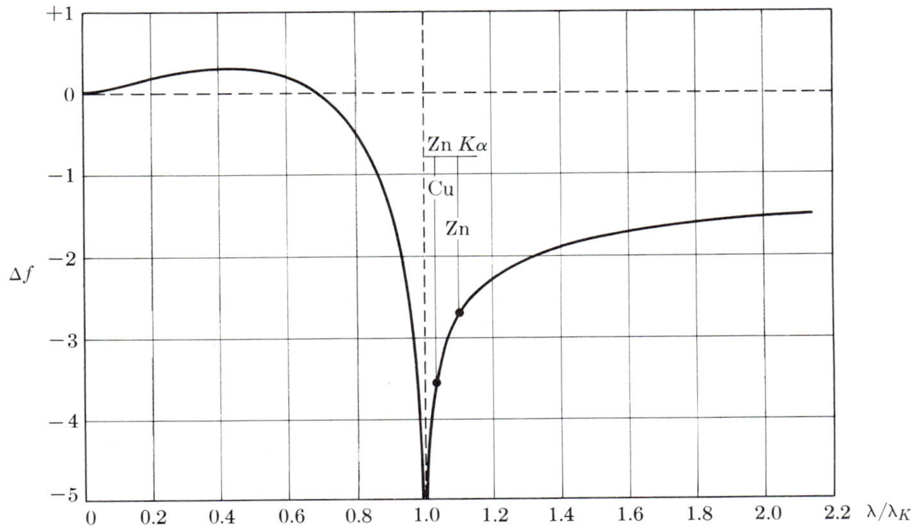

Fig. 13-8 Variation of the scattering-factor correction Δf with λ/λ_K. See text for details. The two points on the curve show the corrections for the scattering of Zn $K\alpha$ radiation by copper and zinc atoms.

When λ/λ_K is less than about 0.8, the correction is practically negligible. When λ/λ_K exceeds about 1.6, the correction is practically constant and independent of small variations in λ_K. But when λ is near λ_K, the slope of the correction curve is quite steep, which means that the Δf correction can be quite different for two elements of nearly the same atomic number. By taking advantage of this fact, we can often increase the intensity of a superlattice line above its normal value.

For example, if ordered CuZn is examined with Mo $K\alpha$ radiation, λ/λ_K is 0.52 for the copper atom and 0.55 for the zinc atom. The value of Δf is then about +0.3 for either atom, and the intensity of a superlattice line would be proportional to $[(29 + 0.3) - (30 + 0.3)]^2 = 1$ at low values of 2θ. Under these circumstances the line would be invisible in the presence of the usual background. But if Zn $K\alpha$ radiation is used, λ/λ_K becomes 1.04 and 1.11 for the copper and zinc atoms, respectively, and Fig. 13-8 shows that the corrections are -3.6 and -2.7, respectively. The superlattice-line intensity is now proportional to $[(29 - 3.6) - (30 - 2.7)]^2 = 3.6$, which is large enough to permit detection of the line. It was by means of Zn $K\alpha$ radiation that Jones and Sykes [13.5] first detected ordering in CuZn. Cu $K\alpha$ radiation also offers some advantage over Mo $K\alpha$, but not so large an advantage as Zn $K\alpha$, and order in CuZn can be detected with Cu $K\alpha$ only if crystal-monochromated radiation is used or if the specimen is a single crystal.

To a very good approximation, the change in atomic scattering factor Δf is independent of scattering angle and therefore a constant for all lines on the diffraction pattern.

By thus taking advantage of this anomalous change in scattering factor near an absorption edge, we are really pushing the x-ray method about as far as it will go. A better tool for the detection of order in alloys of metals of nearly the same atomic number is neutron diffraction (Appendix 2). Two elements may differ in atomic number by only one unit and yet their neutron scattering powers may be entirely different, a situation conducive to high superlattice-line intensity.

13-5 SHORT-RANGE ORDER AND CLUSTERING

Above the critical temperature T_c long-range order disappears and the atomic distribution becomes more or less random. This is indicated by the absence of superlattice lines from the powder pattern. But careful analysis of the diffuse scattering which forms the background of the pattern shows that perfect randomness is not attained. Instead, there is a greater than average tendency for *unlike* atoms to be nearest neighbors. This condition is known as *short-range order*.

For example, when perfect long-range order exists in $AuCu_3$, a gold atom located at 0 0 0 is surrounded by 12 copper atoms at $\frac{1}{2}\,\frac{1}{2}\,0$ and equivalent positions (see Fig. 13–1), and any given copper atom is surrounded by 8 copper and 4 gold atoms. This kind of grouping is a direct result of the existing long-range order, which also requires that gold atoms be on corner sites and copper atoms on face-centered sites. Above T_c this order breaks down and, if the atomic distribution became truly random, a given gold atom might be found on either a corner or face-centered site. It would then have only $\frac{3}{4}(12) = 9$ copper atoms as nearest neighbors, since on the average 3 out of 4 atoms in the solution are copper. Actually, it is observed that some short-range order exists above T_c: at 405°C, for example, which is 15°C above T_c, there are on the average about 11.0 copper atoms around any given gold atom [13.6].

This is a quite general effect. Any solid solution which exhibits long-range order below a certain temperature exhibits some short-range order above that temperature. Above T_c the degree of short-range order decreases as the temperature is raised; i.e., increasing thermal agitation tends to make the atomic distribution more and more random. One interesting fact about short-range order is that it has also been found to exist in solid solutions which do *not* undergo long-range ordering at low temperatures, such as gold-silver and gold-nickel solutions.

We can imagine another kind of departure from randomness in a solid solution, namely, a tendency of *like* atoms to be close neighbors. This effect is known as *clustering*, and it has been observed in aluminum-silver and aluminum-zinc solutions. In fact, there is probably no such thing as a perfectly random solid solution. All real solutions probably exhibit either short-range ordering or clustering to a greater or lesser degree, simply because they are composed of unlike atoms with particular forces of attraction or repulsion operating between them. Short-range order, however, is far more common than clustering.

The degree of short-range order or clustering may be defined in terms of a suitable parameter, just as long-range order is, and the value of this parameter may be related to the diffraction effects produced. The general nature of these effects is illustrated in Fig. 13–9, where the intensity of the diffuse scattering is

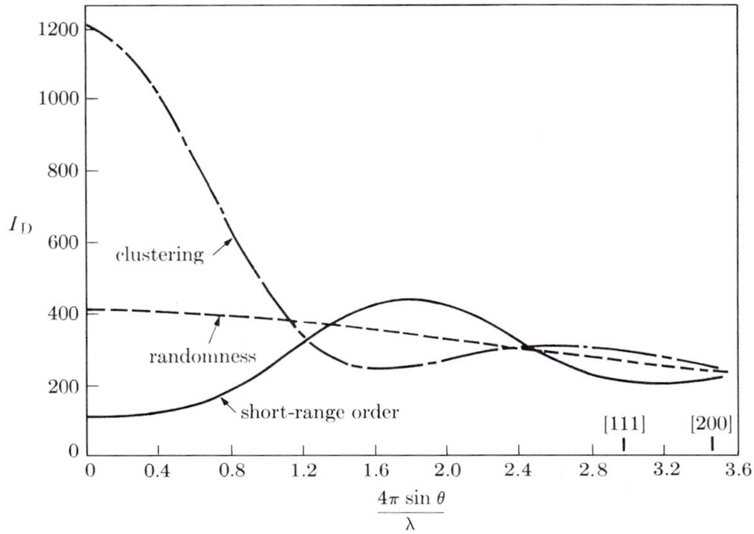

Fig. 13–9 Calculated intensity I_D of diffuse scattering in powder patterns of solid solutions (here, the face-centered cubic alloy Ni_4Au) which exhibit complete randomness, short-range order, and clustering. The short-range order curve is calculated on the basis of one additional unlike neighbor over the random configuration, and the clustering curve on the basis of one less unlike neighbor. Warren and Averbach [13.7].

plotted, not against 2θ, but against a function of $\sin \theta$. (The fundamental lines are not included in Fig. 13–9 because their intensity is much too high compared with the diffuse scattering shown, but the positions of two of them, 111 and 200, are indicated on the abscissa.) If the atomic distribution is perfectly random, the scattered intensity decreases gradually as 2θ or $\sin \theta$ increases from zero, in accordance with Eq. (13–4). If short-range order exists, the scattering at small angles becomes less intense and low broad maxima occur in the scattering curve; these maxima are usually located at the same angular positions as the sharp superlattice lines formed by long-range ordering. Clustering causes strong scattering at low angles.

These effects, however, are all very weak and are masked by the other forms of diffuse scattering which are always present. As a result, the details shown in Fig. 13–9 are never observed in an ordinary powder pattern made with filtered radiation. To disclose these details and so learn something about the structure of the solid solution, it is necessary to use strictly monochromatic radiation and, preferably, single-crystal specimens, and to make allowances for the other forms of diffuse scattering, chiefly temperature-diffuse and Compton modified, that are always present.

PROBLEMS

13–1 A Debye–Scherrer pattern is made with Cu $K\alpha$ radiation of $AuCu_3$ quenched from a temperature T_1. The ratio of the integrated intensity of the 420 line to that of the 421 line is found to be 4.38. Calculate the value of the long-range order parameter S at temperature T_1. (Take the lattice parameter of $AuCu_3$ as 3.75 Å. Ignore the small difference between the Lorentz-polarization factors for these two lines and the corrections to the atomic scattering factors mentioned in Sec. 13–4.)

***13–2** Calculate the ratio of the integrated intensity of the 100 superlattice line to that of the 110 fundamental line for fully ordered β-brass, if Cu $K\alpha$ radiation is used. Estimate the corrections to the atomic scattering factors from Fig. 13–8. The lattice parameter of β-brass (CuZn) is 2.95 Å.

13–3 (a) What is the Bravais lattice of AuCu(I), the ordered tetragonal modification?

b) Calculate the structure factors for the disordered and ordered (tetragonal) forms of AuCu.

c) On the basis of the calculations made in (b) and a consideration of the change in the c/a ratio, describe the qualitative differences between the powder patterns of the ordered (tetragonal) and disordered forms of AuCu.

14

Chemical Analysis by X-ray Diffraction

14-1 INTRODUCTION

A given substance always produces a characteristic diffraction pattern, whether that substance is present in the pure state or as one constituent of a mixture of substances. This fact is the basis for the diffraction method of chemical analysis. *Qualitative analysis* for a particular substance is accomplished by identification of the pattern of that substance. *Quantitative analysis* is also possible, because the intensities of the diffraction lines due to one phase of a mixture depend on the proportion of that phase in the specimen.

The particular advantage of diffraction analysis is that it discloses the presence of a substance *as that substance actually exists in the sample*, and not in terms of its constituent chemical elements. For example, if a sample contains the compound A_xB_y, the diffraction method will disclose the presence of A_xB_y as such, whereas ordinary chemical analysis would show only the presence of elements A and B. Furthermore, if the sample contained both A_xB_y and A_xB_{2y}, both of these compounds would be disclosed by the diffraction method, but chemical analysis would again indicate only the presence of A and B.* To consider another example, chemical analysis of a plain carbon steel reveals only the amounts of iron, carbon, manganese, etc., which the steel contains, but gives no information regarding the phases present. Is the steel in question wholly martensitic, does it contain both martensite and austenite, or is it composed only of ferrite and cementite? Questions such as these can be answered by the diffraction method. Another rather obvious application of diffraction analysis is in distinguishing between different allotropic modifications of the same substance: solid silica, for example, exists in one amorphous and six crystalline modifications, and the diffraction patterns of these seven forms are all different.

Diffraction analysis is therefore useful whenever it is necessary to know the state of chemical combination of the elements involved or the particular phases in which they are present. As a result, the diffraction method has been widely applied

---

* Of course, if the sample contains only A and B, and if it can be safely assumed that each of these elements is wholly in a combined form, then the presence of A_xB_y and A_xB_{2y} can be demonstrated by calculations based on the amounts of A and B in the sample. But this method is not generally applicable, and it usually involves a prior assumption as to the constitution of the sample. For example, a determination of the total amounts of A and B present in a sample composed of A, A_xB_y, and B cannot, in itself, disclose the presence of A_xB_y, either qualitatively or quantitatively.

for the analysis of such materials as ores, clays, refractories, alloys, corrosion products, wear products, industrial dusts, etc. Compared with ordinary chemical analysis, the diffraction method has the additional advantages that it is usually much faster, requires only a very small sample, and is nondestructive.

Detailed treatments of chemical analysis by x-ray diffraction are given by Klug and Alexander [G.39] and Zwell and Danko [14.1]. Nenadic and Crable [14.24] have reviewed diffraction methods of determining quartz, asbestos, and talc in industrial dusts; all of these minerals can cause lung disease.

QUALITATIVE ANALYSIS

14-2 BASIC PRINCIPLES

The powder pattern of a substance is characteristic of that substance and forms a sort of fingerprint by which the substance may be identified. If we had on hand a collection of diffraction patterns for a great many substances, we could identify an unknown by preparing its diffraction pattern and then locating in our file of known patterns one which matched the pattern of the unknown exactly. The collection of known patterns has to be fairly large, if it is to be at all useful, and then pattern-by-pattern comparison in order to find a matching one becomes out of the question.

What is needed is a system of classifying the known patterns so that the one which matches the unknown can be located quickly. Such a system was devised by Hanawalt in 1936. Any one powder pattern is characterized by a set of *line positions* 2θ and a set of relative line *intensities* I. But the angular positions of the lines depend on the wavelength used, and a more fundamental quantity is the spacing d of the lattice planes forming each line. Hanawalt therefore decided to describe each pattern by listing the d and I values of its diffraction lines, and to arrange the known patterns in decreasing values of d for the strongest line in the pattern. This arrangement made possible a search procedure which would quickly locate the desired pattern. In addition, the problem of solving the pattern was avoided and the method could be used even when the crystal structure of the substance concerned was unknown.

14-3 POWDER DIFFRACTION FILE

The task of building up a collection of known patterns was initiated by Hanawalt, Rinn, and Frevel [14.2] at the Dow Chemical Company; they obtained and classified diffraction data on some 1000 different substances. It soon became apparent that these data were of great potential value to a wide range of industries and, beginning in 1941, several technical societies, including the American Society for Testing and Materials, began to cooperate in acquiring and disseminating diffraction data. From 1941 to 1969 the ASTM published and sold an increasing volume of data in the form of 3 × 5 in. file cards, one card for each pattern. Since 1969 this activity has been carried out by the Joint Committee on Powder Diffraction Standards (JCPDS) [14.3] with the cooperation of ten American, Canadian, British, and French societies. In 1976 the Powder Diffraction File (PDF) comprised some 26,000 diffraction patterns in 26 sets, with a new set of about 2000 patterns being added each year. The substances included are elements, alloys, inorganic compounds, minerals, organic compounds, and organometallic compounds.

Hanawalt Method

Since more than one substance can have the same, or nearly the same, d value for its strongest line and even its second strongest line, Hanawalt decided to characterize each substance by the d values of its *three* strongest lines, namely d_1, d_2, and d_3 for the strongest, second-strongest, and third-strongest line, respectively. The values of d_1, d_2, and d_3, together with relative intensities, are usually sufficient to characterize the pattern of an unknown and enable the corresponding pattern in the file to be located. Originally, in each section of the ASTM file, the cards were arranged in groups characterized by a certain range of d_1 spacings. Within each group, e.g., the group covering d_1 values from 2.29 to 2.25 Å, the cards were arranged in decreasing order of d_2 values, rather than d_1 values. When several substances in the same group had identical d_2 values, the order of decreasing d_3 values was followed. The groups themselves were arranged in decreasing order of their d_1 ranges. However, as the number of cards grew, the direct card search became unwieldy, and it was recognized that it would become more cumbersome as the file became larger. Accordingly, the JCPDS decided to number the cards randomly, or chronologically, in each set, rather than according to the value of d_1, and to use the Hanawalt Search Manual to replace the group-arranged card sets.

A typical card from the JCPDS file is reproduced in Fig. 14-1. These cards, grouped in inorganic and organic sections, are available in the following forms:

Cards. 3 × 5 in. (8 × 13 cm) cards for storage in ordinary file drawers.

Microfiche. Cards are photographically reduced and printed on 4 × 6 in. (10 × 15 cm) sheets of photographic film (microfiche). With 116 cards on each microfiche, storage space is much reduced but a microfiche reader is required.

Books. Cards from Sets 1 to 18 are reproduced, three to a page, in book form. All cards from these sets, inorganic and organic, are included in eight 6 × 9 in. (15 × 23 cm) volumes. A separate volume of cards for minerals is available.

Because a particular pattern is almost impossible to locate by a direct search of the total card file, it is necessary to use the search manuals that accompany the file. The same manuals are also necessary to access cards in microfiche and book form, because all data cards are now arranged in order of their file number and not in terms of d spacings. The manuals are of two kinds, according to the method of arranging the data:

1. *Alphabetical.* Substances are listed alphabetically by name. After the name are given the chemical formula, the d values and relative intensities (as subscripts) of the three strongest lines, the file number of the card, and the fiche number. (Starting with Set 26, the fiche number will be omitted.) All entries are fully cross-indexed; i.e., both "sodium chloride" and "chloride: sodium" are listed. This

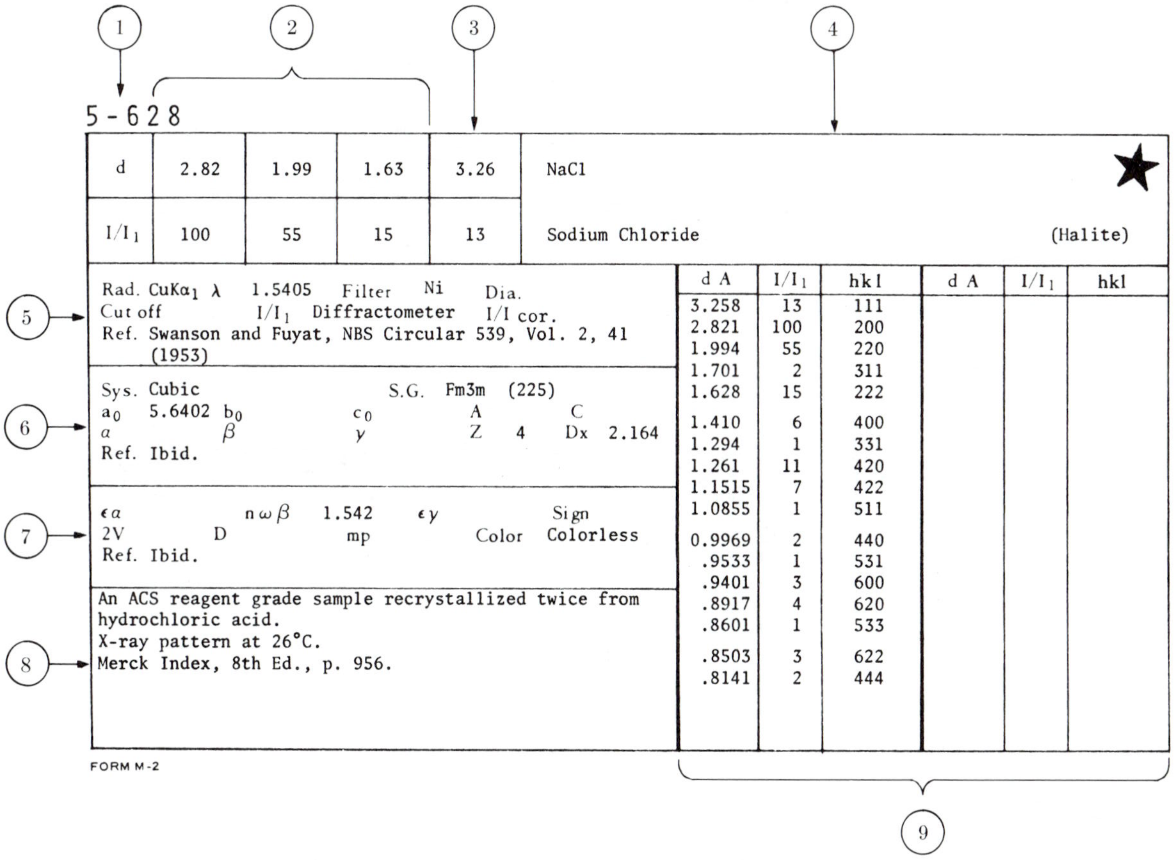

Fig. 14–1 Standard 3 × 5 in. JCPDS diffraction data card (card 628 from Set 5) for sodium chloride. Appearing on the card are 1 (file number), 2 (three strongest lines), 3 (lowest-angle line), 4 (chemical formula and name of substance), 5 (data on diffraction method used), 6 (crystallographic data), 7 (optical and other data), 8 (data on specimen), and 9 (diffraction pattern). Intensities are expressed as percentages of I_1, the intensity of the strongest line on the pattern. Most cards have a symbol in the upper right corner indicating the quality of the data: ✶ (high quality), *i* (lines indexed, intensities fairly reliable), *c* (calculated pattern), and *o* (low reliability). (Courtesy of Joint Committee on Powder Diffraction Standards.)

manual is useful if the investigator has any knowledge of the chemical composition of the sample.

2. *Numerical.* Substances are listed in terms of the *d* spacings of their three strongest lines, and *d* spacings and intensities are given for a total of eight lines for each substance. Also included are the chemical formula, file number, and fiche number. Each substance is listed three times, once with the three strongest lines listed in the usual order $d_1d_2d_3$, again in the order $d_2d_3d_1$, and finally in the order $d_3d_1d_2$. All entries are divided into groups according to the first spacing listed; the arrangement within each group is in decreasing order of the second spacing listed, The purpose of these additional listings (second-strongest line first and third-strongest line first) is to enable the user to match an unknown with an entry in the search manual even when complicating factors have altered the relative intensities of the three strongest lines of the unknown. These complications are usually due to the presence of more than one phase in the sample. This leads to additional lines and even superimposed lines. Use of the numerical search manual requires no knowledge of the chemical composition of the sample.

Fink Method

The Fink method of numerical searching of the data file relies more on *d* spacings than on intensities. It was originally designed for use with electron diffraction patterns, where observed line intensities are not always directly related to structure and therefore not always a reliable guide to identification.

In the Fink-method search manual, which covers inorganic compounds only, *d* spacings and intensities are listed for the eight strongest lines of each substance. The order of listing is determined by the four strongest lines. Each substance is listed four times in Set 24 and later (six times in earlier sets), with one of the four strongest lines listed first and the others following in descending order of *d* values.

14-4 PROCEDURE

Identification of the unknown begins with making its diffraction pattern. Sample preparation should result in fine grain size and in a minimum of preferred orientation, which can cause relative line intensities to differ markedly from their normal values. Note also that relative line intensities depend to some extent on wavelength; this should be kept in mind if the observed pattern is compared with one in the data file made with a different wavelength. Most of the patterns in the file were made with Cu $K\alpha$ radiation, except those for iron-bearing substances.

The pattern may be recorded with a Debye–Scherrer camera, Guinier camera, or diffractometer. Here again, line intensities depend on the apparatus. In particular, absorption effects cause high-angle lines on a Debye–Scherrer pattern to be stronger, relative to low-angle lines, than on a diffractometer recording, as shown in Sec. 4–10.

After the pattern of the unknown is prepared, the plane spacing *d* corresponding to each line on the pattern is calculated, or obtained from tables which give *d* as a function of 2θ for various characteristic wavelengths. Alternatively, a scale may be constructed which gives d directly as a function of line position when laid on the film or diffractometer chart; the accuracy obtainable by such a scale, although not very high, is generally sufficient for identification purposes. If the diffraction pattern has been obtained on film, relative line intensities are usually estimated by eye, on a scale running from 100 for the strongest line down to 10 or 5 for the weakest. On a diffractometer recording the intensity is taken as the maximum intensity measured above background.

After the experimental values of d and I/I_1 are tabulated, the unknown can be identified by the following procedure:

1. Locate the proper d_1 group in the numerical search manual.
2. Read down the second column of d values to find the closest match to d_2. (In comparing experimental and tabulated d values, always allow for the possibility that either set of values may be in error by ± 0.01 Å.)
3. After the closest match has been found for d_1, d_2, and d_3, compare their relative intensities with the tabulated values.
4. When good agreement has been found for the lines listed in the search manual, locate the proper data card in the file, and compare the d and I/I_1 values of all the observed lines with those tabulated. When full agreement is obtained, identification is complete.

14-5 EXAMPLES OF ANALYSIS

The unknown may consist of one or more phases, but the search procedure is initially the same for either case.

Single Phase

When the unknown is a single phase, the search procedure is relatively straightforward. Consider, for example, the pattern described in Table 14-1. It was obtained with Cu $K\alpha$ radiation and a Debye-Scherrer camera; line intensities were estimated. The experimental values of d_1, d_2, and d_3 are 2.82, 1.99, and 1.63 Å,

Table 14-1

Pattern of Unknown

d(Å)	I/I_1	d(Å)	I/I_1
3.25	10	1.00	20
2.82	100	0.95	5
2.18	5	0.94	20
1.99	60	0.89	20
1.71	5	0.86	5
1.63	30	0.85	20
1.42	20	0.82	10
1.25	30	0.79	10
1.15	30	0.78	20
1.09	5		

respectively. By examining the numerical search manual we find that the strongest line falls within the 2.84–2.80 Å group of d_1 values. Inspection of the listed d_2 values in this group discloses seventeen substances having d_2 = 1.99 Å, but only four of these have d_1 = 2.82 Å. The data on these four are shown in Table 14–2 in the form given in the manual. Of these four only NaCl has d_3 = 1.63 Å, and we also note that the intensities listed for the three strongest lines of this substance agree fairly well with the observed intensities; so do the data for the other five lines listed in the manual. We then turn to file card 5-628, reproduced in Fig. 14–1, and compare the complete pattern given there with the observed one. In general the agreement is good, but there are some discrepancies, and these must be resolved before the identification as NaCl is accepted. These discrepancies are:

1. We observe a very weak line with d = 2.18 Å, not listed on the file card. Calculation by Eq. (10–1) shows that this line is a 220 reflection of Cu $K\beta$ radiation not removed by the filter. (We would then expect that the even stronger 200 reflection would also produce a $K\beta$ line. But calculation shows that this $K\beta$ line falls on the 111 $K\alpha$ line and is therefore not seen separately.)

2. The 331 line (d = 1.294 Å) listed on the file card is not observed. However, its intensity is very low.

3. Two high-angle lines (d = 0.79 and 0.78 Å) are observed but are not listed on the file card. However, these can be indexed as 711-551 and 640 lines, which are the next expected lines after 444 in the FCC sequence.

4. After the fifth observed line, observed intensities on the Debye–Scherrer pattern are all higher than those on the file pattern, which was obtained with a diffractometer. These differences are due to absorption effects, as mentioned in Sec. 14–4.

Because the discrepancies are now understood, we can conclude that the specimen is NaCl. However, not all identifications are as positive as this example. A few unexplained weak lines are often assumed to be due to an unknown impurity; the identification of the main constituent is then somewhat tentative, and its

Table 14–2

Portion of JCPDS Search Manual

Spacings and Intensities	Substance	File Number	Fiche Number
2.82_9 1.99_9 $2.26x$ 1.61_9 1.51_9 1.49_9 3.57_8 2.66_8	$(ErSe_2)Q$	19-443	I-106-F 6
* **$2.82x$ 1.99_6 1.63_2** 3.26_1 1.26_1 1.15_1 1.41_1 0.89_1	NaCl	5-628	I- 18-F 8
i **2.82_4 1.99_4 $1.54x$** 1.20_4 1.19_4 2.44_3 5.62_2 4.89_2	$(NH_4)_2WO_2Cl_4$	22- 65	I-145-D 12
$2.82x$ 1.99_8 1.26_3 1.63_2 1.15_2 0.94_1 0.89_1 1.41_1	$(BePd)2C$	18-225	I- 90-D 1

Extracted from the numerical search manual (Inorganic Compounds, Hanawalt Method). The three strongest lines appear in bold-face type. Intensities (rounded off) are shown by the suffix x = 100 and the subscripts 9, 8, 7... = 90, 80, 70.... Certain symbols sometimes follow the chemical formula to describe crystal structure; Q above means base-centered orthorhombic, and $2C$ means simple cubic with two atoms per cell. (Courtesy of the Joint Committee for Powder Diffraction Standards.)

reliability depends on the judgement and experience of the investigator. See also Sec. 14–6.

Because small d spacings correspond to back-reflection lines, they are measurable with greater precision than large d spacings (Sec. 11–1). Spacings smaller than about 1 Å should therefore be computed to the nearest 0.001 Å, rather than to 0.01 Å as in Table 14–1, in order to permit a better match with data in the file.

Mixture of Phases

The analysis is now more complex, but not impossible. Consider the pattern in Table 14–3, obtained with Cu $K\alpha$ radiation and a diffractometer, for which $d_1 = 2.09$ Å, $d_2 = 2.47$ Å, and $d_3 = 1.80$ Å. Examination of the numerical index in the d_1 group 2.09 to 2.05 Å reveals several substances having d_2 values near 2.47 Å, but in no case do the three strongest lines, taken together, agree with those of the unknown. This impasse suggests that the unknown is actually a mixture of phases, and that we are incorrect in assuming that the three strongest lines in the pattern of the unknown are all due to the same substance. Suppose we assume that the strongest line ($d = 2.09$ Å) and the second-strongest line ($d = 2.47$ Å) are formed by two different phases, and that the third-strongest line ($d = 1.80$ Å) is due to, say, the first phase. In other words, we will assume that $d_1 = 2.09$ Å and $d_2 = 1.80$ Å for one phase. A search of the same group of d_1 values, but now in the vicinity of $d_2 = 1.80$ Å, discloses agreement between the three strongest lines of the pattern of copper, file number 4-0836, and three lines in the pattern of our unknown. Turning to card 4-0836, we find good agreement between all lines of the copper pattern, described in Table 14–4, with the starred lines in Table 14–3, the pattern of the unknown.

One phase of the mixture is thus shown to be copper, providing we can account for the remainder of the lines as due to some other substance. These remaining lines are listed in Table 14–5. By multiplying all the observed intensities by a normalizing factor of 1.39, we increase the intensity of the strongest line to 100. We then search the index and card file in the usual way and find that these remaining lines agree with the pattern of cuprous oxide, Cu_2O, which is given at the right of Table 14–5. The unknown is thus shown to be a mixture of copper and cuprous oxide.

Table 14–3

Pattern of Unknown

d(Å)	I/I_1	d(Å)	I/I_1
3.01	5	1.22	4
2.47	72	1.08*	20
2.13	28	1.04*	3
2.09*	100	0.98	5
1.80*	52	0.91*	4
1.50	20	0.83*	8
1.29	9	0.81*	10
1.28*	18		

Table 14–4

Pattern of Copper

d(Å)	I/I_1
2.088	100
1.808	46
1.278	20
1.0900	17
1.0436	5
0.9038	3
0.8293	9
0.8083	8

Table 14-5

Remainder of pattern of unknown			Pattern of Cu_2O	
	I/I_1			
d(Å)	Observed	Normalized	d(Å)	I/I_1
3 01	5	7	3.020	9
2.47	72	100	2.465	100
2.13	28	39	2.135	37
			1.743	1
1.50	20	28	1.510	27
1.29	9	13	1.287	17
1.22	4	6	1.233	4
			1.0674	2
0.98	5	7	0.9795	4
			0.9548	3
			0.8715	3
			0.8216	3

The analysis of mixtures becomes still more difficult when a line from one phase is superimposed on a line from another, and when this composite line is one of the three strongest lines in the pattern of the unknown. The usual procedure then leads only to a very tentative identification of one phase, in the sense that agreement is obtained for some d values but not for all the corresponding intensities. This in itself is evidence of line superposition. Such patterns can be untangled by separating out lines which agree in d value with those of phase X, the observed intensity of any superimposed lines being divided into two parts. One part is assigned to phase X, and the balance, together with the remaining unidentified lines, is treated as in the previous example.

Computer Searching

The procedure for searching the data file for a pattern or patterns matching that of the unknown is essentially one of matching numbers. This task is ideally suited to the computer. As the number of phases in the unknown increases beyond two, manual searching becomes very difficult and time consuming; computer searching is then more efficient.

Johnson and Vand [14.4] have prepared a widely adopted search program written in FORTRAN. It has successfully identified up to six phases in a mixture in less than two minutes.

Computer search procedures may be initiated in two ways:

1. Users with in-house computers may lease from the JCPDS [14.3] magnetic tapes on which are stored all the diffraction data of the current card file. Johnson–Vand search programs for use on IBM 360/370, Univac 1100 series, or CDC 6000/7000 series computers are supplied with the tapes.

2. Those who wish to access a remote computer by teletype may lease from the JCPDS the Diffraction Data Tele-Search called *2dTS*, which provides access to the magnetic-tape version of the data file.

Computer searching is not immune to errors originating in the pattern of the unknown or in the data file. If the unknown contains 3 to 6 phases, the computer may produce from 10 to 50 matching patterns, depending on the width Δd of the "window" selected by the user; this window is the range of d values about a given experimental value within which a match is judged acceptable. From the excessive number of possible matches provided by the computer, the user must select the most probable ones, usually on the basis of some knowledge of the unknown's chemical composition.

Note the possibility of completely automatic sample identification by using the tape output of an automatic diffractometer as the input to a computer searching the data file [7.6].

14-6 PRACTICAL DIFFICULTIES

In theory, the Hanawalt method should lead to the positive identification of any substance whose diffraction pattern is included in the card file. In practice, various difficulties arise, and these are usually due either to errors in the diffraction pattern of the unknown or to errors in the card file.

Errors of the first kind, those affecting the observed positions and intensities of the diffraction lines, have been discussed in various parts of this book and need not be reexamined here. However, the possibility of abnormal intensities due to preferred orientation needs continual emphasis. This condition is particularly common in surface deposits on metals, such as oxide or sulphide layers. If the deposit is very thin, its removal may not yield enough material for examination. The diffractometer is then an ideal instrument for examining the deposit *in situ*, because of the shallow penetration of x-rays. Very thin deposits can yield good diffraction lines and the analyst can often make a positive identification, even in the presence of pronounced preferred orientation, if he has some knowledge of crystal morphology and of the chemical composition of the deposit. Thus, a very thin deposit on sheet steel was identified as graphite by a single strong basal-plane reflection and the knowledge that graphite crystals are normally oriented with their basal planes more or less parallel to the substrate.

These remarks on abnormal intensities are not meant to suggest that successful use of the Hanawalt method requires extremely accurate intensity measurements. If reasonable care has been taken to minimize preferred orientation, then it is often enough to be able to list the diffraction lines in the correct *order* of decreasing intensity.

Errors in the card file itself are generally more serious, because they may go undetected by the investigator and lead to mistaken identifications. Many such errors existed in the earlier editions of the file, but critical editing by members of the JCPDS has resulted in the removal of incorrect data and in the labeling of doubtful data as being of low reliability. Moreover, the JCPDS supports continuing work at the National Bureau of Standards [14.5] and elsewhere, in which the diffraction patterns of pure, well-characterized specimens are carefully prepared and the data added to the file. Most of the high-quality, "starred" patterns in the card file, such as that in Fig. 14–1, have been obtained at the NBS.

Whenever any doubt exists in the investigator's mind as to the validity of a particular identification, he should prepare his own standard pattern. Thus, if the unknown has been tentatively identified as substance X, the pattern of pure X should be prepared under exactly the same experimental conditions used for the pattern of the unknown. Comparison of the two patterns will furnish positive proof, or disproof, of identity.

The Hanawalt method fails completely, of course, when the unknown is a substance not listed in the card file, or when the unknown is a mixture and the component to be identified is not present in sufficient quantity to yield a good diffraction pattern. The latter effect can be quite troublesome, and, as mentioned in Sec. 12-4, mixtures may be encountered which contain more than 50 percent of a particular component without the pattern of that component being visible in the pattern of the mixture.

QUANTITATIVE ANALYSIS (SINGLE PHASE)

14-7 CHEMICAL ANALYSIS BY PARAMETER MEASUREMENT

The lattice parameter of a binary solid solution of B in A depends only on the percentage of B in the alloy, as long as the solution is unsaturated. This fact can be made the basis for chemical analysis by parameter measurement. All that is needed is a parameter vs. composition curve, such as curve *bc* of Fig. 12-8(b), which can be established by measuring the lattice parameter of a series of previously analyzed alloys. This method has been used in diffusion studies to measure the change in concentration of a solution with distance from the original interface. Its accuracy depends entirely on the slope of the parameter-composition curve. In alpha brasses, which can contain from 0 to about 40 percent zinc in copper, an accuracy of ± 1 percent zinc can be achieved without difficulty.

The best source of data on lattice parameters is Pearson's compilation [G.16]. Applications of the parameter method are described by Zwell and Danko [14.1].

This method is applicable only to binary alloys. In ternary solid solutions, for example, the percentage of two components can be independently varied. The result is that two ternary solutions of quite different compositions can have the same lattice parameter.

QUANTITATIVE ANALYSIS (MULTIPHASE)

14-8 BASIC PRINCIPLES

Quantitative analysis by diffraction is based on the fact that the intensity of the diffraction pattern of a particular phase in a mixture of phases depends on the concentration of that phase in the mixture. The relation between intensity and concentration is not generally linear, because the diffracted intensity depends markedly on the absorption coefficient of the mixture and this itself varies with the concentration.

To find the relation between diffracted intensity and concentration, we must go back to the basic equation for the intensity diffracted by a powder specimen. The form of this equation depends on the kind of apparatus used, namely, camera or diffractometer; we shall consider only the diffractometer here. The exact expression for the intensity diffracted by a single-phase powder specimen in a diffractometer is

$$I = \left(\frac{I_0 A \lambda^3}{32\pi r}\right)\left[\left(\frac{\mu_0}{4\pi}\right)^2 \frac{e^4}{m^2}\right]\left(\frac{1}{v^2}\right)\left[|F|^2 p \left(\frac{1 + \cos^2 2\theta}{\sin^2 \theta \cos \theta}\right)\right]\left(\frac{e^{-2M}}{2\mu}\right). \tag{14-1}$$

Here I = integrated intensity per unit length of diffraction line (joules sec^{-1} m^{-1}), I_0 = intensity of incident beam (joules sec^{-1} m^{-2}), A = cross-sectional area of incident beam (m^2), λ = wavelength of incident beam (m), r = radius of diffractometer circle (m), $\mu_0 = 4\pi \times 10^{-7}$ m kg C^{-2}, e = charge on electron (C), m = mass of electron (kg), v = volume of unit cell (m^3), F = structure factor, p = multiplicity factor, θ = Bragg angle, e^{-2M} = temperature factor, and μ = linear absorption coefficient (m^{-1}), which enters as the absorption factor $1/2\mu$.

The derivation of this equation can be found in various advanced texts, for example, those of Warren [G.30] and James [G.7]. It applies to a polycrystalline specimen, made up of randomly oriented grains, in the form of a flat plate of effectively infinite thickness, making equal angles with the incident and diffracted beams and completely filling the incident beam at all angles θ. The second factor in square brackets, containing F, p, and θ, will be recognized as Eq. (4–19), the approximate equation for relative line intensities in a Debye–Scherrer pattern.

A digression on names and symbols is in order here. The quantity I_0 in Eq. (14–1) is a true intensity, i.e., energy per unit area per unit time (joules m^{-2} sec^{-1}), and $I_0 A$ is the *power* of the incident beam. This means that I is the power per unit length of diffraction line in the diffracted beam. If this beam is then incident on a film or counter for a certain time t, then the response of the film or counter is a measure of the *energy* It in unit length of the diffraction line. This is the quantity commonly called integrated intensity. A more descriptive term would be "total diffracted energy," but the term "integrated intensity," symbolized by I, has been too long entrenched in the vocabulary of x-ray diffraction to be changed now.

We can simplify Eq. (14–1) considerably for special cases. As it stands, it applies only to a pure substance. But suppose that we wish to analyze a mixture of two phases, α and β. Then we can concentrate on a particular line of the α phase and rewrite Eq. (14–1) in terms of that phase alone. I now becomes I_α, the intensity of the selected line of the α phase, and the right side of the equation must be multiplied by c_α, the volume fraction of α in the mixture, to allow for the fact that the diffracting volume of α in the mixture is less than it would be if the specimen were pure α. Finally, we must substitute μ_m for μ, where μ_m is the linear absorption coefficient of the mixture. In this new equation, all factors are constant and independent of the concentration of α except c_α and μ_m, and we can write

$$I_\alpha = \frac{K_1 c_\alpha}{\mu_m}, \tag{14-2}$$

where K_1 is a constant. The value of K_1 is unknown, because I_0 is generally unknown. However, K_1 will cancel out if we measure the ratio of I_α to the intensity of some standard reference line. The concentration of α can then be found from this ratio.

The three main methods of analysis differ in what is used as a reference line: (1) *external standard method* (a line from pure α), (2) *direct comparison method* (a line from another phase in the mixture), and (3) *internal standard method* (a line from a foreign material mixed with the specimen).

In all methods, the absorption coefficient μ_m of the mixture is itself a function of c_α and can have a large effect on the measured intensity I_α. Alexander and Klug [14.6] were the first to clearly recognize this effect and to work out the equations needed in analysis.

14-9 EXTERNAL STANDARD METHOD

To put Eq. (14-2) in a useful form, we must express μ_m in terms of the concentration. From Eq. (1-12) we have

$$\frac{\mu_m}{\rho_m} = w_\alpha\left(\frac{\mu_\alpha}{\rho_\alpha}\right) + w_\beta\left(\frac{\mu_\beta}{\rho_\beta}\right), \tag{14-3}$$

where w denotes the weight fraction and ρ the density. Consider unit volume of the mixture. Its weight is ρ_m and the weight of contained α is $w_\alpha \rho_m$. Therefore, the volume of α is $w_\alpha \rho_m/\rho_\alpha$, which is equal to c_α, and a similar expression holds for c_β. Equation (14-3) then becomes

$$\mu_m = c_\alpha \mu_\alpha + c_\beta \mu_\beta = c_\alpha \mu_\alpha + (1 - c_\alpha)\mu_\beta$$
$$= c_\alpha(\mu_\alpha - \mu_\beta) + \mu_\beta;$$

$$I_\alpha = \frac{K_1 c_\alpha}{c_\alpha(\mu_\alpha - \mu_\beta) + \mu_\beta}. \tag{14-4}$$

This equation relates the intensity of a diffraction line from one phase to the volume fraction of that phase and the linear absorption coefficients of both phases.

We can put Eq. (14-4) on a weight basis by considering unit mass of the mixture. The volume of the contained α is w_α/ρ_α and the volume of β is w_β/ρ_β. Therefore,

$$c_\alpha = \frac{w_\alpha/\rho_\alpha}{w_\alpha/\rho_\alpha + w_\beta/\rho_\beta} \tag{14-5}$$

$$= \frac{w_\alpha/\rho_\alpha}{w_\alpha(1/\rho_\alpha - 1/\rho_\beta) + 1/\rho_\beta}. \tag{14-6}$$

Combining Eqs. (14–4) and (14–6) and simplifying, we obtain

$$I_\alpha = \frac{K_1 w_\alpha}{\rho_\alpha[w_\alpha(\mu_\alpha/\rho_\alpha - \mu_\beta/\rho_\beta) + \mu_\beta/\rho_\beta]}. \tag{14–7}$$

For the pure α phase, either Eq. (14–2) or (14–7) gives, for the same line,

$$I_{\alpha p} = \frac{K_1}{\mu_\alpha}, \tag{14–8}$$

where the subscript p denotes diffraction from the pure phase. Division of Eq. (14–7) by Eq. (14–8) eliminates the unknown constant K_1 and gives

$$\frac{I_\alpha}{I_{\alpha p}} = \frac{w_\alpha(\mu_\alpha/\rho_\alpha)}{w_\alpha(\mu_\alpha/\rho_\alpha - \mu_\beta/\rho_\beta) + \mu_\beta/\rho_\beta}. \tag{14–9}$$

This equation permits quantitative analysis of a two-phase mixture, provided that the mass absorption coefficients of each phase are known. If they are not known, a calibration curve can be prepared by using mixtures of known composition. In each case, a specimen of pure α must be available as a reference material, and the measurements of I_α and $I_{\alpha p}$ must be made under identical conditions.

In general, the variation of the intensity ratio $I_\alpha/I_{\alpha p}$ with w_α is not linear, as shown by the curves of Fig. 14–2. The experimental points were obtained by measurements on synthetic binary mixtures of powdered quartz, cristobalite, beryllium oxide, and potassium chloride; the curves were calculated by Eq. (14–9). The agreement is excellent. The line obtained for the quartz-cristobalite mixture is straight because these substances are two allotropic forms of silica and hence have identical mass absorption coefficients. When the mass absorption coefficients of the two phases are equal, Eq. (14–9) becomes simply

$$\frac{I_\alpha}{I_{\alpha p}} = w_\alpha.$$

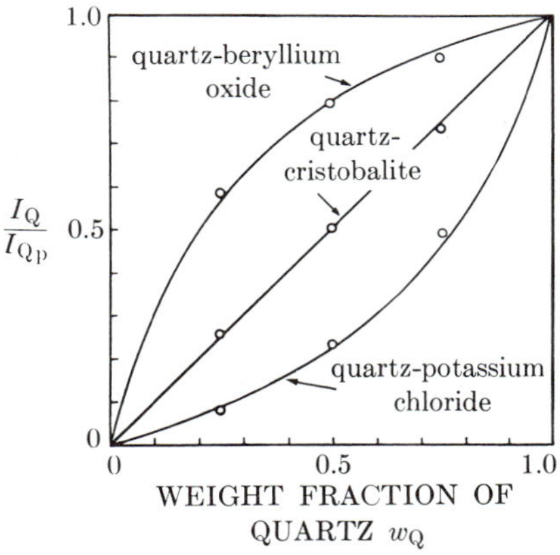

Fig. 14–2 Diffractometer measurements made with Cu $K\alpha$ radiation on binary mixtures. I_Q is the intensity of the reflection from the d = 3.34 Å planes of quartz in a mixture. I_{Qp} is the intensity of the same reflection from pure quartz. Alexander and Klug [14.6].

Fig. 14-2 illustrates very clearly how the intensity of a particular diffraction line from one phase depends on the absorption coefficient of the other phase. For Cu $K\alpha$ radiation, the mass absorption coefficient of BeO is 8.6, of SiO_2 is 34.9, and of KCl is 124 cm^2/gm.

14-10 DIRECT COMPARISON METHOD

This method does not require a sample of the pure phase whose composition is being determined because the required reference line comes from another phase in the mixture.

The direct comparison method is of greatest metallurgical interest because it can be applied directly to polycrystalline aggregates. Since its development by Averbach and Cohen [14.7], it has been widely used for measuring the amount of retained austenite in hardened steel and will be described here in terms of that specific problem, although the method itself is quite general.

The hardening of steel requires two operations: (1) heating to a high temperature to form a homogeneous, face-centered-cubic solid solution called austenite, and (2) quenching the austenite to room temperature to transform it to a hard, metastable, body-centered-tetragonal solid solution called martensite. (These two phases were described in Sec. 12-3.) In practice, the quenched steel may contain some undissolved carbides and, because of incomplete transformation, some austenite is often retained at room temperature. The effect of this austenite on the service behavior of the steel is usually detrimental, but sometimes beneficial. At any rate there is considerable interest in methods of determining the exact amount of austenite present. Quantitative microscopic examination is fairly satisfactory as long as the austenite content is fairly high, but becomes unreliable below about 15 percent austenite in many steels. The x-ray method, on the other hand, is quite accurate in this low-austenite range, often the range of greatest practical interest.

Assume that a hardened steel contains only two phases, martensite and austenite. The problem is to determine the composition of the mixture, when the two phases have the same composition but different crystal structure. The external standard method cannot be used, because it is usually impossible to obtain a reference sample of pure austenite, or of known austenite content, of the same chemical composition as the austenite in the unknown. Instead, we proceed as follows. In the basic intensity equation, Eq. (14-1), we put

$$K_2 = \left(\frac{I_0 A \lambda^3}{32\pi r}\right)\left[\left(\frac{\mu_0}{4\pi}\right)^2 \frac{e^4}{m^2}\right]$$

and

$$R = \left(\frac{1}{v^2}\right)\left[|F|^2 p \left(\frac{1 + \cos^2 2\theta}{\sin^2 \theta \cos \theta}\right)\right](e^{-2M}). \tag{14-10}$$

The diffracted intensity is then given by

$$I = \frac{K_2 R}{2\mu}, \tag{14-11}$$

where K_2 is a constant, independent of the kind and amount of the diffracting substance, and R depends on θ, hkl, and the kind of substance. Designating austenite by the subscript γ and martensite by the subscript α, we can write Eq. (14–11) for a particular diffraction line of each phase:

$$I_\gamma = \frac{K_2 R_\gamma c_\gamma}{2\mu_m},$$

$$I_\alpha = \frac{K_2 R_\alpha c_\alpha}{2\mu_m}.$$

Division of these equations yields

$$\frac{I_\gamma}{I_\alpha} = \frac{R_\gamma c_\gamma}{R_\alpha c_\alpha}. \tag{14–12}$$

The value of c_γ/c_α can therefore be obtained from a measurement of I_γ/I_α and a calculation of R_γ and R_α. (Note that the calculation of R values requires a knowledge of the crystal structures and lattice parameters of both phases.) Once c_γ/c_α is found, the value of c_γ can be obtained from the additional relationship:

$$c_\gamma + c_\alpha = 1.$$

We can thus make an absolute measurement of the austenite content of the steel by direct comparison of the integrated intensity of an austenite line with the integrated intensity of a martensite line.* By comparing several pairs of austenite-martensite lines, we can obtain several independent values of the austenite content.

If the steel contains a third phase, namely, Fe_3C (cementite), we can determine the cementite concentration either by quantitative microscopic examination or by diffraction. If we measure I_C, the integrated intensity of a particular cementite line, and calculate R_C, then we can set up an equation similar to Eq. (14–12) from which c_γ/c_C can be obtained. The value of c_γ is then found from the relation

$$c_\gamma + c_\alpha + c_C = 1.$$

In choosing diffraction lines to measure, we must be sure to avoid overlapping or closely adjacent lines from different phases. Figure 14–3 shows the calculated patterns of austenite and martensite in a 1.0 percent carbon steel, made with Cr $K\alpha$ radiation. Unfortunately, the strong 111 austenite line is too close to the 101–110 martensite line for separate measurement of its integrated intensity. Suitable austenite lines are the 200 and 220; these may be compared with the 002–200 and 112–211 martensite doublets. These "doublets," due to the tetragonality of the martensite unit cell (Fig. 12–5), are not usually resolved into separate lines because all lines are usually quite broad, both from the martensite and austenite, as shown by the pattern in Fig. 14–4. (The unresolved martensite lines are then indexed as a cubic line; for example, the 002–200 doublet is called the 200 line.) The line broadening is due to the nonuniform microstrain in both phases of the quenched steel and, very often, the fine grain size.

* Recalling the earlier discussion of the disappearing-phase x-ray method of locating a solvus line (Sec. 12–4), we note from Eq. (14–12) that the intensity ratio I_γ/I_α is not a linear function of the volume fraction c_γ, or, for that matter, of the weight fraction w_γ.

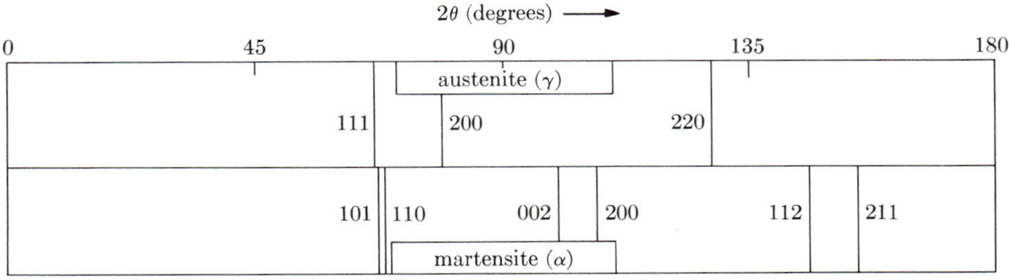

Fig. 14-3 Calculated powder patterns of austenite and martensite, each containing 1.0 percent carbon in solution. Cr $K\alpha$ radiation.

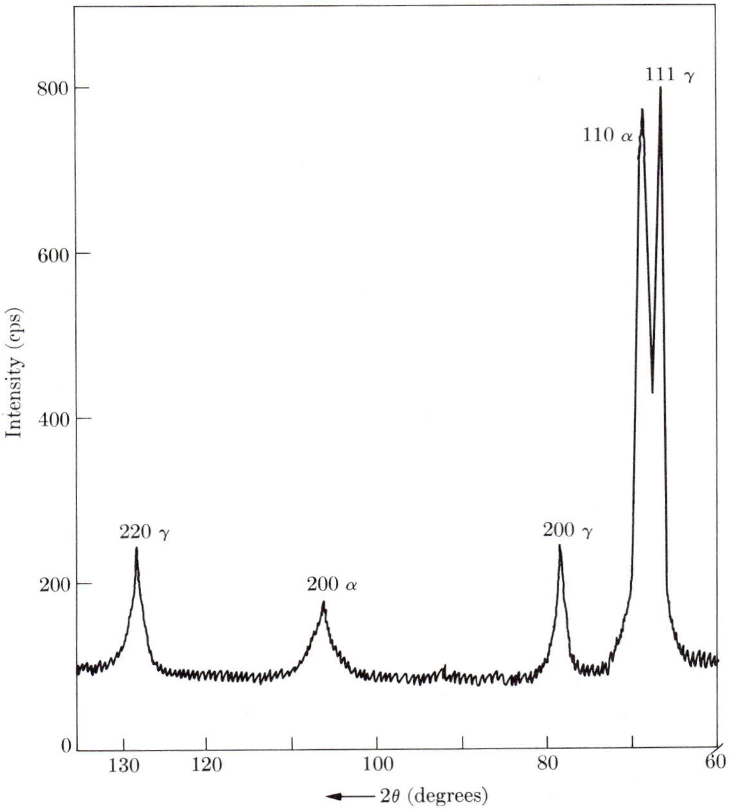

Fig. 14-4 Partial diffractometer pattern of an oil-quenched Ni-V steel, containing about 30 volume percent austenite (γ). Chromium radiation, V filter.

If substantial amounts of carbide are present, as in some tool steels, carbide lines may overlap α and γ lines and cause difficulties in measurement. Durnin and Ridal [14.8] list the α and γ lines that do or do not overlap the lines of Fe_3C and six alloy carbides.

While chromium radiation is the most popular, shorter wavelengths such as Fe $K\alpha$, Co $K\alpha$, and Mo $K\alpha$ will increase the number of lines on the pattern and thus provide more measurable pairs. The low resolution of energy-dispersive diffractometry (Sec. 7–10) is no hindrance here, because the diffraction lines of steel are well separated, and Voskamp [7.22] has described the application of that technique to austenite measurement.

Automatic diffractometers can be easily programmed to measure integrated intensity, and Kelly and Short [7.2] have described automated austenite measurements.

In calculating the value of R for a particular diffraction line, various factors should be kept in mind. The unit cell volume v is calculated from the measured lattice parameters, which are a function of carbon and alloy content. When the martensite doublets are unresolved, the structure factor and multiplicity of the martensite are calculated on the basis of a body-centered *cubic* cell; this procedure, in effect, adds together the integrated intensities of the two lines of the doublet, which is exactly what is done experimentally when the integrated intensity of an unresolved doublet is measured. For greatest accuracy in the calculation of F, the atomic scattering factor f should be corrected for anomalous scattering by an amount Δf (see Sec. 13–4), particularly when Co $K\alpha$ radiation is used. The value of the temperature factor e^{-2M} can be taken from the curve of Fig. 4–20.

Specimen preparation involves wet grinding to remove the surface layer, which may be decarburized or otherwise nonrepresentative of the bulk of the specimen, followed by standard metallographic polishing and etching. This procedure ensures a flat, reproducible surface for the x-ray examination, and allows a preliminary examination of the specimen to be made with the microscope. In grinding and polishing, care should be taken not to produce excessive heat or plastic deformation which would cause partial decomposition of both the martensite and austenite.

In the measurement of diffraction line intensity, it is essential that the *integrated intensity*, not the maximum intensity, be measured. Large variations in line shape can occur because of variations in microstrain and grain size. These variations in line shape will not affect the integrated intensity, but they can make the values of maximum intensity absolutely meaningless.

The sensitivity of the x-ray method in determining small amounts of retained austenite is limited chiefly by the intensity of the continuous background present. The lower the background, the easier it is to detect and measure weak austenite lines. With filtered radiation the minimum detectible amount is about 2 volume percent austenite, and with crystal-monochromated radiation probably about 0.2 percent. The error in the austenite content, originating in the calculation of R and the measurement of I, is probably about 5 percent of the amount present, in the absence of preferred orientation (Sec. 14–12).

In 1971 the National Bureau of Standards issued a standard reference material

(SRM-485) containing a specified amount of austenite [14.9]. The supply of this material is now exhausted and a new one is being prepared. Such a standard material is useful to the investigator who wishes to check his experimental and computational technique. For other standard materials from NBS see [15.2].

Other analytical problems to which the direct comparison method has been applied include the determination of mixed iron oxides in the oxide scale on steel [14.10], the beta phase in titanium alloys [14.11], and mixed uranium and plutonium carbides [14.12].

14-11 INTERNAL STANDARD METHOD

In this method a diffraction line from the phase being determined is compared with a line from a standard substance mixed with the sample in known proportions. The internal standard method is therefore restricted to samples in powder form.

Suppose we wish to determine the amount of phase A in a mixture of phases A, B, C, . . . , where the relative amounts of the other phases present (B, C, D, . . .) may vary from sample to sample. With a known amount of original sample we mix a known amount of a standard substance S to form a new composite sample. Let c_{A} and c'_A be the volume fractions of phase A in the original and composite samples, respectively, and let c_{S} be the volume fraction of S in the composite sample. If a diffraction pattern is now prepared from the composite sample, then from Eq. (14-2) the intensity of a particular line from phase A is given by

$$I_{\mathrm{A}} = \frac{K_3 c'_A}{\mu_m},$$

and the intensity of a particular line from the standard S by

$$I_{\mathrm{S}} = \frac{K_4 c_{\mathrm{S}}}{\mu_m}.$$

Division of one expression by the other gives

$$\frac{I_{\mathrm{A}}}{I_{\mathrm{S}}} = \frac{K_3 c'_A}{K_4 c_{\mathrm{S}}}. \tag{14-13}$$

(Note that μ_m, the linear absorption coefficient of the mixture and an unknown quantity, drops out. Physically, this means that variations in absorption, due to variations in the relative amounts of B, C, D, . . . , have no effect on the ratio $I_{\mathrm{A}}/I_{\mathrm{S}}$ since they affect I_{A} and I_{S} in the same proportion.)

By extending Eq. (14-5) to a number of components, we can write

$$c'_A = \frac{w'_A/\rho_{\mathrm{A}}}{w'_A/\rho_{\mathrm{A}} + w'_B/\rho_{\mathrm{B}} + w'_C/\rho_{\mathrm{C}} + \cdots + w_{\mathrm{S}}/\rho_{\mathrm{S}}}$$

and a similar expression for c_{S}. Therefore,

$$\frac{c'_A}{c_{\mathrm{S}}} = \frac{w'_A \rho_{\mathrm{S}}}{\rho_{\mathrm{A}} w_{\mathrm{S}}}.$$

Substitution of this relation into Eq. (14–13) gives

$$\frac{I_A}{I_S} = K_5 w'_A, \tag{14–14}$$

if w_S is kept constant in all the composite samples. The relation between the weight fractions of A in the original and composite samples is:

$$w'_A = w_A(1 - w_S). \tag{14–15}$$

Combination of Eqs. (14–14) and (14–15) gives

$$\frac{I_A}{I_S} = K_6 w_A. \tag{14–16}$$

The intensity ratio of a line from phase A and a line from the standard S is therefore a linear function of w_A, the weight fraction of A in the original sample. A calibration curve can be prepared from measurements on a set of synthetic samples, containing known concentrations of A and a constant concentration of a suitable standard. Once the calibration curve is established, the concentration of A in an unknown sample is obtained simply by measuring the ratio I_A/I_S for a composite sample containing the unknown and the same proportion of standard as was used in the calibration.

The internal standard method has been widely used for the measurement of the quartz content of industrial dusts. In this analysis, fluorite (CaF_2) has been found to be a suitable internal standard. Figure 14–5 shows a calibration curve prepared from mixtures of quartz and calcium carbonate, of known composition, each mixed with enough fluorite to make the weight fraction of fluorite in each composite sample equal to 0.20. The curve is linear and through the origin, as predicted by Eq. (14–16). Bumsted [14.25] describes the determination of quartz in dust by this method, with particular attention to the problems posed by very small samples.

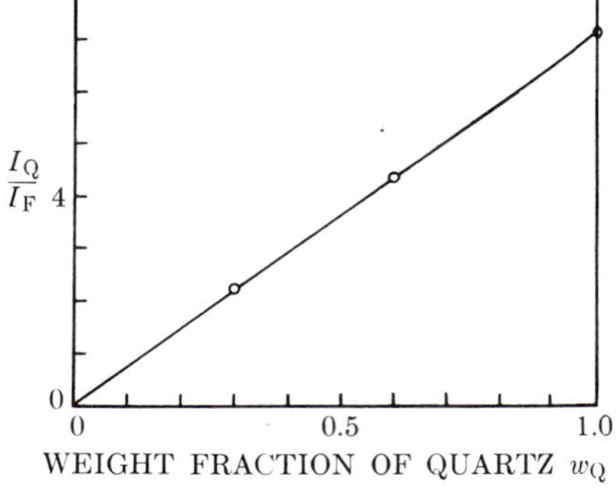

Fig. 14–5 Calibration curve for quartz analysis, with fluorite as internal standard. I_Q is the intensity of the $d = 3.34$ Å line of quartz, and I_F is the intensity of the $d = 3.16$ Å line of fluorite. Alexander and Klug [14.6].

Strictly speaking, Eq. (14-16) is valid only for integrated intensities, and the same is true of all other intensity equations in this chapter. Yet it has been found possible to determine the quartz content of dusts with satisfactory accuracy by simply measuring maximum intensities. This short cut is permissible here only because the shape of the diffraction lines is found to be essentially constant from sample to sample. There is therefore a constant proportionality between maximum and integrated intensity and, as long as all patterns are made under identical experimental conditions, the measurement of maximum intensities gives satisfactory results. Quite erroneous results would be obtained by this procedure if the particle size of the samples were very small and variable, since then a variable amount of line broadening would occur, and this would cause a variation in maximum intensity independent of sample composition.

Other applications of the internal standard method include the analysis of cement [14.13], the analysis of clay minerals [14.14], and the determination of chrysotile asbestos in airborne dust [14.15].

The Joint Committee on Powder Diffraction Standards has generalized the internal standard method by comparing the strongest line on the pattern of a large number of substances with the strongest line from a single standard reference material. The material chosen was α—Al_2O_3, a synthetic corundum commercially available as "Linde A" powder. The current alphabetical (inorganic) search manual of the JCPDS lists I/I_{cor} values for almost 800 compounds. Here I is the maximum intensity of the strongest line from the compound, I_{cor} the same quantity for the corundum, and I/I_{cor} was determined from a mixture, of equal parts by weight, of the compound and corundum. The value of I/I_{cor}, which is reported to two significant figures, for any particular compound A establishes a single point on the calibration curve, like Fig. 14-5, of compound A; if an equi-weight mixture of corundum and the unknown is then made, the weight fraction of A in the unknown is given simply by one half the ratio of I_A/I_{cor} for the unknown-corundum mixture to the tabulated value I_A/I_{cor} for the A-corundum mixture. This method is fast, because the calibration has already been done, but its accuracy is probably low, in view of the use of maximum, rather than integrated, intensities. Variations in line broadening would be expected to introduce errors.

In any method involving powders, accurate sampling and homogeneous mixing can present problems. They are not trivial, and errors in these operations can produce large errors in the final result. Such matters are discussed by Klug and Alexander [G.39] and Grant and Pelton [14.16].

14-12 PRACTICAL DIFFICULTIES

Certain effects can cause great difficulty in quantitative analysis because they cause observed intensities to depart widely from the theoretical. The most important of these complicating factors are:

1. *Preferred orientation.* The basic intensity equation, Eq. (14-1), is derived on the premise of random orientation of the constituent crystals in the sample and is not valid if any preferred orientation exists. It follows that, in the preparation of powder samples for the diffractometer, every effort should be made to avoid preferred orientation. If the sample is a solid polycrystalline aggregate, the analyst has no control over the distribution of orientations in it, but he should at least be aware of the possibility of error due to preferred orientation (texture). The texture problem has probably received most attention in connection with determining austenite in steel by the direct comparison method. Here we have a direct check on texture, because the calculated R values for one phase are simply the theoretical line intensities, in arbitrary units, for that phase in the absence of texture. If the measured intensities of the various lines of a particular phase, say α, are not in the same ratio as their R values, then texture exists; the austenite content determined from a particular pair of α and γ lines will then differ from the value found from another pair. Two approaches have been made to the problem of measuring austenite in the presence of texture:

a) *Averaging intensities* [14.17–14.21]. The basic idea here is simple. If certain lines from, say, austenite are abnormally weak because of texture, then other austenite lines will be abnormally strong. Only by measuring all the lines and averaging them in a particular way can we get valid data. This method, here crudely termed "intensity averaging," has had considerable success, at the cost of increased measurement and computation time. The number of diffraction lines required for the analysis increases with the degree of texture. For strongly textured materials Mo $K\alpha$ radiation is needed to provide enough lines.

b) *Averaging orientations* [14.22, 14.23]. Here the specimen is rotated in particular ways during the measurement of line intensity in order to present more crystal orientations to the incident beam.

2. *Microabsorption.* Consider diffraction from a given crystal of α in a mixture of α and β crystals. The incident beam passes through both α and β crystals on its way to a particular diffracting α crystal, and so does the diffracted beam on its way out of the sample. Both beams are decreased in intensity by absorption, and the decrease can be calculated from the total path length and μ_m, the linear absorption coefficient of the mixture. But a small part of the total path lies entirely within the diffracting α crystal, and for this portion μ_α is the applicable absorption coefficient. If μ_α is much larger than μ_β, or if the particle size of α is much larger than that of β, then the total intensity of the beam diffracted by the α crystals will be much less than that calculated, since the effect of microabsorption in each diffracting α crystal is not included in the basic intensity equation. Evidently, the microabsorption effect is negligible when $\mu_\alpha \approx \mu_\beta$ and both phases have the same particle size, or when the particle size of both phases is very small. Powder samples should therefore be finely ground before analysis.

3. *Extinction.* This effect, described in Sec. 4–12, is a reduction in diffracted intensity as a crystal becomes more nearly perfect. Equation (14–1) is derived for the ideally imperfect crystal, one in which extinction is absent. Samples for chemical analysis should therefore be free of extinction, and this can be accomplished, for powder samples, by grinding or filing. If a solid aggregate must be analyzed directly, the possibility of some extinction in the individual grains of the aggregate should be kept in mind.

Microabsorption and extinction, if present, can seriously decrease the accuracy of the direct comparison method, because this is an absolute method. Fortunately, both effects are negligible in the case of hardened steel. Inasmuch as both the austenite and martensite have the same composition and only a 4 percent difference in density, their linear absorption coefficients are practically identical. Their average particle sizes are also roughly the same. Therefore, microabsorption does not occur. Extinction is absent because of the very nature of hardened steel. The change in specific volume accompanying the transformation of austenite to martensite sets up nonuniform strains in both phases so severe that both kinds of crystals can be considered highly imperfect. If these fortunate circumstances do not exist, and they do not in most other alloy systems, the direct comparison method should be used with caution and checked by some independent method.

On the other hand, the presence of microabsorption and extinction does not invalidate the internal standard method, provided these effects are constant from sample to sample, including the calibration samples. Microabsorption and extinction affect only the values of the constants K_3 and K_4 in Eq. (14–13), and therefore the constant K_6 in Eq. (14–16), and the latter constant determines only the slope of the calibration curve. Therefore, microabsorption and extinction, if present, will have no effect on the accuracy of the internal standard method as long as the crystals of the phase being determined, and those of the standard substance, do not vary in degree of perfection or particle size from one sample to another.

PROBLEMS

The d and I/I_1 values tabulated in Probs. 14–1 to 14–4 represent the diffraction patterns of various unknown substances. Identify the substances involved by reference to the JCPDS powder diffraction file.

***14–1**

d(Å)	I/I_1	d(Å)	I/I_1	d(Å)	I/I_1
3.66	50	1.46	10	1.06	10
3.17	100	1.42	50	1.01	10
2.24	80	1.31	30	0.96	10
1.91	40	1.23	10	0.85	10
1.83	30	1.12	10		
1.60	20	1.08	10		

14–2

d(Å)	I/I_1	d(Å)	I/I_1	d(Å)	I/I_1
5.85	60	2.08	10	1.47	20
3.05	30	1.95	20	1.42	10
2.53	100	1.80	60	1.14	20
2.32	10	1.73	20	1.04	10

*14–3

d(Å)	I/I_1	d(Å)	I/I_1	d(Å)	I/I_1
2.40	50	1.25	20	0.85	10
2.09	50	1.20	10	0.81	20
2.03	100	1.06	20	0.79	20
1.75	40	1.02	10		
1.47	30	0.92	10		
1.26	10				

14–4

d(Å)	I/I_1	d(Å)	I/I_1	d(Å)	I/I_1
3.02	100	2.11	10	1.46	10
2.79	10	1.90	20	1.17	10
2.52	10	1.65	10		
2.31	30	1.62	10		

***14–5** Microscopic examination of a hardened 1.0 percent carbon steel shows no undissolved carbides. X-ray examination of this steel in a diffractometer with filtered cobalt radiation shows that the integrated intensity of the 311 austenite line is 2.33 and the integrated intensity of the unresolved 112–211 martensite doublet is 16.32, both in arbitrary units. Calculate the volume percent austenite in the steel. (Take lattice parameters from Fig. 12–5, Δf corrections from Fig. 13–8, and temperature factors e^{-2M} from Fig. 4–20.)

15

Chemical Analysis by X-ray Spectrometry

15–1 INTRODUCTION

We saw in Chap. 1 that any element, if made the target in an x-ray tube and bombarded with electrons of high enough energy, would emit a characteristic line spectrum. The most intense lines of this spectrum are the $K\alpha$ and $K\beta$ lines. They are always called "characteristic lines" to emphasize the fact that their wavelengths are fixed and characteristic of the emitting element. We also saw that these same lines would be emitted if the element were bombarded with x-rays of high enough energy (fluorescence).

In these phenomena we have the basis for a method of chemical analysis. If the various elements in the sample to be analyzed are made to emit their characteristic lines by electron or x-ray bombardment, then these elements may be identified by analyzing the emitted radiation and showing that these specific wavelengths are present. The analysis is carried out in an x-ray spectrometer of either of the following kinds:

1. *Wavelength-dispersive.* The radiation emitted by the sample is diffracted by lattice planes of known d spacing in a single crystal. In accordance with the Bragg law, radiation of only a single wavelength is reflected for each angular setting of the crystal, and the intensity of this radiation can be measured with a suitable counter, as in Fig. 15–1(a). Because radiation of various wavelengths is physically dispersed in different directions in space, this method is sometimes simply called *dispersive*. A wavelength-dispersive spectrometer is also called a *crystal spectrometer*.

2. *Energy-dispersive.* In this spectrometer, diffraction is not involved. The various wavelengths in the radiation emitted by the sample are separated on the basis of their energies by means of a Si(Li) counter and a multichannel analyzer (MCA); this counter produces pulses proportional in height to the energies in the incident beam, and the MCA then sorts out the various pulse heights, as in Fig. 15–1(b). Because there is no physical separation in space of the various wavelengths (energies), such a spectrometer is often simply called *nondispersive*. It is more recent and less common than the crystal spectrometer.

Two kinds of x-ray spectrometry are possible, depending on the means used to excite the characteristic radiation of the elements in the sample:

1. *X-ray excitation.* The sample is bombarded with x-rays from an x-ray tube

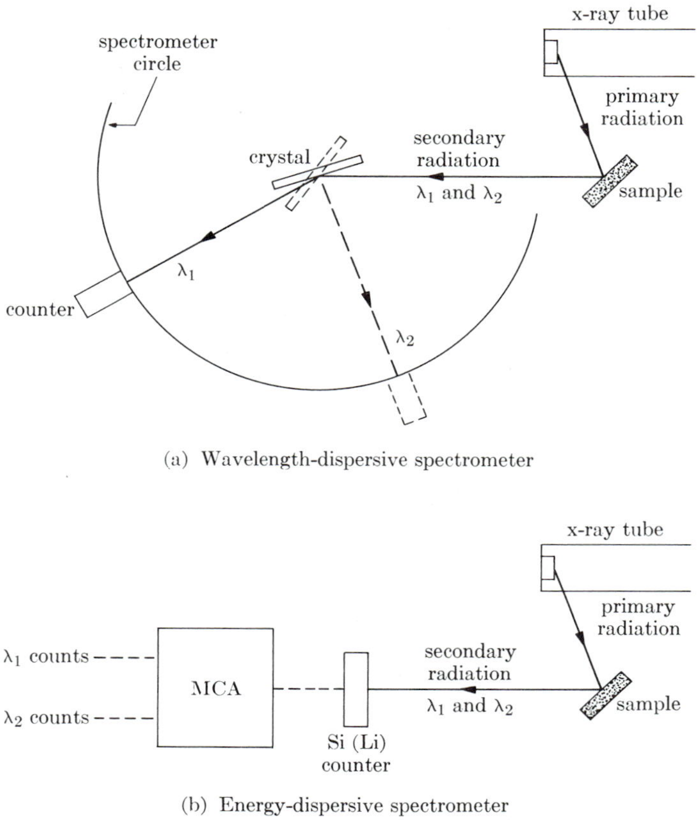

Fig. 15–1 X-ray fluorescence spectrometers. In this example, elements 1 and 2 in the sample emit characteristic wavelengths λ_1 and λ_2. These wavelengths are separately measured by crystal diffraction in (a) or by pulse-height analysis in (b), where MCA = multichannel analyzer.

(Fig. 15–1). The primary radiation causes the sample to emit secondary fluorescent radiation, which is then analyzed in a spectrometer. This method, often called *fluorescence analysis*, is very widely used in industry for chemical analysis. The phenomenon of fluorescence, which is just a nuisance in diffraction experiments, is here made to serve a useful purpose.

2. *Electron excitation.* The sample is bombarded with high-speed electrons in an evacuated apparatus. Historically, this was the first method and was used by Moseley in his work on the relation between characteristic wavelength and atomic number. It is not a practical method for the rapid analysis of many samples, because the apparatus must be evacuated after the insertion of each sample. However, x-ray spectrometers with electron excitation are used in certain instruments of a research nature: in the electron probe microanalyzer (Sec. 15–11) and, as an optional accessory, in the transmission electron microscope and the scanning electron microscope. These devices, incidentally, are often called *electron-column instruments*.

Chemical analysis by x-ray spectrometry can be either *qualitative*, if the various characteristic lines in the emitted spectrum are simply identified, or *quantitative*, if the intensities of these lines are compared with the intensities of lines from a suitable standard. Note that x-ray spectrometry gives information about the chemical *elements* present in the sample, irrespective of their state of chemical combination or the phases in which they exist. X-ray diffraction, on the other hand, as we saw in the previous chapter, discloses the various compounds and phases present in the sample. Spectrometry and diffraction therefore complement one another in the kind of information they provide.

X-ray spectrometry by fluorescence analysis is nondestructive and much more rapid than the ordinary wet methods of chemical analysis. Automatic instruments exist and are widely used in industry for the analysis of such disparate materials as alloys, ores, cements, and petroleum products.

X-ray spectrometry is of such wide interest today as a means of chemical analysis that it has been given book-length treatment by several authors: Jenkins and DeVries [G.29], Birks [G.31], Jenkins [G.41], and Bertin [G.46]. In addition, a book by Jenkins [G.37] presents worked-out problems arising in practical spectrometry, and [7.18] contains many papers on energy-dispersive methods and microanalysis.

15-2 GENERAL PRINCIPLES

Except for Sec. 15-11, this chapter is devoted entirely to fluorescence analysis, i.e., to x-ray excitation of the elements in the sample. We must therefore examine the excitation process itself in some detail.

The fluorescent radiation emitted by the sample should be as intense as possible, so that it will be accurately measurable in a short counting time. The intensity of this emitted radiation depends on both the wavelength and the intensity of the incident primary radiation from the x-ray tube. Suppose that monochromatic radiation of constant intensity and of wavelength λ is incident on an element which has a K absorption edge at λ_K, and that we can continuously vary λ. As we decrease λ from a value larger than λ_K, no K fluorescence occurs until λ is just shorter than λ_K. The fluorescent intensity is then a maximum. Further decrease in λ causes the fluorescent intensity to decrease, in much the same manner as the absorption coefficient. This is natural since, as mentioned in Sec. 1-5, fluorescence and true absorption are but two aspects of the same phenomenon. At any one value of λ, the fluorescent intensity is directly proportional to the incident intensity.

The best exciting agent would therefore be a strong characteristic line of wavelength just shorter than λ_K. It is clearly impossible to satisfy this requirement for more than one fluorescing element at a time, and in practice we use a tube with a tungsten, or other heavy metal, target or a chromium target, with as high a power rating as possible. The exciting radiation is then that part of the continuous spectrum and such characteristic lines of the target as have shorter wavelengths than the absorption edge of the fluorescing element. Tube choice depends on the elements to be most often determined: a W tube will give higher fluorescent intensities from heavy elements and a Cr tube from light elements. One manufacturer supplies a dual-target tube, from which either W or Cr radiation can be obtained at the turn of a switch without breaking the tube vacuum.

The fluorescent yields ω of the elements in the sample are also relevant here. As we saw in Sec. 1–5, fluorescent radiation and the ejection of an Auger electron are competing processes, and the Auger effect is more important for light elements. For elements lighter than Cl (atomic number $Z = 17$) the probability of K fluorescence is less than 10 percent.

The beam of secondary radiation issuing from the sample consists largely of fluorescent radiation, but there are some other weak components present as well. These are coherent scattered radiation, coherent diffracted radiation, and incoherent (Compton modified) radiation. These components appear as a background on which the spectral lines are superimposed. This background is normally low (see Fig. 15–3), but it may become rather high if the sample contains a large proportion of elements of low atomic number, because the sample will then emit a large amount of Compton modified radiation.

The wavelength range used in fluorescence analysis extends from about 0.2 Å to 20 Å. The lower limit is imposed mainly by the maximum voltage that can be applied to the x-ray tube, which lies in the range 50–100 kV in commercial instruments. At 100 kV the short-wavelength limit of the continuous spectrum from the tube is $12.4/100 = 0.12$ Å. The maximum intensity occurs at about 1.5 times this value, or 0.18 Å. Incident radiation of this wavelength would cause K fluorescence in Hf ($Z = 72$), and the emitted $K\alpha$ radiation would have a wavelength of 0.22 Å. For heavier elements we can use L rather than K lines; thus the lower limit on λ does not impose any upper limit on the atomic number Z of detectible elements. Figure 15–2 shows how the wavelength of the strongest K and L lines varies with atomic number.

The upper limit on wavelength depends on the equipment available and is imposed by the large absorption of long-wavelength fluorescent radiation by anything it encounters, such as air and the counter window. Absorption therefore puts an unfortunate lower limit on the detectible light elements. If the spectrometer operates in air, Ti ($Z = 22$, $K\alpha = 2.75$ Å) is about the lightest element detectible. (Ti $K\alpha$ radiation is decreased to one half its original intensity by passage through only 10 cm of air.) If a path filled with helium is provided for the x-rays traversing the spectrometer, absorption is decreased to such an extent that Al is measurable ($Z = 13$, $K\alpha = 8.3$ Å). In commerical vacuum spectrometers the usual lower limit is F ($Z = 9$, $K\alpha = 18.3$ Å). (The vacuum requirement in the sample chamber of a fluorescent spectrometer is not nearly as severe as in an electron-excitation instrument. Therefore, the pump-down time after sample insertion is much shorter in the former.)

Another important factor which limits the detection of light elements is absorption in the sample itself. Fluorescent radiation is produced not only at the surface of the sample but also in its interior, to a depth depending on the depth of

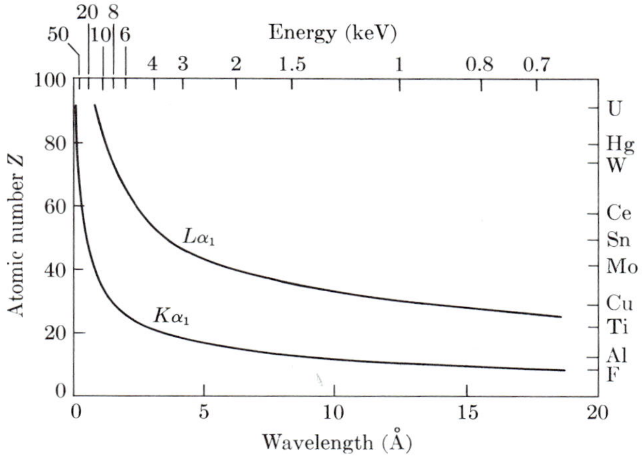

Fig. 15–2 Variation with atomic number Z of the wavelength and energy of the strongest emission lines of the K and L series.

effective penetration by the primary beam, which in turn depends on the overall absorption coefficient of the sample. The fluorescent radiation produced within the sample then undergoes absorption on its way out. Because long-wavelength fluorescent radiation will be highly absorbed by the sample, the fluorescent radiation outside the sample comes only from a thin surface skin and its intensity is accordingly low. It follows that detection of small amounts of a light element in a heavy-element matrix is practically impossible. On the other hand, even a few parts per million of a heavy element in a light-element matrix can be detected.

WAVELENGTH DISPERSION

15–3 SPECTROMETERS

In one type of spectrometer, called single-channel, the analyzing crystal and counter are mechanically coupled, as in a diffractometer. Thus, when the crystal is set at a particular Bragg angle θ, the counter is automatically set at the corresponding angle 2θ. With the counter connected to a ratemeter and recorder, the whole spectrum can be continuously scanned and recorded.

Figure 15–3(a) shows an example of such a scan. It looks a lot like a powder pattern made with a diffractometer, for example, Fig. 7–5. However, the lines of a powder pattern are reflections of the same wavelength from planes of different indices *hkl*, whereas the lines in Fig. 15–3(a) all have the same indices (those of the (200) reflecting planes of the analyzing crystal) but each is formed by a different wavelength. That wavelength is calculable from the Bragg angle and the interplanar spacing of the crystal.

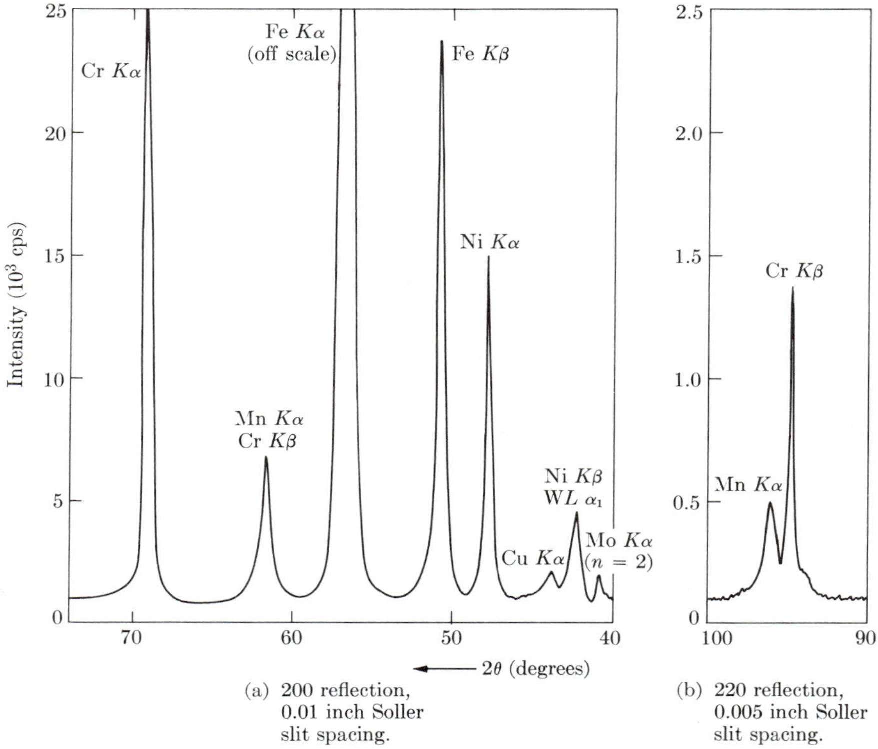

Fig. 15–3 Recording of fluorescent spectrum of a stainless steel containing 19.4 Cr, 9.5 Ni, 1.5 Mo, 1.4 W, 1.0 Mn (in weight percent), balance mainly Fe. Flat LiF crystal analyzers. Platinum-target x-ray tube, 50 kV, 40 mA. (Courtesy of Diano Corporation.)

The spectrum in Fig. 15–3(a) is that of a stainless steel, excited by the primary beam from a platinum-target x-ray tube. The *K* lines of all the major constituents (Fe, Cr, and Ni) and of some of the minor constituents (Mo and Mn) are apparent. One of the *L* lines of W is also present. The Cu $K\alpha$ line originates from copper existing as an impurity in the tube target. Figure 15–3(b), a recording made at higher resolution, is discussed later.

X-ray Optics

The analyzing crystal may be flat or curved, with resulting differences in spectrometer design.

The *flat-crystal* type, illustrated in Fig. 15–4, has the simpler design. The x-ray tube is placed as close as possible to the sample, so that the primary radiation on it, and the fluorescent radiation it emits, will be as intense as possible. For the operator's protection against scattered radiation, the sample is enclosed in a thick metal box, which contains a single opening through which the fluorescent beam leaves. The sample area irradiated is of the order of 2 cm square. Fluorescent

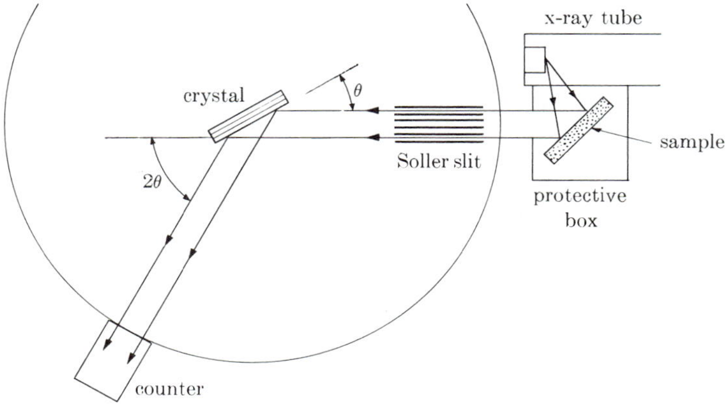

Fig. 15-4 X-ray spectrometer, flat-crystal type.

radiation is emitted in all directions by this area, which acts as a source of radiation for the spectrometer proper. Because of the large size of this source, the beam of fluorescent radiation issuing from the protective box contains a large proportion of widely divergent and convergent radiation. Collimation of this beam before it strikes the analyzing crystal is therefore absolutely necessary, if any resolution at all is to be obtained. This collimation is achieved by passing the beam through a Soller slit whose plates are at right angles to the plane of the spectrometer circle, because it is the divergence (and convergence) in this plane that we want to eliminate.

Essentially parallel radiation from the collimator is then incident on the flat crystal, and a portion of it is diffracted into the counter by lattice planes parallel to the crystal face. Because no focusing occurs, the beam diffracted by the crystal is fairly wide and the counter receiving slit must also be wide.

The *curved-crystal* spectrometer is illustrated in Fig. 15-5. Radiation from the sample passes through the narrow slit S and diverges to the analyzing crystal, which has its reflecting planes bent to a radius of $2R$ and its surface ground to a radius R. Diffracted radiation of a single wavelength is brought to a focus at the counter receiving slit, located on the focusing circle passing through S and the face of the crystal, as described in Sec. 6-13. But now the radius R of the focusing circle is fixed, for a crystal of given curvature, and the slit-to-crystal and crystal-to-focus distances must both be varied as θ is varied. The focusing relation, found from Eq. (6-13), is

$$D = 2R \sin \theta,$$

where D stands for both the slit-to-crystal and crystal-to-focus distances, which must be kept equal to one another. This is accomplished by rotation of both the crystal and the counter about the center O of the focusing circle, in such a manner that rotation of the crystal through an angle x (about O) is accompanied by rotation of the counter through an angle $2x$. At the same time the counter is rotated about

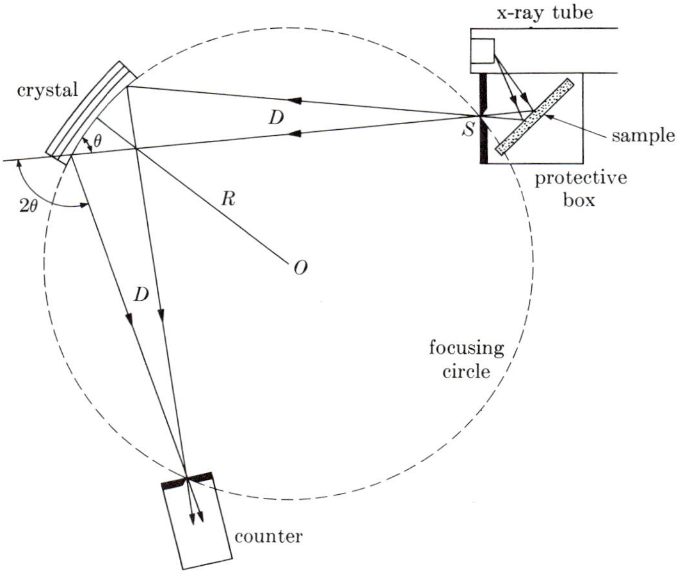

Fig. 15–5 X-ray spectrometer, curved-crystal type.

a vertical axis through its slit, by means of another coupling, so that it always points at the crystal.

D increases as θ increases and may become inconveniently large, for a crystal of given radius of curvature R_1, at large θ values. In order to keep D within reasonable limits, it is necessary to change to another crystal, of smaller radius R_2, for this high-θ (long-wavelength) range.

Crystals

The angle 2θ at which a particular wavelength is reflected depends only on the d spacing of the analyzing crystal. The Bragg law shows that the longest wavelength that can be reflected is equal to $2d$. We therefore need crystals with small d spacings for short wavelengths (high-Z elements) and large d spacings for long wavelengths (low-Z elements). Table 15-1, after Jenkins [G.41], lists the most commonly used crystals, and Bertin [G.46] gives a much longer list.

Other factors affecting the choice of a crystal are its reflecting power and resolving power. These properties are discussed later.

Counters

The scintillation counter (Sec. 7–7) and the sealed gas proportional counter (Sec. 7–5) are both used in spectrometry. The scintillation counter is better for the very short wavelength region because of its greater efficiency there (Fig. 7–12); in the 1 Å–2 Å range either counter is suitable. In the long-wavelength region a gas-flow proportional counter is required, because of its thin low-absorption window.

Table 15-1

Analyzing Crystals

Crystal	Reflecting plane	$2d(\theta)$	Usual atomic number range
Lithium fluoride (LiF)	(420)	1.80	high Z, high resolution
Lithium fluoride (LiF)	(200)	4.03	all $Z > 19$
Germanium (Ge)	(111)	6.53	(no 222 reflection)*
Pyrolytic graphite (PG)	(002)	6.72	P(15), S(16), and Cl(17)
Pentaerythritol (PET)	(002)	8.74	Al(13) through K(19)
Ammonium dihydrogen phosphate (ADP)	(101)	10.64	Mg(12)
Rubidium acid phthalate (RAP)	(001)	26.1	F(9) and Na(11)

* Ge has no 222 (second-order 111) reflection (Sec. 6-13). Second-order reflections sometimes cause line-overlap problems. See text.

Some counters have two windows in line. The beam from the crystal passes through a gas-flow counter, with two side windows opposite one another, and then strikes a scintillation counter. By collecting pulses from both counters, high sensitivity to both long and short wavelengths is achieved.

A single-channel pulse-height analyzer (Sec. 7-9) in the counting circuit may be useful in dealing with interfering high-order reflections (harmonics) [G.29]. If the crystal is set to diffract spectral line λ_A from element A in the sample, it will also diffract, at the same angle 2θ, wavelengths $\lambda_A/2$ in the second order, $\lambda_A/3$ in the third order, etc. If the sample also emits line λ_B from element B, and if λ_B happens to equal $\lambda_A/2$, then the condition called *harmonic overlap* exists and the counter is receiving two spectral lines simultaneously. As a specific example, suppose a sample contains phosphorus (P $K\alpha$ = 6.16 Å) and calcium (Ca $K\beta$ = 3.09 Å). When the crystal is set to diffract P $K\alpha$, it will also diffract Ca $K\beta$ in the second order at very nearly the same angle. But the counter pulses from these two wavelengths will be of different sizes, and the window of the pulse-height analyzer can be set to pass the P $K\alpha$ pulses and exclude the Ca $K\beta$ pulses. (An alternate solution would be to use a germanium crystal as analyzer; as mentioned in Table 15-1, its (111) planes do not produce a second-order reflection.)

The escape-pulse phenomenon described in Sec. 7-9 is also relevant here and should be kept in mind when using a pulse-height analyzer.

Single-Channel Spectrometers

The spectrometers shown in Figs. 15-1(a), 15-4, and 15-5 are *single-channel* instruments, in the sense that they have only one counter, which is regarded as a "channel" through which information is received. The various spectral lines are measured sequentially by moving the counter from one line to another, either manually or by a mechanical drive.

In automatic single-channel spectrometers, the angular positions $2\theta_A$, $2\theta_B$, . . . , at which lines λ_A, λ_B, . . . will be reflected, are preset; the counter moves rapidly to position $2\theta_A$, remains there long enough to make an accurate intensity measurement, swings rapidly to $2\theta_B$, etc. After each measurement the total counts received are recorded on a print-out device or sent to a computer (Sec. 15–6) for conversion to percent concentration of the element involved. A large number of line positions 2θ can be programmed, and automatic sample changers can sequentially put up to 60 samples into position. Unattended analysis of many samples for many elements is therefore possible.

Multichannel Spectrometers

These are automatic instruments which have as many channels (crystals and counters) as there are spectral lines to be measured. In each channel a crystal and counter are fixed at the correct angular positions to measure a selected spectral line, such as line λ_A from element A. A number of channels are arranged in a circle around a centrally located x-ray tube (Fig. 15–6). All the analyzing channels receive the same fluorescent radiation from the sample, while one nondispersive control channel receives fluorescent radiation directly from a standard. The control channel serves to monitor the output of the x-ray tube. Some instruments have as many as 30 channels.

When a sample is being analyzed, all counters are started simultaneously. When the control channel has accumulated a predetermined number of counts, all counters are automatically stopped and the number of counts collected by each counter is recorded. Because the total fluorescent energy received in each analyzing counter is related to a fixed amount of energy entering the control counter, variations in the x-ray tube output do not affect the accuracy of the results.

The kind of crystal and kind of counter in each channel are selected to be best suited to the wavelength to be measured in that channel. No compromise has to be made in order to cover a certain wavelength range, as in a single-channel instrument.

Because all channels operate simultaneously, a multichannel instrument is fast.

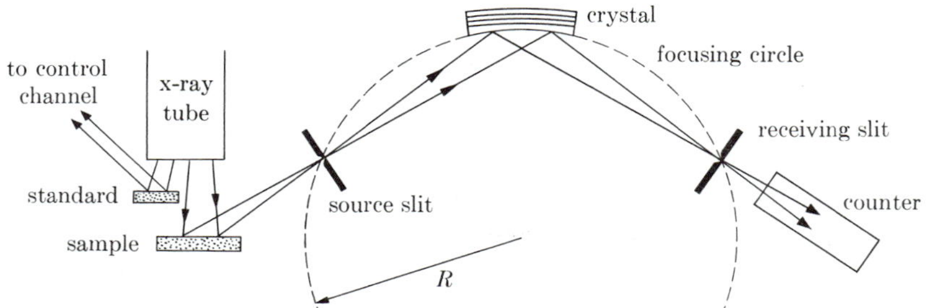

Fig. 15–6 Multichannel spectrometer, curved-crystal type, with relative arrangement of x-ray tube, sample, and one analyzing channel. (The tube shown is of the end-window type: the face of the target is inclined to the tube axis and the x-rays produced escape through a window in the end of the tube. Other spectrometers of this kind use a side-window x-ray tube.)

(a)

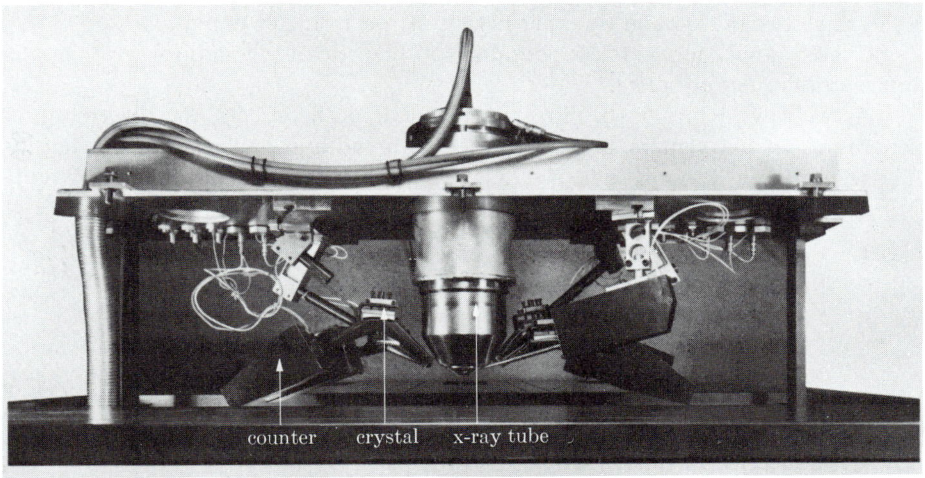

(b)

Fig. 15–7 Multichannel crystal spectrometer. (a) Overall view. (b) Close-up view of end-window x-ray tube and several analyzing channels; the crystal and counter of one channel are indicated; this entire enclosure is evacuated. (Courtesy of Applied Research Laboratories.)

A complete analysis for many elements, including print-out of the results, takes about one minute.

Multichannel spectrometers may contain, in addition to the fixed channels, a scanning channel in which a moving crystal and counter can sequentially scan the entire spectrum, just as in a single-channel instrument. This feature is useful for qualitative analysis or for quantitative determination of an element for which a fixed channel has not been preset.

Multichannel spectrometers are made by several manufacturers. Figure 15–7 shows one example.

15-4 INTENSITY AND RESOLUTION

In any spectrometer the attainment of adequate intensity and adequate resolution are important problems. The intensity of the fluorescent radiation emitted by the sample is very much less than that of the primary radiation incident on it, and can become very low indeed when the fluorescing element is only a minor constituent of the sample. This fluorescent radiation is then diffracted by the analyzing crystal, and another large loss of intensity occurs, because diffraction is such an inefficient process. The diffracted beam entering the counter may therefore be very weak, and a long counting time will be necessary to measure its intensity with acceptable accuracy. Spectrometer design must therefore ensure maximum intensity of the radiation entering the counter. At the same time, the spectrometer must be capable of high resolution, if the sample contains elements which have characteristic lines of very nearly the same wavelength and which must be separately identified. Both these factors, intensity and resolution, are affected by the kind of analyzing crystal used and by other details of spectrometer design.

Curved analyzing crystals, because of their focusing action, give greater intensity than flat crystals.

As we saw in Fig. 6–10, resolution depends both on $\Delta 2\theta$, the dispersion, or separation, of line centers, and on B, the line breadth at half-maximum intensity. The resolution will be adequate if $\Delta 2\theta$ is equal to or greater than $2B$. By differentiating the Bragg law, we obtain

$$\frac{\lambda}{\Delta\lambda} = \frac{2 \tan \theta}{\Delta 2\theta}. \tag{15-1}$$

When the minimum value of $\Delta 2\theta$, namely $2B$, is inserted, this becomes

$$\frac{\lambda}{\Delta\lambda} = \frac{\tan \theta}{B}. \tag{15-2}$$

The left-hand side of this equation gives the resolving power required to separate two lines of mean wavelength λ and wavelength difference $\Delta\lambda$. The right-hand side gives the resolving power available, and this involves both the mean Bragg angle of the lines and their breadth. Note that the available resolving power increases rapidly with θ, for a given line breadth. This means that, of two crystals producing the same line breadth, the one with the smaller plane spacing d will have the greater resolving power, because it will reflect to higher 2θ angles. For a given crystal, second-order reflections provide greater resolving power than first-order reflections, because they occur at larger angles, but their intensity is less than a fifth of that of first-order reflections.

The factors affecting the line width B can be discussed only with reference to particular spectrometers. In the flat crystal type (Fig. 15-4), the value of B depends partly on the collimation of the beam striking the crystal and partly on the perfection of the crystal itself. The beam reflected by the crystal into the counter is fairly wide, in a linear sense, but almost parallel; its *angular* width is measured by its divergence, and this is equal, if the crystal is perfect, to the divergence of the beam striking the crystal. The latter divergence is controlled by the Soller slit. If l is the length of the slit and s the spacing between plates, then the maximum divergence allowed is

$$\alpha = \frac{2s}{l} \text{ radian.}$$

For a typical slit with $l = 10$ cm and $s = 0.025$ cm, $\alpha = 0.3°$. But further divergence is produced by the mosaic structure of the analyzing crystal: this divergence is related to the extent of disorientation of the mosaic blocks, and has a value of about 0.2° for the crystals normally used. The line width B is the sum of these two effects and is typically of the order of 0.5°. The line width can be decreased by increasing the degree of collimation, but the intensity will also be decreased. Conversely, if the problem at hand does not require fine resolution, a more "open" collimator is used in order to increase intensity. Normally, the collimation is designed to produce a line width of about 0.5°, which will provide adequate resolution for most work.

When a curved reflecting crystal (Fig. 15-5) is used, the line width depends mainly on the width of the source slit S and the precision with which the crystal is ground and bent. The line width is normally about the same as that obtained with a flat crystal, namely, about 0.5°.

Knowing the d spacing of an analyzing crystal and the line width B it produces, we can calculate its resolving power and compare it with the resolving power required to separate closely spaced spectral lines. The smallest wavelength difference in the K series occurs between the $K\beta$ line of an element of atomic number Z and the $K\alpha$ line of an element of atomic number $(Z + 1)$. This difference itself varies with atomic number and is least for the $K\beta$ line of vanadium ($Z = 23$) and the $K\alpha$ line of chromium ($Z = 24$); these two wavelengths are 2.284 and 2.291 Å, respectively, and their difference is only 0.007 Å. A more common problem is the separation of the $K\beta$ line of chromium ($Z = 24$) from the $K\alpha$ line of manganese ($Z = 25$), since both of these elements occur in all stainless steels. The wavelength difference here is 0.018 Å and the mean wavelength 2.094 Å. The required resolving power $\lambda/\Delta\lambda$ is therefore 2.094/0.018 or 116. The available resolving power for a LiF crystal (200 reflection, $2d = 4.03$ Å, and $B = 0.5°$) is 70, which is insufficient to resolve Cr $K\beta$ and Mn $K\alpha$, as shown by Fig. 15-3(a). Increasing the resolving power to 124 by using a crystal with smaller d spacing (LiF crystal, 220 reflection,

$2d = 2.85$ Å, $B = 0.5°$) separates the lines enough for identification. If, in addition, the collimation is improved to an estimated B of 0.4°, the resolving power becomes 155, and the lines are well separated [Fig. 15–3(b)]. [An equivalent way of estimating resolution is to calculate the dispersion $\Delta 2\theta$ produced by a given crystal and compare it with the dispersion required, namely, $2B$. The value of $\Delta 2\theta$ is given by $(2 \tan \theta)(\Delta\lambda/\lambda)$, from Eq. (15–1).]

To sum up, high intensity is desirable in fluorescent analysis in order that the counting time required to obtain good accuracy be reasonably short; if the element to be detected is present only in small concentrations and a crystal of low reflecting power is used, the required counting times will be prohibitively long. High resolution is desirable whenever the analysis requires use of a spectral line having very nearly the same wavelength as another line from the sample or the x-ray tube target.

15–5 QUALITATIVE ANALYSIS

In qualitative work sufficient accuracy can be obtained by automatic scanning of the spectrum, with the ratemeter output fed to a chart recorder. Interpretation of the recorded spectrum will be facilitated if the analyst has on hand (a) a table of corresponding values of λ and 2θ for the particular analyzing crystal used, and (b) a single table of the principal K and L lines of all the elements arranged in numerical order of wavelength. Such a wavelength table is given in Vol. 4 of [G.11] and is much more useful than one, like that in Appendix 7, in which K and L lines are listed separately in order of atomic number.

Since it is important to know whether an observed line is due to an element in the sample or to an element in the x-ray tube target, a preliminary investigation should be made of the spectrum emitted by the target alone. For this purpose a substance like carbon or a plastic is placed in the sample holder and irradiated in the usual way; such a substance merely scatters part of the primary radiation into the spectrometer, and does not contribute any observable fluorescent radiation of its own. The spectrum so obtained will disclose the characteristic lines of the target and of any impurities it may contain.

15–6 QUANTITATIVE ANALYSIS

In the absence of interfering effects one would expect that the intensity I_A of a fluorescent line from element A in the sample would be directly proportional to the atomic fraction of A present. But interfering effects do exist; they are not trivial; and the fluorescent intensity can depart widely from proportionality to the amount present. Examples are shown in Fig. 15–8 for three binary mixtures containing iron. These curves demonstrate that the fluorescent intensity from a given element depends markedly on the other element or elements present.

This behavior is due chiefly to two effects:

1. *Matrix absorption.* As the composition of the sample changes, so does its absorption coefficient. As a result there are changes both in the absorption of the

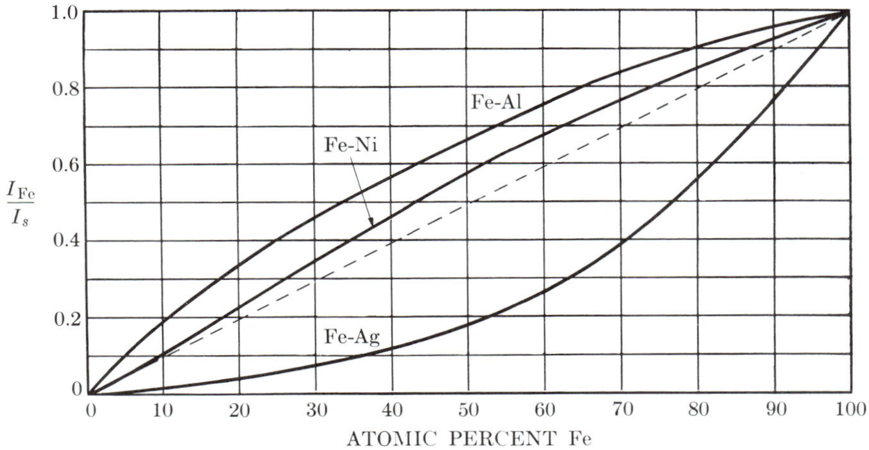

Fig. 15-8 Effect of iron concentration on the intensity of Fe $K\alpha$ radiation fluoresced by various mixtures. I_{Fe} and I_s are the Fe $K\alpha$ intensities from the mixture and from pure iron, respectively. Friedman and Birks [15.1].

primary radiation traveling into the sample and in the absorption of the fluorescent radiation traveling out. The absorption of the primary radiation is difficult to calculate, because the part of that radiation effective in causing K fluorescence, for example, in A has wavelengths extending from λ_{SWL}, the short-wavelength limit of the continuous spectrum, to λ_{KA}, the K absorption edge of A. To each of these incident wavelengths corresponds a different incident intensity and a different matrix absorption coefficient. The absorption of the fluorescent radiation, of wavelength λ_A, depends only on the absorption coefficient of the specimen for that particular wavelength. (Absorption effects are particularly noticeable in the Fe-Al and Fe-Ag curves of Fig. 15-8. The absorption coefficient of an Fe-Al alloy is less than that of an Fe-Ag alloy of the same iron content, with the result that the depth of effective penetration of the incident beam is greater for the Fe-Al alloy. A larger number of iron atoms can therefore contribute to the fluorescent beam, and this beam itself will undergo less absorption than in the Fe-Ag alloy. The over-all result is that the intensity of the fluorescent Fe $K\alpha$ radiation outside the specimen is greater for the Fe-Al alloy.)

2. *Enhancement* (*multiple excitation*). If the primary radiation causes element B in the specimen to emit its characteristic radiation, of wavelength λ_B, and if λ_B is less than λ_{KA}, then fluorescent K radiation from A will be excited not only by the incident beam but also by fluorescent radiation from B. (This effect is evident in the Fe-Ni curve of Fig. 15-8. Ni $K\alpha$ radiation can excite Fe $K\alpha$ radiation, and the result is that the observed intensity of the Fe $K\alpha$ radiation from an Fe-Ni alloy is closer to that for an Fe-Al alloy of the same iron content than one would expect from a simple comparison of the absorption coefficients of the two alloys. In the case of an Fe-Ag alloy, the observed Fe $K\alpha$ intensity is much lower, even though Ag $K\alpha$ can excite Fe $K\alpha$, because of the very large absorption in the specimen.)

These effects so complicate the calculation of fluorescent intensities that quantitative analysis is usually performed on an empirical basis, i.e., by the use of standard samples of known composition. These samples need not cover the 0–100 percent range, as in Fig. 15–8, but only quite limited ranges, because the greatest use of fluorescent analysis is in control work, where a great many samples of approximately the same composition have to be analyzed to see if their composition falls within specified limits. Standard samples of known composition, established by wet chemical analysis, may be purchased from the National Bureau of Standards [15.2] or from various commerical sources [15.3].

Three methods are used for quantitative analysis: calibration curves, empirical coefficients, and fundamental parameters.

Calibration Curves

When only a single element is to be determined and its concentration range is narrow, the matrix composition is essentially constant and so are the effects of matrix absorption and enhancement. The analytical problem is then reduced to the preparation of a single calibration curve, which is often linear. Figure 15–9 shows an example, for analysis in the parts-per-million (ppm) range. In this particular case, the background, represented by the intercept on the ordinate, is very high.

Empirical-Coefficient Method

This method, the most widely used of all, is required for such materials as alloys, cements, and ores, where typically five or more elements must be determined and where each element may vary in concentration over a substantial range. The

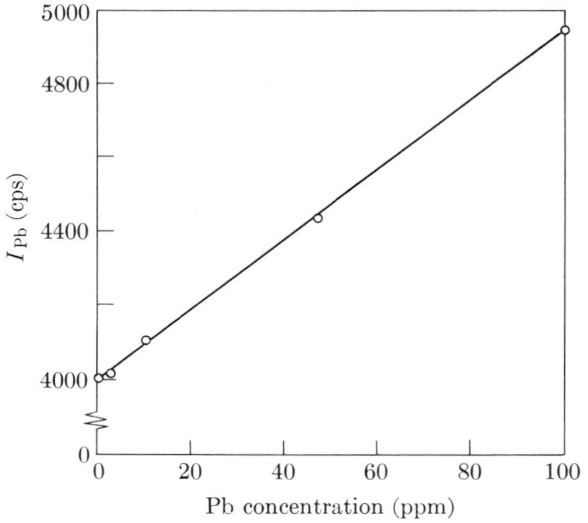

Fig. 15–9 Calibration curve for the determination of lead in oil. I_{Pb} is the intensity of the Pb $L\alpha$ line. Tungsten-target tube (40 kV, 20 mA), LiF analyzing crystal, scintillation counter. Jenkins and DeVries [G.37].

intensity I_A of a line from element A is no longer proportional to the concentration of A but depends also on the concentrations of elements B, C, D, . . . , because of variations in matrix absorption and enhancement.

A graphical method, as represented by a calibration curve, is no longer adequate because a whole family of such curves would be needed. Instead, an analytical approach is adopted. A set of simultaneous equations is written, involving measured line intensities, the desired concentrations, and empirical coefficients determined from previous measurements on standard specimens.

Equations of various forms are in use, and the whole subject is much too complex for detailed treatment here. The books referenced earlier (Sec. 15-1) should be consulted for particulars, as well as various papers in [15.11].

We will consider here only the approach of Birks [G.31], which is perhaps the simplest. Let I_{Ap} be the intensity of a line from pure A. Then $I_A/I_{Ap} = R_A$, the relative intensity of the line from A. To a first approximation and ignoring interfering effects, we might write

$$w_A = R_A \tag{15-3}$$

where w_A is the weight fraction of A in the sample. To allow for interferences we then modify the equation to

$$w_A = R_A(\alpha_{AA}w_A + \alpha_{AB}w_B + \alpha_{AC}w_C + \cdots). \tag{15-4}$$

Here a coefficient like α_{AB}, sometimes called an influence coefficient, measures the absorption-enhancement effect of element B on I_A and hence R_A. The coefficient is weighted by the amount of B present; although it ostensibly reflects only the effect of B on R_A, its value depends on the other elements present and is therefore valid only for a particular kind of matrix.

Suppose we wish to analyze ternary alloys containing elements A, B, and C. We then need three standard samples, at least as many as the number of elements to be determined, in order to find the coefficients α_{ij}. These standards should cover the same composition range as the unknowns to be analyzed. From measurements of R_A, R_B, and R_C on each of the three standards, we can determine the nine coefficients α_{ij} from three sets of three equations of the form of Eq. (15-4). To analyze an unknown, three equations of the same form, involving the values of R_A, R_B, and R_C for the unknown and the nine known coefficients, together with the auxiliary equation

$$w_A + w_B + w_C = 1, \tag{15-5}$$

are solved for w_A, w_B, and w_C by multiple regression.

Other analysis equations, some more complex than Eq. (15-4), are also in use, namely, those developed by Lucas-Tooth and Pyne [15.4, 15.5], Lachance and Traill [15.6], Claisse and Quintin [15.7], and Rasberry and Heinrich [15.8, 15.9].

Whichever analysis equation is chosen, a small computer will be necessary to solve it. Automatic spectrometers usually include a computer, and the manufacturer supplies programs for chemical analysis. Once the instrument is calibrated by means of standard samples, the operator has only to insert samples and read their chemical composition on a teletyped print-out.

Fundamental-Parameter method

In this method fluorescent intensities are calculated from first principles. The data needed for the calculation are (1) the spectral distribution (intensity vs. wavelength) of the primary beam from the x-ray tube, (2) mass absorption coefficients μ/ρ of all elements in the sample, and (3) fluorescent yields ω of all elements. The calculations require that measured line intensities I from the unknown be converted into relative intensities R by comparison with pure metal standards. (Use of R rather than I values eliminates the need to know the wavelength-dependent efficiencies of reflection by the analyzing crystal and detection by the counter.)

The advantage of the fundamental-parameter method is that only pure element standards are required. The disadvantage is that a large computer is needed, because the intensity calculations must be integrated over all the wavelengths in the primary beam; in addition, μ/ρ and ω values are not known very exactly, especially for the lighter elements.

The working equations relating line intensity to concentration that emerge from fundamental-parameter calculations are similar in form to the equations used in the empirical-coefficient method. These calculations therefore yield the α_{ij} coefficients directly. Jenkins *et al.* [15.10] discuss the use of a large computer to calculate coefficients for a particular class of samples and the subsequent application of these coefficients to the analysis of particular samples of that class by means of a small on-line computer. They also compare experimental and calculated coefficients. See also various papers in [15.11].

Sample Preparation

Solid samples must be surface ground or machined in a reproducible way, because variations in surface roughness can affect line intensities. Liquid samples can be contained in various kinds of cells.

Powder samples, after fine grinding and mixing, are pressed into special holders. These samples are the most difficult to prepare. Adequate mixing is essential, because only a thin surface layer is actually analyzed and this must be representative of the whole sample. To improve homogeneity, the powder sample is sometimes fused with a flux, such as sodium tetraborate, in a small platinum crucible; the result is a smooth, glassy pellet that may be analyzed directly. In monitoring air pollution by solid particles, the dust and the filter material on which it is caught are exposed together to the primary x-ray beam.

Bertin [G.46] describes sample preparation in considerable detail.

Precision and Accuracy

Precision (reproducibility) is governed by counting statistics, stability of the spectrometer components, and sample homogeneity.

Accuracy (approach to the "true" value) is governed by the homogeneity of the unknown and standard samples and the accuracy with which the compositions of the standards are known.

When the ratio of peak intensity to background for a particular fluorescent line is low, as it will be for low concentrations, the counting error is given by Eq. (7-10) and not by Eq. (7-9) applied to the peak alone.

ENERGY DISPERSION

15-7 SPECTROMETERS

The essential parts of an energy-dispersive spectrometer were shown in Fig. 15-1(b). It consists of a Si(Li) counter and a FET preamplifier, both cooled by liquid nitrogen, and a multichannel analyzer (MCA). It is mechanically simple, because it has no analyzing crystal, and electronically complex, because of the MCA.

The utility of this kind of spectrometer is based on two properties: (1) the excellent energy resolution of the Si(Li) counter (Sec. 7-8), which is far better than that of any other type of "proportional" counter, and (2) the ability of the MCA to perform rapid pulse-height analysis (Sec. 7-9). The latter feature makes this spectrometer very much faster than a single-channel crystal spectrometer. The MCA can measure the intensities of all the spectral lines from the sample in about a minute, unless elements in very low concentrations are to be determined.

Figure 15-10 shows an energy-dispersive spectrometer, operating, in this particular example, off the same x-ray tube as the diffractometer shown on the

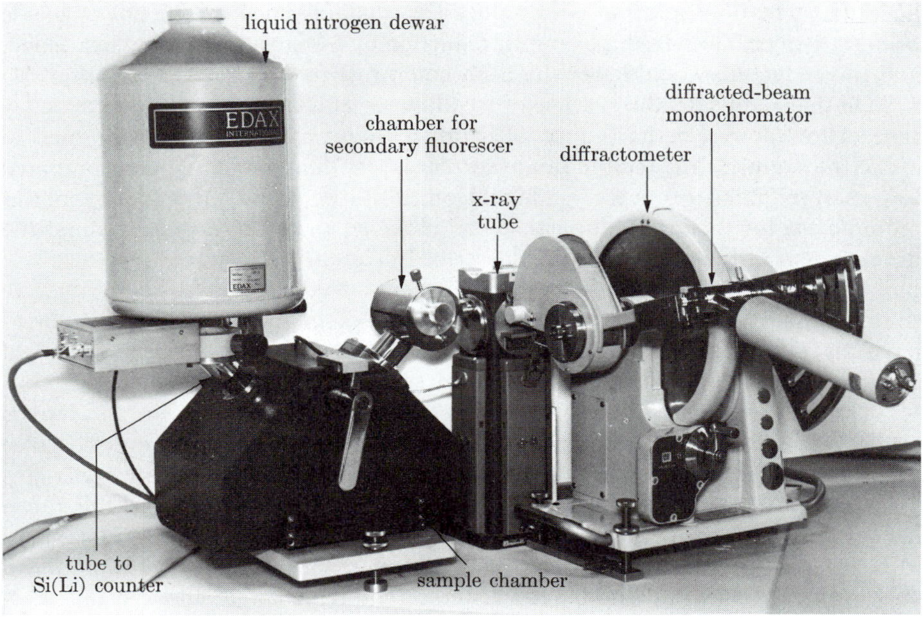

Fig. 15-10 Energy-dispersive fluorescence spectrometer (to left of x-ray tube) and diffractometer (at right). (Courtesy of Philips Electronic Instruments, Inc.)

right. The various components are identified, including the large dewar (cryostat) to hold a supply of liquid nitrogen. The role of the secondary fluorescer is described later (Sec. 15–9). Other essential parts of the spectrometer, not shown here, are the MCA and the video display. The latter is described in Sec. 15–10.

15–8 INTENSITY AND RESOLUTION

In a crystal spectrometer the intensity of the fluorescent radiation from the sample is greatly reduced before it reaches the counter by (a) the collimator, which is necessary for good resolution but which blocks all rays except those nearly parallel to the collimator axis, and (b) the crystal, which diffracts very inefficiently. Neither of these losses exists in the energy-dispersive spectrometer, so that the fluorescent radiation incident on the counter can be quite intense.

However, this apparent advantage cannot be fully exploited because of limitations on the *total* count rate. The counter of a crystal spectrometer counts pulses of only a single size, corresponding to x-rays of a single wavelength (single energy). The counter of an energy-dispersive instrument must count pulses of all sizes at the same time. Suppose a sample of A and B contains only a small amount of B and that x-rays are arriving at the counter at rates equivalent to 200,000 cps for A $K\alpha$ radiation and 2000 cps for B $K\alpha$. Suppose also that the counting rate varies linearly with x-ray intensity only up to 10,000 cps, because losses occur in the counter-electronics system at higher rates. Then even the weaker B $K\alpha$ radiation is being measured inaccurately, because the total count rate exceeds the range of linearity. The only solution is to reduce the excitation power (the power of the primary x-ray beam). In this example, reduction by a factor of 20 would be needed, and this reduction would lower the B $K\alpha$ count rate to 100 cps.

On the credit side, the absence of a collimator and crystal means that an x-ray tube of low power, perhaps air-cooled rather than water-cooled, can be used to excite the sample. The required tube current is less than a milliampere, compared to tens of milliamperes in a crystal-spectrometer tube. In fact, it is the low power requirement for this application that has provided the main impetus behind the development of the miniature tubes described in Sec. 1–7.

Another result of the low power requirement is that a low-intensity radioactive source, rather than an x-ray tube, can be used to excite the sample. Such sources have made possible various types of simple portable spectrometers, of limited capability but useful for such tasks as alloy sorting and ore prospecting.

The resolution of an energy-dispersive spectrometer is better, the shorter the wavelength or the greater the energy, as shown by Eq. (7–5). The reverse is true of the crystal spectrometer, as may be concluded from Eq. (15–2). For wavelengths greater than about 1 Å, the crystal spectrometer has better resolution; this range includes the $K\alpha$ lines of elements lighter than Kr ($Z = 36$). (This conclusion applies to a LiF crystal with $d = 2$ Å, a Si(Li) counter with pulse-distribution width $W = 160$ eV for Mn $K\alpha$ radiation, and the requirement that the line or pulse-distribution separation be twice the breadth.) Incidentally, much verbal confusion is possible in this area. *Resolution* may be defined quantitatively as $\Delta\lambda/\lambda$, which is the reciprocal of *resolving power* $\lambda/\Delta\lambda$. "Good" resolution then means a small numerical value of $\Delta\lambda/\lambda$. Not all writers agree with, or adhere to, these distinctions. A further difficulty is that almost everyone also uses the word "resolution" in a completely nonnumerical sense, as in the first sentence of this paragraph.

15-9 EXCITATION AND FILTRATION

Nondispersive spectrometry in its most developed form is based on *selective counting* by pulse-height analysis. But selective counting can be supplemented, or even replaced, by *selective excitation* and/or *selective filtration.*

Selective Excitation

Instead of exciting the sample by the general radiation from the x-ray tube, one can use the almost monochromatic fluorescent radiation from a secondary fluorescer, as in Fig. 15-11. Proper selection of the secondary fluorescer can then restrict the number of fluorescing elements in the sample, because only those with Z less than that of the secondary fluorescer are excited. Thus an Fe secondary fluorescer will excite Cr in a stainless steel sample but not Fe or Ni. As a result, the sensitivity for Cr detection is increased, because selective excitation has decreased the load (total count rate) on the detector/analyzer system.

The spectrometer in Fig. 15-10 operates from a secondary fluorescer. If the general radiation from an x-ray tube is preferred, for a particular analysis, the secondary fluorescer may be replaced by a piece of plastic, which will scatter the general radiation toward the sample. In either case, an x-ray tube operating at a current normal for diffraction experiments, a few tens of milliamperes, will give adequate intensity.

Selective Filtration

The proper filter, inserted between sample and counter, can prevent much unwanted radiation from reaching the counter. For example, in the analysis of brasses (Cu-Zn alloys), a Ni filter will pass much of the Cu $K\alpha$ radiation and absorb most of the Zn $K\alpha$. Selective filtration is most effective when the wavelengths to be separated are close together or widely apart, because, in either case, a filter can be chosen with quite different absorption coefficients for the two wavelengths. Balanced filters (Sec. 7-13) have also been used.

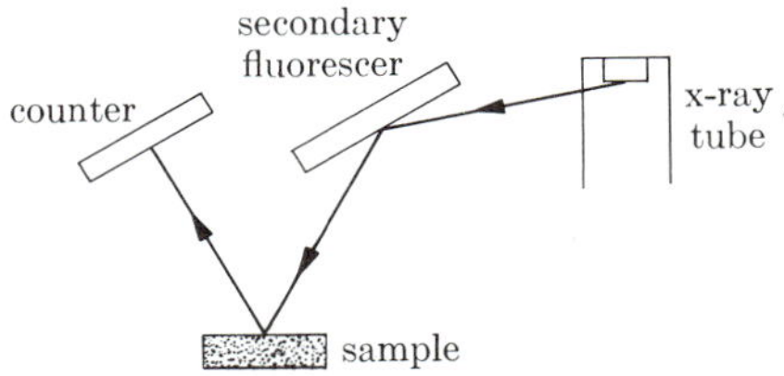

Fig. 15-11 Sample excitation by radiation from a secondary fluorescer.

In some portable spectrometers selective filtration entirely replaces pulse-height analysis as a means of wavelength separation, and only a simple counting system is needed.

15-10 CHEMICAL ANALYSIS

Energy-dispersive analysis is more often qualitative or semiquantitative than quantitative.

Qualitative analysis depends on identification of the peaks on the video display of the MCA, which shows the counts accumulated vs. x-ray energy. The analyst should have a table or chart of the energies of all K and L lines arranged consecutively, as in Vol. 4 of [G.11]. (Only $K\alpha$-line energies are listed in Appendix 7.) Or these data may be stored in the memory of the MCA, to be retrieved when needed.

Figure 15-12 shows an example of an energy-dispersed spectrum; the abscissa is energy in keV (from about 5 to almost 9 keV) and the ordinate is the number of counts (full scale = 50,000 counts) received in each channel in a certain time. This spectrum is the analog of the wavelength-dispersed spectrum of Fig. 15-3.

The display of Fig. 15-12 may be altered in various ways by the operator and various kinds of information may be presented. The horizontal and/or vertical scales may be expanded in order to examine particular parts of the spectrum. (The energy scale shown has 20 eV per channel, so that 200 channels of the MCA are being used for the energy range of 4 keV shown.) If some of the peaks in this spectrum are thought to be due to the K lines of iron, pushing a few buttons will (a) display on the top line the atomic number and chemical symbol of the suspected element (Z26, Fe), the line designation (K), and the energy of the $K\alpha$ line (6400 eV), and (b) superimpose "marker lines" (bright vertical lines) on the spectrum, showing the $K\alpha$ and $K\beta$ energies and relative intensities. Two of the peaks in this spectrum are thus identified as Fe $K\alpha$ (6400 eV) and Fe $K\beta$ (7060 eV).

Other controls enable the operator to determine the number of counts in a given peak by first selecting an energy window of the pulse-height analyzer to straddle the peak and then counting the total pulses within the width of that window.

The reader may wonder why an energy-dispersed spectrum as shown in Fig. 15-12 does not also contain the various *hkl* diffraction lines of the specimen, like the energy-dispersive diffraction pattern of Fig. 7-23. After all, the experimental arrangements for diffraction and spectroscopy are similar. Spectra sometimes *do* contain diffraction lines. Ordinarily, however, they are not observed because spectrometers do not contain collimators. The resulting "crossfire" of diffracted beams means that the angle θ is not defined; the diffracted radiation is therefore not peaked at particular energies but is merely spread over the background.

Quantitative analysis is done by the methods described earlier for wavelength dispersion. However, problems with partially overlapped peaks are more common

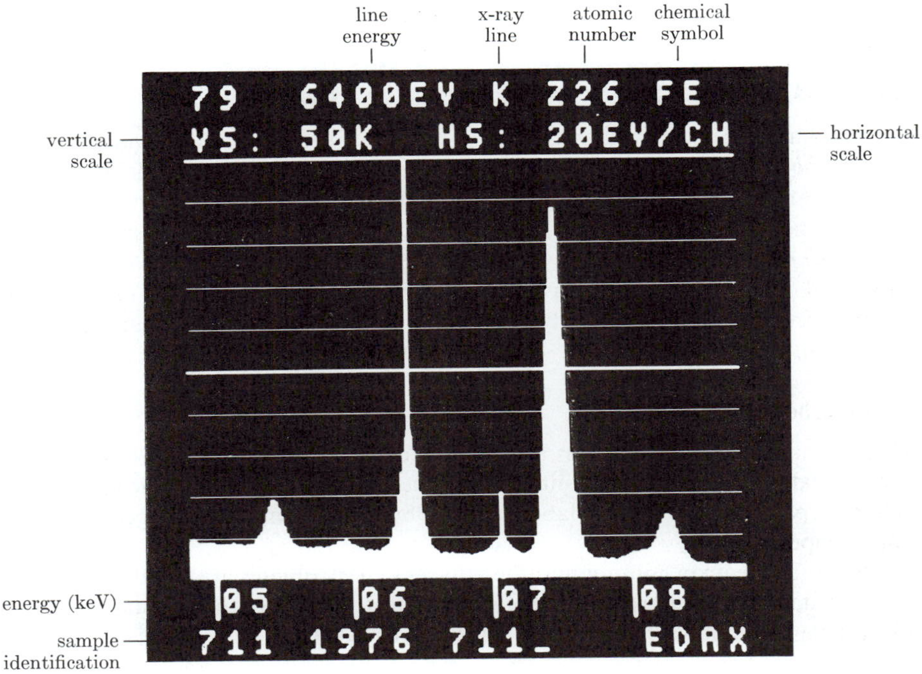

Fig. 15-12 Video display of an energy-dispersive spectrometer. (Courtesy of Philips Electronic Instruments, Inc.)

in energy dispersion because of the poorer resolution in the long-wavelength (lighter-element) region. Peak overlap makes it difficult to measure the area under each peak, and it is the area, rather than peak height, which is proportional to x-ray intensity, because the peak width W varies with energy.

Such problems are usually attacked by computer "unfolding" of the overlapped peaks. This operation amounts to the subtraction of the counts in the assumed pulse distribution of one peak from those in the combined peak; the result is two separate peaks which, if added together, would give the observed, overlapped peak. The computer that performs this operation and other kinds of "spectrum manipulation," such as background subtraction, is part of the same small computer that acts as an MCA.

Summary. The various differences between wavelength-dispersive and energy-dispersive spectrometry have been described in this chapter. All in all, wavelength dispersion by a crystal spectrometer is superior, chiefly because of its better resolution for most elements of interest, for the quantitative determination of several elements in a complex sample with high accuracy. Energy dispersion has a special place in microanalysis, in portable spectrometers, or wherever fast, semi-quantitative analyses are required.

MICROANALYSIS

15-11 MICROANALYSIS

Chemical analysis of a very small region in a large sample is called *microanalysis*. The diameter of the analyzed region is of the order of a micron or less. Such analysis can be done with an *x-ray microprobe* (*x-ray microanalyzer*) or with various forms of the *electron microscope*, if the latter is equipped with an x-ray spectrometer.

These devices are collectively known as *electron-column instruments*. They all may be likened to elaborate x-ray tubes, in which the specimen is the target and in which extreme measures have been taken to focus the electron beam from the filament into a very small spot.

Microprobe

This instrument always includes an x-ray spectrometer, because its whole purpose is chemical analysis. It came into general use in the early 1950s but is currently being replaced in many laboratories by spectrometer-equipped electron microscopes.

Both the electron column and the adjacent spectrometer are highly evacuated, a circumstance that improves light element detectability. With a crystal spectrometer all elements down to boron ($Z = 5$) can be detected.

Electron Microscope

In this instrument the primary purpose of the electron beam is to produce an image, and sometimes an electron diffraction pattern, of the sample. However, the electrons hitting the sample also generate x-rays, and a chemical analysis of the electron-irradiated area is therefore possible if a spectrometer is attached to the microscope.

The electron focal spot on the sample in a scanning electron microscope is at least ten times smaller than it is in the microprobe, in order to provide good spatial resolution in the electron image. As a result, the current in the electron beam hitting the sample is very small, of the order of 10^{-9} ampere or less, and the intensity of the emitted x-rays very low. Under these circumstances an energy-dispersive spectrometer is preferred, because it is far more efficient than a crystal spectrometer in detecting x-rays from the specimen, as well as being very much faster. In fact, the largest single application of the energy-dispersive spectrometer is as an attachment to the scanning electron microscope.

General

The electrons striking the sample inevitably generate unwanted white radiation as well as characteristic radiation. The observed x-ray lines are therefore accompanied by a much higher background than lines generated by fluorescence. This circumstance makes it harder to detect elements that are present in small amounts.

Both qualitative and quantitative analyses are possible. However, more than

90 percent of the work done with these instruments is qualitative or semi-quantitative, because the analyst is usually seeking an answer to one of the following questions:

1. What elements are present in this small region of the sample?
2. Is the concentration of element A in this region higher or lower than in this other region?

Answers to these questions enable precipitate particles, for example, to be identified and concentration gradients to be studied.

Quantitative microanalysis is more difficult than fluorescence analysis of macroscopic samples, whichever method is used:

1. *Empirical-coefficient method.* This method requires standards, and it is very difficult to obtain standards that are chemically homogeneous *on a micron scale.*

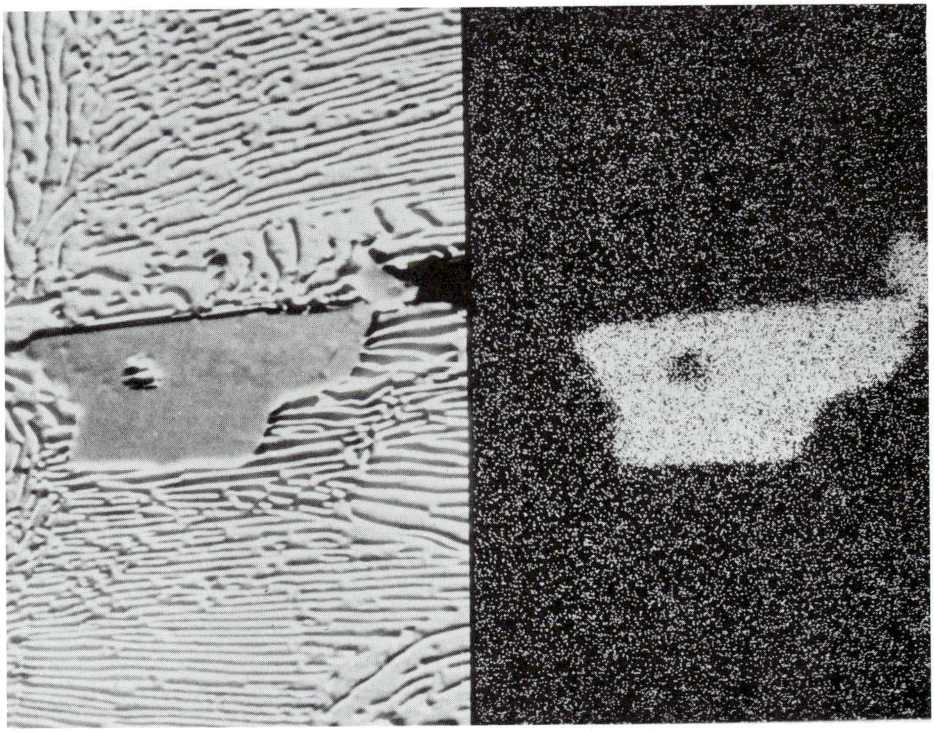

(a) Electron micrograph (b) Sulphur x-ray image

Fig. 15-13 Identification of an inclusion in cast iron by means of a scanning electron microscope and an energy-dispersive spectrometer. (a) Image formed by secondary electrons, 2280X, (b) x-ray image or "map" formed by the $K\alpha$ radiation of sulphur. (Courtesy of Advanced Metals Research Corporation.)

2. *Fundamental-parameter method.* The physics of x-ray production by electron impact is more complex than x-ray fluorescence. The calculation from first principles of line intensity vs. concentration is therefore more difficult.

Very useful "x-ray images," showing the spatial distribution of chemical composition, can be produced with a microprobe or spectrometer-equipped scanning electron microscope, because in either instrument a selected area of the sample can be scanned with the electron beam. Suppose the electron micrograph of Fig. 15–13(a), showing an inclusion in a specimen of cast iron, represents the area to be scanned, and we wish to know the distribution of sulphur in this area. The spectrometer is adjusted to isolate the S $K\alpha$ line from the sample, and the intensity of this line is used to modulate the intensity of the electron beam in the video display of the spectrometer. When the two electron beams (the beam scanning the sample and the beam scanning the video screen) are scanning in synchronism, bright areas on the video display correspond to sulphur-rich regions of the sample and dark areas to sulphur-poor regions, as in (b). The inclusion in this specimen is therefore identified as a sulphide.

PROBLEMS

***15–1** Consider LiF analyzing crystals, with lattice parameter a = 4.028 Å and line width B = 0.5°. Calculate the dispersion $\Delta 2\theta$ in degrees for the 200, 220, and 420 reflections for (a) Co $K\beta$ and Ni $K\alpha$ and (b) Sn $K\beta$ and Sb $K\alpha$. (c) Which of these reflections gives adequate resolution for (a) and for (b)?

15–2 Verify the statement in Sec. 15–8 regarding spectrometer resolution by wavelength dispersion (WD) and energy dispersion (ED) by calculating the percent resolution $\Delta\lambda/\lambda$ for each type and for wavelengths of 0.5, 1.0, and 1.5 Å. For the WD spectrometer, assume a LiF crystal with $2d$ = 4.03 Å (200 reflection) and line width B = 0.5°. For the ED spectrometer, assume a Si(Li)-FET counter and Eq. (7–5). Assume also that the line or pulse-distribution separation must be twice the breadth for adequate resolution.

16

Measurement of Residual Stress

16-1 INTRODUCTION

When a polycrystalline piece of metal is deformed elastically in such a manner that the strain is uniform over relatively large distances, the lattice plane spacings in the constituent grains change from their stress-free value to some new value corresponding to the magnitude of the applied stress, this new spacing being essentially constant from one grain to another for any particular set of planes similarly oriented with respect to the stress. This uniform macrostrain, as we saw in Sec. 9–4, causes a *shift* of the diffraction lines to new 2θ positions. On the other hand, if the metal is deformed plastically, the lattice planes usually become distorted in such a way that the spacing of any particular (hkl) set varies from one grain to another or from one part of a grain to another. This nonuniform microstrain causes a *broadening* of the corresponding diffraction line. Actually, both kinds of strain are usually superimposed in plastically deformed metals, and diffraction lines are both shifted and broadened, because not only do the plane spacings vary from grain to grain but their mean value differs from that of the undeformed metal.

In this chapter we will be concerned with the line shift due to uniform strain. From this shift the strain may be calculated and, knowing the strain, we can determine the stress present, either by a calculation involving the mechanically measured elastic constants of the material, or by a calibration procedure involving measurement of the strains produced by known stresses. X-ray diffraction can therefore be used as a method of "stress" measurement. Note, however, that stress is not measured directly by the x-ray method or, for that matter, by any other method of "stress" measurement. It is always strain that is measured; the stress is determined indirectly, by calculation or calibration.

The various methods of "stress" measurement differ only in the kind of strain gauge used. In the electric-resistance method for the measurement of applied stress, the gauge is a short length of fine wire or foil cemented to the surface of the metal being tested; any strain in the metal is shared by the gauge, and any extension or contraction of the gauge is accompanied by a change in its resistance, which can therefore be used as a measure of strain. In the x-ray method, the strain gauge is the spacing of lattice planes.

In principle, the x-ray method is applicable to any crystalline material. While it has had some application to stress measurement in ceramics and rocks, its major use is the measurement of residual stress in metals and alloys.

The x-ray method has been reviewed by Barrett and Massalski [G.25], Klug and Alexander [G.39], Taylor [G.19], and Norton [16.1, 16.2]. A detailed treatment has been published by the Society of Automotive Engineers [16.3].

16-2 APPLIED STRESS AND RESIDUAL STRESS

Before the x-ray method is examined in any detail, it is advisable to consider first a more general subject, namely, the difference between applied stress and residual stress, and to gain a clear idea of what these terms mean. Consider a metal bar deformed elastically, for example in uniform tension. The *applied stress* is given simply by the applied force per unit area of cross section. If the external force is removed, the stress disappears, and the bar regains its initial stress-free dimensions. On the other hand, there are certain operations that can be performed on a metal part, which will leave it in a stressed condition even after all external forces have been removed. This stress, which persists in the absence of external force, is called *residual stress*.

For example, consider the assembly shown in Fig. 16-1(a). It consists of a hollow section through which is passed a loosely fitting bolt with threaded ends. If nuts are screwed on these ends and tightened, the sides of the assembly are compressed and the bolt is placed in tension. The stresses present are residual, inasmuch as there are no external forces acting on the assembly as a whole. Notice also that the tensile stresses in one part of the assembly are balanced by compressive stresses in other parts. This balance of opposing stresses, required by the fact that the assembly as a whole is in equilibrium, is characteristic of all states of residual stress.

An exactly equivalent condition of residual stress can be produced by welding a cross bar into an open section, as shown in Fig. 16-1(b). We can reasonably

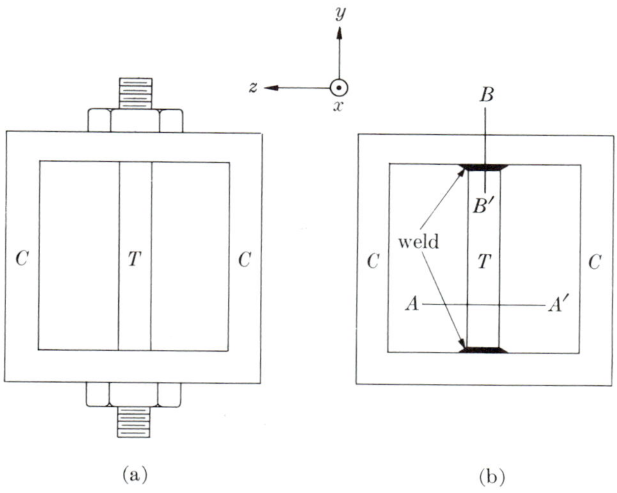

Fig. 16-1 Examples of residual stress. T = tension, C = compression.

assume that, at the instant the second weld is completed, a substantial portion of the central bar is hot but that the two side members are far enough from the heated zone to be at room temperature. On cooling, the central bar tries to contract thermally but is restrained by the side members. It does contract partially, but not as much as it would if it were free, and the end result is that the side members are placed in compression and the central rod in tension when the whole assembly is at room temperature. Residual stress is quite commonly found in welded structures, which are often called *weldments.* Actually, the residual stress state in the weldment of Fig. 16–1(b) is more complex than might at first appear. Across the section AA' the stress σ_y is entirely tensile and constant. But on BB', which crosses the weld itself, the stress σ_z in the z direction varies both in magnitude and sign with the position y of the point considered. Numerous studies of welds have shown that σ_z is tensile (positive) in the weld and compressive (negative) on either side.

Plastic flow can also set up residual stresses. The beam shown in Fig. 16–2(a) is supported at two points and loaded by two equal forces F applied near each end. At any point between the two supports the stress in the outside fibers is constant, tensile on the top of the beam and compressive on the bottom. These stresses are a maximum on the outside surfaces and decrease to zero at the neutral axis, as indicated by the stress diagram at the right of (a). This diagram shows how the longitudinal stress varies across the section AA', when all parts of the beam are

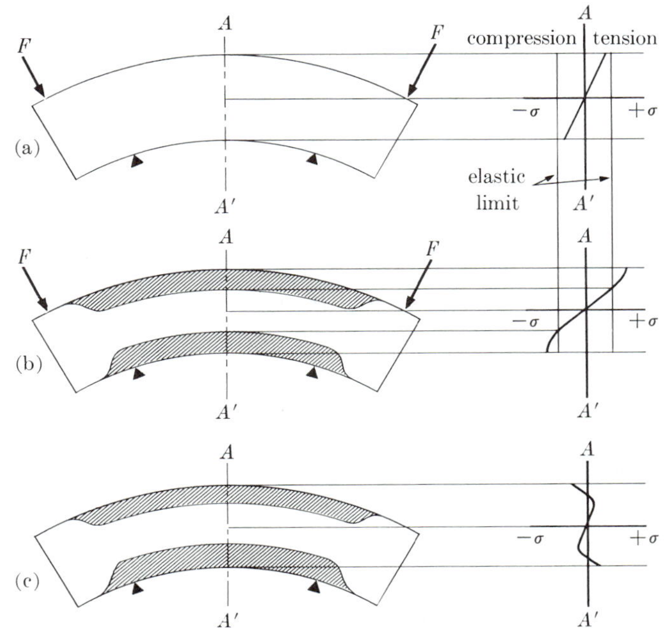

Fig. 16–2 Residual stress induced by plastic flow in bending: (a) loaded below elastic limit; (b) loaded beyond elastic limit; (c) unloaded. Shaded regions have been plastically strained.

below the elastic limit. Suppose the load on the beam is now increased to the point where the elastic limit is exceeded, not only in the outer fibers but to a considerable depth. Then plastic flow will take place in the outer portions of the beam, indicated by shading in (b), but there will be an inner region still only elastically strained, because the stress there is still below the elastic limit. The stresses above the neutral axis are still entirely tensile, both in the elastically and plastically strained portions, and those below entirely compressive. If the load is now removed, these stresses try to relieve themselves by straightening the beam. Under the action of these internal forces, the beam does partially straighten itself, and to such an extent that the stress in the outer regions is not only reduced to zero but is actually changed in sign, as indicated in (c). The end result is that the unloaded beam contains residual compressive stress in its top outside portion and residual tensile stress in its lower outside portion. It is quite common to find residual stress in metal parts which have been plastically deformed, not only by bending but by rolling, drawing, machining, grinding, etc.

Note that the service stress in any loaded machine or structure is the algebraic sum of the applied stress, due to the service load, and any residual stress that may have existed before the service load was applied. If the residual stress is not known, neither is the service stress. When the service stress reaches dangerous levels, failure occurs. Interest in residual stress stems mainly from the role it plays in three kinds of metal failure: fatigue failure, brittle fractures in general, and stress-corrosion cracking.

Applied stress is easy to measure. An electric-resistance gauge is cemented to the unloaded part, and gauge readings are made before and after the load is applied. The difference in the gauge readings gives the strain, and multiplication of the strain by Young's modulus E gives the applied stress.

Residual stress is more difficult to measure, and there are only two practical methods of doing it:

1. *X-ray diffraction.* This method is nondestructive for the measurement of surface stress. If the stress is to be measured at some point below the surface, material must be removed down to that point to expose a new surface for x-ray examination; the x-ray method then becomes destructive. Ordinarily, however, one is most interested in the stress at the surface, where the applied stress is usually highest and where failures usually originate.

2. *Dissection (mechanical relaxation).* This method is inherently destructive, even for the measurement of surface stress, and slow. Part of the residually-stressed object is removed, by cutting or some other method. This removal upsets the pre-existing balance of forces, with the result that the remainder of the object mechanically relaxes (strains) more or less completely. If this strain is measured, the stress originally existing can be computed. For example, suppose we wish to know the stress in the side members of the weldment in Fig. 16–1(b). We fix a gauge, to measure strain in the y direction, on one of these members. We then cut through the central rod along AA'. This cut allows the side members, originally in compression, to elongate to their stress-free length, and the gauge shows a strain of

$+\varepsilon_y$. Before the cut was made, the strain relative to the stress-free state must have been $-\varepsilon_y$, so that the original stress σ_y was $-\varepsilon_y E$. Similarly, the residual stress at various depths of the bent beam of Fig. 16-2(c) may be calculated from the change in curvature that results from successive removal of layers parallel to the neutral plane.

The x-ray method, being nondestructive, has the great advantage that it makes possible repeated measurements on the same specimen. For example, we may measure stress before and after some treatment designed to produce or modify residual stress. Or we may measure residual stress on a machine component at various stages in its service life.

Note also that the x-ray method measures the *existing* stress, whether it be solely residual or the sum of residual and applied. It therefore has the capability of measuring the actual service stress in a machine or structure under a service load.

16-3 GENERAL PRINCIPLES

The x-ray method is best approached by first considering the case of uniaxial stress, where the stress acts only in a single direction, even though this condition is rare in practice. The more general case of biaxial stress will be dealt with later.

Uniaxial stress

Consider a cylindrical rod of cross-sectional area A stressed elastically in tension by a force F (Fig. 16-3). There is a stress $\sigma_y = F/A$ in the y direction but none in the x or z directions. (This stress is the only *normal stress* acting; there are also *shear stresses* present, but these are not directly measurable by x-ray diffraction.) The stress σ_y produces a strain ε_y in the y direction given by

$$\varepsilon_y = \frac{\Delta L}{L} = \frac{L_f - L_0}{L_0},$$

where L_0 and L_f are the original and final lengths of the bar. This strain is related to the stress by

$$\sigma_y = E\varepsilon_y. \tag{16-1}$$

The elongation of the bar is accompanied by a decrease in its diameter D. The strains in the x and z directions are therefore given by

$$\varepsilon_x = \varepsilon_z = \frac{D_f - D_0}{D_0},$$

where D_0 and D_f are the original and final diameters of the bar. If the material of the bar is isotropic, these strains are related by the equation

$$\varepsilon_x = \varepsilon_z = -\nu\varepsilon_y, \tag{16-2}$$

where ν is Poisson's ratio for the material of the bar. The value of ν ranges from about 0.25 to about 0.45 for most metals and alloys.

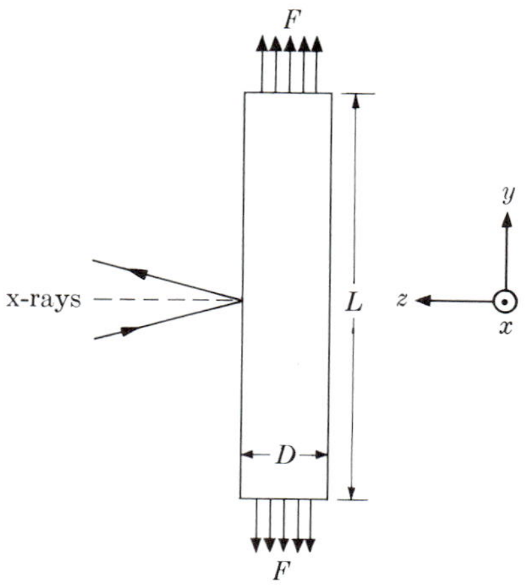

Fig. 16–3 Bar in pure tension, with x-rays reflected from planes parallel to axis.

To measure ε_y by x-rays would require diffraction from planes perpendicular to the axis of the bar. Since this is usually physically impossible, we use instead reflecting planes which are parallel to the axis of the bar by making the back-reflection x-ray measurement indicated in Fig. 16–3. (It is essential that a back-reflection technique be used, in order to gain sufficient precision in the measurement of plane spacing. Even quite large stresses cause only a very small change in d.) In this way we obtain a measurement of the strain in the z direction since this is given by

$$\varepsilon_z = \frac{d_n - d_0}{d_0}, \tag{16–3}$$

where d_n is the spacing of the planes parallel to the bar axis under stress, and d_0 is the spacing of the same planes in the absence of stress. (The subscript n describes the fact that the reflecting-plane normal is *normal* to the specimen surface.) Combining Eqs. (16–1), (16–2), and (16–3), we obtain the relation

$$\sigma_y = -\frac{E}{\nu}\left(\frac{d_n - d_0}{d_0}\right), \tag{16–4}$$

which gives the required stress.

It should be noted that only a particular set of grains contributes to a particular hkl reflection. These are grains whose (hkl) planes are parallel to the surface of the bar, as indicated in Fig. 16–4, and which are compressed by the applied stress, that is, d_n is less than d_0. Grains whose (hkl) planes are normal to the surface have these planes extended, as shown in an exaggerated fashion in the drawing. The spacing d_{hkl} therefore varies with crystal orientation, and there is thus no possibility of using any of the extrapolation procedures described in Chap. 11 to measure d_{hkl} accurately. Instead we must determine this spacing from the 2θ position of a single diffraction line.

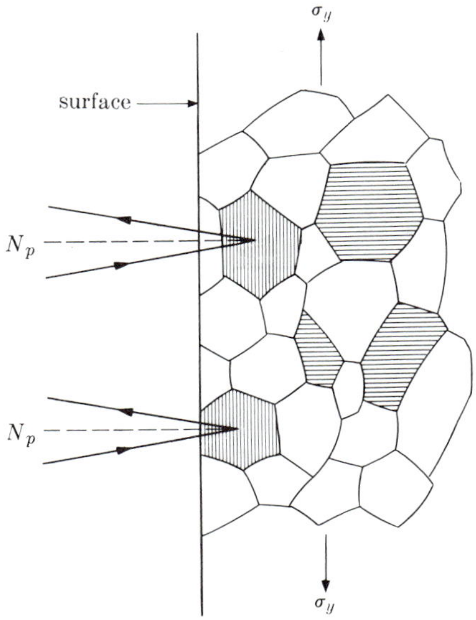

Fig. 16-4 Diffraction from strained aggregate, tension axis vertical. Lattice planes shown belong to the same (hkl) set. N_p = reflecting-plane normal.

Equation (16-4) shows that knowledge of d_0 is required. If the specimen contains only applied stress, d_0 can be obtained from a measurement on the unloaded specimen. (Such a stress measurement is rarely made, and then only for certain research purposes; it is far easier to measure applied stress with an electric-resistance gauge.) If the specimen contains residual stress, d_0 must be measured on a small stress-free portion cut out of the specimen; the method then becomes destructive and of no interest.

Note that the value of d_0 cannot be obtained from measurements on a "similar" stress-free material. If the specimen is iron, for example, it is not sufficiently accurate to look up the lattice parameter of "iron" in a handbook and calculate d_0 from that parameter. The specimen may contain more or less impurities than the material for which the parameter is given, and impurities can change the parameter.

Equation (16-4) is therefore not a practical basis for the measurement of residual stress. We will see later that *two* measurements of plane spacing on the *stressed* specimen are required for a nondestructive determination of stress.

Biaxial Stress

In a bar subject to pure tension the normal stress acts only in a single direction. But in general there will be stress components in two or three directions at right angles to one another, forming so-called biaxial or triaxial stress systems. However, the stress at right angles to a free surface is always zero at that surface, so that at the surface of a body, which is the only place where we can measure stress, we never have to deal with more than two stress components and these lie in the plane of the surface. Only in the interior of a body can the stresses be triaxial.

Within a stressed body, whatever the stress system, three mutually perpendicular directions (1, 2, and 3) can be found which are normal to planes on which no shear stress acts. These are called the *principal directions*, and the stresses acting in these directions (σ_1, σ_2, and σ_3) are called the *principal stresses*.

Consider now a portion of the surface of a stressed body (Fig. 16–5). Principal stresses σ_1 and σ_2 are parallel to the surface, and σ_3 is zero. However, ε_3, the strain normal to the surface, is not zero. It has a finite value, given by the Poisson contractions due to σ_1 and σ_2:

$$\varepsilon_3 = -\nu(\varepsilon_1 + \varepsilon_2) = -\frac{\nu}{E}(\sigma_1 + \sigma_2). \tag{16–5}$$

The value of ε_3 can be found by measuring the spacing d of planes parallel to the surface and is given by Eq. (16–3). Substituting this value into (16–5), we obtain

$$\frac{d_n - d_0}{d_0} = -\frac{\nu}{E}(\sigma_1 + \sigma_2). \tag{16–6}$$

Therefore, in the general case, only the sum of the principal stresses can be obtained from such a measurement, and the value of d_0 is again required.

What we wish to measure, however, is the single stress σ_ϕ acting in some chosen direction in the surface, say the direction OC of Fig. 16–5, where OC makes an angle ϕ with principal direction 1. We do this by making two measurements, one of the strain ε_3 along the surface normal and one of the strain ε_ψ along OB. The direction OB lies in the vertical plane $OABC$ through OC at an angle ψ, usually chosen to be 45°, to the surface normal. The strain ε_3 is derived from the spacing d_n of planes parallel to the surface, and the strain ε_ψ from the spacing d_i of planes whose normal is *inclined* along OB. The spacings of these planes are shown diagrammatically in Fig. 16–6(a). This is a plot in polar coordinates d,ψ in the plane $OABC$ containing the stress σ_ϕ to be measured. The length and direction

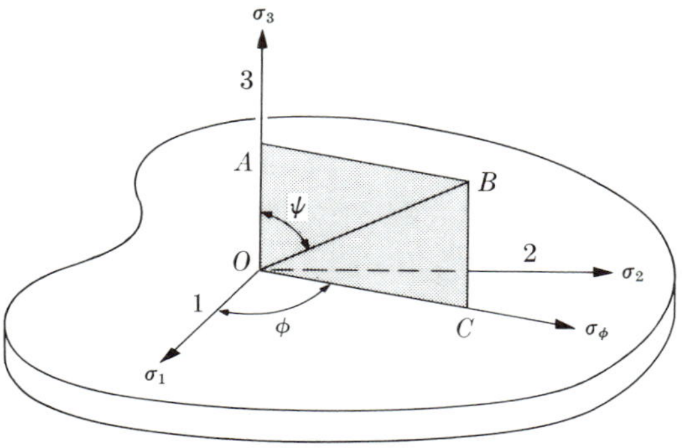

Fig. 16–5 Stresses at the surface of a stressed body. $\sigma_3 = 0$. The stress to be measured is σ_ϕ.

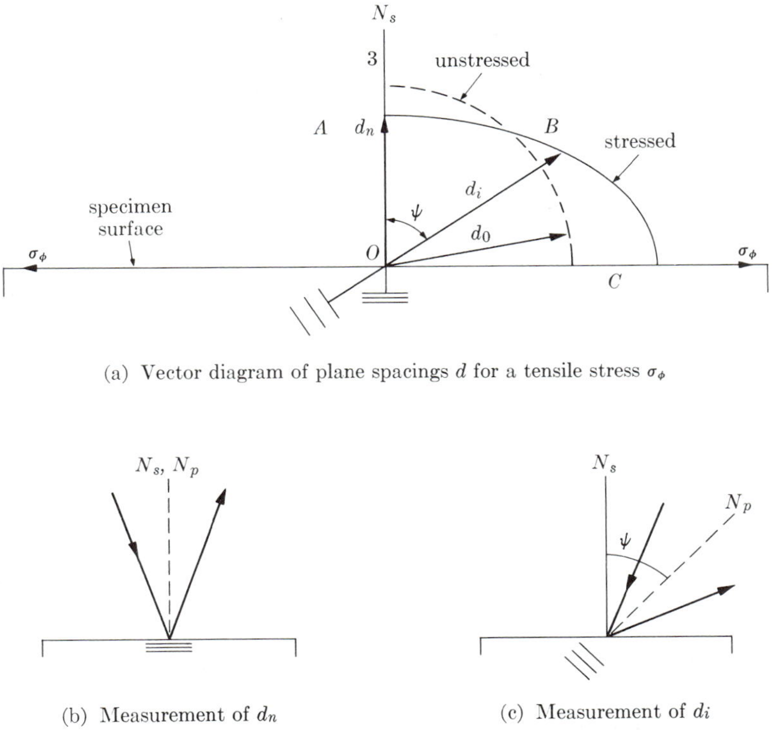

Fig. 16-6 (a) Plane-spacing diagram. (b) and (c) Orientations of x-ray beams relative to specimen. N_s = normal to specimen surface, N_p = normal to reflecting planes.

of any vector show the spacing and plane-normal direction, respectively, of any selected (*hkl*) set of planes. If the specimen were unstressed, the end of the d_0 vector would describe the dashed circle shown, because plane spacing is then independent of plane orientation. This is not true when stress is present; if the stress is tensile, d_i increases with ψ along the curve shown by the full line. Figures 16-6(b) and (c) show the orientations of the x-ray beam required to measure d_n and d_i.

Elasticity theory for an isotropic solid shows that the strain along the inclined line OB is

$$\varepsilon_\psi = \frac{1}{E}\left[\sigma_\phi(1 + \nu)\sin^2\psi - \nu(\sigma_1 + \sigma_2)\right]. \tag{16-7}$$

Subtraction of Eq. (16-5) from (16-7) yields

$$\boxed{\varepsilon_\psi - \varepsilon_3 = \frac{\sigma_\phi}{E}(1 + \nu)\sin^2\psi.} \tag{16-8}$$

This equation is the basis for the x-ray measurement of stress. It shows, as first recognized by Glocker *et al.* in 1936 [16.4], that the *difference* between two strains in a stressed specimen depends only on the stress acting in the plane of those strains. Expressing the strains in terms of plane spacings, we have

$$\frac{d_i - d_0}{d_0} - \frac{d_n - d_0}{d_0} = \frac{d_i - d_n}{d_0} = \frac{\sigma_\phi}{E}(1 + \nu)\sin^2\psi. \tag{16-9}$$

An ingenious approximation is now made [16.4]; this approximation, coupled with Eq. (16–8), is at the heart of the x-ray method. Because d_i, d_n, and d_0 are very nearly equal to one another, $(d_i - d_n)$ is small compared to d_0. The unknown spacing d_0 can therefore be replaced in the denominator above by d_n or d_i with negligible error (Prob. 16–1). If d_0 is replaced by d_n, Eq. (16–9) can be written

$$\sigma_\phi = \frac{E}{(1 + \nu)\sin^2\psi}\left(\frac{d_i - d_n}{d_n}\right). \tag{16-10}$$

This equation allows us to calculate the stress in any chosen direction from plane spacings determined from two measurements, made in a plane normal to the surface and containing the direction of the stress to be measured. Note that the angle ϕ does not appear in this equation and fortunately so, because we do not generally know the directions of the principal stresses *a priori*. Nor is it necessary to know the unstressed plane spacing d_0; the measurement is therefore non-destructive, because there is no need to cut out part of the specimen to obtain a stress-free sample for comparison. (Although the basis of the x-ray method, Eq. (16–8), involves a difference of two strains, we see that in the final analysis we measure, not strains, but *lengths*, namely, the d spacings of differently oriented planes.)

Note also that Eq. (16–10) is valid whether the stress system is uniaxial or biaxial, i.e., whether σ_ϕ is the only stress or is part of a biaxial system. If a transverse stress $\sigma_{\phi+90°}$ exists at right angles to σ_ϕ, it will alter d_i and d_n by the same factor but will not change the value of $(d_i - d_n)/d_n$. This other stress can be detected, if present, by a d_i measurement at an azimuth angle of $(\phi + 90°)$.

The magnitudes and directions of the principal stresses σ_1 and σ_2, if needed, may be found by making four measurements: the normal spacing d_n and three inclined spacings d_i, determined at azimuth angles of ϕ, $\phi + 60°$, and $\phi - 60°$ [G.25]. However, the principal directions may often be inferred from the symmetry of the specimen. For example, quenching a solid cylinder from a high temperature produces a residual stress system in which the longitudinal stress is σ_1 and the circumferential (hoop) stress is σ_2.

16–4 DIFFRACTOMETER METHOD

Until about 1950, x-ray stress measurements were made only by photographic methods. Today the diffractometer is preferred because it is faster and more precise. The photographic method, which still has its place, will be described later.

Steel which has been hardened by quenching and tempering produces very broad diffraction lines. Measuring the positions of such lines on a photographic film is difficult and can only be done by making a microphotometer record of the film. However, such measurements can be made in a straightforward fashion with a diffractometer, as was first shown by Christenson and Rowland [16.5]. Their demonstration that residual stress could be accurately measured in hardened steel, an important industrial material, greatly enlarged the scope of the x-ray method.

The standard diffractometer method is often called the *two-exposure method*, because two plane-spacing measurements are made, one of d_n at $\psi = 0$ and one of d_i at $\psi = 45°$. The word "exposure" is here only a relic of the earlier photographic method.

Most diffractometers must be modified to meet two requirements of stress measurement:

1. The specimen holder must be robust enough to support fairly large and heavy specimens, and it must be possible to rotate the holder about the diffractometer axis independently of the counter rotation, in order to change the angle ψ.

2. In the usual technique, the counter must be mounted on a radial slide that allows counter movement along a radius of the diffractometer circle. (Some prefer to leave the counter in its usual position and move the counter slit. In the following, when counter movement is mentioned, either practice is meant.)

The second requirement arises from the need to preserve focusing conditions, as shown in Fig. 16–7. In (a) the specimen is equally inclined to the incident and diffracted beams; ψ is zero and the specimen normal N_s coincides with the reflecting-

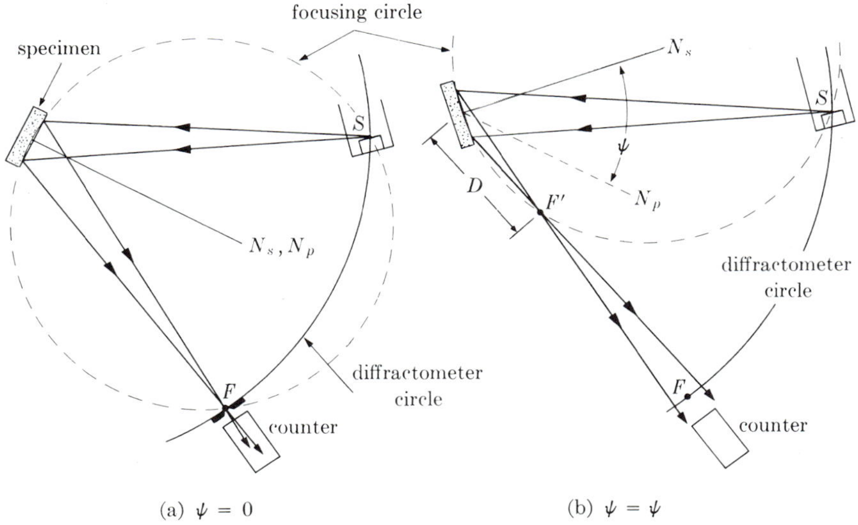

Fig. 16–7 Use of a diffractometer for stress measurement.

plane normal N_p. Radiation divergent from the source S is diffracted to a focus at F on the diffractometer circle. In (b) the specimen has been turned through an angle ψ for the inclined measurement. Because the focusing circle is always tangent to the specimen surface, rotation of the specimen alters the focusing circle both in position and radius, and the diffracted rays now come to a focus at F', located a distance D from the diffractometer axis. If R is the radius of the diffractometer circle, then it may be shown that

$$\frac{D}{R} = \frac{\sin(\theta - \psi)}{\sin(\theta + \psi)}. \qquad (16\text{–}11)$$

If ψ is 45°, then D/R is 0.70 for $\theta = 80°$ and 0.47 for $\theta = 70°$.

When ψ is not zero, the focal point of the diffracted beam therefore lies between F, the usual position of the counter receiving slit, and the specimen. To preserve focusing, the counter should be moved so that the slit is at F'.

Because the angular position 2θ of the diffracted beam is measured directly with a diffractometer, it is convenient to write the stress equation in terms of 2θ rather than plane spacings. Differentiating the Bragg law, we obtain

$$\frac{\Delta d}{d} = -\frac{\cot\theta\,\Delta 2\theta}{2}.$$

Combining this relation with Eq. (16–10) gives

$$\sigma_\phi = \frac{E\cot\theta(2\theta_n - 2\theta_i)}{2(1+\nu)\sin^2\psi}. \qquad (16\text{–}12)$$

Put

$$K_1 = \frac{E\cot\theta}{2(1+\nu)\sin^2\psi}. \qquad (16\text{–}13)$$

Then

$$\sigma_\phi = K_1(2\theta_n - 2\theta_i) = K_1(\Delta 2\theta), \qquad (16\text{–}14)$$

where $2\theta_n$ is the observed value (in radians) of the diffraction angle in the "normal" measurement ($\psi = 0$) and $2\theta_i$ its value in the inclined measurement ($\psi = \psi$). The constant K_1 is called the *stress constant*. For greatest sensitivity K_1 should be as small as possible, which is why θ should be as large as possible. For measurements made on the 211 line of steel (body-centered cubic) with Cr $K\alpha$ radiation, $\theta = 78°$. If we put $\psi = 45°$ and assume $E = 30 \times 10^3$ ksi* and $\nu = 0.29$, then $K_1 = 86.3$ ksi/deg $\Delta 2\theta$. If the standard deviations of the measurements of $2\theta_n$ and $2\theta_i$ are each 0.02°, the standard deviation of the stress σ_ϕ is 2.4 ksi. This value is independent of the stress level. The largest residual stress measured to date is a compressive stress of 300 ksi in steel; this stress would cause a line shift $\Delta 2\theta$ of 3.5° for the stress constant given above.

Essentially, the quantity measured in the diffractometer method is $\Delta 2\theta = (2\theta_n - 2\theta_i)$, the shift in diffraction-line position due to stress as the angle ψ is

---

* 1 ksi = 10^3 psi = 10^3 lb/in^2 = 0.7031 kg/mm^2 = 6.895 MPa.
1 kg/mm^2 = 1.422 ksi = 9.807 MPa.

changed. But instrument misalignment can introduce small errors which cause a change in 2θ even for a stress-free specimen, when ψ is changed from 0 to 45°. It is therefore necessary to determine this change experimentally and apply it as a correction. If $(\Delta 2\theta)_0$ is the line shift for a stress-free specimen and $(\Delta 2\theta)_m$ the measured shift for a stressed specimen, then the line shift due to stress is

$$\Delta 2\theta = (\Delta 2\theta)_m - (\Delta 2\theta)_0. \tag{16-15}$$

The correction is best determined on a specimen of fine powder, which is necessarily free of macrostress. The powder should have about the same composition as the material in which stress is to be measured in order that its diffraction line occur at about the same 2θ position, because the correction $(\Delta 2\theta)_0$ itself depends to some extent on 2θ.

A value of $(\Delta 2\theta)_0$ greater than about 0.1° points to the need for better instrument alignment. The general alignment procedure stipulated by the diffractometer manufacturer should be carried out. Two further conditions must be satisfied:

1. The specimen surface must lie on the diffractometer axis and remain there as ψ is changed. Failure to achieve this condition is usually the major source of error in stress measurement. Some kind of mechanical pointer (feeler gauge) is often used to position the specimen in the proper place.

2. The radial motion of the counter must be truly radial. If the beam from the x-ray tube is restricted by a very narrow incident-beam slit to pass over the diffractometer axis and the counter is positioned at $2\theta = 0$ to receive the direct beam, nonradiality of the counter slide will be disclosed by an apparent deviation of the beam from 0° as the counter is moved toward the diffractometer axis.

A reasonably strong high-angle diffraction line is needed for the measurement of stress. The combination of (hkl) reflecting planes and wavelength λ that will produce such a line varies from one kind of material to another. These combinations and the approximate 2θ position of the line are listed in Table 16-1 for various materials.

Most stress measurements are based on Eq. (16-14). However, this equation involves a small error because K_1 is not truly constant. Equation (16-13) shows that K_1 is proportional to cot θ, which varies slightly as 2θ changes from $2\theta_n$ to $2\theta_i$. Better accuracy results from the use of Eq. (16-10), which requires the calculation of d values from the observed 2θ values. A shorter calculation will yield sin θ values, which vary inversely as d, as shown by the Bragg law. Therefore,

$$\frac{d_i - d_n}{d_n} = \frac{\sin \theta_n}{\sin \theta_i} - 1. \tag{16-16}$$

Equation (16-10) then becomes

$$\sigma_\phi = \frac{E}{(1 + \nu)\sin^2 \psi}\left(\frac{\sin \theta_n}{\sin \theta_i} - 1\right) = K_2\left(\frac{\sin \theta_n}{\sin \theta_i} - 1\right) \tag{16-17}$$

where K_2 is a new stress constant.

Table 16–1

Diffraction Data and Stress Constants

Alloy	Composition (weight percent)	λ, hkl [b]	2θ (deg)	$K_e = \dfrac{E}{1+\nu}$ (10^3 ksi)	K_1 [a] (ksi/deg)	Reference
(A) Ferritic and martensitic steels (body-centered cubic)						
Armco iron	Fe + 0.02 C	Cr, 211	156	27.9		[16.19]
4340 (50 R_C)[c]	Fe + 0.40 C, 1.8 Ni, 0.8 Cr, 0.25 Mo	Cr, 211	156.0	24.5	89.3	[16.20]
4340	Fe + 0.40 C, 1.8 Ni, 0.8 Cr, 0.25 Mo	Cr, 211	156		122	[16.22]
4340	Fe + 0.40 C, 1.8 Ni, 0.8 Cr, 0.25 Mo	Cr, 211	156		(101) [d]	[16.24]
4130	Fe + 0.30 C, 1.0 Cr, 0.20 Mo	Cr, 211	156		87	[16.22]
Railroad steel	Fe + 0.75 C	Cr, 211	156.1		90.9 [e]	[16.23]
D-6AC	Fe + 0.50 C, 1.0 Cr, 0.5 Ni, 1.0 Mo, 0.1 V	Cr, 211	156.2		113.4 [e]	[16.23]
D-6AC	Fe + 0.50 C, 1.0 Cr, 0.5 Ni, 1.0 Mo, 0.1 V	Cr, 211	156		(98) [d]	[16.24]
Vascomax 250	Fe + 18.5 Ni, 8.5 Co, 4.8 Mo, 0.1 Al, 0.4 Ti	Cr, 211	156		(111) [d]	[16.24]
4820 (carburized)	Fe + 0.20 C, 3.5 Ni, 0.25 Mo (core composition)	Cr, 211	156	(24.3) [d,g] (26.2) [d,h]		[16.21]
410 SS[f] (22 R_C)	Fe + 12.5 Cr	Cr, 211	155.1	25.6	98.4	[16.20]
410 SS (42 R_C)	Fe + 12.5 Cr	Cr, 211	155.1	25.1	96.7	[16.20]
422 SS (34 R_C)	Fe + 0.22 C, 12.0 Cr, 0.7 Ni, 1.0 Mo, 1.0 W, 0.25 V	Cr, 211	154.8	26.4	103.2	[16.20]
422 SS (39 R_C)	Fe + 0.22 C, 12.0 Cr, 0.7 Ni, 1.0 Mo, 1.0 W, 0.25 V	Cr, 211	154.8	26.1	103.4	[16 20]
(B) Austenitic alloys (face-centered cubic)						
304 SS	Fe + 0.08 C, 18 Cr, 8 Ni, 2 Mn	Cr, 220	129.0	20.2	170.0	[16.20]
Incoloy 903	Fe + 38 Ni, 15 Co, 3Cb, 1.4 Ti, 0.7 Al	Cr, 220	128.0	31.2	264.0	[16.20]
Incoloy 903	Fe + 38 Ni, 15 Co, 3 Cb, 1.4 Ti, 0.7 Al	Cu, 331	146.3		74.6 [e]	[16.23]
Incoloy 800	Fe + 32.5 Ni, 21 Cr, 0.4 Ti, 0.4 Al	Cr, 220	129.0	23.4	196.0	[16.20]
Incoloy 800	Fe + 32.5 Ni, 21 Cr, 0.4 Ti, 0.4 Al	Cu, 420	147.0	21.5	110.0	[16.20]

Measurement of Line Position

Line positions cannot be measured with sufficient precision on a chart recording made with a ratemeter. Instead, a scaler is used to determine the count rate at several positions on the line profile, and from these data the position of the line center is calculated. This procedure is particularly necessary when the lines are broad, as they are from hardened steel; the line width at half-maximum intensity is then 5°–10° 2θ. If the line is 8° wide and the stress constant K_1 is 86.3 ksi/deg $\Delta 2\theta$, as given above, a stress of 50 ksi will cause the line to shift by only 7 percent of its width when the specimen is turned through 45°. Measurement of such a small shift requires that the line center be accurately located at each angle ψ.

The standard method of finding the center of a diffraction line, broad or narrow, is to fit a parabola to the top of the line and take the axis of the parabola

Table 16-1 (*Continued*)

Alloy	Composition (weight percent)	λ, hkl^b	2θ (deg)	$K_e = \frac{E}{1+\nu}$ (10^3 ksi)	$K_1{}^a$ (ksi/deg)	Reference
(C) Nickel alloys (face-centered cubic)						
Monel K 500	66.5 Ni, 29.5 Cu, 2.7 Al, 1.0 Fe	Cu, 420	150.0	21.0	98.4	[16.20]
Inconel 600	76 Ni, 15.5 Cr, 8 Fe	Cu, 420	151.0	23.1	105.0	[16.20]
Inconel 600	76 Ni, 15.5 Cr, 8 Fe	Cr, 220	131.0	21.1	174.0	[16.20]
Inconel 718	52.5 Ni, 19 Cr, 18.5 Fe, 3.1 Mo, 5.0 Cb, 0.9 Ti, 0.4 Al	Cu, 331	145.0	19.7–20.3	109.0–112.0	[16.20]
Inconel 718	52.5 Ni, 19 Cr, 18.5 Fe, 3.1 Mo, 5.0 Cb, 0.9 Ti, 0.4 Al	Cu, 331	146.1		122.4 [e]	[16.23]
Inconel 718	52.5 Ni, 19 Cr, 18.5 Fe, 3.1 Mo, 5.0 Cb, 0.9 Ti, 0.4 Al	Cr, 220	128.0	31.2–31.4	263.0–265.0	[16.20]
Inconel X750	73 Ni, 15.5 Cr, 7 Fe, 2.5 Ti, 0.9 Cb, 0.8 Al	Cr, 220	131.0	36.8	301.0	[16.20]
(D) Aluminum alloys (face-centered cubic)						
2024	Al + 4.4 Cu, 1.5 Mg, 0.6 Mn	Cu, 511	163	8.51		[16.19]
2024-T3	Al + 4.4 Cu, 1.5 Mg, 0.6 Mn	Cr, 311	139		44	[16.22]
7075	Al + 1.6 Cu, 2.5 Mg, 0.3 Cr, 5.5 Zn	Cr, 311	139.0	8.83	56.9	[16.20]
7079-T611	Al + 0.6 Cu, 3.7 Mg, 0.2 Cr, 4.7 Zn	Cr, 311	139		50	[16.22]
2219-T87	Al + 6.3 Cu, 0.3 Mn, 0.18 Zr, 0.1 V, 0.06 Ti	Cr, 311	139.5		55.0 [e]	[16.23]
(E) Copper alloy (face-centered cubic)						
Cu-Ni	85 Cu, 15 Ni	Cu, 420	146.0	18.6	98.8	[16.20]
(F) Titanium alloys (hexagonal close-packed)						
Ti-6-4	Ti + 6 Al, 4 V	Cu, 213	142.0	12.2	74.0	[16.20]
Ti-6-4	Ti + 6 Al, 4 V	Cu, 213	142		70	[16.22]
Ti-6-2-4-2	Ti + 6 Al, 2 Sn, 4 Zr, 2 Mo	Cu, 213	140.7	14.8	92.3	[16.20]

(a) Value of K_1 from Eq. (16–13) with $\psi = 45°$.
(b) In this column only: "Cr" means Cr $K\alpha$; "Cu" means Cu $K\alpha$.
(c) R_C = Rockwell C hardness number.
(d) Calculated from data in referenced paper.
(e) Measured with counter in same radial position, between F and F' of Fig. 16–7(b), for $\psi = 0$ and 45°.
(f) SS = stainless steel.
(g) As quenched.
(h) Average value after tempering at 200° F and above.

1 ksi = 10^3 psi = 10^3 lb/in^2 = 0.7031 kg/mm^2 = 6.895 MPa.
1 kg/mm^2 = 1.422 ksi = 9.807 MPa.

as the line center (Fig. 16–8). This method was first used for stress measurement by Ogilvie [16.6].

The equation of a parabola with its axis parallel to the y axis and vertex at (h, k) is

$$(x - h)^2 = p(y - k). \tag{16–18}$$

If we put $x = 2\theta$ and $y = I$, this equation represents the shape of the diffraction line near its peak. We substitute several pairs of 2θ, I values into the equation and solve for h by the method of least squares. The value of h equals the 2θ position of the line center. Only two or three points on either side of the peak near its

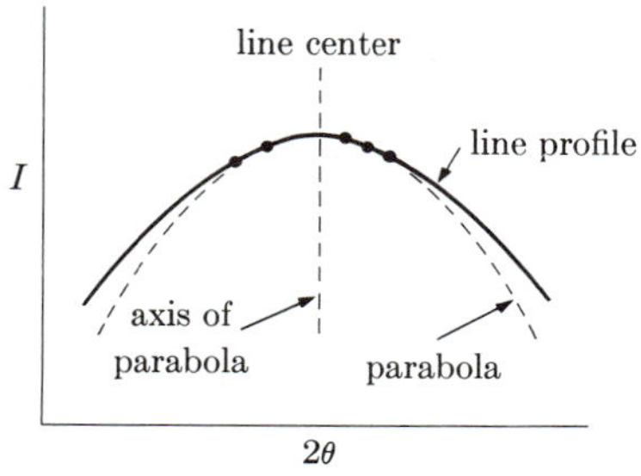

Fig. 16–8 Method of locating the center of a diffraction line.

maximum are sufficient to locate the parabola with surprising accuracy. The positions of diffraction lines as broad as 8° 2θ at half-maximum intensity can be reproducibly determined to within 0.02° by this method.

A simpler method, much used, for locating the parabola axis was suggested by Koistinen and Marburger [16.7]. Only three points on the line profile need be measured but they must be separated by the same angular interval c, as shown in Fig. 16–9. The central point should be near the maximum and the other two have intensities of about 85 percent of the maximum. Once the intensity differences $a = y_2 - y_1$ and $b = y_2 - y_3$ are found, the center of the line is given by

$$h = x_1 + \frac{c}{2}\left(\frac{3a + b}{a + b}\right). \tag{16–19}$$

The y coordinate may be intensity I in counts/sec, counts n for a fixed time, or reciprocal times $1/t$ for a fixed count.

As mentioned in Sec. 9–4, the breadth of an x-ray line often correlates well with the hardness of the specimen. Marburger and Koistinen [16.30] also showed that the hardness of certain quenched and tempered steels is related to the breadth

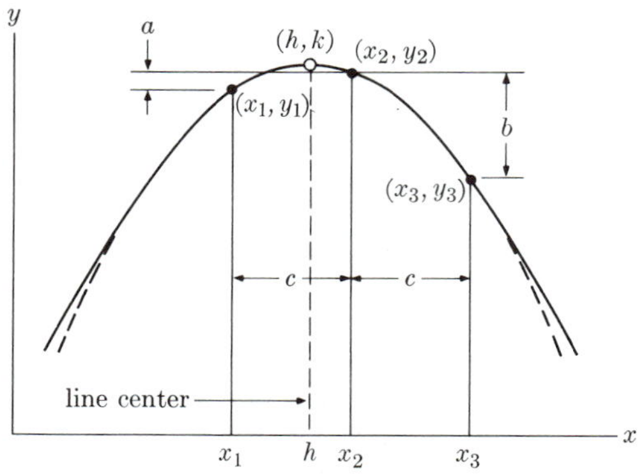

Fig. 16–9 Three-point method for fitting a parabola. $x = 2\theta$, $y = I$.

B_P of the *parabola* used to fit the x-ray line for stress measurements. The breadth at half-maximum intensity ($y = k/2$) is

$$B_P = 2c[k/(a + b)]^{1/2}. \qquad (16\text{–}20)$$

This equation requires that $k = I_{max}$ be measured at the position $2\theta = h$ given by Eq. (16–19). This additional measurement may be avoided and B_P determined solely from the data obtained in the stress measurements, if the following more complex expression is used:

$$B_P = c\left[\frac{(3a + b)^2 + 8y_1(a + b)}{2(a + b)^2}\right]^{1/2}.$$

Inasmuch as B_P (breadth of the parabola) and B (breadth of the x-ray line) both correlate well with hardness, the relation between B_P and B must be a simple one, perhaps linear or nearly so.

When the lines are broad, certain corrections should be applied to the intensity data *before* finding the line center, as pointed out by Koistinen and Marburger. In calculating the relative intensities of the various lines on a powder pattern, we saw in Sec. 4–9 that one factor controlling these intensities was the Lorentz-polarization (L-P) factor $(1 + \cos^2 2\theta)/(\sin^2 \theta \cos \theta)$. This factor can vary considerably over the width of a single line, when the line is broad and in the high-angle region. However, the L-P factor applies to integrated intensities. To obtain an expression governing intensities at particular values of 2θ within a single line, we drop the $(1/\cos \theta)$ term, which relates to line breadth, and obtain the modified L-P factor $(1 + \cos^2 2\theta)/\sin^2 \theta$. The variation of this factor with 2θ makes a high-angle line asymmetrical about its center. Absorption in the specimen has a similar effect when ψ is not zero, because the absorption factor is then $(1 - \tan \psi \cot \theta)$. If we combine these two factors into one and call it the *LPA* factor, then

$$\begin{aligned} LPA &= \text{(modified L-P factor)(absorption factor)} \\ &= \left(\frac{1 + \cos^2 2\theta}{\sin^2 \theta}\right)(1 - \tan \psi \cot \theta). \qquad (16\text{–}21) \end{aligned}$$

Measured intensities are to be divided by *LPA* in order to make the lines more nearly symmetrical, before determining the line center by the least-squares or three-point method. The function *LPA* is tabulated in [16.3].

If the background is high, because of fluorescence by the specimen, better accuracy is attainable by subtracting the background, assumed linear across the line, before applying the *LPA* correction and finding the line center.

Kelly and Eichen [7.7] and James and Cohen [16.25] have described computer-controlled stress measurements. The computer controls the necessary 2θ and ψ rotations, corrects line asymmetry by the *LPA* correction, finds the line center by curve fitting, and prints out the value of the stress.

Specimen Preparation

Ideally, the specimen surface should be clean and smooth. Dirt and scale, if present, must be removed, but any removal of material must be done by a process

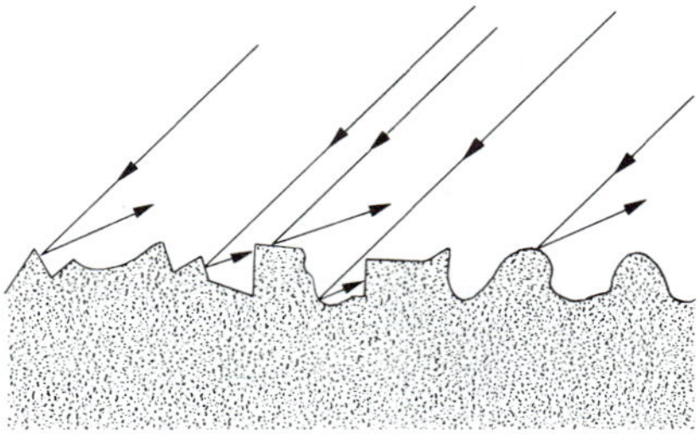

Fig. 16–10 Diffraction from a rough surface when the incident beam is inclined.

that does not itself produce residual stress and thereby change the stress to be measured.

Grinding and machining are known to introduce large stresses to depths of at least 0.005 inch (125 μm). The effect of acid etching is less certain; some investigators have reported that etching produces some stress, others that the effect is negligible; at any rate, deep etching is objectionable because of the surface roughness it produces (see below). The only sure method of removing metal without introducing stress is by electrolytic polishing, i.e., the process in which the specimen is made the anode in an electrolytic cell; details are given in [16.3] and by Tegart [16.31].

Surface roughness should be strictly avoided, because the high points in a rough surface are not stressed in the same way as the bulk of the material and yet they contribute most to the diffraction pattern, especially the one made at inclined incidence, as indicated in Fig. 16–10. Of course, the surface should not be touched at all prior to the stress measurement, if the object is to measure residual surface stresses caused by some treatment such as machining, grinding, shot peening, etc. Such treatments produce steep stress gradients normal to the surface, and the removal of any material by polishing or etching would defeat the purpose of the measurement.

Subsurface Measurements

When the stress inside a specimen is to be determined, material is removed in layers and the stress is measured at each new exposed surface. These measured stresses are not those previously existing, because the removal of a stressed layer changes the stress in the remaining material. The measured stresses have therefore to be corrected. The layer-removal correction is described by Moore and Evans [16.8] and is also given in [16.3].

If the stress varies rapidly with distance below the surface, allowance must also be made for the finite penetration of the x-ray beam into the specimen. The incident beam is not only measuring the stress at the surface then under study but is also sampling different stresses within the depth of penetration of the x-ray beam. The method for correcting measured stresses for beam penetration is given in [16.3].

An example of stress measurements in depth is given in Fig. 16–11, which shows the residual stress produced in hardened steel by grinding. The extremely

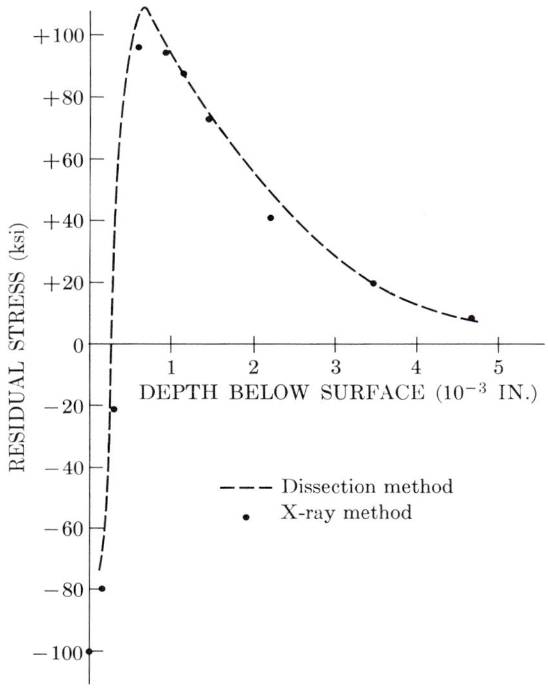

Fig. 16–11 Longitudinal residual stress, parallel to grinding direction, as a function of depth below the ground surface in a steel heat treated to a Rockwell C hardness of 59. X-ray measurements by Koistinen and Marburger [16.7]; dissection measurements by Letner [16.9]. Note inability of dissection method to measure stress right at the surface.

steep stress gradient normal to the surface should be noted; the stress changes from $+100$ ksi to -100 ksi in about 0.0006 inch (15 μm), which means that the gradient is about 3×10^5 ksi/in. Beam penetration corrections were necessary here, as they generally are when measurements are made on ground or machined surfaces.

Variant Techniques

The following variations of the standard method described above are also in use:

1. The inclined measurement is made at $\psi = 60°$ rather than 45°. Sensitivity therefore improves, but any roughness of the specimen surface will have a more serious effect.

2. The counter is left in its usual position on the diffractometer circle (no radial movement) for the inclined measurement. The diffracted beam is then poorly focused at the counter and the intensity at the counter slit is decreased, but there is no need for a radial slide on the diffractometer.

3. *$\sin^2 \psi$ method.* Inclined measurements are made at a number of angles rather than at only one. Values of d_i or $2\theta_i$ are plotted against $\sin^2 \psi$, and the stress σ_ϕ

is derived from the slope of the line. The more points, the more precisely is the line established.

4. *Parallel-beam method.* If the incident x-ray beam is composed of parallel, rather than divergent, rays, no focusing occurs, and there is no need to move the counter radially as ψ is changed. Nearly parallel radiation is obtained by placing Soller slits in the incident and diffracted beams, with the plates of the slits parallel to the diffractometer axis so as to minimize divergence in the plane normal to that axis. The parallel-beam method is the standard method in Japan [G.44]. Compared to the divergent-beam method of Fig. 16–7, the parallel-beam method is mechanically simpler and much more tolerant of specimen positioning errors (Sec. 16–7), but it yields lower diffracted intensity at the counter.

5. *Side-inclination method.* The geometry of the Schulz method for pole-figure determination (Fig. 9–16) is adopted. For the measurement of d_n the specimen is positioned so that the direction in which stress is to be measured is parallel to the diffractometer axis. The ψ rotation (shown as an α rotation in Fig. 9–16) is then made about the axis AA' for the measurement of d_i. The focal point of the diffracted beam remains on the diffractometer circle as ψ is changed, so that radial motion of the counter is not needed. This method has another advantage for some specimens in which stress has to be measured at the bottom of a concave region, such as in the space between the teeth of a gear; depending on the shape of the concavity, it may be physically impossible to get an x-ray beam in and out with the standard method, because either the incident or diffracted beam is blocked by the specimen at one ψ angle or another; with the slide-inclination method this problem may not exist. On the debit side, the height of the beam is severely limited.

Special Diffractometers

It is sometimes necessary to measure residual stress in an object that is too large and/or heavy to be accommodated on a standard diffractometer. Or one may wish to measure stress in a large structure, as distinct from a machine component. In either case, it is necessary to bring the diffractometer to the object to be measured rather than vice versa.

Compact, mobile diffractometers have been made for this purpose, especially by Japanese manufacturers [G.44]. Because they are intended only for stress measurement, the angular 2θ range of counter movement covers only the high-angle region. The diffractometer is usually mounted at the end of a boom which is fixed to a wheeled cart or a mobile platform. The ψ angle is changed by rotating the whole diffractometer (x-ray tube and counter) with respect to the specimen about the diffractometer axis, which is an imaginary axis lying in the surface of the specimen.

A more recent, and very promising, development is due to Cohen and James, who adapted a position-sensitive proportional counter to stress measurement. This counter has the great advantage that it can measure the 2θ position of a diffracted beam without a 2θ movement of the counter (Sec. 7–5). Preliminary work was done with the counter mounted on a standard diffractometer [16.10]. Later a portable instrument was made, in which the counter and a miniature, air-cooled x-ray tube are fixed to a single support that moves in a curved guide through an arc of 45° [16.32]. Three rods fixed to the guide allow the unit to be properly positioned relative to the specimen. Stress is measured by determining the line position at $\psi = 0$, swinging the counter and tube together through 45°, and determining the line position again. This instrument has the following features:

1. It is fast. All the points on the diffraction-line profile are measured simultaneously and stored as I, 2θ data in an MCA, as in time-analysis diffractometry (Sec. 7–10). A minicomputer then establishes the line centers for $\psi = 0$ and 45° by least-squares parabola fitting, computes the stress, and displays it. Stress can be measured with a standard deviation of about 5 ksi in 20 seconds or less, depending on line width.

2. It is truly portable. The prototype weighs only 23 lb (10 kg) and can be carried by one man, with the aid of a neck strap and two handles, and held in position for a stress measurement. As a fixed instrument, it should be useful for monitoring stress in material coming off a production line.

3. The special counter and associated electronics are rather expensive.

Fastress Stress Analyzer

This instrument (registered trademark: Fastress) is based on a design by Weinman *et al.* [16.11]. It has two x-ray tubes and four counters.

The two x-ray tubes T_1 and T_2 [Fig. 16–12(a)] are positioned so that two incident beams strike the fixed specimen at the proper angles for simultaneous measurement at $\psi = 0$ and 45°. The diffracted beam for the d_n measurement ($\psi = 0$) is located automatically by a pair C_1 of narrow, side-by-side counters; the beam for the d_i measurement ($\psi = 45°$) is located by a similar pair C_2. Figure 16–13 is an overall view.

Figure 16–12(b) shows the front view of counter pair C_1 (C_2 is identical) and the method of locating the center of a diffraction line. The two counters A and B are connected to two separate ratemeters. The difference between the ratemeter outputs actuates a servomechanism which drives A and B as a unit back and forth along an arc centered at O on the specimen surface. When this difference is zero, both ratemeters are measuring equal intensities and C_1 is centered on the diffraction line, as indicated on the drawing. (Line asymmetry, not shown, can be allowed for automatically.) When C_1 and C_2 are each centered on the diffraction lines, a voltage proportional to the difference γ between their angular positions is sent to a strip-chart recorder. Because γ is a linear function of $(2\theta_n - 2\theta_i)$, the deflection of the recorder pen indicates stress directly. Stresses may be measured with a precision variously reported as 2 to 10 ksi in about one minute, which is the time required to establish a good average of the pen deflection.

The Fastress instrument is not a diffractometer but rather a "stress gauge" that must be calibrated with specimens containing known stresses. It is specifically designed for measurements on ferritic and martensitic steels (BCC) by observation of the 211 reflection with Cr $K\alpha$ radiation at a 2θ angle of about 156°. Counter movements are restricted to a range of 152° to 159°, and only materials that produce

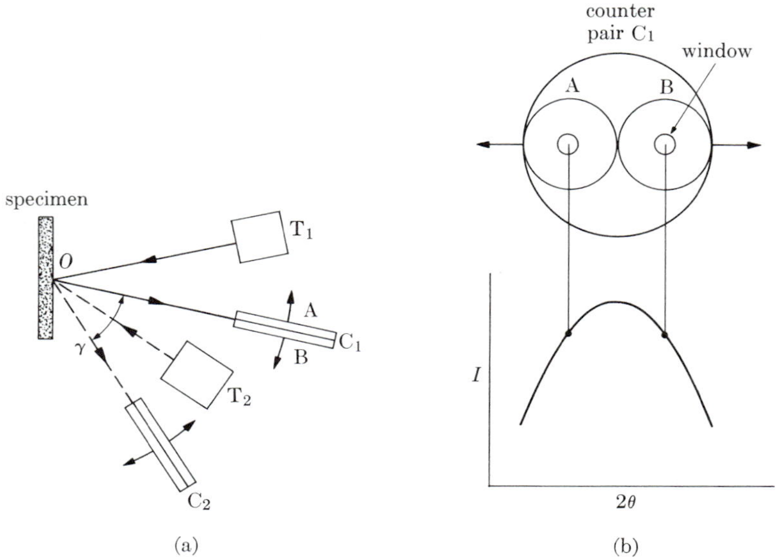

Fig. 16–12 Fastress stress analyzer. (a) Arrangement of x-ray tubes T and counters C. (b) Front view of one counter pair.

Fig. 16–13 Fastress stress analyzer. Close-up of x-ray tubes and counters. The table supporting the specimen shown here is removable, so that stress may be measured in much larger objects. (Courtesy of American Analytical Corporation.)

a diffraction line in this range can be investigated. The equipment is mobile and can be used either in the plant or laboratory.

Both incident x-ray beams are collimated by pinholes, and the irradiated area on the specimen is a spot about $\frac{1}{8}$ inch (3 mm) in diameter, or smaller. The instrument is therefore well suited to making a rapid exploration of possible stress variations from point to point on a surface. Such variations, on an apparently uniform surface, are often surprisingly large. (With a diffractometer the irradiated area is usually about 1 cm^2, in order to obtain adequate intensity at the counter, and the observed stress is an average over that area.)

16-5 PHOTOGRAPHIC METHOD

This method of stress measurement is not often used today, because it is slower and less precise than the diffractometer method. However, the photographic method still has some advantages. The required apparatus is very simple: a small back-reflection camera fixed to the head of a shockproof x-ray tube energized through a shockproof cable. This apparatus is smaller, lighter, more robust, and much cheaper than a mobile diffractometer. It is also more mobile and better suited to work in a confined space.

Facilities for film development are needed, either a standard darkroom or, for field work, a portable darkroom in the form of a light-tight box with gloved portholes [16.12]. Rapid-process film of the Polaroid Land type does not appear to have been adapted to stress cameras.

Two experimental techniques have been used:

1. Two exposures are made, with the incident beam inclined at a different angle to the specimen for each exposure [G.25, 16.3, 16.13, 6.8]. Only one side of the Debye ring on the film is measured in each exposure. This method is entirely analogous to the standard diffractometer method.

2. A single inclined exposure is made, but both sides of the Debye ring are measured.

Norton [16.13] has analyzed the errors in each technique and has concluded that the single-exposure technique is, in practice, just as precise as the two-exposure technique, besides being twice as fast, even though the precision of the two-exposure technique is theoretically better.

Only the single-exposure technique will be described here. The incident beam (Fig. 16-14) is collimated by a pinhole, not shown, and is inclined at an angle β to the normal N_s to the specimen surface. The reflected rays register on two strips of film held in a curved holder of radius R centered on O. (Flat-film cameras have also been used.) The two rays shown do not lie on a true Debye cone because they are reflected by sets of planes, 1 and 2, differently oriented with respect to the stress σ_ϕ and therefore with different spacings. The two sets of planes have slightly different Bragg angles θ_1 and θ_2, and their normals N_{p1} and N_{p2} make slightly different angles α_1 and α_2 with the incident beam. Equation (16-12) can then be written for each ψ angle, ψ_1 and ψ_2:

$$\sigma_\phi = \frac{E \cot \theta\,(\theta_n - \theta_{i1})}{(1 + \nu) \sin^2 \psi_1}, \tag{16-22}$$

$$\sigma_\phi = \frac{E \cot \theta\,(\theta_n - \theta_{i2})}{(1 + \nu) \sin^2 \psi_2}. \tag{16-23}$$

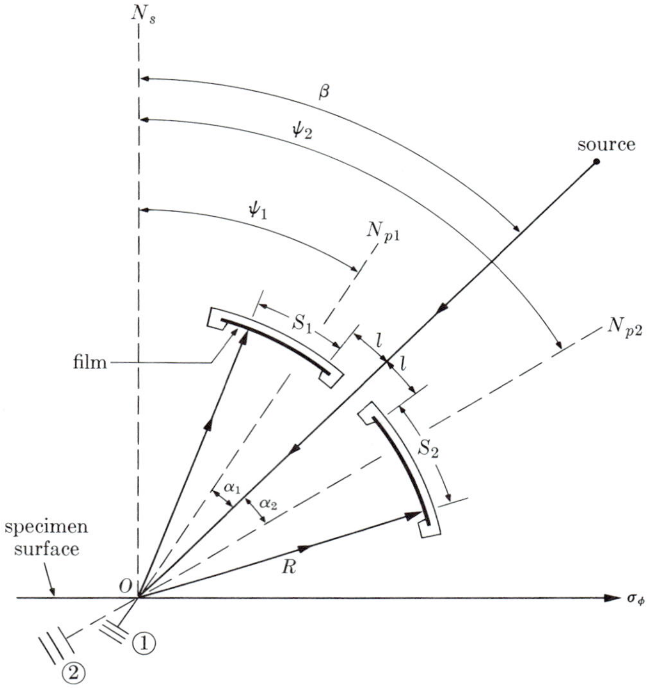

Fig. 16–14 Photographic method of stress measurement (single-exposure technique). After Norton [16.13].

The quantities measured are the distances S_1 and S_2 between the diffraction lines and the shadows of the inside knife edges, which are at a distance l from the incident beam. Therefore,

$$2\alpha_1 R = 2[(\pi/2) - \theta_{i1}]R = l + S_1,$$

$$2\alpha_2 R = 2[(\pi/2) - \theta_{i2}]R = l + S_2.$$

By subtraction we find

$$2R(\theta_{i2} - \theta_{i1}) = S_1 - S_2. \tag{16–24}$$

By eliminating θ_n from Eqs. (16–22) and (16–23) and combining the result with (16–24), we have

$$\sigma_\phi = \frac{E \cot \theta \,(S_1 - S_2)}{2R(1 + \nu)(\sin^2 \psi_1 - \sin^2 \psi_2)}. \tag{16–25}$$

But $\psi_1 = \beta - \alpha_1$ and $\psi_2 = \beta + \alpha_2$. In evaluating the sine terms in Eq. (16–25), it is sufficiently accurate to put $\alpha_1 = \alpha_2 = 90° - \theta$. The result is the working equation

$$\sigma_\phi = K_3(S_2 - S_1), \tag{16–26}$$

where the stress constant is given by

$$K_3 = \frac{E}{4R(1 + \nu)\sin^2\theta \sin 2\beta}. \tag{16-27}$$

If β is chosen to be 45°, the sin 2β term drops out.

The stress camera shown in Fig. 16–15 has a radius of 70.0 mm. It takes two standard packets of dental x-ray film which slide into curved slots in the film holder; thus film can be inserted and removed for processing without disturbing the camera alignment. Knife edges in the film holder cast sharp x-ray shadows at each end of each film strip; these serve as fiducial marks for the measurement of line position and for the correction of film shrinkage. The correct distance between specimen and film is set by replacing the collimator aperture with a retractable pointer, not shown; the tip of this pointer lies on the axis of the incident beam at the right distance from the film, i.e., at the point O of Fig. 16–14. Line positions on the film are measured with a conventional viewing box and scale, such as that of Fig. 6–18, or with a microphotometer. If the standard deviation of the quantity $(S_2 - S_1)$ is 0.04 mm, which can be achieved by careful, repeated measurements on each film, then the standard deviation in stress for steel is 4.5 ksi [16.13].

The small diameter of the incident beam from a stress camera or a Fastress unit is an advantage when one wishes to measure stress variations from point to point on a surface, as in the region near a weld. The stress distribution shown in Fig. 16–16, determined by a photographic method, simulates the residual stresses due to spot welding. The specimen was a steel strip 10 × 3 × $\frac{1}{4}$ inch (25 × 8 × 0.6 cm). A circular area of about $\frac{3}{8}$ inch (1 cm) diameter, whose size is indicated on the graph, was heated locally to about 700°C for a few seconds by clamping the strip at its center between the two electrodes

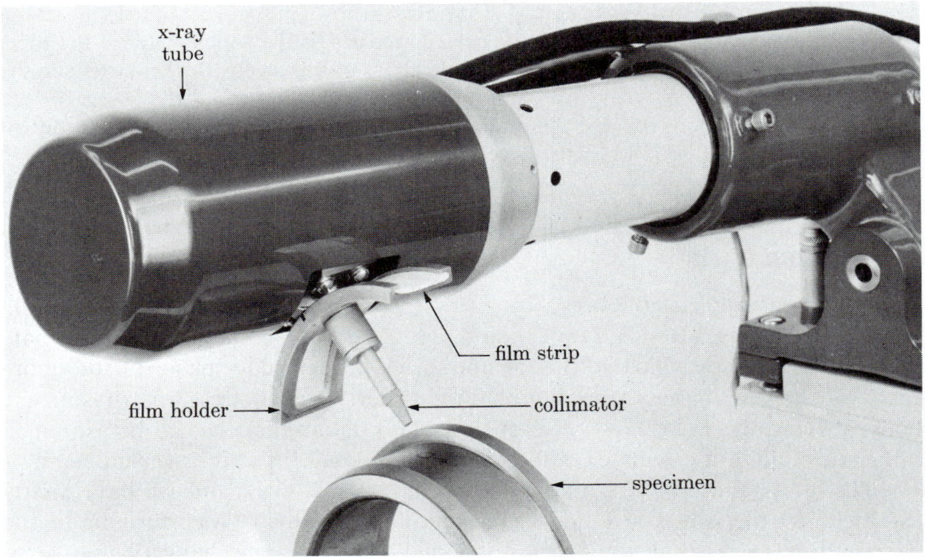

Fig. 16–15 Stress camera in position for a stress measurement by the single-exposure technique. The head of the x-ray tube is enclosed in a protective cover. (Courtesy of Advanced Metals Research Corporation.)

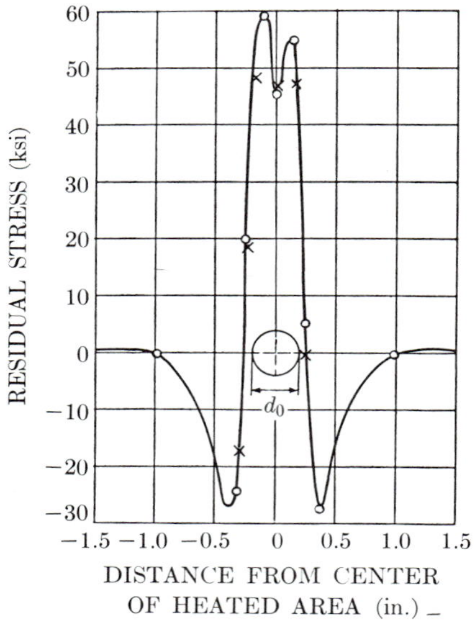

Fig. 16–16 Longitudinal residual stress in the surface of a steel strip, due to localized heating. d_0 is diameter of heated area. Norton and Rosenthal [16.14].

of a spot welding machine. The central area rapidly expanded but was constrained by the relatively cold metal around it. Plastic flow occurred in the hot central region, and residual stress was set up during cooling. The stresses shown in Fig. 16–16 are longitudinal stresses at points along a line through the heated area across the 3-inch width of the strip. Tensile residual stress, almost equal to the yield stress of this steel (60.5 ksi), exists in the heated area.

Other investigations by the photographic method include measurements on Al [16.12, 16.15, 16.17], Ti [16.12], and steel [16.12, 16.16, 16.18].

16–6 CALIBRATION

For the measurement of stress by x-rays we have developed three working equations, namely, (16–14), (16–17), and (16–26). Each contains an appropriate stress constant K, by which diffraction line shift is converted to stress. Furthermore each was derived on the assumption that the material under stress was an isotropic body obeying the usual laws of elasticity. This assumption has to be examined rather carefully if a calculated value of K is to be used for stress measurement.

The stress constant K contains the quantity $E/(1 + \nu)$, and we have tacitly assumed that the values of E and ν measured in the ordinary way during a tensile test are to be used in calculating the value of K. But these mechanically measured values are not necessarily the correct ones to apply to a diffraction measurement. In the latter, strains are measured in particular crystallographic directions, namely, the directions normal to the (hkl) reflecting planes, and we know that both E and

ν vary with crystallographic direction. This anisotropy of elastic properties varies from one metal to another: for example, measurements on single crystals of α-iron show that E has a value of 41.2×10^6 psi in the direction [111] and 19.2×10^6 psi in [100], whereas the values of E for aluminum show very little variation, being 10.9×10^6 psi in [111] and 9.1×10^6 psi in [100]. The mechanically measured values are 30×10^6 and 10×10^6 psi for polycrystalline iron and aluminum, respectively. These latter values are evidently average values for aggregates of contiguous grains having random orientation. In the x-ray method, however, only grains having a particular orientation relative to the incident beam, and therefore a particular orientation with respect to the measured stress, are able to reflect. There is therefore no good reason why the mechanically measured values of E and ν should be applied to these particular grains. Stated alternately, an aggregate of randomly oriented grains may behave isotropically but individual grains of particular orientations in that aggregate may not.

These considerations are amply supported by experiment. By making x-ray measurements on materials subjected to known stresses, we can determine the stress constant K experimentally. The values of K so obtained can differ substantially from the values calculated from the mechanically measured elastic constants. Moreover, for the same material the measured values of K usually vary with the indices (hkl) of the reflecting planes.

Methods have been proposed for calculating the proper values of E and ν to use with x-ray stress measurements from values measured in various directions in single crystals. Such calculations usually rest on assumptions of dubious validity, and the results are not in good agreement with experiment. The safest procedure is to measure K on a specimen subjected to known stresses, and specific examples of calibration procedures may be found in [16.3, 16.19, 16.20, 16.21, 16.22, 16.24].

The usual practice is to set up known stresses in a body by bending. The specimen is a flat strip, seen edge-on in Fig. 16-17, supported at two points and loaded by the forces F at two other points. This four-point bending produces a tensile stress in the front surface, and this stress is constant in magnitude between the two inner supports. If the calibration is performed on a diffractometer, the bending fixture must be so designed that the front surface of the specimen coincides with the diffractometer axis at any ψ angle and at any degree of bending.

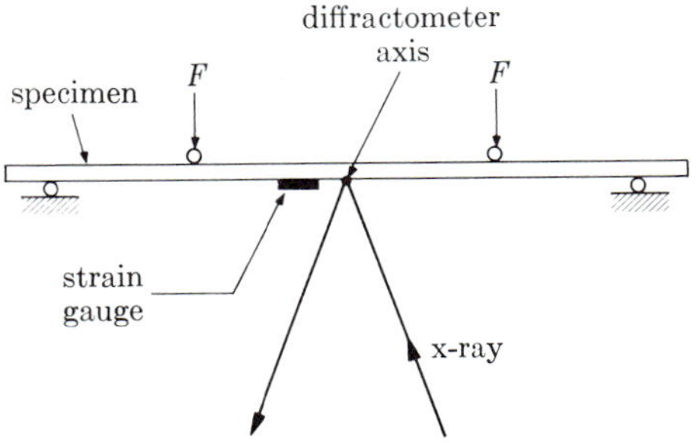

Fig. 16-17 Specimen loaded in bending for calibration of the x-ray method.

The strain due to bending is measured by an electrical-resistance strain gauge mounted near the region examined by x-rays. The product of this strain and the mechanically measured elastic modulus E of the material is the surface longitudinal stress σ_ϕ needed for the calibration. (Any uncertainty in E will cause a corresponding uncertainty in the stress constant K. The best approach is to measure E on the same specimen by direct loading of the same kind as that used in the x-ray calibration, e.g., by four-point bending.)

After a range of known stresses σ_ϕ, not exceeding the elastic limit, are measured by x-rays, the data are plotted in the form of a calibration curve. If, for example, a diffractometer is used and the measurement is based on Eq. (16–14), the calibration curve is a plot of σ_ϕ vs. $\Delta 2\theta = (2\theta_n - 2\theta_i)$, and the slope of the line is equal to K_1. (If the line does not go through the origin, it indicates that the calibrating member itself contained residual stress before a bending moment was applied.)

The x-ray technique used in calibration should match the technique to be used later for the measurement of unknown stresses. For example, the stress constant determined with the focusing technique of Fig. 16–7 is not expected to agree with the stress constant determined by the parallel-beam technique.

The stress constant K appropriate to any technique can be usefully divided into two parts, as suggested by Norton [16.13]:

$$K = K_e K_g \tag{16–28}$$

where K_e equals $E/(1 + \nu)$ and involves only the elastic constants of the material and where K_g contains the geometrical quantities appropriate to a particular technique. Thus the K_g factor for stress constant K_1 is, from Eq. (16–13),

$$K_{g1} = \frac{\cot \theta}{2 \sin^2 \psi}. \tag{16–29}$$

The K_e part of any stress constant is the important part, because it is determined only by the properties of the material, and it is the quantity that should be reported as the result of a calibration experiment. Similarly, any report of x-ray stress measurements should include the value of K_e used in the calculations.

Experimentally determined stress constants are given in Table 16–1 for a variety of industrial alloys, ferrous and nonferrous. All have been determined with the diffractometer.

16–7 PRECISION AND ACCURACY

As in most measurements, precision (reproducibility) is easier to attain than accuracy (approach to the "true" value).

Precision

High precision requires that the random errors involved in measuring diffraction line position be minimized, and these errors are larger, the wider the line.

Diffractometer

Precision in establishing the profile, and hence the position, of a diffraction line is governed mainly by the statistical error in counting. As shown by Eq. (7–8), this error is proportional to $1/\sqrt{N}$, where N is the number of counts. Many investigators use a fixed-count technique and count long enough at each position 2θ to accumulate 10^5 counts. The relative standard deviation of such a count is 0.3 percent. Ways of estimating the standard deviation of the measured stress from the errors involved in counting and curve fitting are treated in [16.3]. James and Cohen [16.25] have made an experimental study of the precision attainable by different techniques; they show, for example, that better precision is attained by fitting a parabola to a number of points, seven or more, by least squares than by fitting it to three points, as in Fig. 16-9, even when the total time spent in counting is made the same for both techniques. It is difficult to give a single figure for the precision attainable in a stress measurement in a reasonable measuring time without hedging it about with many qualifications, but, for measurements on steel, standard deviations of 2 ksi–3 ksi (for narrow diffraction lines) and 4 ksi–5 ksi (broad lines) are probably conservative estimates.

Stress camera

Precision is increased by making repeated measurements on each film and averaging the results. Using the single-exposure technique, Norton [16.13] obtained a standard deviation on steel (narrow lines) of 4.5 ksi.

Accuracy

High accuracy requires that systematic errors be minimized. These errors are chiefly of two kinds, those of a geometrical nature and those involved in the elastic-constant portion K_e of the stress constant. The systematic errors cause an error in the measured stress which is proportional to the magnitude of the stress.

In a diffractometer measurement by the standard focusing technique, the chief geometrical error is a displacement of the specimen surface from the diffractometer axis. If the specimen is displaced by an amount x along the reflecting-plane normal, then a line shift will occur even for a stress-free specimen, as it is rotated from $\psi = 0$ to $\psi = \psi$. This shift, in radians, is given by [16.29]

$$2\theta_n - 2\theta_i = \Delta 2\theta = \frac{2x \cos \theta}{R}\left[1 - \frac{\sin \theta}{\sin (\theta - \psi)}\right], \tag{16-30}$$

where R is the diffractometer radius and x is taken as positive when the displacement is in front of the axis. For $x = 10^{-2}$ inch (250 μm), $\theta = 78°$, $\psi = 45°$, and $R = 5.73$ in. (14.55 cm), $\Delta 2\theta = -0.033$ deg.

In the single-exposure camera method, faulty positioning of the camera relative to the specimen causes errors in R and β.

Given careful experimental technique, an inaccurate value of K_e remains as the largest source of error. Inspection of the data in Table 16-1 shows rather poor agreement between certain investigators for the value of K_e or K_1 for the same or similar material. The investigator who does not have the time or facilities to measure K_e before making stress measurements has two options: (1) he can select a value, or a mean value, from a list such as that of Table 16-1, with a possible error of up to about 10 percent, or (2) he can calculate K_e from the mechanically measured values of E and ν, with a possible error of about 30 percent. While such errors would be intolerably large in some measurements, they do not necessarily impair the value of residual stress measurements. As Norton [16.13] points out, the investigator is usually more concerned with precision than accuracy, in that he wants to know how the stress varies from one point to another on the specimen or how the stress at a particular point varies with the treatment of the specimen. An error of even 30 percent in the absolute magnitude of the measured stresses would seldom change the conclusions reached in the investigation.

(In certain nickel alloys Prevey [16.20] has found abnormally large differences between the x-ray and mechanical values of K_e; for these alloys his x-ray values are up to 80 percent larger than the mechanical values reported in the literature.)

16-8 PRACTICAL DIFFICULTIES

Certain conditions inherent in the specimen can affect the accuracy, and even the possibility, of x-ray stress measurements.

Large Grain Size

If the grains are large, the diffraction line is spotty and its position not well defined. This condition is obvious with a photographic, but not with a diffractometric, technique. If this condition is suspected, a back-reflection pinhole photograph should be made before stress measurements with a diffractometer are attempted. Or the line may be recorded on a dental film placed in front of the counter slit.

Some industrial products have grains so large that x-ray stress measurements are impossible. In borderline cases, oscillation of the specimen during the measurement, by ± 2–5 degrees about an axis lying in the specimen surface, will increase the number of reflecting grains enough to produce a measurable line.

Preferred Orientation

A moderate degree of preferred orientation causes no difficulty, but a fairly sharp texture has two effects:

1. The diffraction line selected for the stress measurement may be strong at $\psi = 0$ and virtually absent at $\psi = 45°$, or vice versa. The texture will then control the possible ψ angles.

2. If the material is elastically very anisotropic, the mechanically measured value of E will depend markedly on direction in the specimen. This effect is automatically allowed for if K_e is determined experimentally, but not if K_e is calculated from E determined on a non-textured specimen.

Plastic Deformation

If the specimen has been plastically deformed in a particular way in the region examined by x-rays, the x-ray method does not indicate the true macrostress. (If the plastic deformation causing the residual stress is remote from the region examined, the accuracy of the x-ray method is not affected; this condition exists, for example, in surface measurements on quenched specimens, where residual stress in the outer surface is due to plastic flow in the interior.)

When a specimen is stretched plastically a few percent and then unloaded, x-ray measurements show a line shift indicating residual compressive macrostress in the direction of prestrain. The effect is symmetrical: after plastic compression, x-rays indicate residual tensile stress. It is not a surface effect, because x-ray measurements made after successive removal of surface layers show that the stress persists throughout the specimen. On the other hand, dissection measurements show that a true macrostress does not exist, and, in fact, none would be expected after uniform deformation. The stress indicated by x-rays is called *pseudo-macrostress*, "pseudo" because it is not a true macrostress causing strain on dissection and "macro" because it causes an x-ray line shift. Pseudo-macrostress is actually an unusual kind of microstress, in which the portions of the material that are in tension and in compression are unequal in volume. It has been discussed in various reviews [16.26–16.28].

The effect was first observed after uniaxial deformation, but such deformation is not restricted to pure tension and compression. Plastic bending, for example, causes true macrostress (Fig. 16–2), but the deformation mode is predominantly a tension or compression of layers parallel to the neutral axis of the beam. The longitudinal residual stress indicated by x-rays is therefore the sum of true macrostress and pseudo-macrostress, and the x-ray result will be numerically larger at either surface than the result obtained by dissection.

Deformation by rolling or die drawing has a certain uniaxial character, but the forces on the material at the roll or die surface are inclined to the specimen axis. Macrostress is produced, superimposed on a complex system of microstresses; the latter probably include a pseudo-macrostress. As a result, the x-ray method does not measure the true macrostress, but rather the sum of macrostress and some component of the microstress that causes an additional line shift.

On the other hand, plastic deformation by grinding or shot peening produces macrostresses that are accurately measurable by x-rays, as shown by excellent agreement between x-ray and dissection measurements. Figure 16–11 shows an example. Deformation by these processes appears to be multiaxial, rather than uniaxial, and pseudo-macrostress is accordingly absent.

In summary, the x-ray method does not reveal the true macrostress in specimens that have been plastically deformed by tension, compression bending, rolling, or die drawing. It would be wrong to conclude, however, that the x-ray method is in error. Pseudo-macrostress is just as real as macrostress. X-rays indicate the sum of the two, dissection only the latter.

(When the $\sin^2 \psi$ method (Sec. 16–4) is used, some specimens that have been plastically deformed in the region examined yield values of d_i that vary with $\sin^2 \psi$ in an oscillatory manner, rather than linearly [16.33, 16.34, 16.26]. These oscillations in d_i are not fully understood. They must be caused by a system of microstresses more complex than pseudo-macrostress, because pseudo-macrostress, like true macrostress, yields a linear variation of d_i with $\sin^2 \psi$. When oscillations occur, the standard two-exposure method of stress measurement (Sec. 16.4) can be seriously in error.)

PROBLEMS

***16–1** A uniaxial tensile stress of 100 ksi is being measured in iron by reflecting x-rays from the (211) planes at $\psi = 0$ and 45°. Take the lattice parameter a of the stress-free iron to be 2.8665 Å, $E = 30 \times 10^6$ psi, and $\nu = 0.29$. (a) Calculate d_0, d_n, and d_i to five significant figures. (b) What percent error is made in replacing $(d_i - d_n)/d_0$ by $(d_i - d_n)/d_n$ in Eq. (16–9)?

16–2 Derive Eq. (16–11).

16–3 Derive Eq. (16–19).

16–4 Show that the absorption factor for the diffractometer method is $(1 - \tan\psi \cot\theta)$.

***16–5** The circumferential (hoop) stress in a cylinder of 1045 steel, due to water quenching followed by glass bead peening, is measured with a diffractometer (Cr $K\alpha$ radiation, 211 reflection). The line shift $\Delta 2\theta$ when a stress-free specimen is rotated from $\psi = 0$ to $\psi = 45°$ is $-0.10°$. Take the stress constant K_1 as 90 ksi/deg. The time t given below is the time required to accumulate 20,000 counts at each angle.

ψ	2θ	t(sec)	LPA
0	155.00°	69.20	1.910912
	155.80	54.47	1.916158
	156.60	71.64	1.921281
45°	156.00	35.84	1.509882
	156.50	32.35	1.521150
	157.00	33.83	1.532391

Calculate the residual stress with and without correction for line asymmetry by the LPA factor.

Appendices

APPENDIX 1
THE RECIPROCAL LATTICE

A1–1 INTRODUCTION

All the diffraction phenomena described in this book have been discussed in terms of the Bragg law. This simple law, admirable for its very simplicity, is in fact applicable to a very wide range of phenomena and is all that is needed for an understanding of a great many applications of x-ray diffraction. Yet there are diffraction effects which the Bragg law is totally unable to explain, notably those involving diffuse scattering at non-Bragg angles, and these effects demand a more general theory of diffraction for their explanation. The reciprocal lattice provides the framework for such a theory. This powerful concept was introduced into the field of diffraction by the German physicist Ewald in 1921 and has since become an indispensable tool in the solution of many problems.

Although the reciprocal lattice may at first appear rather abstract or artificial, the time spent in grasping its essential features is time well spent, because the reciprocal-lattice theory of diffraction, being general, is applicable to all diffraction phenomena from the simplest to the most intricate. Familiarity with the reciprocal lattice will therefore not only provide the student with the necessary key to complex diffraction effects but will deepen his understanding of even the simplest.

A1–2 VECTOR MULTIPLICATION

Since the reciprocal lattice is best formulated in terms of vectors, we shall first review a few theorems of vector algebra, namely, those involving the multiplication of vector quantities.

The *scalar product* (or dot product) of two vectors* $\mathbf{a}$ and $\mathbf{b}$, written $\mathbf{a} \cdot \mathbf{b}$, is a scalar quantity equal in magnitude to the product of the absolute values of the two vectors and the cosine of the angle α between them, or

$$\mathbf{a} \cdot \mathbf{b} = ab \cos \alpha.$$

Geometrically, Fig. A1–1 shows that the scalar product of two vectors may be regarded as the product of the length of one vector and the projection of the other upon the first. If one of the vectors, say $\mathbf{a}$, is a unit vector (a vector of unit length), then $\mathbf{a} \cdot \mathbf{b}$ gives immediately the length of the projection of $\mathbf{b}$ on $\mathbf{a}$. The scalar product of sums or differences of vectors is formed simply by term-by-term multiplication:

$$(\mathbf{a} + \mathbf{b}) \cdot (\mathbf{c} - \mathbf{d}) = (\mathbf{a} \cdot \mathbf{c}) - (\mathbf{a} \cdot \mathbf{d}) + (\mathbf{b} \cdot \mathbf{c}) - (\mathbf{b} \cdot \mathbf{d}).$$

* Bold-face symbols stand for vectors. The same symbol in italic stands for the absolute value of the vector.

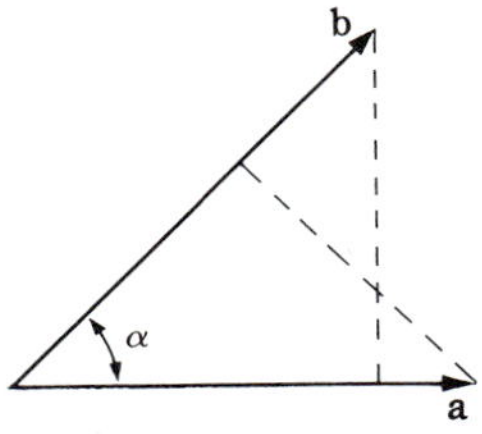

Fig. A1-1 Scalar product of two vectors.

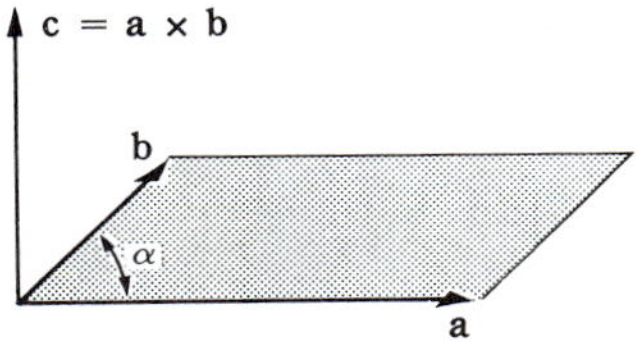

Fig. A1-2 Vector product of two vectors.

The order of multiplication is of no importance; i.e.,

$$\mathbf{a} \cdot \mathbf{b} = \mathbf{b} \cdot \mathbf{a}.$$

The *vector product* (or cross product) of two vectors **a** and **b**, written $\mathbf{a} \times \mathbf{b}$, is *a vector* **c** at right angles to the plane of **a** and **b**, and equal in magnitude to the product of the absolute values of the two vectors and the sine of the angle α between them, or

$$\mathbf{c} = \mathbf{a} \times \mathbf{b},$$

$$c = ab \sin \alpha.$$

The magnitude of **c** is simply the area of the parallelogram constructed on **a** and **b**, as suggested by Fig. A1-2. The direction of **c** is that in which a right-hand screw would move if rotated in such a way as to bring **a** into **b**. It follows from this that the direction of the vector product **c** is reversed if the order of multiplication is reversed, or that

$$\mathbf{a} \times \mathbf{b} = -(\mathbf{b} \times \mathbf{a}).$$

A1-3 THE RECIPROCAL LATTICE

Corresponding to any crystal lattice, we can construct a *reciprocal lattice*, so called because many of its properties are reciprocal to those of the crystal lattice. Let the crystal lattice have a unit cell defined by the vectors $\mathbf{a}_1$, $\mathbf{a}_2$, and $\mathbf{a}_3$. Then the corresponding reciprocal lattice has a unit cell defined by the vectors $\mathbf{b}_1$, $\mathbf{b}_2$, and $\mathbf{b}_3$, where

$$\mathbf{b}_1 = \frac{1}{V}(\mathbf{a}_2 \times \mathbf{a}_3), \tag{1}$$

$$\mathbf{b}_2 = \frac{1}{V}(\mathbf{a}_3 \times \mathbf{a}_1), \tag{2}$$

$$\mathbf{b}_3 = \frac{1}{V}(\mathbf{a}_1 \times \mathbf{a}_2), \tag{3}$$

and V is the volume of the crystal unit cell. This way of defining the vectors $\mathbf{b}_1$, $\mathbf{b}_2$, $\mathbf{b}_3$ in terms of the vectors $\mathbf{a}_1$, $\mathbf{a}_2$, $\mathbf{a}_3$ gives the reciprocal lattice certain useful properties which we will now investigate.

Consider the general triclinic unit cell shown in Fig. A1–3. The reciprocal-lattice axis $\mathbf{b}_3$ is, according to Eq. (3), normal to the plane of $\mathbf{a}_1$ and $\mathbf{a}_2$, as shown. Its length is given by

$$\begin{aligned} b_3 &= \frac{|\mathbf{a}_1 \times \mathbf{a}_2|}{V} \\ &= \frac{(\text{area of parallelogram } OACB)}{(\text{area of parallelogram } OACB)(\text{height of cell})} \\ &= \frac{1}{OP} = \frac{1}{d_{001}}, \end{aligned}$$

since OP, the projection of $\mathbf{a}_3$ on $\mathbf{b}_3$, is equal to the height of the cell, which in turn is simply the spacing d of the (001) planes of the crystal lattice. Similarly, we find that the reciprocal lattice axes $\mathbf{b}_1$ and $\mathbf{b}_2$ are normal to the (100) and (010) planes, respectively, of the crystal lattice, and are equal in length to the reciprocals of the spacings of these planes.

By extension, similar relations are found for all the planes of the crystal lattice. The whole reciprocal lattice is built up by repeated translations of the unit cell by the vectors $\mathbf{b}_1$, $\mathbf{b}_2$, $\mathbf{b}_3$. This produces an array of points each of which is labeled with its coordinates in terms of the basic vectors. Thus, the point at the end of the

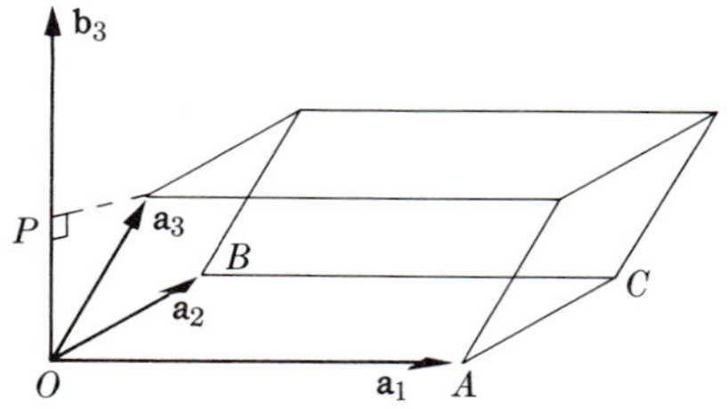

Fig. A1–3 Location of the reciprocal-lattice axis $\mathbf{b}_3$.

$\mathbf{b}_1$ vector is labeled 100, that at the end of the $\mathbf{b}_2$ vector 010, etc. This extended reciprocal lattice has the following properties:

1. A vector $\mathbf{H}_{hkl}$ drawn from the origin of the reciprocal lattice to any point in it having coordinates hkl is perpendicular to the plane in the crystal lattice whose Miller indices are hkl. This vector is given in terms of its coordinates by the expression

$$\mathbf{H}_{hkl} = h\mathbf{b}_1 + k\mathbf{b}_2 + l\mathbf{b}_3.$$

2. The length of the vector $\mathbf{H}_{hkl}$ is equal to the reciprocal of the spacing d of the (hkl) planes, or

$$H_{hkl} = \frac{1}{d_{hkl}}.$$

The important thing to note about these relations is that the reciprocal-lattice array of points completely describes the crystal, in the sense that each reciprocal-lattice point is related to a set of planes in the crystal and represents the orientation and spacing of that set of planes.

Before proving these general relations, we might consider particular examples of the reciprocal lattice as shown in Figs. A1-4 and A1-5 for cubic and hexagonal crystals. In each case, the reciprocal lattice is drawn from any convenient origin, not necessarily that of the crystal lattice, and to any convenient scale of reciprocal angstroms. Note that Eqs. (1) through (3) take on a very simple form for any crystal whose unit cell is based on mutually perpendicular vectors, i.e., cubic, tetragonal, or orthorhombic. For such crystals, $\mathbf{b}_1$, $\mathbf{b}_2$, and $\mathbf{b}_3$ are parallel, respectively, to $\mathbf{a}_1$, $\mathbf{a}_2$, and $\mathbf{a}_3$, while b_1, b_2, and b_3 are simply the reciprocals of a_1, a_2, and a_3. In Figs. A1-4 and A1-5, four cells of the reciprocal lattice are shown, together with two $\mathbf{H}$ vectors in each case. By means of the scales shown, it may be verified that each $\mathbf{H}$ vector is equal in length to the reciprocal of the spacing of the corresponding planes and normal to them. Note that reciprocal lattice points such as

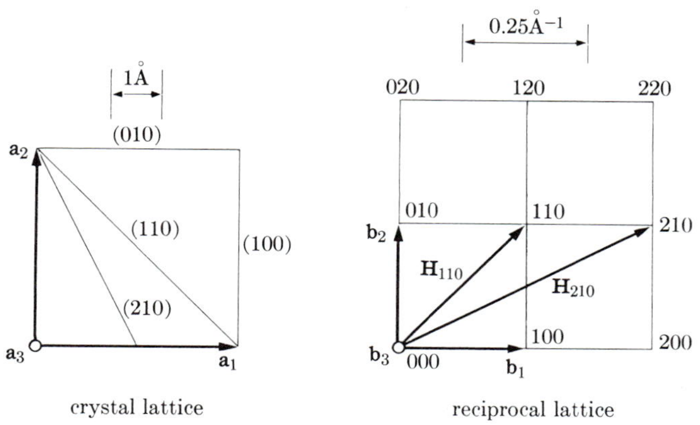

Fig. A1-4 The reciprocal lattice of a cubic crystal which has $a_1 = 4$ Å. The axes $\mathbf{a}_3$ and $\mathbf{b}_3$ are normal to the drawing.

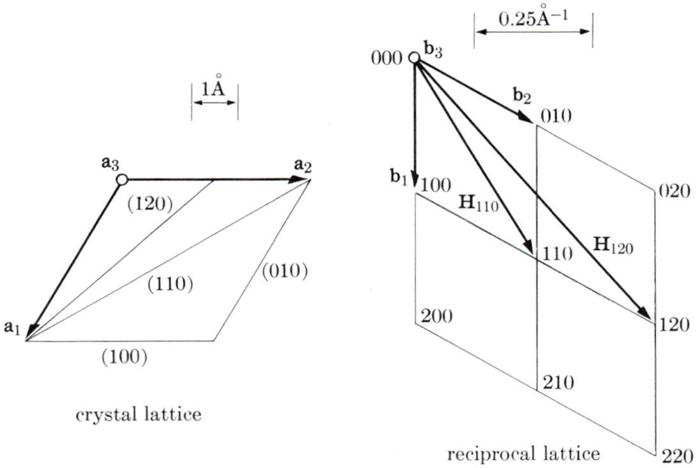

Fig. A1–5 The reciprocal lattice of a hexagonal crystal which has $a_1 = 4$ Å. (Here the three-symbol system of plane indexing is used and $\mathbf{a}_3$ is the axis usually designated **c**.) The axes $\mathbf{a}_3$ and $\mathbf{b}_3$ are normal to the drawing.

nh, *nk*, *nl*, where *n* is an integer, correspond to planes parallel to (*hkl*) and having $1/n$ their spacing. Thus, $\mathbf{H}_{220}$ is perpendicular to (220) planes and therefore parallel to $\mathbf{H}_{110}$, since (110) and (220) are parallel, but $\mathbf{H}_{220}$ is twice as long as $\mathbf{H}_{110}$ since the (220) planes have half the spacing of the (110) planes.

Other useful relations between the crystal and reciprocal vectors follow from Eqs. (1) through (3). Since $\mathbf{b}_3$, for example, is normal to both $\mathbf{a}_1$ and $\mathbf{a}_2$, its dot product with either one of these vectors is zero, or

$$\mathbf{b}_3 \cdot \mathbf{a}_1 = \mathbf{b}_3 \cdot \mathbf{a}_2 = 0.$$

The dot product of $\mathbf{b}_3$ and $\mathbf{a}_3$, however, is unity, since (see Fig. A1–3)

$$\begin{aligned} \mathbf{b}_3 \cdot \mathbf{a}_3 &= (b_3)(\text{projection of } \mathbf{a}_3 \text{ on } \mathbf{b}_3) \\ &= \left(\frac{1}{OP}\right)(OP) \\ &= 1. \end{aligned}$$

In general,

$$\begin{aligned} \mathbf{a}_m \cdot \mathbf{b}_n &= 1, \quad \text{if } m = n, &(4) \\ &= 0, \quad \text{if } m \neq n. &(5) \end{aligned}$$

The fact that $\mathbf{H}_{hkl}$ is normal to (*hkl*) and H_{hkl} is the reciprocal of d_{hkl} may be proved as follows. Let *ABC* of Fig. A1–6 be part of the plane nearest the origin in the set (*hkl*). Then, from the definition of Miller indices, the vectors from the origin to the points *A*, *B*, and *C* are $\mathbf{a}_1/h$, $\mathbf{a}_2/k$, and $\mathbf{a}_3/l$, respectively. Consider the vector **AB**, that is, a vector drawn from *A* to *B*, lying in the plane (*hkl*). Since

$$\frac{\mathbf{a}_1}{h} + \mathbf{AB} = \frac{\mathbf{a}_2}{k},$$

then

$$\mathbf{AB} = \frac{\mathbf{a}_2}{k} - \frac{\mathbf{a}_1}{h}.$$

Forming the dot product of **H** and **AB**, we have

$$\mathbf{H} \cdot \mathbf{AB} = (h\mathbf{b}_1 + k\mathbf{b}_2 + l\mathbf{b}_3) \cdot \left(\frac{\mathbf{a}_2}{k} - \frac{\mathbf{a}_1}{h}\right).$$

Evaluating this with the aid of Eqs. (4) and (5), we find

$$\mathbf{H} \cdot \mathbf{AB} = 1 - 1 = 0.$$

Since this product is zero, **H** must be normal to **AB**. Similarly, it may be shown that **H** is normal to **AC**. Since **H** is normal to two vectors in the plane (hkl), it is normal to the plane itself.

To prove the reciprocal relation between H and d, let **n** be a unit vector in the direction of **H**, i.e., normal to (hkl). Then

$$d = ON = \frac{\mathbf{a}_1}{h} \cdot \mathbf{n}.$$

But

$$\mathbf{n} = \frac{\mathbf{H}}{H}.$$

Therefore

$$\begin{aligned} d &= \frac{\mathbf{a}_1}{h} \cdot \frac{\mathbf{H}}{H} \\ &= \frac{\mathbf{a}_1}{h} \cdot \frac{(h\mathbf{b}_1 + k\mathbf{b}_2 + l\mathbf{b}_3)}{H} \\ &= \frac{1}{H}. \end{aligned}$$

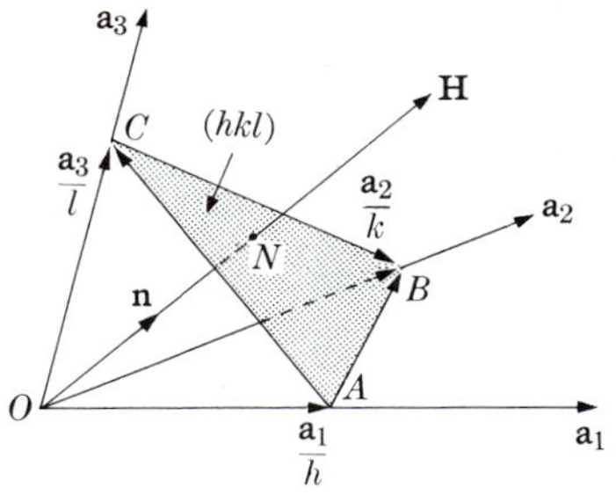

Fig. A1–6 Relation between reciprocal-lattice vector **H** and crystal plane (hkl).

Used purely as a geometrical tool, the reciprocal lattice is of considerable help in the solution of many problems in crystal geometry. Consider, for example, the relation between the planes of a zone and the axis of that zone. Since the planes of a zone are all parallel to one line, the zone axis, their normals must be coplanar. This means that planes of a zone are represented, in the reciprocal lattice, by a set of points lying on a plane passing through the origin of the reciprocal lattice. If the plane (hkl) belongs to the zone whose axis is $[uvw]$, then the normal to (hkl), namely, **H**, must be perpendicular to $[uvw]$. Express the zone axis as a vector in the crystal lattice and **H** as a vector in the reciprocal lattice:

$$\text{Zone axis} = u\mathbf{a}_1 + v\mathbf{a}_2 + w\mathbf{a}_3,$$

$$\mathbf{H} = h\mathbf{b}_1 + k\mathbf{b}_2 + l\mathbf{b}_3.$$

If these two vectors are perpendicular, their dot product must be zero:

$$(u\mathbf{a}_1 + v\mathbf{a}_2 + w\mathbf{a}_3) \cdot (h\mathbf{b}_1 + k\mathbf{b}_2 + l\mathbf{b}_3) = 0,$$

$$hu + kv + lw = 0.$$

This is the relation given without proof in Sec. 2–6. By similar use of reciprocal-lattice vectors, other problems of crystal geometry, such as the derivation of the plane-spacing equations given in Appendix 3, may be greatly simplified.

A1–4 DIFFRACTION AND THE RECIPROCAL LATTICE

The great utility of the reciprocal lattice, however, lies in its connection with diffraction problems. We shall consider how x-rays scattered by the atom O at the origin of the crystal lattice (Fig. A1–7) are affected by those scattered by any other atom A whose coordinates with respect to the origin are $p\mathbf{a}_1$, $q\mathbf{a}_2$ and $r\mathbf{a}_3$, where p, q, and r are integers. Thus,

$$\mathbf{OA} = p\mathbf{a}_1 + q\mathbf{a}_2 + r\mathbf{a}_3.$$

Let the incident x-rays have a wavelength λ, and let the incident and diffracted beams be represented by the unit vectors $\mathbf{S}_0$ and $\mathbf{S}$, respectively. $\mathbf{S}_0$, $\mathbf{S}$, and $\mathbf{OA}$ are, in general, not coplanar.

To determine the conditions under which diffraction will occur, we must determine the phase difference between the rays scattered by the atoms O and A. The lines Ou and Ov in Fig. A1–7 are wave fronts perpendicular to the incident beam $\mathbf{S}_0$ and the scattered beam $\mathbf{S}$, respectively. Let δ be the path difference for rays scattered by O and A. Then

$$\begin{aligned}\delta &= uA + Av \\ &= Om + On \\ &= \mathbf{S}_0 \cdot \mathbf{OA} + (-\mathbf{S}) \cdot \mathbf{OA} \\ &= -\mathbf{OA} \cdot (\mathbf{S} - \mathbf{S}_0).\end{aligned}$$

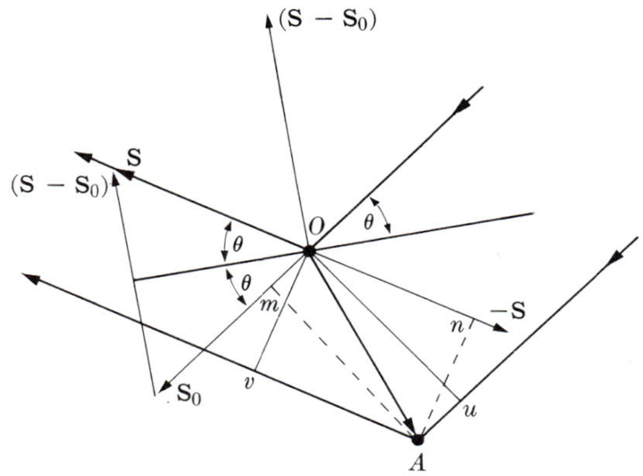

Fig. A1–7 X-ray scattering by atoms at O and A. After Guinier [G.10].

The corresponding phase difference, in radians, is given by

$$\phi = \frac{2\pi\delta}{\lambda}$$

$$= -2\pi\left(\frac{\mathbf{S} - \mathbf{S}_0}{\lambda}\right) \cdot \mathbf{OA}. \qquad (6)$$

Diffraction is now related to the reciprocal lattice by expressing the vector $(\mathbf{S} - \mathbf{S}_0)/\lambda$ as a vector in that lattice. Let

$$\frac{\mathbf{S} - \mathbf{S}_0}{\lambda} = h\mathbf{b}_1 + k\mathbf{b}_2 + l\mathbf{b}_3.$$

This is now in the form of a vector in reciprocal space but, at this point, no particular significance is attached to the parameters h, k, and l. They are continuously variable and may assume any values, integral or nonintegral. Equation (6) now becomes

$$\phi = -2\pi(h\mathbf{b}_1 + k\mathbf{b}_2 + l\mathbf{b}_3) \cdot (p\mathbf{a}_1 + q\mathbf{a}_2 + r\mathbf{a}_3) = -2\pi(hp + kq + lr).$$

A diffracted beam will be formed only if reinforcement occurs, and this requires that ϕ be an integral multiple of 2π. This can happen only if h, k, and l are integers. Therefore the condition for diffraction is that the vector $(\mathbf{S} - \mathbf{S}_0)/\lambda$ end on a *point* in the reciprocal lattice, or that

$$\boxed{\frac{\mathbf{S} - \mathbf{S}_0}{\lambda} = \mathbf{H} = h\mathbf{b}_1 + k\mathbf{b}_2 + l\mathbf{b}_3}, \qquad (7)$$

where h, k, and l are now restricted to integral values.

Both the Laue equations and the Bragg law can be derived from Eq. (7). The former are obtained by forming the dot product of each side of the equation and the three crystal-lattice vectors $\mathbf{a}_1$, $\mathbf{a}_2$, $\mathbf{a}_3$ successively. For example,

$$\mathbf{a}_1 \cdot \left(\frac{\mathbf{S} - \mathbf{S}_0}{\lambda}\right) = \mathbf{a}_1 \cdot (h\mathbf{b}_1 + k\mathbf{b}_2 + l\mathbf{b}_3)$$

$$= h,$$

or

$$\mathbf{a}_1 \cdot (\mathbf{S} - \mathbf{S}_0) = h\lambda. \tag{8}$$

Similarly,

$$\mathbf{a}_2 \cdot (\mathbf{S} - \mathbf{S}_0) = k\lambda, \tag{9}$$

$$\mathbf{a}_3 \cdot (\mathbf{S} - \mathbf{S}_0) = l\lambda. \tag{10}$$

Equations (8) through (10) are the vector form of the equations derived by von Laue in 1912 to express the necessary conditions for diffraction. They must be satisfied simultaneously for diffraction to occur.

As shown in Fig. A1–7, the vector $(\mathbf{S} - \mathbf{S}_0)$ bisects the angle between the incident beam $\mathbf{S}_0$ and the diffracted beam $\mathbf{S}$. The diffracted beam $\mathbf{S}$ can therefore be considered as being reflected from a set of planes perpendicular to $(\mathbf{S} - \mathbf{S}_0)$. In fact, Eq. (7) states that $(\mathbf{S} - \mathbf{S}_0)$ is parallel to $\mathbf{H}$, which is in turn perpendicular to the planes (hkl). Let θ be the angle between $\mathbf{S}$ (or $\mathbf{S}_0$) and these planes. Then, since $\mathbf{S}$ and $\mathbf{S}_0$ are unit vectors,

$$(S - S_0) = 2 \sin \theta.$$

Therefore

$$\frac{2 \sin \theta}{\lambda} = \frac{S - S_0}{\lambda} = H = \frac{1}{d},$$

or

$$\lambda = 2d \sin \theta.$$

The conditions for diffraction expressed by Eq. (7) may be represented graphically by the "Ewald construction" shown in Fig. A1–8. The vector $\mathbf{S}_0/\lambda$ is drawn parallel to the incident beam and $1/\lambda$ in length. The terminal point O of this vector is taken as the origin of the reciprocal lattice, drawn to the same scale as the vector $\mathbf{S}_0/\lambda$. A sphere of radius $1/\lambda$ is drawn about C, the initial point of the incident-beam vector. Then the condition for diffraction from the (hkl) planes is that the point hkl in the reciprocal lattice (point P in Fig. A1–8) touch the surface of the sphere, and the direction of the diffracted-beam vector $\mathbf{S}/\lambda$ is found by joining C to P. When this condition is fulfilled, the vector $\mathbf{OP}$ equals both $\mathbf{H}_{hkl}$ and $(\mathbf{S} - \mathbf{S}_0)/\lambda$, thus satisfying Eq. (7). Since diffraction depends on a reciprocal-lattice point touching the surface of the sphere drawn about C, this sphere is known as the "sphere of reflection."

Our initial assumption that p, q, and r are integers apparently excludes all crystals except those having only one atom per cell, located at the cell corners. For if the unit cell contains more than one atom, then the vector $\mathbf{OA}$ from the origin to "any atom" in the crystal may have nonintegral coordinates. However, the

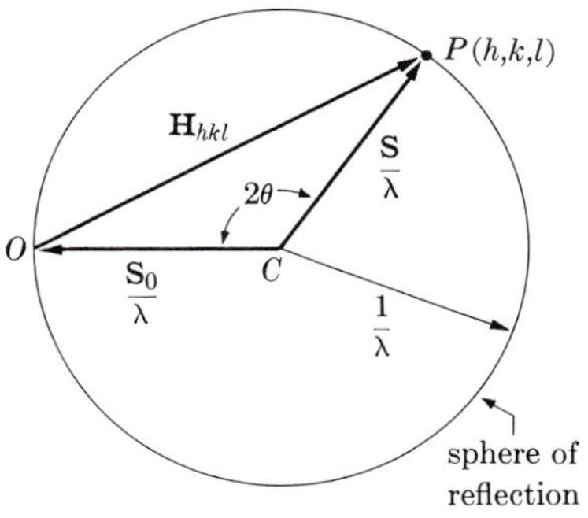

Fig. A1-8 The Ewald construction. Section through the sphere of reflection containing the incident and diffracted beam vectors.

presence of these additional atoms in the unit cell affects only the intensities of the diffracted beams, not their directions, and it is only the diffraction directions which are predicted by the Ewald construction. Stated in another way, the reciprocal lattice depends only on the shape and size of the unit cell of the crystal lattice and not at all on the arrangement of atoms within that cell. If we wish to take atom arrangement into consideration, we may weight each reciprocal-lattice point hkl with the appropriate value of the scattering power ($= |F|^2$, where F is the structure factor) of the particular (hkl) planes involved. Some planes may then have zero scattering power, thus eliminating some reciprocal-lattice points from consideration, e.g., all reciprocal-lattice points having odd values of $(h + k + l)$ for body-centered crystals.

The common methods of x-ray diffraction are differentiated by the methods used for bringing reciprocal-lattice points into contact with the surface of the sphere of reflection. The radius of the sphere may be varied by varying the incident wavelength (Laue method), or the position of the reciprocal lattice may be varied by changes in the orientation of the crystal (rotating-crystal and powder methods).

A1-5 THE ROTATING-CRYSTAL METHOD

As stated in Sec. 3-6, when monochromatic radiation is incident on a single crystal rotated about one of its axes, the reflected beams lie on the surface of imaginary cones coaxial with the rotation axis. The way in which this reflection occurs may be shown very nicely by the Ewald construction. Suppose a simple cubic crystal is rotated about the axis [001]. This is equivalent to rotation of the reciprocal lattice about the $\mathbf{b}_3$ axis. Figure A1-9 shows a portion of the reciprocal lattice oriented in this manner, together with the adjacent sphere of reflection.

All crystal planes having indices ($hk1$) are represented by points lying on a plane (called the "$l = 1$ layer") in the reciprocal lattice, normal to $\mathbf{b}_3$. When the reciprocal lattice rotates, this plane cuts the reflection sphere in the small circle shown, and any points on the $l = 1$ layer which touch the sphere surface must

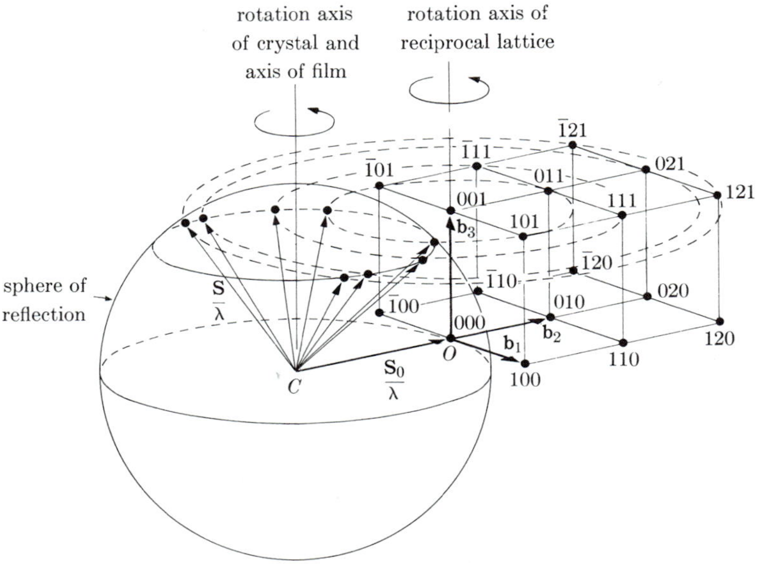

Fig. A1–9 Reciprocal-lattice treatment of rotating-crystal method.

touch it on this circle. Therefore all diffracted-beam vectors $\mathbf{S}/\lambda$ must end on this circle, which is equivalent to saying that the diffracted beams must lie on the surface of a cone. In this particular case, all the *hk*1 points shown intersect the surface of the sphere sometime during their rotation about the $\mathbf{b}_3$ axis, producing the diffracted beams shown in Fig. A1–9. In addition many *hk*0 and *hk*$\bar{1}$ reflections would be produced, but these have been omitted from the drawing for the sake of clarity.

This simple example may suggest how the rotation photograph of a crystal of unknown structure, and therefore having an unknown reciprocal lattice, can yield clues as to the distribution in space of reciprocal-lattice points. By taking a number of photographs with the crystal rotated successively about various axes, the crystallographer gradually discovers the complete distribution of reflecting points. Once the reciprocal lattice is known, the crystal lattice is easily derived, because it is a corollary of Eqs. (1) through (3) that the reciprocal of the reciprocal lattice is the crystal lattice.

A1–6 THE POWDER METHOD

The random orientations of the individual crystals in a powder specimen are equivalent to the rotation of a single crystal about all possible axes during the x-ray exposure. The reciprocal lattice therefore takes on all possible orientations relative to the incident beam, but its origin remains fixed at the end of the $\mathbf{S}_0/\lambda$ vector.

Consider any point *hkl* in the reciprocal lattice, initially at P_1 (Fig. A1–10). This point can be brought into a reflecting position on the surface of the reflection

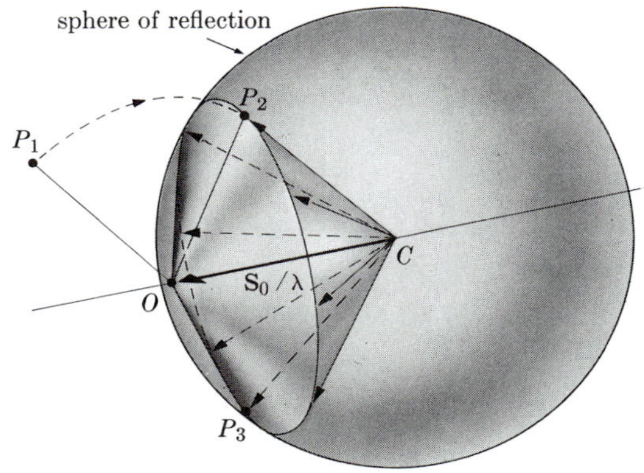

Fig. A1–10 Formation of a cone of diffracted rays in the powder method.

sphere by a rotation of the lattice about an axis through O and normal to OC, for example. Such a rotation would move P_1 to P_2. But the point hkl can still remain on the surface of the sphere [i.e., reflection will still occur from the same set of planes (hkl)] if the reciprocal lattice is then rotated about the axis OC, since the point hkl will then move around the small circle P_2P_3. During this motion, the **H** vector sweeps out a cone whose apex is at O, and the diffracted beams all lie on the surface of another cone whose apex is at C. The axes of both cones coincide with the incident beam.

The number of different hkl reflections obtained on a powder photograph depends, in part, on the relative magnitudes of the wavelength and the crystal-lattice parameters or, in reciprocal-lattice language, on the relative sizes of the sphere of reflection and the reciprocal-lattice unit cell. To find the number of reflections we may regard the reciprocal lattice as fixed and the incident-beam vector $\mathbf{S}_0/\lambda$ as rotating about its terminal point through all possible positions. The reflection sphere therefore swings about the origin of the reciprocal lattice and sweeps out a sphere of radius $2/\lambda$, called the "limiting sphere" (Fig. A1–11). All reciprocal-lattice points within the limiting sphere can touch the surface of the reflection sphere and cause reflection to occur.

It is also a corollary of Eqs. (1) through (3) that the volume v of the reciprocal-lattice unit cell is the reciprocal of the volume V of the crystal unit cell. Since there is one reciprocal-lattice point per cell of the reciprocal lattice, the number of reciprocal-lattice points within the limiting sphere is given by

$$n = \frac{(4\pi/3)(2/\lambda)^3}{v} = \frac{32\pi V}{3\lambda^3}. \tag{11}$$

Not all of these n points will cause a separate reflection: some of them may have a zero structure factor, and some may be at equal distances from the reciprocal-lattice origin, i.e., correspond to planes of the same spacing. (The latter effect is taken care of by the multiplicity factor, since this gives the number of different planes in

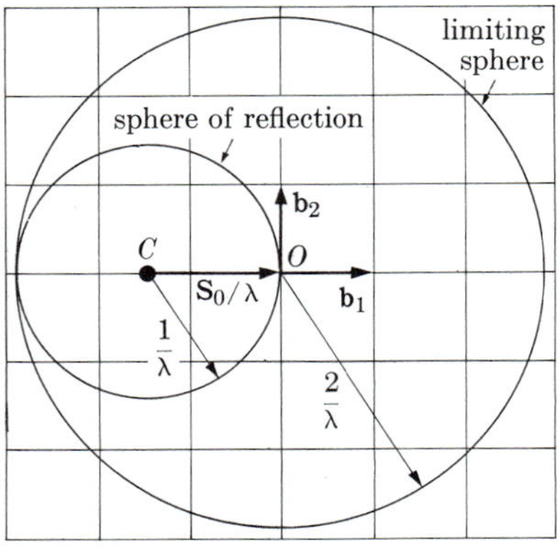

Fig. A1–11 The limiting sphere for the powder method.

a form having the same spacing.) However, Eq. (11) may always be used directly to obtain an upper limit to the number of possible reflections. For example, if $V = 50$ Å^3 and $\lambda = 1.54$ Å, then $n = 460$. If the specimen belongs to the triclinic system, this number will be reduced by a factor of only 2, the multiplicity factor, and the powder photograph will contain 230 separate diffraction lines! As the symmetry of the crystal increases, so does the multiplicity factor and the fraction of reciprocal-lattice points which have zero structure factor, resulting in a decrease in the number of diffraction lines. For example, the powder pattern of a diamond cubic element has only 5 lines, for the same values of V and λ assumed above.

A1–7 THE LAUE METHOD

Diffraction occurs in the Laue method because of the continuous range of wavelengths present in the incident beam. Stated alternatively, contact between a fixed reciprocal-lattice point and the sphere of reflection is produced by continuously varying the radius of the sphere. There is therefore a whole set of reflection spheres, not just one; each has a different center, but all pass through the origin of the reciprocal lattice. The range of wavelengths present in the incident beam is of course not infinite. It has a sharp lower limit at λ_{SWL}, the short-wavelength limit of the continuous spectrum; the upper limit is less definite but is often taken as the wavelength of the K absorption edge of the silver in the emulsion (0.48 Å), because the effective photographic intensity of the continuous spectrum drops abruptly at that wavelength [see Fig. 1–19(c)].

To these two extreme wavelengths correspond two extreme reflection spheres, as shown in Fig. A1–12, which is a section through these spheres and the $l = 0$ layer of a reciprocal lattice. The incident beam is along the $\mathbf{b}_1$ vector, i.e., perpendicular to the (h00) planes of the crystal. The larger sphere shown is centered at B and has a radius equal to the reciprocal of λ_{SWL}, while the smaller sphere is centered at A and has a radius equal to the reciprocal of the wavelength of the silver K absorption edge.

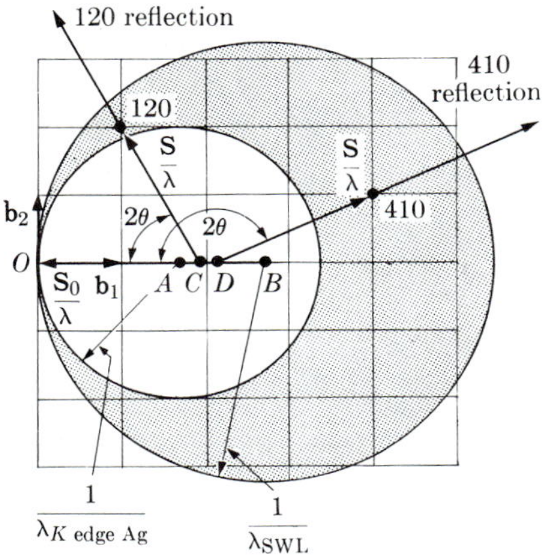

Fig. A1–12 Reciprocal-lattice treatment of the Laue method. $(\mathbf{S} - \mathbf{S}_0)/\lambda = \mathbf{H}$.

There is a whole series of spheres lying between these two and centered on the line segment AB. Therefore any reciprocal-lattice point lying in the shaded region of the diagram is on the surface of one of these spheres and corresponds to a set of crystal planes oriented to reflect one of the incident wavelengths. In the forward direction, for example, a 120 reflection will be produced. To find its direction, we locate a point C on AB which is equidistant from the origin O and the reciprocal-lattice point 120; C is therefore the center of the reflection sphere passing through the point 120. Joining C to 120 gives the diffracted-beam vector $\mathbf{S}/\lambda$ for this reflection. The direction of the 410 reflection, one of the many backward-reflected beams, is found in similar fashion; here the reciprocal-lattice point in question is situated on a reflection sphere centered at D.

There is another way of treating the Laue method which is more convenient for many purposes. The basic diffraction equation, Eq. (7), is rewritten in the form

$$\boxed{\mathbf{S} - \mathbf{S}_0 = \lambda \mathbf{H}}. \tag{12}$$

Both sides of this equation are now dimensionless and the radius of the sphere of reflection is simply unity, since $\mathbf{S}$ and $\mathbf{S}_0$ are unit vectors. But the position of the reciprocal-lattice points is now dependent on the wavelength used, since their distance from the origin of the reciprocal lattice is now given by λH.

In the Laue method, each reciprocal-lattice point (except 000) is drawn out into a line segment directed to the origin, because of the range of wavelengths present in the incident beam. The result is shown in Fig. A1–13,* which is drawn to correspond to Fig. A1–12. The point nearest the origin on each line segment has

* In this figure, as well as in Figs. A1–11 and A1–12, the size of the reciprocal lattice, relative to the size of the reflection sphere, has been exaggerated for clarity.

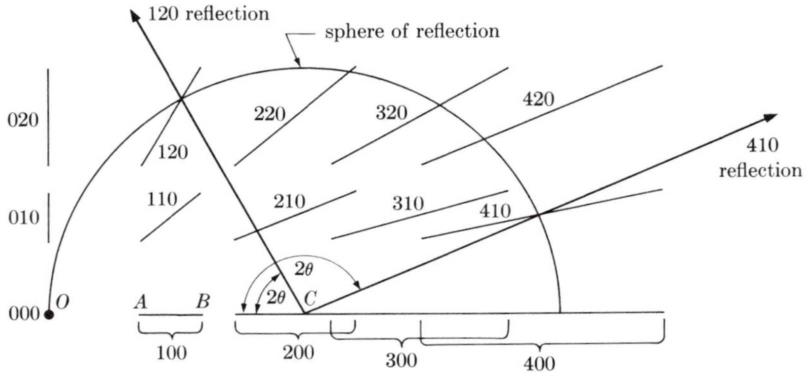

Fig. A1–13 Alternative reciprocal-lattice treatment of the Laue method. $\mathbf{S} - \mathbf{S}_0 = \lambda\mathbf{H}$.

a value of λH corresponding to the shortest wavelength present, while the point on the other end has a value of λH corresponding to the longest effective wavelength. Thus the 100 reciprocal-lattice line extends from A to B, where $OA = \lambda_{\min}H_{100}$ and $OB = \lambda_{\max}H_{100}$. Since the length of any line increases as H increases, for a given range of wavelengths, overlapping occurs for the higher orders, as shown by 200, 300, 400, etc. The reflection sphere is drawn with unit radius, and reflection occurs whenever a reciprocal-lattice line intersects the sphere surface. Graphically, the advantage of this construction over that of Fig. A1–12 is that all diffracted beams are now drawn from the same point C, thus facilitating the comparison of the diffraction angles 2θ for different reflections.

This construction also shows why the diffracted beams from planes of a zone are arranged on a cone in the Laue method. All reciprocal-lattice lines representing the planes of one zone lie on a plane passing through the origin of the reciprocal lattice. This plane cuts the reflection sphere in a circle, and all the diffracted beam vectors **S** must end on this circle, thus producing a conical array of diffracted beams, the axis of the cone coinciding with the zone axis.

Another application of this construction, to the problem of temperature-diffuse scattering, will illustrate the general utility of the reciprocal-lattice method in treating diffuse scattering phenomena. The reciprocal lattice of any crystal may be regarded as a distribution of "scattered intensity" in reciprocal space, in the sense that a scattered beam will be produced whenever the sphere of reflection intersects a point in reciprocal space where the "scattered intensity" is not zero. If the crystal is perfect, the scattered intensity is concentrated at points in reciprocal space, the points of the reciprocal lattice, and is zero everywhere else. But if anything occurs to disturb the regularity of the crystal lattice, then these points become smeared out, and appreciable scattered intensity exists in regions of reciprocal space where h, k, and l are nonintegral. For example, if the atoms of the crystal are undergoing thermal vibration, then each point of the reciprocal lattice spreads out into a region which may be considered, to a first approximation, as roughly spherical in shape, as suggested by Fig. A1–14(a). In other words, the

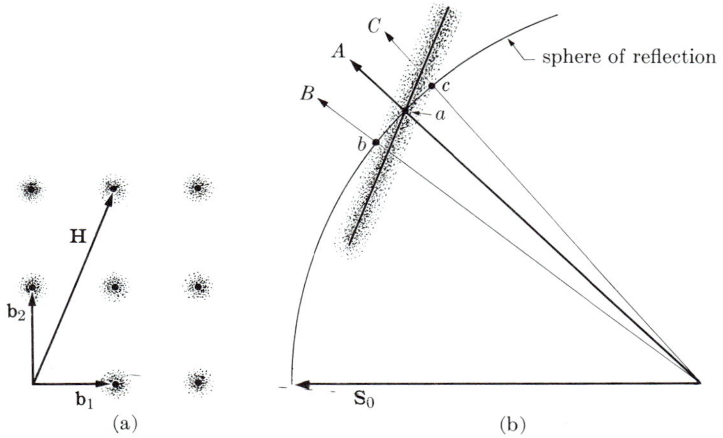

Fig. A1–14 The effect of thermal vibration on the reciprocal lattice.

thermally produced elastic waves which run through the crystal lattice so disturb the regularity of the atomic planes that the corresponding **H** vectors end, not on points, but in small spherical regions. The scattered intensity is not distributed uniformly within each region: it remains very high at the central point, where h, k, and l are integral, and is very weak and diffuse in the surrounding volume, as indicated in the drawing.

What then will be the effect of thermal agitation on, for example, a transmission Laue pattern? If we use the construction of Fig. A1–13, i.e., if we make distances in the reciprocal lattice equal to λH, then each spherical volume in the reciprocal lattice will be drawn out into a rod, roughly cylindrical in shape and directed to the origin, as indicated in Fig. A1–14(b), which is a section through the reflection sphere and one such rod. The axis of each rod is a line of high intensity and this is surrounded by a low-intensity region. This line intersects the reflection sphere at a and produces the strong diffracted beam A, the ordinary Laue reflection. But on either side of A there are weak scattered rays, extending from B to C, due to the intersection, extending from b to c, of the diffuse part of the rod with the sphere of reflection. In a direction normal to the drawing, however, the diffuse rod intersects the sphere in an arc equal only to the rod diameter, which is much shorter than the arc bc. We are thus led to expect, on a film placed in the transmission position, a weak and diffuse streak running *radially* through the usual sharp, intense Laue spot.

Figure A1–15 shows an example of this phenomenon, often called *thermal asterism* because of the radial direction of the diffuse streaks. This photograph was obtained from aluminum at 280°C in 5 minutes. Actually, thermal agitation is quite pronounced in aluminum even at room temperature, and thermal asterism is usually evident in overexposed room-temperature photographs. Even in Fig. 3–6(a), which was given a normal exposure of about 15 minutes, radial streaks are

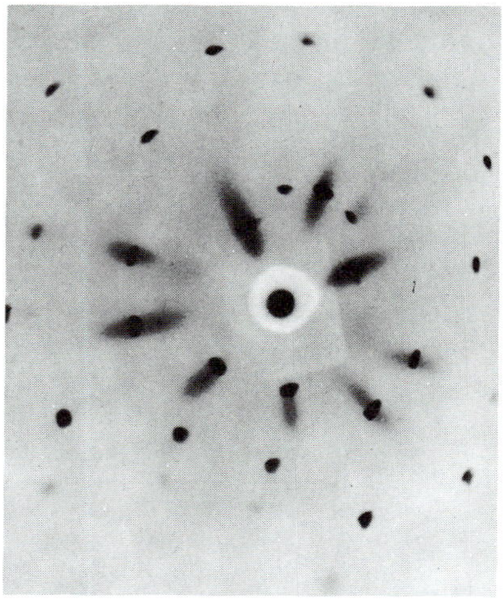

Fig. A1–15 Transmission Laue pattern showing thermal asterism. Aluminum crystal, 280°C, 5 min exposure.

faintly visible. In this latter photograph, there is a streak near the center that does not pass through any Laue spot: it is due to a reciprocal-lattice rod so nearly tangent to the reflection sphere that the sphere intersects only the diffuse part of the rod and not its axis.

The enlargement of reciprocal-lattice points caused by thermal vibration, depicted in Fig. A1–14(a), is not observed in powder patterns. It is so weak and diffuse that it is lost in the background.

APPENDIX 2
ELECTRON AND NEUTRON DIFFRACTION

A2-1 INTRODUCTION

Just as a beam of x-rays has a dual wave-particle character so, inversely, does a stream of particles have certain properties peculiar to wave motion. In particular, such a stream of particles can be diffracted by a periodic arrangement of scattering centers. This was first predicted theoretically by de Broglie in 1924 and demonstrated experimentally by Davisson and Germer in 1927 (for electrons) and by Von Halban and Preiswerk in 1936 (for neutrons).

If a stream of particles can behave like wave motion, it must have a wavelength associated with it. The theory of wave mechanics indicates that this wavelength is given by the ratio of Planck's constant h to the momentum of the particle, or

$$\lambda = \frac{h}{mv}, \tag{1}$$

where m is the mass and v the velocity of the particle. If a stream of particles is directed at a crystal under the proper conditions, diffraction will occur in accordance with the Bragg law just as for x-rays, and the directions of diffraction can be predicted by the use of that law and the wavelength calculated from Eq. (1). Both electrons and neutrons have proved to be useful particles for the study of crystalline structure by diffraction and numerous applications of these techniques have been found in metallurgy and in solid state physics and chemistry. The differences between x-ray, electron, and neutron diffraction by crystals are such that these three techniques supplement one another to a remarkable degree, each giving a particular kind of information which the others are incapable of supplying.

A2-2 ELECTRON DIFFRACTION

A stream of fast electrons is obtained in a tube operating on much the same principles as an x-ray tube. The wavelength associated with the electrons depends on the applied voltage, since the kinetic energy of the electrons is given by

$$\tfrac{1}{2}mv^2 = eV, \tag{2}$$

where e is the charge on the electron and V the applied voltage. Combination of Eqs. (1) and (2) shows the inverse relation between wavelength and voltage:

$$\lambda = \sqrt{\frac{150}{V}},$$

where λ is in angstroms and the applied voltage V is in volts. This equation requires small relativistic corrections at high voltages, due to the variation of electron mass with velocity. At an operating voltage of 100 kV, the electron wavelength is about 0.04 Å, or considerably shorter than the wavelength of x-rays used in diffraction.

Electron diffraction differs from x-ray diffraction in the following ways:

1. Electrons are much less penetrating than x-rays. They are easily absorbed by air, which means that the specimen and the photographic film on which the diffraction pattern is recorded must both be enclosed within the evacuated tube in which the electron beam is produced. Transmission patterns can be made only of specimens so thin as to be classified as foils or films. Reflection patterns from thick specimens are recorded by a glancing-angle technique, somewhat like that shown in Fig. 6–13, on a film placed behind the specimen; such a pattern will be representative only of a thin surface layer, because diffraction occurs over a depth of only a few hundred angstroms or less. Electron diffraction is therefore well suited to the study of thin surface layers.

2. Electrons are scattered much more intensely than x-rays, so that even a very thin layer of material gives a strong diffraction pattern in a short time.

3. The intensity of electron scattering decreases as 2θ increases, as with x-rays, but much more rapidly. This circumstance, coupled with the very short wavelength of the electron beam, causes the entire observable diffraction pattern to be confined to an angular region of about $\pm 4^\circ\ 2\theta$.

Experimentally, electron diffraction has developed in two stages:

1. The earlier work, before the invention of the electron microscope, was done with homemade apparatus called "electron diffraction cameras." Many important studies were made of the structures of metal foils, electrodeposits, films deposited by evaporation, oxide films on metals, and surface layers due to polishing.

2. Virtually all current work is done with the electron microscope. Here, microscopy and diffraction go hand in hand. The diffraction pattern discloses the orientation of the crystal (grain) under examination and thus the proper imaging conditions required to disclose what the microscopist wishes to see. Another application is in the determination of the structure of very small crystals revealed by the microscope.

Still another application is low-energy electron diffraction, called LEED, carried out in special apparatus with an operating voltage of the order of 100 volts. The electron beam then has such low energy that it can penetrate only a monolayer or so of atoms at the specimen surface. The resulting diffraction pattern reveals the arrangement of these surface atoms, an arrangement often quite different from that of the underlying material. Such studies are important in the understanding of phenomena such as catalysis.

A2–3 NEUTRON DIFFRACTION

By making a small opening in the wall of a nuclear reactor, a beam of neutrons can be obtained. The neutrons in such a beam have kinetic energies extending over a considerable range, but a "monochromatic" beam, i.e., a beam composed of neutrons with a single energy, can be obtained by diffraction from a single crystal and this diffracted beam can be used in diffraction experiments. If E is the kinetic energy of the neutrons, then

$$E = \tfrac{1}{2}mv^2, \tag{3}$$

where m is the mass of the neutron (1.68×10^{-27} kg) and v is its velocity. Combination of Eqs. (1) and (3) gives the wavelength of the neutron beam:

$$\lambda = \frac{h}{\sqrt{2mE}}. \tag{4}$$

The neutrons issuing from a reactor have their kinetic energies distributed in much the same way as those of gas molecules in thermal equilibrium; i.e., they follow the Maxwell distribution law. The largest fraction of these so-called "thermal neutrons" therefore has kinetic energy equal to kT, where k is Boltzmann's constant and T the absolute temperature. If this fraction is selected by the monochromating crystal, then we can insert $E = kT$ in Eq. (4) and find

$$\lambda = \frac{h}{\sqrt{2mkT}}. \tag{5}$$

T is of the order of 300° to 400° K, which means that λ is about 1 or 2 Å, i.e., of the same order of magnitude as x-ray wavelengths. Diffraction experiments are performed with a neutron diffractometer, in which the intensity of the beam diffracted by the specimen is measured with a proportional counter filled with BF_3 gas.

Neutron diffraction differs markedly from x-ray or electron diffraction in several ways:

1. A neutron beam is highly penetrating. An iron plate, 1 cm thick, is opaque to electrons, virtually opaque to 1.5 Å x-rays, but transmits 35 percent of 1.5 Å neutrons.

2. The intensity of neutron scattering varies quite irregularly with the atomic number Z of the scattering atom. Elements with almost the same values of Z may have quite different neutron-scattering powers, and elements with widely separated values of Z may scatter neutrons equally well. Furthermore, some light elements, such as carbon, scatter neutrons more intensely than some heavy elements, such as tungsten. It follows that structure analyses can be carried out with neutron diffraction that are impossible, or possible only with great difficulty, with x-ray or electron diffraction. In a compound of hydrogen or carbon, for example, with a heavy metal, x-rays will not "see" the light hydrogen or carbon atom because of its relatively low scattering power, whereas its position in the lattice can be determined with ease by neutron diffraction. Neutrons can also distinguish in many cases between elements differing by only one atomic number, elements which scatter x-rays with almost equal intensity; neutron diffraction, for example, shows strong superlattice lines from ordered FeCo, whereas with x-rays they are practically invisible.

3. Neutrons have a small magnetic moment. If the scattering atom also has a net magnetic moment, the two interact and modify the total scattering. In substances that have an ordered arrangement of atomic moments (antiferromagnetic, ferrimagnetic, and ferromagnetic) neutron diffraction can disclose both the magnitude and direction of the moments. Only neutron diffraction can furnish such information, and it has had a major impact on studies of magnetic structure.

Diffuse scattering at small angles (in transmission), mentioned in regard to x-rays at the end of Sec. 9–3, also occurs with neutrons. Neutron small-angle scattering has certain advantages over x-rays as a means of studying inhomogeneities in materials, particularly because thick specimens, rather than thin foils, can be examined.

Neutron diffraction would doubtless have wider application if all potential investigators had easy access to high-intensity neutron sources, but the number of such sources is very limited.

APPENDIX 3
LATTICE GEOMETRY

A3–1 PLANE SPACINGS

The value of d, the distance between adjacent planes in the set (hkl), may be found from the following equations.

Cubic:
$$\frac{1}{d^2} = \frac{h^2 + k^2 + l^2}{a^2}$$

Tetragonal:
$$\frac{1}{d^2} = \frac{h^2 + k^2}{a^2} + \frac{l^2}{c^2}$$

Hexagonal:
$$\frac{1}{d^2} = \frac{4}{3}\left(\frac{h^2 + hk + k^2}{a^2}\right) + \frac{l^2}{c^2}$$

Rhombohedral:
$$\frac{1}{d^2} = \frac{(h^2 + k^2 + l^2)\sin^2\alpha + 2(hk + kl + hl)(\cos^2\alpha - \cos\alpha)}{a^2(1 - 3\cos^2\alpha + 2\cos^3\alpha)}$$

Orthorhombic:
$$\frac{1}{d^2} = \frac{h^2}{a^2} + \frac{k^2}{b^2} + \frac{l^2}{c^2}$$

Monoclinic:
$$\frac{1}{d^2} = \frac{1}{\sin^2\beta}\left(\frac{h^2}{a^2} + \frac{k^2\sin^2\beta}{b^2} + \frac{l^2}{c^2} - \frac{2hl\cos\beta}{ac}\right)$$

Triclinic:
$$\frac{1}{d^2} = \frac{1}{V^2}(S_{11}h^2 + S_{22}k^2 + S_{33}l^2 + 2S_{12}hk + 2S_{23}kl + 2S_{13}hl)$$

In the equation for triclinic crystals,

$$V = \text{volume of unit cell (see below)},$$

$$S_{11} = b^2c^2\sin^2\alpha,$$

$$S_{22} = a^2c^2\sin^2\beta,$$

$$S_{33} = a^2b^2\sin^2\gamma,$$

$$S_{12} = abc^2(\cos\alpha\cos\beta - \cos\gamma),$$

$$S_{23} = a^2bc(\cos\beta\cos\gamma - \cos\alpha),$$

$$S_{13} = ab^2c(\cos\gamma\cos\alpha - \cos\beta).$$

A3-2 CELL VOLUMES

The following equations give the volume V of the unit cell.

Cubic:
$$V = a^3$$

Tetragonal:
$$V = a^2c$$

Hexagonal:
$$V = \frac{\sqrt{3}\,a^2c}{2} = 0.866a^2c$$

Rhombohedral:
$$V = a^3\sqrt{1 - 3\cos^2\alpha + 2\cos^3\alpha}$$

Orthorhombic:
$$V = abc$$

Monoclinic:
$$V = abc\sin\beta$$

Triclinic:
$$V = abc\sqrt{1 - \cos^2\alpha - \cos^2\beta - \cos^2\gamma + 2\cos\alpha\cos\beta\cos\gamma}$$

A3-3 INTERPLANAR ANGLES

The angle ϕ between the plane $(h_1k_1l_1)$, of spacing d_1, and the plane $(h_2k_2l_2)$, of spacing d_2, may be found from the following equations. (V is the volume of the unit cell.)

Cubic:
$$\cos\phi = \frac{h_1h_2 + k_1k_2 + l_1l_2}{\sqrt{(h_1^2 + k_1^2 + l_1^2)(h_2^2 + k_2^2 + l_2^2)}}$$

Tetragonal:
$$\cos\phi = \frac{\dfrac{h_1h_2 + k_1k_2}{a^2} + \dfrac{l_1l_2}{c^2}}{\sqrt{\left(\dfrac{h_1^2 + k_1^2}{a^2} + \dfrac{l_1^2}{c^2}\right)\left(\dfrac{h_2^2 + k_2^2}{a^2} + \dfrac{l_2^2}{c^2}\right)}}$$

Hexagonal:
$$\cos\phi = \frac{h_1h_2 + k_1k_2 + \frac{1}{2}(h_1k_2 + h_2k_1) + \dfrac{3a^2}{4c^2}\,l_1l_2}{\sqrt{\left(h_1^2 + k_1^2 + h_1k_1 + \dfrac{3a^2}{4c^2}\,l_1^2\right)\left(h_2^2 + k_2^2 + h_2k_2 + \dfrac{3a^2}{4c^2}\,l_2^2\right)}}$$

Rhombohedral:
$$\cos\phi = \frac{a^4d_1d_2}{V^2}\,[\sin^2\alpha(h_1h_2 + k_1k_2 + l_1l_2)$$
$$+ (\cos^2\alpha - \cos\alpha)(k_1l_2 + k_2l_1 + l_1h_2 + l_2h_1 + h_1k_2 + h_2k_1)]$$

Orthorhombic:
$$\cos\phi = \frac{\dfrac{h_1h_2}{a^2} + \dfrac{k_1k_2}{b^2} + \dfrac{l_1l_2}{c^2}}{\sqrt{\left(\dfrac{h_1^2}{a^2} + \dfrac{k_1^2}{b^2} + \dfrac{l_1^2}{c^2}\right)\left(\dfrac{h_2^2}{a^2} + \dfrac{k_2^2}{b^2} + \dfrac{l_2^2}{c^2}\right)}}$$

Monoclinic:

$$\cos\phi = \frac{d_1 d_2}{\sin^2\beta}\left[\frac{h_1 h_2}{a^2} + \frac{k_1 k_2 \sin^2\beta}{b^2} + \frac{l_1 l_2}{c^2} - \frac{(l_1 h_2 + l_2 h_1)\cos\beta}{ac}\right]$$

Triclinic:

$$\cos\phi = \frac{d_1 d_2}{V^2}\,[S_{11}h_1h_2 + S_{22}k_1k_2 + S_{33}l_1l_2 + S_{23}(k_1l_2 + k_2l_1) + S_{13}(l_1h_2 + l_2h_1) + S_{12}(h_1k_2 + h_2k_1)]$$

APPENDIX 4
THE RHOMBOHEDRAL-HEXAGONAL TRANSFORMATION

The lattice of points shown in Fig. A4–1 is rhombohedral, that is, it possesses the symmetry elements characteristic of the rhombohedral system. The primitive rhombohedral cell has axes $\mathbf{a}_1$(R), $\mathbf{a}_2$(R), and $\mathbf{a}_3$(R). The same lattice of points, however, may be referred to a hexagonal cell having axes $\mathbf{a}_1$(H), $\mathbf{a}_2$(H), and **c**(H). The hexagonal cell is no longer primitive, since it contains three lattice points per unit cell (at 0 0 0, $\frac{2}{3}\ \frac{1}{3}\ \frac{1}{3}$, and $\frac{1}{3}\ \frac{2}{3}\ \frac{2}{3}$), and it has three times the volume of the rhombohedral cell.

If one wishes to know the indices $(HK \cdot L)$, referred to hexagonal axes, of a plane whose indices (hkl), referred to rhombohedral axes, are known, the following equations may be used:

$$H = h - k,$$
$$K = k - l,$$
$$L = h + k + l.$$

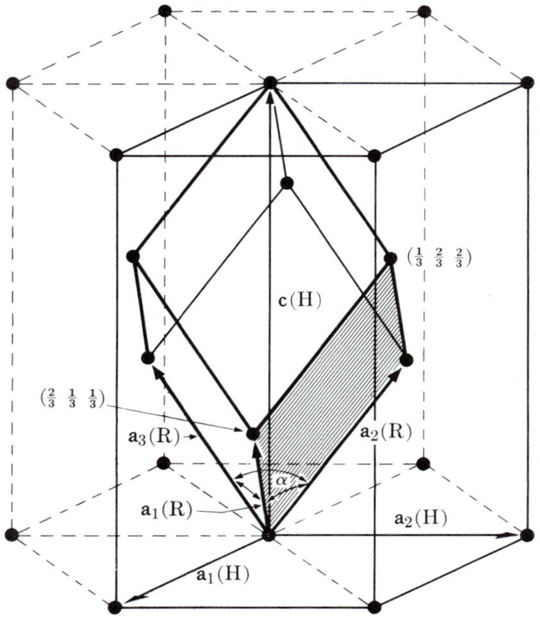

Fig. A4–1 Rhombohedral and hexagonal unit cells in a rhombohedral lattice.

Thus, the (001) face of the rhombohedral cell (shown shaded in the figure) has indices $(0\bar{1} \cdot 1)$ when referred to hexagonal axes.

Since a rhombohedral lattice may be referred to hexagonal axes, it follows that the powder pattern of a rhombohedral substance can be indexed on a hexagonal Hull–Davey or Bunn chart. How then can we recognize the true nature of the lattice? From the equations given above, it follows that

$$-H + K + L = 3k.$$

If the lattice is really rhombohedral, then k is an integer and the only lines appearing in the pattern will have hexagonal indices $(HK \cdot L)$ such that the sum $(-H + K + L)$ is always an integral multiple of 3. If this condition is not satisfied, the lattice is hexagonal.

When the pattern of a rhombohedral substance has been so indexed, i.e., with reference to hexagonal axes, and the true nature of the lattice determined, we usually want to know the indices (hkl) of the reflecting planes when referred to rhombohedral axes. The transformation equations are

$$h = \tfrac{1}{3}(2H + K + L),$$

$$k = \tfrac{1}{3}(-H + K + L),$$

$$l = \tfrac{1}{3}(-H - 2K + L).$$

There is then the problem of determining the lattice parameters a_R and α of the rhombohedral unit cell. But the dimensions of the rhombohedral cell can be determined from the dimensions of the hexagonal cell, and this is an easier process than solving the rather complicated plane-spacing equation for the rhombohedral system. The first step is to index the pattern on the basis of hexagonal axes. Then the parameters a_H and c of the hexagonal cell are calculated in the usual way. Finally, the parameters of the rhombohedral cell are determined from the following equations:

$$a_R = \tfrac{1}{3}\sqrt{3a_H^2 + c^2},$$

$$\sin \frac{\alpha}{2} = \frac{3}{2\sqrt{3 + (c/a_H)^2}}.$$

Finally, it should be noted that, if the c/a ratio of the hexagonal cell in Fig. A4–1 takes on the special value of 2.45, then the angle α of the rhombohedral cell will equal 60° and the lattice of points will be face-centered cubic. Compare Fig. A4–1 with Fig. 2–7.

Further information on the rhombohedral-hexagonal relationship and on unit cell transformations in general may be obtained from the *International Tables for X-Ray Crystallography* [G.11], Vol. 1, pp. 15–21.

APPENDIX 5
CRYSTAL STRUCTURES OF SOME ELEMENTS

Most of the following data are from Pearson [G.16, Vol. 2], who should be consulted for data on certain high-temperature or high-pressure crystal forms not given below. The data on carbon and iodine, and the distance of closest approach in samarium, are from Barrett and Massalski [G.25]. Lattice parameters in both of these sources are given in Å units; they have been multiplied by 1.002056/1.00202, and rounded off to the same number of significant figures, in order to convert them to Å* units. See note on wavelength tables in Appendix 7.

Element	Type of Structure	Temp. (°C)	Lattice parameters (Å*) *a*	*b*	*c* or axial angle	Distance of closest approach (Å*)
Ac Actinium	FCC, A1		5.311			3.755
Al Aluminum	FCC, A1	25	4.0497			2.8636
Am Americium, α*	Hex., La type	20	3.4681		11.240	3.4505
Sb Antimony	Rhomb., A7	25	4.5069		$\alpha = 57° 6'27''$	2.906
As Arsenic	Rhomb., A7	22.5	4.1319		$\alpha = 54° 8'$	2.507
Ba Barium	BCC, A2	25	5.013			4.341
Be Beryllium, α*	HCP, A3	R.T.	2.286		3.584	2.2257
Bi Bismuth	Rhomb., A7	25	4.736		$\alpha = 57° 14'$	3.071
B Boron*	Tetrag.	R.T.	8.80		5.05	
Cd Cadmium	HCP, A3	21	2.9789		5.6169	2.9789
Ca Calcium, α*	FCC, A1	26	5.5886			3.9517
C Carbon, diamond	Cubic, A4	20	3 5671			1.544
Carbon, graphite*	Hex., A9	20	2.4613		6.7080	1.421
Ce Cerium*	FCC, A1	23	5.1603			3.6488
Cs Cesium	BCC, A2	173°K	6.0797			5.265
Cr Chromium	BCC, A2	20	2.8847			2.498
Co Cobalt, α*	HCP, A3	R.T.	2.507		4.070	2.497
Cobalt, β	FCC, A1	R.T.	3.544			2.506
Cu Copper	FCC, A1	20	3.6148			2.5561
Dy Dysprosium, α*	HCP, A3	R.T.	3.5904		5.6477	3.5030
Er Erbium, α*	HCP, A3	R.T.	3.5589		5.5876	3.4681
Eu Europium	BCC, A2	25	4.5822			3.9682
Gd Gadolinium, α*	HCP, A3	20	3.6361		5.7828	3.5731
Ga Gallium	Orthorh.	R.T.	4.523	7.661	4.524	2.484
Ge Germanium	Cubic, A4	25	5.6577			2.4498
Au Gold	FCC, A1	25	4.0786			2.8840
Hf Hafnium, α*	HCP, A3	24	3.1947		5.0513	3.1274
Ho Holmium, α*	HCP, A3	R.T.	3.5774		5.6160	3.4858
In Indium	Tetrag., A6	R.T.	4.5981		4.9469	3.2513
I Iodine	Orthorh.	26	4.79	7.25	9.78	3.54
Ir Iridium	FCC, A1	R.T.	3.8390			2.7146
Fe Iron, α*	BCC, A2	20	2.8665			2.4824
Iron, γ	FCC, A1	916	3.6469			2.5787
Iron, δ	BCC, A2	1394	2.9323			2.5394
La Lanthanum, α*	Hex.	R.T.	3.770		12.159	3.739
Pb Lead	FCC, A1	25	4.9504			3.5004

Element	Type of Structure	Temp. (°C)	Lattice parameters (Å*) *a*	*b*	*c* or axial angle	Distance of closest approach (Å*)
Li Lithium*	BCC, A2	25	3.5101			3.0398
Lu Lutetium*	HCP, A3	R.T.	3.5032		5.5511	3.4345
Mg Magnesium	HCP, A3	25	3.2095		5.2107	3.1971
Mn Manganese, α*	Cubic, A12	R.T.	8.9142			
Hg Mercury	Rhomb., A10	227°K	3.005		$\alpha = 70° 32'$	3.005
Mo Molybdenum	BCC, A2	20	3.1469			2.7253
Nd Neodynium, α*	Hex., La type	R.T.	3.6580		11.7996	3.6280
Np Neptunium, α*	Orthorh.	20	6.663	4.723	4.887	2.60
Ni Nickel	FCC, A1		3.5239			2.4920
Nb Niobium	BCC, A2	25	3.3067			2.8637
Os Osmium	HCP, A3	20	2.7354		4.3193	2.6755
Pd Palladium	FCC, A1	22	3.8908			2.7511
P Phosphorous, black*	Orthorh.	22	3.3137	10.478	4.3765	2.224
Pt Platinum	FCC, A1	20	3.9240			2.7747
Pu Plutonium, α*	Monocl.	21	6.183	4.822	10.963 $\beta = 101.79°$	2.57
Po Polonium, α*	Cubic	~10	3.345			3.345
K Potassium	BCC, A2	78°K	5.247			4.524
Pr Praseodymium, α*	Hex., La type	R.T.	3.6726		11.8358	3.6402
Pa Protactinium	Tetrag.		3 925		3.238	3.212
Re Rhenium	HCP, A3	R.T.	2.760		4.458	2.741
Rh Rhodium	FCC, A1	20	3.8045			2.6902
Rb Rubidium	BCC, A2	20	5.70			4.94
Ru Ruthenium	HCP, A3	25	2.7059		4.2818	2.6503
Sm Samarium	Rhomb.		8.996		$\alpha = 23° 13'$	3.588
Sc Scandium, α*	HCP, A3	R.T.	3.3091		5.2735	3.2561
Se Selenium*	Hex., A8	25	4.3658		4.9592	2.321
Si Silicon	Cubic, A4	25	5.4309			2.3517
Ag Silver	FCC, A1	25	4.0863			2.8895
Na Sodium	BCC, A2	20	4.2908			3.7159
Sr Strontium, α*	FCC, A1	25	6.0851			4.3029
S Sulphur*	Orthorh.	24.8	10.4650	12.8665	24.4869	2.037 (mean)
Ta Tantalum	BCC, A2	R.T.	3.298			2.856
Tc Technetium	HCP, A3	R.T.	2.735		4.388	2.703
Te Tellurium	Hex., A8	25	4.4568		5.9270	2.864
Tb Terbium, α*	HCP, A3	R.T.	3.6011		5.6938	3.5253
Tl Thallium, α*	HCP, A3	18	3.4567		5.5250	3.4077
Th Thorium, α*	FCC, A1	R.T.	5.0847			3.5951
Tm Thulium, α*	HCP, A3	R.T.	3.5376		5.5548	3.4474
Sn Tin (white), β*	Tetrag., A5	25	5.8317		3.1815	3.022
Tin (grey), α	Cubic, A4	20	6.4894			2.8100
Ti Titanium, α*	HCP, A3	25	2.9512		4.6845	2.8964
Titanium, β	BCC, A2	900	3.3066			2.8636
W Tungsten	BCC, A2	25	3.1653			2.7412
U Uranium, α*	Orthorh., A20	25	2.8538	5.8697	4.9550	2.7540
Uranium, β	Tetrag.	720	10.759		5.656	
Uranium, γ	BCC, A2	805	3.524			3.052
V Vanadium	BCC, A2	R.T.	3.0232			2.6182
Yb Ytterbium*	FCC, A1	R.T.	5.4864			3.8794
Y Yttrium*	HCP, A3	R.T.	3.6475		5.7308	3.5509
Zn Zinc	HCP, A3	25	2.6650		4.9470	2.6650
Zr Zirconium, α*	HCP, A3	25	3.2313		5.1479	3.1790
Zirconium, β	BCC, A2	862	3.6091			3.1256

* Ordinary form of an element that exists, or is thought to exist, in more than one form.

APPENDIX 6
CRYSTAL STRUCTURES OF SOME COMPOUNDS AND SOLID SOLUTIONS

Substance	Type of structure	Lattice parameters (Å)	Spacing of cleavage planes (Å)
NaCl KCl AgBr	FCC, B1 FCC, B1 FCC, B1	$a = 5.639$ $a = 6.290$ $a = 5.77$	2.820
CaF_2 (fluorite)	FCC, C1	$a = 5.46$	
$CaCO_3$ (calcite)	Rhombohedral, G1	$a = 6.37$ $\alpha = 46.1°$	3.036
SiO_2 (α-quartz)	Hexagonal, C8	$a = 4.90$ $c = 5.39$	
$H_2KAl_2(SiO_4)_3$ (mica, muscovite)	Monoclinic	$a = 5.18$ $b = 8.96$ $c = 20.15$ $\beta = 98.6°$	10.08
Fe_3C (cementite)	Orthorhombic	$a = 4.525$ $b = 5.088$ $c = 6.740$	
Austenite	FCC, A1	$a = 3.555 + 0.044x$ (x = weight percent carbon)	
Martensite	BC Tetragonal	$a = 2.867 - 0.013x$ $c = 2.867 + 0.116x$ (x = weight percent carbon)	

APPENDIX 7
X-RAY WAVELENGTHS

All of the following values are extracted from much longer tables on pp. 6–43 of Vol. 4 of the *International Tables for X-Ray Crystallography* [G.11], which are in turn taken from J. A. Bearden, *Rev. Mod. Phys.*, **39**, 78 (1967).

E($K\alpha$) is the energy $h\nu$ of the unresolved $K\alpha$ line to the nearest 0.01 keV. (The values in [G.11] are of higher accuracy.) In computing E($K\alpha$), $K\alpha_1$ is given twice the weight of $K\alpha_2$.

The wavelengths are given in Å* units. This unit is *defined* by the wavelength of the W$K\alpha_1$ line = 0.2090100 Å*. The Å* unit is believed to be equal to the angstrom Å to within 5 parts per million and involves a conversion factor of 1.002056 ± 0.000005 Å/kX. Because of the still remaining uncertainty in this conversion factor it was decided to introduce the Å* unit. The distinction between Å and Å* is negligible except in work of the very highest accuracy.

Wavelengths (in Å* units) and Energies (in keV) of Some Characteristic Emission Lines and Absorption Edges

Element	E($K\alpha$) (keV)	$K\alpha_2$ strong	$K\alpha_1$ very strong	$K\beta_1$ weak	K edge	$L\alpha_1$	L_{III} edge
1 H							
2 He							
3 Li	0.05	228			226.5		
4 Be	0.11	114			111		
5 B	0.18	67.6					
6 C	0.28	44.7			43.68		
7 N	0.39	31.6			30.99		
8 O	0.52	23.62			23.32		
9 F	0.68	18.32					
10 Ne	0.85	14.610		14.452	14.3018		
11 Na	1.04	11.9101		11.575	11.569		405
12 Mg	1.25	9.8900		9.521	9.5122		250.7
13 Al	1.49	8.34173	8.33934	7.960	7.94813		170.4
14 Si	1.74	7.12791	7.12542	6.753	6.738		123
15 P	2.01	6.160	6.157	5.796	5.784		94
16 S	2.31	5.37496	5.37216	5.0316	5.0185		
17 Cl	2.62	4.7307	4.7278	4.4034	4.3971		
18 A	2.96	4.19474	4.19180	3.8860	3.87090		
19 K	3.31	3.7445	3.7414	3.4539	3.4365		42.1
20 Ca	3.69	3.36166	3.35839	3.0897	3.0703	36.33	35.49
21 Sc	4.09	3.0342	3.0309	2.7796	2.762	31.35	
22 Ti	4.51	2.75216	2.74851	2.51391	2.49734	27.42	27.29
23 V	4.95	2.50738	2.50356	2.28440	2.2691	24.25	
24 Cr	5.41	2.293606	2.28970	2.08487	2.07020	21.64	20.7
25 Mn	5.90	2.10578	2.101820	1.91021	1.89643	19.45	
26 Fe	6.40	1.939980	1.936042	1.75661	1.74346	17.59	17.525
27 Co	6.93	1.792850	1.788965	1.62079	1.60815	15.972	15.915
28 Ni	7.47	1.661747	1.657910	1.500135	1.48807	14.561	14.525
29 Cu	8.04	1.544390	1.540562	1.392218	1.38059	13.336	13.288
30 Zn	8.63	1.439000	1.435155	1.29525	1.2834	12.254	12.131

Element	E($K\alpha$) (keV)	$K\alpha_2$ strong	$K\alpha_1$ very strong	$K\beta_1$ weak	K edge	$L\alpha_1$	L_{III} edge
31 Ga	9.24	1.34399	1.340083	1.20789	1.1958	11.292	11.100
32 Ge	9.88	1.258011	1.254054	1.12894	1.11658	10.4361	10.187
33 As	10.53	1.17987	1.17588	1.05730	1.0450	9.6709	9.367
34 Se	11.21	1.10882	1.10477	0.99218	0.97974	8.9900	8.646
35 Br	11.91	1.04382	1.03974	0.93279	0.9204	8.3746	7.984
36 Kr	12.63	0.9841	0.9801	0.8785	0.86552	7.817	7.392
37 Rb	13.38	0.92969	0.925553	0.82868	0.81554	7.3183	6.862
38 Sr	14.14	0.87943	0.87526	0.78292	0.76973	6.8628	6.387
39 Y	14.93	0.83305	0.82884	0.74072	0.72766	6.4488	5.962
40 Zr	15.75	0.79015	0.78593	0.70173	0.68883	6.0705	5.579
41 Nb	16.58	0.75044	0.74620	0.66576	0.65298	5.7243	5.230
42 Mo	17.44	0.713590	0.709300	0.632288	0.61978	5.40655	4.913
43 Tc	18.33	0.67932	0.67502	0.60130	0.58906	5.1148	4.630
44 Ru	19.24	0.647408	0.643083	0.572482	0.56051	4.84575	4.369
45 Rh	20.17	0.617630	0.613279	0.545605	0.53395	4.59743	4.1299
46 Pd	21.12	0.589821	0.585448	0.520520	0.5092	4.36767	3.9074
47 Ag	22.11	0.563798	0.5594075	0.497069	0.48589	4.15443	3.6999
48 Cd	23.11	0.539422	0.535010	0.475105	0.46407	3.95635	3.5047
49 In	24.14	0.516544	0.512113	0.454545	0.44371	3.77192	3.3237
50 Sn	25.20	0.495053	0.490599	0.435236	0.42467	3.59994	3.1557
51 Sb	26.28	0.474827	0.470354	0.417085	0.40668	3.43941	3.0003
52 Te	27.38	0.455784	0.451295	0.399995	0.38974	3.28920	2.8555
53 I	28.51	0.437829	0.433318	0.383905	0.37381	3.14860	2.7196
54 Xe	29.67	0.42087	0.41634	0.36872	0.3584	3.0166	2.5926
55 Cs	30.86	0.404835	0.400290	0.354364	0.34451	2.8924	2.4740
56 Ba	32.07	0.389668	0.385111	0.340811	0.33104	2.77595	2.3629
57 La	33.31	0.375313	0.370737	0.327983	0.31844	2.66570	2.261
58 Ce	34.57	0.361683	0.357092	0.315816	0.30648	2.5615	2.166
59 Pr	35.87	0.348749	0.344140	0.304261	0.29518	2.4630	2.0791
60 Nd	37.19	0.336472	0.331846	0.293299	0.28453	2.3704	1.9967
61 Pm	38.54	0.324803	0.320160	0.28290	0.27431	2.2822	1.9191
62 Sm	39.92	0.313698	0.309040	0.27301	0.26464	2.1998	1.8457
63 Eu	41.33	0.303118	0.298446	0.263577	0.25553	2.1209	1.7761
64 Gd	42.77	0.293038	0.288353	0.25460	0.24681	2.0468	1.7117
65 Tb	44.24	0.283423	0.278724	0.24608	0.23841	1.9765	1.6497
66 Dy	45.73	0.274247	0.269533	0.23788	0.23048	1.90881	1.5916
67 Ho	47.26	0.265486	0.260756	0.23012	0.22291	1.8450	1.5368
68 Er	48.83	0.257110	0.252365	0.22266	0.21567	1.78425	1.4835
69 Tm	50.42	0.249095	0.244338	0.21556	0.20880	1.7268	1.4334
70 Yb	52.04	0.241424	0.236655	0.20884	0.20224	1.67189	1.3862
71 Lu	53.70	0.234081	0.229298	0.20231	0.19585	1.61951	1.3405
72 Hf	55.40	0.227024	0.222227	0.19607	0.18982	1.56958	1.2972
73 Ta	57.11	0.220305	0.215497	0.190089	0.18394	1.52197	1.2553
74 W	58.87	0.213828	0.2090100	0.184374	0.17837	1.47639	1.2155
75 Re	60.67	0.207611	0.202781	0.178880	0.17302	1.43290	1.1773
76 Os	62.50	0.201639	0.196794	0.173611	0.16787	1.39121	1.1408
77 Ir	64.36	0.195904	0.191047	0.168542	0.16292	1.35128	1.1058
78 Pt	66.26	0.190381	0.185511	0.163675	0.15818	1.31304	1.0723
79 Au	68.20	0.185075	0.180195	0.158982	0.153593	1.27640	1.04000
80 Hg	70.18	0.179958	0.175068	0.154487	0.14918	1.24120	1.0091
81 Tl	72.19	0.175036	0.170136	0.150142	0.14495	1.20739	0.9793
82 Pb	74.25	0.170294	0.165376	0.145970	0.140880	1.17501	0.95073
83 Bi	76.34	0.165717	0.160789	0.141948	0.13694	1.14386	0.9234
84 Po	78.48	0.16130	0.15636	0.13807		1.11386	
85 At	80.66	0.15705	0.15210	0.13432		1.08500	
86 Rn	82.88	0.15294	0.14798	0.13069		1.05723	
87 Fr	85.14	0.14896	0.14399	0.12719		1.03049	
88 Ra	87.46	0.14512	0.14014	0.12382		1.00473	0.8028
89 Ac	89.81	0.14141	0.136417	0.12055		0.97993	
90 Th	92.22	0.137829	0.132813	0.117396	0.11307	0.95600	0.7607
91 Pa	94.67	0.134343	0.129325	0.114345		0.93284	
92 U	97.18	0.130968	0.125947	0.111394	0.10723	0.910639	0.7223

Some commonly used K wavelengths

Element	$K\alpha$ (weighted average)*	$K\alpha_2$ strong	$K\alpha_1$ very strong	$K\beta_1$ weak
Cr	2.29100	2.293606	2.28970	2.08487
Fe	1.937355	1.939980	1.936042	1.75661
Co	1.790260	1.792850	1.788965	1.62079
Cu	1.541838	1.544390	1.540562	1.392218
Mo	0.710730	0.713590	0.709300	0.632288

* $K\alpha_1$ is given twice the weight of $K\alpha_2$.

Characteristic L Lines of Tungsten

Line	Relative intensity	Wavelength
$L\alpha_1$	Very strong	1.47639
$L\alpha_2$	Weak	1.48743
$L\beta_1$	Strong	1.281809
$L\beta_2$	Medium	1.24460
$L\beta_3$	Weak	1.26269
$L\gamma_1$	Weak	1.09855

APPENDIX 8
MASS ABSORPTION COEFFICIENTS μ/ρ (cm²/gm) AND DENSITIES ρ

The mass absorption coefficients are extracted from much longer tables on pp. 61–66 of Vol. 4 of the *International Tables for X-Ray Crystallography* [G.11]. Although these coefficients are given to four significant figures, the actual accuracy is much less; the uncertainty ranges from less than 2 percent to more than 15 percent, depending on absorber and wavelength. [G.11] should be consulted for details.

The densities of elements solid at room temperature, except *P*, are x-ray densities, rounded off, from pp. 46–56 of Vol. 3 of [G.11]. Densities of gases are from *Metals Handbook* (Cleveland: American Society for Metals, 1948).

Absorber	Density (gm/cm³)	Mo		Cu		Co		Cr	
		$K\alpha$ 0.711 Å	$K\beta$ 0.632 Å	$K\alpha$ 1.542 Å	$K\beta$ 1.392 Å	$K\alpha$ 1.790 Å	$K\beta$ 1.621 Å	$K\alpha$ 2.291 Å	$K\beta$ 2.085 Å
1 H	0.08375×10^{-3}	0.3727	0.3699	0.3912	0.3882	0.3966	0.3928	0.4116	0.4046
2 He	0.1664×10^{-3}	0.2019	0.1972	0.2835	0.2623	0.3288	0.2966	0.4648	0.4001
3 Li	0.533	0.1968	0.1866	0.4770	0.3939	0.6590	0.5283	1.243	0.9639
4 Be	1.85	0.2451	0.2216	1.007	0.7742	1.522	1.152	3.183	2.388
5 B	2.47	0.3451	0.2928	2.142	1.590	3.357	2.485	7.232	5.385
6 C	2.27 (graphite)	0.5348	0.4285	4.219	3.093	6.683	4.916	14.46	10.76
7 N	1.165×10^{-3}	0.7898	0.6054	7.142	5.215	11.33	8.330	24.42	18.23
8 O	1.332×10^{-3}	1.147	0.8545	11.03	8.062	17.44	12.85	37.19	27.88
9 F	1.696×10^{-3}	1.584	1.154	15.95	11.66	25.12	18.57	53.14	39.99
10 Ne	0.8387×10^{-3}	2.209	1.597	22.13	16.24	34.69	25.72	72.71	54.91
11 Na	0.966	2.939	2.098	30.30	22.23	47.34	35.18	98.48	74.66
12 Mg	1.74	3.979	2.825	40.88	30.08	63.54	47.38	130.8	99.62
13 Al	2.70	5.043	3.585	50.23	37.14	77.54	58.08	158.0	120.7
14 Si	2.33	6.533	4.624	65.32	48.37	100.4	75.44	202.7	155.6
15 P	1.82 (yellow)	7.870	5.569	77.28	57.44	118.0	89.05	235.5	181.6
16 S	2.09	9.625	6.835	92.53	68.90	141.2	106.6	281.9	217.2
17 Cl	3.214×10^{-3}	11.64	8.261	109.2	81.79	164.7	125.3	321.5	250.2
18 A	1.663×10^{-3}	12.62	8.949	119.5	89.34	180.9	137.3	355.5	275.8
19 K	0.862	16.20	11.51	148.4	111.7	222.0	169.9	426.8	334.2
20 Ca	1.53	19.00	13.56	171.4	129.0	257.4	196.4	499.6	389.3
21 Sc	2.99	21.04	15.00	186.0	140.8	275.5	212.2	520.9	410.7
22 Ti	4.51	23.25	16.65	202.4	153.2	300.5	231.0	571.4	449.0
23 V	6.09	25.24	18.07	222.6	168.0	332.7	254.7	75.06	501.0
24 Cr	7.19	29.25	20.99	252.3	191.1	375.0	288.1	85.71	65.79
25 Mn	7.47	31.86	22.89	272.5	206.7	405.1	311.2	96.08	73.75
26 Fe	7.87	37.74	27.21	304.4	233.6	56.25	345.5	113.1	86.77
27 Co	8.8	41.02	29.51	338.6	258.7	62.86	47.71	124.6	96.06
28 Ni	8.91	47.24	34.18	48.83	282.8	73.75	56.05	145.7	112.5
29 Cu	8.93	49.34	35.77	51.54	38.74	78.11	59.22	155.2	119.5
30 Zn	7.13	55.46	40.26	59.51	45.30	88.71	68.00	171.7	133.5
31 Ga	5.91	56.90	41.69	62.13	46.65	94.15	71.39	186.9	144.0
32 Ge	5.32	60.47	44.26	67.92	51.44	102.0	77.79	199.9	154.5
33 As	5.78	65.97	48.57	75.65	57.01	114.0	86.76	224.0	173.3
34 Se	4.81	68.82	51.20	82.89	62.32	125.1	95.11	246.1	190.4
35 Br	3.12 (liquid)	74.68	55.56	90.29	68.07	135.8	103.5	266.2	206.2

Absorber	Density (gm/cm^3)	Mo		Cu		Co		Cr	
		$K\alpha$ 0.711 Å	$K\beta$ 0.632 Å	$K\alpha$ 1.542 Å	$K\beta$ 1.392 Å	$K\alpha$ 1.790 Å	$K\beta$ 1.621 Å	$K\alpha$ 2.291 Å	$K\beta$ 2.085 Å
36 Kr	3.488×10^{-3}	79.10	58.64	97.02	73.22	145.7	111.2	284.6	220.7
37 Rb	1.53	83.00	62.07	106.3	80.16	159.6	121.8	311.7	241.8
38 Sr	2.58	88.04	65.59	115.3	86.77	173.5	132.2	339.3	263.4
39 Y	4.48	97.56	72.57	127.1	96.19	190.2	145.4	368.9	286.9
40 Zr	6.51	16.10	75.20	136.8	103.3	204.9	156.6	398.6	309.7
41 Nb	8.58	16.96	81.22	148.8	112.3	222.9	170.4	431.9	336.4
42 Mo	10.22	18.44	13.29	158.3	119.7	236.6	181.0	457.4	356.5
43 Tc	11.50	19.78	14.30	167.7	126.9	250.8	191.9	485.5	378.0
44 Ru	12.36	21.33	15.40	180.8	137.0	269.4	206.6	517.9	404.4
45 Rh	12.42	23.05	16.65	194.1	147.1	289.0	221.8	555.2	433.7
46 Pd	12.00	24.42	17.63	205.0	155.6	304.3	234.0	580.9	455.1
47 Ag	10.50	26.38	19.10	218.1	165.8	323.5	248.9	617.4	483.5
48 Cd	8.65	27.73	20.13	229.3	174.0	341.8	262.1	658.8	513.5
49 In	7.29	29.13	21.18	242.1	183.3	362.7	277.1	705.8	548.0
50 Sn	7.29	31.18	22.62	253.3	193.1	374.1	288.7	708.8	556.6
51 Sb	6.69	33.01	23.91	266.5	203.6	391.3	303.1	733.4	578.8
52 Te	6.25	33.92	24 67	273.4	208.4	404.4	311.7	768.9	602.7
53 I	4.95	36.33	26.53	291.7	221.9	434.0	333.2	835.2	650.8
54 Xe	5.495×10^{-3}	38.31	27.86	309.8	235.9	459.0	353.4	755.4	685.2
55 Cs	1.91 (−10°C)	40.44	29.51	325.4	247.5	483.8	371.5	802.7	725.1
56 Ba	3.59	42.37	31.00	336.1	256.0	499.0	383.6	587.3	644.6
57 La	6.17	45.34	33.10	353.5	270.8	519.0	401.9	222.9	661.5
58 Ce	6.77	48.56	35.54	378.8	289.6	559.1	431.5	240.4	509.4
59 Pr	6.78	50.78	37.09	402.2	306.8	596.2	458.8	260.5	205.3
60 Nd	7.00	53.28	38.88	417.9	319.8	531.7	475.9	271.3	213.4
61 Pm		55.52	40.52	441.1	336.4	401.4	503.2	284.7	223.8
62 Sm	7.54	57.96	42.40	453.5	346.6	411.8	446.3	295.0	231.5
63 Eu	5.25	61.18	44.74	417.9	369.0	165.2	476.9	312.7	244.9
64 Gd	7.87	62.79	45.95	426.7	377.2	169.5	346.7	318.9	250.3
65 Tb	8.27	66.77	48.88	321.9	399.9	178.7	367.1	338.9	265.2
66 Dy	8.53	68.89	50.38	336.6	360.2	184.9	142.9	351.7	275.0
67 Ho	8.80	72.14	52.76	128.4	272.4	189.8	146.3	363.3	283.4
68 Er	9.04	75.61	55.07	134.3	291.7	198.4	153.0	379.7	296.1
69 Tm	9.33	78.98	57.94	140.2	288.5	207.4	159.8	397.0	309.6
70 Yb	6.97	80.23	59.22	144.7	110.9	214.0	164.9	409.6	319.4
71 Lu	9.84	84.18	62.04	152.0	116.5	224.6	173.2	429.5	335.1
72 Hf	13.28	86.33	64.15	157.7	121.0	232.9	179.6	445.0	347.2
73 Ta	16.67	89.51	66.07	161.5	123.9	238.3	183.9	454.7	355.0
74 W	19.25	95.76	70.57	170.5	131.5	249.7	193.7	470.4	369.1
75 Re	21.02	98.74	72.47	178.3	137.3	261.8	202.7	495.5	388.0
76 Os	22.58	100.2	74.13	183.8	141.3	270.3	209.0	512.4	401.1
77 Ir	22.55	103.4	77.20	192.2	147.6	283.4	218.8	539.6	421.6
78 Pt	21.44	108.6	80.23	198.2	151.2	295.2	226.4	571.6	443.9
79 Au	19.28	111.3	82.33	207.8	160.6	303.3	235.7	568.0	446.7
80 Hg	13.55	114.7	85.30	216.2	166.6	317.0	245.7	597.9	468.9
81 Tl	11.87	119.4	88.25	222.2	171.1	326.3	252.7	616.9	483.3
82 Pb	11.34	122.8	90.55	232.1	178.6	340.8	263.8	644.5	504.9
83 Bi	9.80	125.9	93.50	242.9	187.5	355.3	275.8	667.2	524.2
86 Rn	4.40 (liq., −62°C)	117.2	100.7	263.7	203.0	387.1	299.7	731.4	573.2
90 Th	11.72	99.46	73.34	306.8	236.6	449.0	348.4	844.1	663.0
92 U	19.05	96.67	72.63	305.7	236.2	446.3	346.9	774.0	657.1
94 Pu	19.81	48.84	78.99	352.9	271.2	519.6	401.6	803.2	771.6

APPENDIX 9
VALUES OF $\sin^2 \theta$

$\theta°$	.0	.1	.2	.3	.4	.5	.6	.7	.8	.9	Differences .01	.02	.03	.04	.05
00	.0000	0000	0000	0000	0000	0001	0001	0001	0002	0002					
1	.0003	0004	0004	0005	0006	0007	0008	0009	0010	0011					
2	.0012	0013	0015	0016	0018	0019	0021	0022	0024	0026					
3	.0027	0029	0031	0033	0035	0037	0039	0042	0044	0046					
4	.0049	0051	0054	0056	0059	0062	0064	0067	0070	0073			Interpolate		
5	.0076	0079	0082	0085	0089	0092	0095	0099	0102	0106					
6	.0109	0113	0117	0120	0124	0128	0132	0136	0140	0144					
7	.0149	0153	0157	0161	0166	0170	0175	0180	0184	0189					
8	.0194	0199	0203	0208	0213	0218	0224	0229	0234	0239					
9	.0245	0250	0256	0261	0267	0272	0278	0284	0290	0296					
10	.0302	0308	0314	0320	0326	0332	0338	0345	0351	0358	1	1	2	2	3
1	.0364	0371	0377	0384	0391	0397	0404	0411	0418	0425	1	1	2	3	3
2	.0432	0439	0447	0454	0461	0468	0476	0483	0491	0498	1	1	2	3	4
3	.0506	0514	0521	0529	0537	0545	0553	0561	0569	0577	1	2	2	3	4
4	.0585	0593	0602	0610	0618	0627	0635	0644	0653	0661	1	2	3	3	4
15	.0670	0679	0687	0696	0705	0714	0723	0732	0741	0751	1	2	3	4	4
6	.0760	0769	0778	0788	0797	0807	0816	0826	0835	0845	1	2	3	4	5
7	.0855	0865	0874	0884	0894	0904	0914	0924	0934	0945	1	2	3	4	5
8	.0955	0965	0976	0986	0996	1007	1017	1028	1039	1049	1	2	3	4	5
9	.1060	1071	1082	1092	1103	1114	1125	1136	1147	1159	1	2	3	4	6
20	.1170	1181	1192	1204	1215	1226	1238	1249	1261	1273	1	2	3	5	6
1	.1284	1296	1308	1320	1331	1343	1355	1367	1379	1391	1	2	4	5	6
2	.1403	1415	1428	1440	1452	1464	1477	1489	1502	1514	1	2	4	5	6
3	.1527	1539	1552	1565	1577	1590	1603	1616	1628	1641	1	3	4	5	6
4	.1654	1667	1680	1693	1707	1720	1733	1746	1759	1773	1	3	4	5	7
25	.1786	1799	1813	1826	1840	1853	1867	1881	1894	1908	1	3	4	5	7
6	.1922	1935	1949	1963	1977	1991	2005	2019	2033	2047	1	3	4	6	7
7	.2061	2075	2089	2104	2118	2132	2146	2161	2175	2190	1	3	4	6	7
8	.2204	2219	2233	2248	2262	2277	2291	2306	2321	2336	1	3	4	6	7
9	.2350	2365	2380	2395	2410	2425	2440	2455	2470	2485	2	3	5	6	8
30	.2500	2515	2530	2545	2561	2576	2591	2607	2622	2637	2	3	5	6	8
1	.2653	2668	2684	2699	2715	2730	2746	2761	2777	2792	2	3	5	6	8
2	.2808	2824	2840	2855	2871	2887	2903	2919	2934	2950	2	3	5	6	8
3	.2966	2982	2998	3014	3030	3046	3062	3079	3095	3111	2	3	5	6	8
4	.3127	3143	3159	3176	3192	3208	3224	3241	3257	3274	2	3	5	7	8
35	.3290	3306	3323	3339	3356	3372	3389	3405	3422	3438	2	3	5	7	8
6	.3455	3472	3488	3505	3521	3538	3555	3572	3588	3605	2	3	5	7	8
7	.3622	3639	3655	3672	3689	3706	3723	3740	3757	3773	2	3	5	7	8
8	.3790	3807	3824	3841	3858	3875	3892	3909	3926	3943	2	3	5	7	8
9	.3960	3978	3995	4012	4029	4046	4063	4080	4097	4115	2	3	5	7	9
40	.4132	4149	4166	4183	4201	4218	4235	4252	4270	4287	2	3	5	7	9
1	.4304	4321	4339	4356	4373	4391	4408	4425	4443	4460	2	3	5	7	9
2	.4477	4495	4512	4529	4547	4564	4582	4599	4616	4634	2	3	5	7	9
3	.4651	4669	4686	4703	4721	4738	4756	4773	4791	4808	2	3	5	7	9
4	.4826	4843	4860	4878	4895	4913	4930	4948	4965	4983	2	3	5	7	9

θ°	.0	.1	.2	.3	.4	.5	.6	.7	.8	.9	Differences .01	.02	.03	.04	.05
45	.5000	5017	5035	5052	5070	5087	5105	5122	5140	5157	2	3	5	7	9
6	.5174	5192	5209	5227	5244	5262	5279	5297	5314	5331	2	3	5	7	9
7	.5349	5366	5384	5401	5418	5436	5453	5471	5488	5505	2	3	5	7	9
8	.5523	5540	5557	5575	5592	5609	5627	5644	5661	5679	2	3	5	7	9
9	.5696	5713	5730	5748	5765	5782	5799	5817	5834	5851	2	3	5	7	9
50	.5868	5885	5903	5920	5937	5954	5971	5988	6005	6022	2	3	5	7	9
1	.6040	6057	6074	6091	6108	6125	6142	6159	6176	6193	2	3	5	7	9
2	.6210	6227	6243	6260	6277	6294	6311	6328	6345	6361	2	3	5	7	8
3	.6378	6395	6412	6428	6445	6462	6479	6495	6512	6528	2	3	5	7	8
4	.6545	6562	6578	6595	6611	6628	6644	6661	6677	6694	2	3	5	7	8
55	.6710	6726	6743	6759	6776	6792	6808	6824	6841	6857	2	3	5	7	8
6	.6873	6889	6905	6921	6938	6954	6970	6986	7002	7018	2	3	5	7	8
7	.7034	7050	7066	7081	7097	7113	7129	7145	7160	7176	2	3	5	6	8
8	.7192	7208	7223	7239	7254	7270	7285	7301	7316	7332	2	3	5	6	8
9	.7347	7363	7378	7393	7409	7424	7439	7455	7470	7485	2	3	5	6	8
60	.7500	7515	7530	7545	7560	7575	7590	7605	7620	7635	2	3	5	6	8
1	.7650	7664	7679	7694	7709	7723	7738	7752	7767	7781	2	3	5	6	8
2	.7796	7810	7825	7839	7854	7868	7882	7896	7911	7925	1	3	4	6	7
3	.7939	7953	7967	7981	7995	8009	8023	8037	8051	8065	1	3	4	6	7
4	.8078	8092	8106	8119	8133	8147	8160	8174	8187	8201	1	3	4	6	7
65	.8214	8227	8241	8254	8267	8280	8293	8307	8320	8333	1	3	4	5	7
6	.8346	8359	8372	8384	8397	8410	8423	8435	8448	8461	1	3	4	5	7
7	.8473	8486	8498	8511	8523	8536	8548	8560	8572	8585	1	3	4	5	6
8	.8597	8609	8621	8633	8645	8657	8669	8680	8692	8704	1	2	4	5	6
9	.8716	8727	8739	8751	8762	8774	8785	8796	8808	8819	1	2	4	5	6
70	.8830	8841	8853	8864	8875	8886	8897	8908	8918	8929	1	2	3	5	6
1	.8940	8951	8961	8972	8983	8993	9004	9014	9024	9035	1	2	3	4	6
2	.9045	9055	9066	9076	9086	9096	9106	9116	9126	9135	1	2	3	4	5
3	.9145	9155	9165	9174	9184	9193	9203	9212	9222	9231	1	2	3	4	5
4	.9240	9249	9259	9268	9277	9286	9295	9304	9313	9321	1	2	3	4	5
75	.9330	9339	9347	9356	9365	9373	9382	9390	9398	9407	1	2	3	4	4
6	.9415	9423	9431	9439	9447	9455	9463	9471	9479	9486	1	2	3	3	4
7	.9494	9502	9509	9517	9524	9532	9539	9546	9553	9561	1	2	2	3	4
8	.9568	9575	9582	9589	9596	9603	9609	9616	9623	9629	1	1	2	3	4
9	.9636	9642	9649	9655	9662	9668	9674	9680	9686	9692	1	1	2	3	3
80	.9698	9704	9710	9716	9722	9728	9733	9739	9744	9750	1	1	2	2	3
1	.9755	9761	9766	9771	9776	9782	9787	9792	9797	9801					
2	.9806	9811	9816	9820	9825	9830	9834	9839	9843	9847					
3	.9851	9856	9860	9864	9868	9872	9876	9880	9883	9887					
4	.9891	9894	9898	9901	9905	9908	9911	9915	9918	9921					
													Interpolate		
85	.9924	9927	9930	9933	9936	9938	9941	9944	9946	9949					
6	.9951	9954	9956	9958	9961	9963	9965	9967	9969	9971					
7	.9973	9974	9976	9978	9979	9981	9982	9984	9985	9987					
8	.9988	9989	9990	9991	9992	9993	9994	9995	9996	9996					
9	.9997	9998	9998	9999	9999	9999	1.00	1.00	1.00	1.00					

From Henry, Lipson, and Wooster [G.8].

APPENDIX 10
QUADRATIC FORMS OF MILLER INDICES

Cubic					Hexagonal	
$h^2 + k^2 + l^2$	*hkl*				$h^2 + hk + k^2$	*hk*
	Simple	Face-centered	Body-centered	Diamond		
1	100				1	10
2	110	. . .	110		2	
3	111	111	. . .	111	3	11
4	200	200	200		4	20
5	210				5	
6	211	. . .	211		6	
7					7	21
8	220	220	220	220	8	
9	300, 221				9	30
10	310	. . .	310		10	
11	311	311	. . .	311	11	
12	222	222	222		12	22
13	320				13	31
14	321	. . .	321		14	
15					15	
16	400	400	400	400	16	40
17	410, 322				17	
18	411, 330	. . .	411, 330		18	
19	331	331	. . .	331	19	32
20	420	420	420		20	
21	421				21	41
22	332	. . .	332		22	
23					23	
24	422	422	422	422	24	
25	500, 430				25	50
26	510, 431	. . .	510, 431		26	
27	511, 333	511, 333	. . .	511, 333	27	33
28					28	42
29	520, 432				29	
30	521	. . .	521		30	
31					31	51
32	440	440	440	440	32	
33	522, 441				33	
34	530, 433	. . .	530, 433		34	
35	531	531	. . .	531	35	
36	600, 442	600, 442	600, 442		36	60
37	610				37	43
38	611, 532	. . .	611, 532		38	
39					39	52
40	620	620	620	620	40	
41	621, 540, 443				41	
42	541	. . .	541		42	
43	533	533	. . .	533	43	61
44	622	622	622		44	
45	630, 542				45	
46	631	. . .	631		46	
47					47	
48	444	444	444	444	48	44
49	700, 632				49	70, 53

Cubic					Hexagonal	
$h^2 + k^2 + l^2$	*hkl*				$h^2 + hk + k^2$	*hk*
	Simple	Face-centered	Body-centered	Diamond		
50	710, 550, 543	. . .	710, 550, 543		50	
51	711, 551	711, 551	. . .	711, 551	51	
52	640	640	640		52	62
53	720, 641				53	
54	721, 633, 552	. . .	721, 633, 552		54	
55					55	
56	642	642	642	642	56	
57	722, 544				57	71
58	730	. . .	730		58	
59	731, 553	731, 553	. . .	731, 553	59	

APPENDIX 11

VALUES OF $(\sin\theta)/\lambda$ ($Å^{-1}$)

θ	Radiation				
	Mo $K\alpha$ (0.711 Å)	Cu $K\alpha$ (1.542 Å)	Co $K\alpha$ (1.790 Å)	Fe $K\alpha$ (1.937 Å)	Cr $K\alpha$ (2.291 Å)
0°	0.00	0.00	0.00	0.00	0.00
1	0.02	0.01	0.01	0.01	0.01
2	0.05	0.02	0.02	0.02	0.02
3	0.07	0.03	0.03	0.03	0.02
4	0.10	0.05	0.04	0.04	0.03
5	0.12	0.06	0.05	0.04	0.04
6	0.15	0.07	0.06	0.05	0.05
7	0.17	0.08	0.07	0.06	0.05
8	0.20	0.09	0.08	0.07	0.06
9	0.22	0.10	0.09	0.08	0.07
10	0.24	0.11	0.10	0.09	0.08
11	0.27	0.12	0.11	0.10	0.08
12	0.29	0.13	0.12	0.11	0.09
13	0.32	0.15	0.13	0.12	0.10
14	0.34	0.16	0.14	0.12	0.11
15	0.36	0.17	0.14	0.13	0.11
16	0.39	0.18	0.15	0.14	0.12
17	0.41	0.19	0.16	0.15	0.13
18	0.43	0.20	0.17	0.16	0.13
19	0.46	0.21	0.18	0.17	0.14
20	0.48	0.22	0.19	0.18	0.15
21	0.51	0.23	0.20	0.18	0.15
22	0.53	0.24	0.21	0.19	0.16
23	0.55	0.25	0.22	0.20	0.17
24	0.57	0.26	0.23	0.21	0.18
25	0.60	0.27	0.24	0.22	0.18
26	0.62	0.28	0.24	0.23	0.19
27	0.64	0.29	0.25	0.23	0.20
28	0.66	0.30	0.26	0.24	0.20
29	0.68	0.31	0.27	0.25	0.21

θ	Radiation				
	Mo $K\alpha$ (0.711 Å)	Cu $K\alpha$ (1.542 Å)	Co $K\alpha$ (1.790 Å)	Fe $K\alpha$ (1.937 Å)	Cr $K\alpha$ (2.291 Å)
30	0.70	0.32	0.28	0.26	0.22
31	0.72	0.33	0.29	0.27	0.22
32	0.75	0.34	0.30	0.27	0.23
33	0.77	0.35	0.30	0.28	0.24
34	0.79	0.36	0.31	0.29	0.24
35	0.81	0.37	0.32	0.29	0.25
36	0.83	0.38	0.33	0.30	0.26
37	0.85	0.39	0.34	0.31	0.26
38	0.87	0.40	0.34	0.32	0.27
39	0.89	0.41	0.35	0.32	0.27
40	0.91	0.42	0.36	0.33	0.28
41	0.93	0.43	0.37	0.34	0.29
42	0.94	0.43	0.37	0.35	0.29
43	0.96	0.44	0.38	0.35	0.30
44	0.98	0.45	0.39	0.36	0.30
45	0.99	0.46	0.40	0.36	0.31
46	1.01	0.47	0.40	0.37	0.31
47	1.03	0.47	0.41	0.38	0.32
48	1.05	0.48	0.42	0.38	0.32
49	1.06	0.49	0.42	0.39	0.33
50	1.08	0.50	0.43	0.39	0.33
52	1.11	0.51	0.44	0.41	0.34
54	1.14	0.52	0.45	0.42	0.35
56	1.17	0.54	0.46	0.43	0.36
58	1.20	0.55	0.47	0.44	0.37
60	1.22	0.56	0.48	0.45	0.38
62	1.24	0.57	0.49	0.46	0.39
64	1.26	0.58	0.50	0.46	0.39
66	1.28	0.59	0.51	0.47	0.40
68	1.30	0.60	0.52	0.48	0.40
70	1.32	0.61	0.53	0.48	0.41
72	1.34	0.62	0.53	0.49	0.41
74	1.35	0.62	0.54	0.50	0.42
76	1.37	0.63	0.54	0.50	0.42
78	1.38	0.63	0.55	0.50	0.43
80	1.39	0.64	0.55	0.51	0.43
82	1.39	0.64	0.55	0.51	0.43
84	1.40	0.64	0.56	0.51	0.43
86	1.40	0.65	0.56	0.51	0.43
88	1.41	0.65	0.56	0.52	0.43
90	1.41	0.65	0.56	0.52	0.43

APPENDIX 12
ATOMIC SCATTERING FACTORS

$\frac{\sin\theta}{\lambda}$ (Å^{-1})	0.0	0.1	0.2	0.3	0.4	0.5	0.6	0.7	0.8	0.9	1.0	1.1	1.2
H	1	0.81	0.48	0.25	0.13	0.07	0.04	0.03	0.02	0.01	0.00	0.00	
He	2	1.88	1.46	1.05	0.75	0.52	0.35	0.24	0.18	0.14	0.11	0.09	
Li^{+}	2	1.96	1.8	1.5	1.3	1.0	0.8	0.6	0.5	0.4	0.3	0.3	
Li	3	2.2	1.8	1.5	1.3	1.0	0.8	0.6	0.5	0.4	0.3	0.3	
Be^{+2}	2	2.0	1.9	1.7	1.6	1.4	1.2	1.0	0.9	0.7	0.6	0.5	
Be	4	2.9	1.9	1.7	1.6	1.4	1.2	1.0	0.9	0.7	0.6	0.5	
B^{+3}	2	1.99	1.9	1.8	1.7	1.6	1.4	1.3	1.2	1.0	0.9	0.7	
B	5	3.5	2.4	1.9	1.7	1.5	1.4	1.2	1.2	1.0	0.9	0.7	
C	6	4.6	3.0	2.2	1.9	1.7	1.6	1.4	1.3	1.16	1.0	0.9	
N^{+5}	2	2.0	2.0	1.9	1.9	1.8	1.7	1.6	1.5	1.4	1.3	1.16	
N^{+3}	4	3.7	3.0	2.4	2.0	1.8	1.66	1.56	1.49	1.39	1.28	1.17	
N	7	5.8	4.2	3.0	2.3	1.9	1.65	1.54	1.49	1.39	1.29	1.17	
O	8	7.1	5.3	3.9	2.9	2.2	1.8	1.6	1.5	1.4	1.35	1.26	
O^{-2}	10	8.0	5.5	3.8	2.7	2.1	1.8	1.5	1.5	1.4	1.35	1.26	
F	9	7.8	6.2	4.45	3.35	2.65	2.15	1.9	1.7	1.6	1.5	1.35	
F^{-}	10	8.7	6.7	4.8	3.5	2.8	2.2	1.9	1.7	1.55	1.5	1.35	
Ne	10	9.3	7.5	5.8	4.4	3.4	2.65	2.2	1.9	1.65	1.55	1.5	
Na^{+}	10	9.5	8.2	6.7	5.25	4.05	3.2	2.65	2.25	1.95	1.75	1.6	
Na	11	9.65	8.2	6.7	5.25	4.05	3.2	2.65	2.25	1.95	1.75	1.6	
Mg^{+2}	10	9.75	8.6	7.25	5.95	4.8	3.85	3.15	2.55	2.2	2.0	1.8	
Mg	12	10.5	8.6	7.25	5.95	4.8	3.85	3.15	2.55	2.2	2.0	1.8	
Al^{+3}	10	9.7	8.9	7.8	6.65	5.5	4.45	3.65	3.1	2.65	2.3	2.0	
Al	13	11.0	8.95	7.75	6.6	5.5	4.5	3.7	3.1	2.65	2.3	2.0	
Si^{+4}	10	9.75	9.15	8.25	7.15	6.05	5.05	4.2	3.4	2.95	2.6	2.3	
Si	14	11.35	9.4	8.2	7.15	6.1	5.1	4.2	3.4	2.95	2.6	2.3	
P^{+5}	10	9.8	9.25	8.45	7.5	6.55	5.65	4.8	4.05	3.4	3.0	2.6	
P	15	12.4	10.0	8.45	7.45	6.5	5.65	4.8	4.05	3.4	3.0	2.6	
P^{-3}	18	12.7	9.8	8.4	7.45	6.5	5.65	4.85	4.05	3.4	3.0	2.6	
S^{+6}	10	9.85	9.4	8.7	7.85	6.85	6.05	5.25	4.5	3.9	3.35	2.9	
S	16	13.6	10.7	8.95	7.85	6.85	6.0	5.25	4.5	3.9	3.35	2.9	
S^{-2}	18	14.3	10.7	8.9	7.85	6.85	6.0	5.25	4.5	3.9	3.35	2.9	
Cl	17	14.6	11.3	9.25	8.05	7.25	6.5	5.75	5.05	4.4	3.85	3.35	
Cl^{-}	18	15.2	11.5	9.3	8.05	7.25	6.5	5.75	5.05	4.4	3.85	3.35	
A	18	15.9	12.6	10.4	8.7	7.8	7.0	6.2	5.4	4.7	4.1	3.6	
K^{+}	18	16.5	13.3	10.8	8.85	7.75	7.05	6.44	5.9	5.3	4.8	4.2	
K	19	16.5	13.3	10.8	9.2	7.9	6.7	5.9	5.2	4.6	4.2	3.7	3.3
Ca^{+2}	18	16.8	14.0	11.5	9.3	8.1	7.35	6.7	6.2	5.7	5.1	4.6	
Ca	20	17.5	14.1	11.4	9.7	8.4	7.3	6.3	5.6	4.9	4.5	4.0	3.6
Sc^{+3}	18	16.7	14.0	11.4	9.4	8.3	7.6	6.9	6.4	5.8	5.35	4.85	
Sc	21	18.4	14.9	12.1	10.3	8.9	7.7	6.7	5.9	5.3	4.7	4.3	3.9
Ti^{+4}	18	17.0	14.4	11.9	9.9	8.5	7.85	7.3	6.7	6.15	5.65	5.05	
Ti	22	19.3	15.7	12.8	10.9	9.5	8.2	7.2	6.3	5.6	5.0	4.6	4.2
V	23	20.2	16.6	13.5	11.5	10.1	8.7	7.6	6.7	5.9	5.3	4.9	4.4
Cr	24	21.1	17.4	14.2	12.1	10.6	9.2	8.0	7.1	6.3	5.7	5.1	4.6
Mn	25	22.1	18.2	14.9	12.7	11.1	9.7	8.4	7.5	6.6	6.0	5.4	4.9

$\frac{\sin\theta}{\lambda}$ (Å^{-1})	0.0	0.1	0.2	0.3	0.4	0.5	0.6	0.7	0.8	0.9	1.0	1.1	1.2
Fe	26	23.1	18.9	15.6	13.3	11.6	10.2	8.9	7.9	7.0	6.3	5.7	5.2
Co	27	24.1	19.8	16.4	14.0	12.1	10.7	9.3	8.3	7.3	6.7	6.0	5.5
Ni	28	25.0	20.7	17.2	14.6	12.7	11.2	9.8	8.7	7.7	7.0	6.3	5.8
Cu	29	25.9	21.6	17.9	15.2	13.3	11.7	10.2	9.1	8.1	7.3	6.6	6.0
Zn	30	26.8	22.4	18.6	15.8	13.9	12.2	10.7	9.6	8.5	7.6	6.9	6.3
Ga	31	27.8	23.3	19.3	16.5	14.5	12.7	11.2	10.0	8.9	7.9	7.3	6.7
Ge	32	28.8	24.1	20.0	17.1	15.0	13.2	11.6	10.4	9.3	8.3	7.6	7.0
As	33	29.7	25.0	20.8	17.7	15.6	13.8	12.1	10.8	9.7	8.7	7.9	7.3
Se	34	30.6	25.8	21.5	18.3	16.1	14.3	12.6	11.2	10.0	9.0	8.2	7.5
Br	35	31.6	26.6	22.3	18.9	16.7	14.8	13.1	11.7	10.4	9.4	8.6	7.8
Kr	36	32.5	27.4	23.0	19.5	17.3	15.3	13.6	12.1	10.8	9.8	8.9	8.1
Rb^+	36	33.6	28.7	24.6	21.4	18.9	16.7	14.6	12.8	11.2	9.9	8.9	
Rb	37	33.5	28.2	23.8	20.2	17.9	15.9	14.1	12.5	11.2	10.2	9.2	8.4
Sr	38	34.4	29.0	24.5	20.8	18.4	16.4	14.6	12.9	11.6	10.5	9.5	8.7
Y	39	35.4	29.9	25.3	21.5	19.0	17.0	15.1	13.4	12.0	10.9	9.9	9.0
Zr	40	36.3	30.8	26.0	22.1	19.7	17.5	15.6	13.8	12.4	11.2	10.2	9.3
Nb	41	37.3	31.7	26.8	22.8	20.2	18.1	16.0	14.3	12.8	11.6	10.6	9.7
Mo	42	38.2	32.6	27.6	23.5	20.8	18.6	16.5	14.8	13.2	12.0	10.9	10.0
Tc	43	39.1	33.4	28.3	24.1	21.3	19.1	17.0	15.2	13.6	12.3	11.3	10.3
Ru	44	40.0	34.3	29.1	24.7	21.9	19.6	17.5	15.6	14.1	12.7	11.6	10.6
Rh	45	41.0	35.1	29.9	25.4	22.5	20.2	18.0	16.1	14.5	13.1	12.0	11.0
Pd	46	41.9	36.0	30.7	26.2	23.1	20.8	18.5	16.6	14.9	13.6	12.3	11.3
Ag	47	42.8	36.9	31.5	26.9	23.8	21.3	19.0	17.1	15.3	14.0	12.7	11.7
Cd	48	43.7	37.7	32.2	27.5	24.4	21.8	19.6	17.6	15.7	14.3	13.0	12.0
In	49	44.7	38.6	33.0	28.1	25.0	22.4	20.1	18.0	16.2	14.7	13.4	12.3
Sn	50	45.7	39.5	33.8	28.7	25.6	22.9	20.6	18.5	16.6	15.1	13.7	12.7
Sb	51	46.7	40.4	34.6	29.5	26.3	23.5	21.1	19.0	17.0	15.5	14.1	13.0
Te	52	47.7	41.3	35.4	30.3	26.9	24.0	21.7	19.5	17.5	16.0	14.5	13.3
I	53	48.6	42.1	36.1	31.0	27.5	24.6	22.2	20.0	17.9	16.4	14.8	13.6
Xe	54	49.6	43.0	36.8	31.6	28.0	25.2	22.7	20.4	18.4	16.7	15.2	13.9
Cs	55	50.7	43.8	37.6	32.4	28.7	25.8	23.2	20.8	18.8	17.0	15.6	14.5
Ba	56	51.7	44.7	38.4	33.1	29.3	26.4	23.7	21.3	19.2	17.4	16.0	14.7
La	57	52.6	45.6	39.3	33.8	29.8	26.9	24.3	21.9	19.7	17.9	16.4	15.0
Ce	58	53.6	46.5	40.1	34.5	30.4	27.4	24.8	22.4	20.2	18.4	16.6	15.3
Pr	59	54.5	47.4	40.9	35.2	31.1	28.0	25.4	22.9	20.6	18.8	17.1	15.7
Nd	60	55.4	48.3	41.6	35.9	31.8	28.6	25.9	23.4	21.1	19.2	17.5	16.1
Pm	61	56.4	49.1	42.4	36.6	32.4	29.2	26.4	23.9	21.5	19.6	17.9	16.4
Sm	62	57.3	50.0	43.2	37.3	32.9	29.8	26.9	24.4	22.0	20.0	18.3	16.8
Eu	63	58.3	50.9	44.0	38.1	33.5	30.4	27.5	24.9	22.4	20.4	18.7	17.1
Gd	64	59.3	51.7	44.8	38.8	34.1	31.0	28.1	25.4	22.9	20.8	19.1	17.5
Tb	65	60.2	52.6	45.7	39.6	34.7	31.6	28.6	25.9	23.4	21.2	19.5	17.9
Dy	66	61.1	53.6	46.5	40.4	35.4	32.2	29.2	26.3	23.9	21.6	19.9	18.3
Ho	67	62.1	54.5	47.3	41.1	36.1	32.7	29.7	26.8	24.3	22.0	20.3	18.6
Er	68	63.0	55.3	48.1	41.7	36.7	33.3	30.2	27.3	24.7	22.4	20.7	18.9
Tm	69	64.0	56.2	48.9	42.4	37.4	33.9	30.8	27.9	25.2	22.9	21.0	19.3

$\frac{\sin\theta}{\lambda}$ ($Å^{-1}$)	0.0	0.1	0.2	0.3	0.4	0.5	0.6	0.7	0.8	0.9	1.0	1.1	1.2
Yb	70	64.9	57.0	49.7	43.2	38.0	34.4	31.3	28.4	25.7	23.3	21.4	19.7
Lu	71	65.9	57.8	50.4	43.9	38.7	35.0	31.8	28.9	26.2	23.8	21.8	20.0
Hf	72	66.8	58.6	51.2	44.5	39.3	35.6	32.3	29.3	26.7	24.2	22.3	20.4
Ta	73	67.8	59.5	52.0	45.3	39.9	36.2	32.9	29.8	27.1	24.7	22.6	20.9
W	74	68.8	60.4	52.8	46.1	40.5	36.8	33.5	30.4	27.6	25.2	23.0	21.3
Re	75	69.8	61.3	53.6	46.8	41.1	37.4	34.0	30.9	28.1	25.6	23.4	21.6
Os	76	70.8	62.2	54.4	47.5	41.7	38.0	34.6	31.4	28.6	26.0	23.9	22.0
Ir	77	71.7	63.1	55.3	48.2	42.4	38.6	35.1	32.0	29.0	26.5	24.3	22.3
Pt	78	72.6	64.0	56.2	48.9	43.1	39.2	35.6	32.5	29.5	27.0	24.7	22.7
Au	79	73.6	65.0	57.0	49.7	43.8	39.8	36.2	33.1	30.0	27.4	25.1	23.1
Hg	80	74.6	65.9	57.9	50.5	44.4	40.5	36.8	33.6	30.6	27.8	25.6	23.6
Tl	81	75.5	66.7	58.7	51.2	45.0	41.1	37.4	34.1	31.1	28.3	26.0	24.1
Pb	82	76.5	67.5	59.5	51.9	45.7	41.6	37.9	34.6	31.5	28.8	26.4	24.5
Bi	83	77.5	68.4	60.4	52.7	46.4	42.2	38.5	35.1	32.0	29.2	26.8	24.8
Po	84	78.4	69.4	61.3	53.5	47.1	42.8	39.1	35.6	32.6	29.7	27.2	25.2
At	85	79.4	70.3	62.1	54.2	47.7	43.4	39.6	36.2	33.1	30.1	27.6	25.6
Rn	86	80.3	71.3	63.0	55.1	48.4	44.0	40.2	36.8	33.5	30.5	28.0	26.0
Fr	87	81.3	72.2	63.8	55.8	49.1	44.5	40.7	37.3	34.0	31.0	28.4	26.4
Ra	88	82.2	73.2	64.6	56.5	49.8	45.1	41.3	37.8	34.6	31.5	28.8	26.7
Ac	89	83.2	74.1	65.5	57.3	50.4	45.8	41.8	38.3	35.1	32.0	29.2	27.1
Th	90	84.1	75.1	66.3	58.1	51.1	46.5	42.4	38.8	35.5	32.4	29.6	27.5
Pa	91	85.1	76.0	67.1	58.8	51.7	47.1	43.0	39.3	36.0	32.8	30.1	27.9
U	92	86.0	76.9	67.9	59.6	52.4	47.7	43.5	39.8	36.5	33.3	30.6	28.3
Np	93	87	78	69	60	53	48	44	40	37	34	31	29
Pu	94	88	79	69	61	54	49	44	41	38	34	31	29
Am	95	89	79	70	62	55	50	45	42	38	35	32	30
Cm	96	90	80	71	62	55	50	46	42	39	35	32	30
Bk	97	91	81	72	63	56	51	46	43	39	36	33	30
Cf	98	92	82	73	64	57	52	47	43	40	36	33	31
	99	93	83	74	65	57	52	48	44	40	37	34	31
	100	94	84	75	66	58	53	48	44	41	37	34	31

From Peiser, Rooksby, and Wilson [G.13]. More extensive tables, at smaller intervals of $(\sin\theta)/\lambda$, are given on pp. 72–98 of Vol. 4 of [G.11].

APPENDIX 13
MULTIPLICITY FACTORS FOR THE POWDER METHOD

Cubic:	hkl / 48*	hhl / 24	$0kl$ / 24*	$0kk$ / 12	hhh / 8	$00l$ / 6	
Hexagonal and Rhombohedral:	$hk \cdot l$ / 24*	$hh \cdot l$ / 12*	$0k \cdot l$ / 12*	$hk \cdot 0$ / 12*	$hh \cdot 0$ / 6	$0k \cdot 0$ / 6	$00 \cdot l$ / 2
Tetragonal:	hkl / 16*	hhl / 8	$0kl$ / 8	$hk0$ / 8*	$hh0$ / 4	$0k0$ / 4	$00l$ / 2
Orthorhombic:	hkl / 8	$0kl$ / 4	$h0l$ / 4	$hk0$ / 4	$h00$ / 2	$0k0$ / 2	$00l$ / 2
Monoclinic:	hkl / 4	$h0l$ / 2	$0k0$ / 2				
Triclinic:	hkl / 2						

Note that, in cubic crystals, for example, hhl stands for such indices as 112 (or 211), $0kl$ for such indices as 012 (or 210), $0kk$ for such indices as 011 (or 110), etc.

* These are the usual multiplicity factors. In some crystals, planes having these indices comprise two forms with the same spacing but different structure factor, and the multiplicity factor for each form is half the value given above. In the cubic system, for example, there are some crystals in which permutations of the indices (hkl) produce planes which are not structurally equivalent; in such crystals (AuBe, discussed in Sec. 2–7, is an example), the plane (123), for example, belongs to one form and has a certain structure factor, while the plane (321) belongs to another form and has a different structure factor. There are $\frac{48}{2} = 24$ planes in the first form and 24 planes in the second. This question is discussed more fully by Henry, Lipson, and Wooster [G.8].

APPENDIX 14

LORENTZ-POLARIZATION FACTOR $\left(\frac{1+\cos^2 2\theta}{\sin^2\theta\cos\theta}\right)$

$\theta°$	.0	.1	.2	.3	.4	.5	.6	.7	.8	.9
2	1639	1486	1354	1239	1138	1048	968.9	898.3	835.1	778.4
3	727.2	680.9	638.8	600.5	565.6	533.6	504.3	477.3	452.3	429.3
4	408.0	388.2	369.9	352.7	336.8	321.9	308.0	294.9	282.6	271.1
5	260.3	250.1	240.5	231.4	222.9	214.7	207.1	199.8	192.9	186.3
6	180.1	174.2	168.5	163.1	158.0	153.1	148.4	144.0	139.7	135.6
7	131.7	128.0	124.4	120.9	117.6	114.4	111.4	108.5	105.6	102.9
8	100.3	97.80	95.37	93.03	90.78	88.60	86.51	84.48	82.52	80.63
9	78.79	77.02	75.31	73.66	72.05	70.49	68.99	67.53	66.12	64.74
10	63.41	62.12	60.87	59.65	58.46	57.32	56.20	55.11	54.06	53.03
11	52.04	51.06	50.12	49.19	48.30	47.43	46.58	45.75	44.94	44.16
12	43.39	42.64	41.91	41.20	40.50	39.82	39.16	38.51	37.88	37.27
13	36.67	36.08	35.50	34.94	34.39	33.85	33.33	32.81	32.31	31.82
14	31.34	30.87	30.41	29.96	29.51	29.08	28.66	28.24	27.83	27.44
15	27.05	26.66	26.29	25.92	25.56	25.21	24.86	24.52	24.19	23.86
16	23.54	23.23	22.92	22.61	22.32	22.02	21.74	21.46	21.18	20.91
17	20.64	20.38	20.12	19.87	19.62	19.38	19.14	18.90	18.67	18.44
18	18.22	18.00	17.78	17.57	17.36	17.15	16.95	16.75	16.56	16.36
19	16.17	15.99	15.80	15.62	15.45	15.27	15.10	14.93	14.76	14.60
20	14.44	14.28	14.12	13.97	13.81	13.66	13.52	13.37	13.23	13.09
21	12.95	12.81	12.68	12.54	12.41	12.28	12.15	12.03	11.91	11.78
22	11.66	11.54	11.43	11.31	11.20	11.09	10.98	10.87	10.76	10.65
23	10.55	10.45	10.35	10.24	10.15	10.05	9.951	9.857	9.763	9.671
24	9.579	9.489	9.400	9.313	9.226	9.141	9.057	8.973	8.891	8.810
25	8.730	8.651	8.573	8.496	8.420	8.345	8.271	8.198	8.126	8.054
26	7.984	7.915	7.846	7.778	7.711	7.645	7.580	7.515	7.452	7.389
27	7.327	7.266	7.205	7.145	7.086	7.027	6.969	6.912	6.856	6.800
28	6.745	6.692	6.637	6.584	6.532	6.480	6.429	6.379	6.329	6.279
29	6.230	6.183	6.135	6.088	6.042	5.995	5.950	5.905	5.861	5.817
30	5.774	5.731	5.688	5.647	5.605	5.564	5.524	5.484	5.445	5.406
31	5.367	5.329	5.292	5.254	5.218	5.181	5.145	5.110	5.075	5.040
32	5.006	4.972	4.939	4.906	4.873	4.841	4.809	4.777	4.746	4.715
33	4.685	4.655	4.625	4.595	4.566	4.538	4.509	4.481	4.453	4.426
34	4.399	4.372	4.346	4.320	4.294	4.268	4.243	4.218	4.193	4.169
35	4.145	4.121	4.097	4.074	4.052	4.029	4.006	3.984	3.962	3.941
36	3.919	3.898	3.877	3.857	3.836	3.816	3.797	3.777	3.758	3.739
37	3.720	3.701	3.683	3.665	3.647	3.629	3.612	3.594	3.577	3.561
38	3.544	3.527	3.513	3.497	3.481	3.465	3.449	3.434	3.419	3.404
39	3.389	3.375	3.361	3.347	3.333	3.320	3.306	3.293	3.280	3.268
40	3.255	3.242	3.230	3.218	3.206	3.194	3.183	3.171	3.160	3.149
41	3.138	3.127	3.117	3.106	3.096	3.086	3.076	3.067	3.057	3.048
42	3.038	3.029	3.020	3.012	3.003	2.994	2.986	2.978	2.970	2.962
43	2.954	2.946	2.939	2.932	2.925	2.918	2.911	2.904	2.897	2.891
44	2.884	2.878	2.872	2.866	2.860	2.855	2.849	2.844	2.838	2.833

θ°	.0	.1	.2	.3	.4	.5	.6	.7	.8	.9
45	2.828	2.824	2.819	2.814	2.810	2.805	2.801	2.797	2.793	2.789
46	2.785	2.782	2.778	2.775	2.772	2.769	2.766	2.763	2.760	2.757
47	2.755	2.752	2.750	2.748	2.746	2.744	2.742	2.740	2.738	2.737
48	2.736	2.735	2.733	2.732	2.731	2.730	2.730	2.729	2.729	2.728
49	2.728	2.728	2.728	2.728	2.728	2.728	2.729	2.729	2.730	2.730
50	2.731	2.732	2.733	2.734	2.735	2.737	2.738	2.740	2.741	2.743
51	2.745	2.747	2.749	2.751	2.753	2.755	2.758	2.760	2.763	2.766
52	2.769	2.772	2.775	2.778	2.782	2.785	2.788	2.792	2.795	2.799
53	2.803	2.807	2.811	2.815	2.820	2.824	2.828	2.833	2.838	2.843
54	2.848	2.853	2.858	2.863	2.868	2.874	2.879	2.885	2.890	2.896
55	2.902	2.908	2.914	2.921	2.927	2.933	2.940	2.946	2.953	2.960
56	2.967	2.974	2.981	2.988	2.996	3.004	3.011	3.019	3.026	3.034
57	3.042	3.050	3.059	3.067	3.075	3.084	3.092	3.101	3.110	3.119
58	3.128	3.137	3.147	3.156	3.166	3.175	3.185	3.195	3.205	3.215
59	3.225	3.235	3.246	3.256	3.267	3.278	3.289	3.300	3.311	3.322
60	3.333	3.345	3.356	3.368	3.380	3.392	3.404	3.416	3.429	3.441
61	3.454	3.466	3.479	3.492	3.505	3.518	3.532	3.545	3.559	3.573
62	3.587	3.601	3.615	3.629	3.643	3.658	3.673	3.688	3.703	3.718
63	3.733	3.749	3.764	3.780	3.796	3.812	3.828	3.844	3.861	3.878
64	3.894	3.911	3.928	3.946	3.963	3.980	3.998	4.016	4.034	4.052
65	4.071	4.090	4.108	4.127	4.147	4.166	4.185	4.205	4.225	4.245
66	4.265	4.285	4.306	4.327	4.348	4.369	4.390	4.412	4.434	4.456
67	4.478	4.500	4.523	4.546	4.569	4.592	4.616	4.640	4.664	4.688
68	4.712	4.737	4.762	4.787	4.812	4.838	4.864	4.890	4.916	4.943
69	4.970	4.997	5.024	5.052	5.080	5.109	5.137	5.166	5.195	5.224
70	5.254	5.284	5.315	5.345	5.376	5.408	5.440	5.471	5.504	5.536
71	5.569	5.602	5.636	5.670	5.705	5.740	5.775	5.810	5.846	5.883
72	5.919	5.956	5.994	6.032	6.071	6.109	6.149	6.189	6.229	6.270
73	6.311	6.352	6.394	6.437	6.480	6.524	6.568	6.613	6.658	6.703
74	6.750	6.797	6.844	6.892	6.941	6.991	7.041	7.091	7.142	7.194
75	7.247	7.300	7.354	7.409	7.465	7.521	7.578	7.636	7.694	7.753
76	7.813	7.874	7.936	7.999	8.063	8.128	8.193	8.259	8.327	8.395
77	8.465	8.536	8.607	8.680	8.754	8.829	8.905	8.982	9.061	9.142
78	9.223	9.305	9.389	9.474	9.561	9.649	9.739	9.831	9.924	10.02
79	10.12	10.21	10.31	10.41	10.52	10.62	10.73	10.84	10.95	11.06
80	11.18	11.30	11.42	11.54	11.67	11.80	11.93	12.06	12.20	12.34
81	12.48	12.63	12.78	12.93	13.08	13.24	13.40	13.57	13.74	13.92
82	14.10	14.28	14.47	14.66	14.86	15.07	15.28	15.49	15.71	15.94
83	16.17	16.41	16.66	16.91	17.17	17.44	17.72	18.01	18.31	18.61
84	18.93	19.25	19.59	19.94	20.30	20.68	21.07	21.47	21.89	22.32
85	22.77	23.24	23.73	24.24	24.78	25.34	25.92	26.52	27.16	27.83
86	28.53	29.27	30.04	30.86	31.73	32.64	33.60	34.63	35.72	36.88
87	38.11	39.43	40.84	42.36	44.00	45.76	47.68	49.76	52.02	54.50

From Henry, Lipson, and Wooster [G.8].

APPENDIX 15
DATA FOR CALCULATION OF THE TEMPERATURE FACTOR

Values of $\phi(x) = \dfrac{1}{x}\displaystyle\int_0^x \dfrac{\xi}{e^{\xi} - 1}\, d\xi$ as a Function of x

x	.0	.1	.2	.3	.4	.5	.6	.7	.8	.9
0	1.000	0.975	0.951	0.928	0.904	0.882	0.860	0.839	0.818	0.797
1	0.778	0.758	0.739	0.721	0.703	0.686	0.669	0.653	0.637	0.622
2	0.607	0.592	0.578	0.565	0.552	0.539	0.526	0.514	0.503	0.491
3	0.480	0.470	0.460	0.450	0.440	0.431	0.422	0.413	0.404	0.396
4	0.388	0.380	0.373	0.366	0.359	0.352	0.345	0.339	0.333	0.327
5	0.321	0.315	0.310	0.304	0.299	0.294	0.289	0.285	0.280	0.276
6	0.271	0.267	0.263	0.259	0.255	0.251	0.248	0.244	0.241	0.237

For x greater than 7, $\phi(x)$ is given to a good approximation by $(1.642/x)$. (From Vol. 2, p. 264 of [G.11]).

Debye Temperatures

James [G.7, p. 221] gives the following values of the characteristic Debye temperature Θ for some cubic metals.

Metal	Θ(°K)	Metal	Θ(°K)
Al	390	Ta	245
Ca	230	Pb	88
Cu	320	Fe	430
Ag	210	Co	410
Au	175	Ni	400
Cr	485	Pd	275
Mo	380	Ir	285
W	310	Pt	230

APPENDIX 16

ATOMIC WEIGHTS Based on the assigned relative atomic mass of ^{12}C = 12.

Element	Symbol	Atomic number	Atomic weight	Element	Symbol	Atomic number	Atomic weight
Actinium	Ac	89	(227)	Mercury	Hg	80	200.59
Aluminium	Al	13	26.9815	Molybdenum	Mo	42	95.94
Americium	Am	95	(243)	Neodymium	Nd	60	144.24
Antimony	Sb	51	121.75	Neon	Ne	10	20.179
Argon	Ar	18	39.948	Neptunium	Np	93	237.0482
Arsenic	As	33	74.9216	Nickel	Ni	28	58.71
Astatine	At	85	(210)	Niobium	Nb	41	92.9064
Barium	Ba	56	137.34	Nitrogen	N	7	14.0067
Berkelium	Bk	97	(247)	Nobelium	No	102	(254)
Beryllium	Be	4	9.01218	Osmium	Os	76	190.2
Bismuth	Bi	83	208.9806	Oxygen	O	8	15.9994
Boron	B	5	10.81	Palladium	Pd	46	106.4
Bromine	Br	35	79.904	Phosphorus	P	15	30.9738
Cadmium	Cd	48	112.40	Platinum	Pt	78	195.09
Calcium	Ca	20	40.08	Plutonium	Pu	94	(242)
Californium	Cf	98	(249)	Polonium	Po	84	(210)
Carbon	C	6	12.011	Potassium	K	19	39.102
Cerium	Ce	58	140.12	Praseodymium	Pr	59	140.9077
Cesium	Cs	55	132.9055	Promethium	Pm	61	(147)
Chlorine	Cl	17	35.453	Protactinium	Pa	91	231.0359
Chromium	Cr	24	51.996	Radium	Ra	88	226.0254
Cobalt	Co	27	58.9332	Radon	Rn	86	(222)
Copper	Cu	29	63.546	Rhenium	Re	75	186.2
Curium	Cm	96	(247)	Rhodium	Rh	45	102.9055
Dysprosium	Dy	66	162.50	Rubidium	Rb	37	85.4678
Einsteinium	Es	99	(254)	Ruthenium	Ru	44	101.07
Erbium	Er	68	167.26	Samarium	Sm	62	150.4
Europium	Eu	63	151.96	Scandium	Sc	21	44.9559
Fermium	Fm	100	(253)	Selenium	Se	34	78.96
Fluorine	F	9	18.9984	Silicon	Si	14	28.086
Francium	Fr	87	(223)	Silver	Ag	47	107.868
Gadolinium	Gd	64	157.25	Sodium	Na	11	22.9898
Gallium	Ga	31	69.72	Strontium	Sr	38	87.62
Germanium	Ge	32	72.59	Sulfur	S	16	32.06
Gold	Au	79	196.9665	Tantalum	Ta	73	180.9479
Hafnium	Hf	72	178.49	Technetium	Tc	43	98.9062
Helium	He	2	4.00260	Tellurium	Te	52	127.60
Holmium	Ho	67	164.9303	Terbium	Tb	65	158.9254
Hydrogen	H	1	1.0080	Thallium	Tl	81	204.37
Indium	In	49	114.82	Thorium	Th	90	232.0381
Iodine	I	53	126.9045	Thulium	Tm	69	168.9342
Iridium	Ir	77	192.22	Tin	Sn	50	118.69
Iron	Fe	26	55.847	Titanium	Ti	22	47.90
Krypton	Kr	36	83.80	Tungsten	W	74	183.85
Lanthanum	La	57	138.9055	Uranium	U	92	238.029
Lawrencium	Lr	103	(257)	Vanadium	V	23	50.9414
Lead	Pb	82	207.2	Xenon	Xe	54	131.30
Lithium	Li	3	6.941	Ytterbium	Yb	70	173.04
Lutetium	Lu	71	174.97	Yttrium	Y	39	88.9059
Magnesium	Mg	12	24.305	Zinc	Zn	30	65.37
Manganese	Mn	25	54.9380	Zirconium	Zr	40	91.22
Mendelevium	Md	101	(256)				

Values in parentheses represent the most stable known isotopes.

APPENDIX 17
PHYSICAL CONSTANTS

Charge on electron	e	1.602×10^{-19}	coulomb
Mass of electron	m	9.109×10^{-31}	kg
Mass of neutron		1.675×10^{-27}	kg
Velocity of light	c	2.998×10^{8}	m/sec
Planck's constant	h	6.626×10^{-34}	joule sec
Boltzmann's constant	k	1.381×10^{-23}	joule/deg K
Avogadro's number	N	6.023×10^{26}	per kg mol
Gas constant ($=Nk$)	R	8.314×10^{3}	joules/deg K, kg mol
		1.986×10^{3}	cal/deg K, kg mol

1 cal = 4.186 joule
1 erg = 10^{-7} joule
1 eV = 1.602×10^{-19} joule

General References

The following books are listed mainly in order of publication. Starred entries are reference books.

***G.1** **Strukturbericht** (Leipzig: Akademische Verlagsgesellschaft, 1931–1943. Also available from Ann Arbor, Mich.: Edwards Brothers, 1943). A series of seven volumes describing crystal structures whose solutions were published in the years 1913 to 1939, inclusive.

***G.2** Ralph W. G. Wyckoff. *The Structure of Crystals*, 2nd ed. (New York: Chemical Catalog Co., 1931. Supplement for 1930–34, New York: Reinhold, 1935). Crystallography (including space-group theory) and x-ray diffraction. In addition, full descriptions are given of a large number of known crystal structures.

G.3 W. L. Bragg. *The Crystalline State. Vol. I: A General Survey* (London: George Bell, 1933). This book, together with [G.7, G.12, and G.23], form a continuing series, edited by W. L. Bragg, to which this book is an introduction. It is a very readable survey of the field by the father of structure analysis. Contains very clear accounts in broad and general terms of crystallography (including space-group theory), diffraction, and structure analysis. An historical account of the development of x-ray crystallography is also included.

G.4 Arthur H. Compton and Samuel K. Allison. *X-Rays in Theory and Experiment* (New York: D. Van Nostrand, 1935). A standard treatise on the physics of x-rays and x-ray diffraction, with emphasis on the former.

G.5 M. J. Buerger. *X-Ray Crystallography* (New York: Wiley, 1942). Theory and practice of rotating and oscillating crystal methods. Space-group theory.

G.6 Wayne T. Sproull. *X-Rays in Practice* (New York: McGraw-Hill, 1946). X-ray diffraction and radiography, with emphasis on industrial applications.

G.7 R. W. James. *The Crystalline State. Vol. II: The Optical Principles of the Diffraction of X-Rays* (London: George Bell, 1948). Excellent book on advanced theory of x-ray diffraction. Includes thorough treatments of diffuse scattering (due to thermal agitation, small particle size, crystal imperfections, etc.), the use of Fourier series in structure analysis, and scattering by gases, liquids, and amorphous solids.

G.8 N. F. M. Henry; H. Lipson; and W. A. Wooster. *The Interpretation of X-Ray Diffraction Photographs* (London: Macmillan, 1951). Rotating and oscillating crystal methods, as well as powder methods, are described. Good section on analytical methods of indexing powder photographs.

***G.9** *Structure Reports* (Utrecht: Oosthoek, 1951 to date). A continuation, sponsored by the International Union of Crystallography, of *Strukturbericht* [G.1]. The volume numbers begin with Vol. 8, where *Strukturbericht* left off. The results of structure determinations are usually given in sufficient detail that the reader has no need to consult the original paper.

G.10 André Guinier. *X-Ray Crystallographic Technology* (London: Hilger and Watts, 1952). Excellent treatment of the theory and practice of x-ray diffraction. The title is not fair to the book, which includes a considerable body of theory and detailed experimental technique. The theory and applications of the reciprocal lattice are very well described. Includes treatments of focusing monochromators, small-angle scattering, and diffraction by amorphous substances.

***G.11** *International Tables for X-Ray Crystallography* (Birmingham, England: Kynoch Press, for the International Union of Crystallography).

Vol. I. Symmetry Groups (1952). Tables of point groups and space groups. The student should not overlook the interesting Historical Introduction written by Max von Laue.

Vol. II. Mathematical Tables (1959). In addition to mathematical tables useful in crystallography and x-ray diffraction, there is considerable textual material on crystal geometry, diffraction methods, temperature and absorption corrections, etc.

Vol. III. Physical and Chemical Tables (1962). Includes data on characteristic wavelengths, absorption coefficients, atomic scattering factors, Compton scattering, etc. Also treatments of intensity measurements, texture determination, particle size broadening, small angle scattering, and radiation hazards.

Vol. IV. Revised and Supplementary Tables (1974). Contains revised data on characteristic wavelengths, absorption coefficients, and atomic scattering factors. Also some new material useful in structure determination.

G.12 H. Lipson and W. Cochrane. *The Crystalline State. Vol. III: The Determination of Crystal Structures* (London: George Bell, 1953). Advanced structure analysis by means of space group theory and Fourier series. Experimental methods are not included; i.e., the problem of structure analysis is covered from the point at which $|F|^2$ values have been determined by experiment to the final solution. Contains many illustrative examples.

G.13 H. S. Peiser; H. P. Rooksby; and A. J. C. Wilson, editors. *X-Ray Diffraction by Polycrystalline Materials* (London: Institute of Physics, 1955). Contains some thirty chapters, by some thirty different authors, on the theory and practice of the powder method in its many variations. These chapters are grouped into three major sections: experimental technique, interpretation of data, and applications in specific fields of science and industry. Much useful information.

G.14 George L. Clark. *Applied X-Rays*, 4th ed. (New York: McGraw-Hill, 1955). Very comprehensive and oriented toward applications of x-ray diffraction, medical and industrial radiography and microradiography, and chemical and biological effects of x-rays. A wide range of crystal structures is also described.

G.15 André Guinier and Gerard Fournet. *Small-Angle Scattering of X-Rays* (New York: Wiley, 1955). A full description of small-angle scattering phenomena, including theory, experimental technique, interpretation of results, and applications.

***G.16** W. B. Pearson. *A Handbook of Lattice Spacings and Structures of Metals and Alloys* (New York: Pergamon Press, 1958). A most useful source of information. Gives the crystal structures of intermediate phases, and the variation of lattice parameter with composition in solid solutions, of binary and ternary alloys. Also gives the crystal structures of metal borides, carbides, hydrides, nitrides, and binary oxides.

Vol. 2 (1967). An updated and expanded version of the 1958 book, now referred to as Vol. 1. The investigator should first consult Vol. 2, because it contains virtually all the data of Vol. 1, together with later data, and need consult Vol. 1 only when he or she wants a description of the earlier work, up to 1956.

G.17 Leonid V. Azaroff and Martin J. Buerger. *The Powder Method in X-Ray Crystallography* (New York: McGraw-Hill, 1958). The making and interpretation of Debye–Scherrer photographs, including precise parameter measurements.

G.18 H. Lipson and C. A. Taylor. *Fourier Transforms and X-Ray Diffraction* (London: George Bell, 1958). On the interpretation of x-ray scattering by Fourier transforms.

G.19 A. Taylor. *X-Ray Metallography* (New York: Wiley, 1961). X-ray diffraction, radiography, and microradiography. Structures of metals and alloys; determination of phase diagrams, texture, grain size, and residual stress; chemical analysis and studies of ceramics.

G.20 G. E. Bacon. *Neutron Diffraction*, 2nd ed. (Oxford: Clarendon Press, 1962). Theory and practice of neutron diffraction, with applications to magnetic materials, structure determination, small-angle scattering, and amorphous materials.

G.21 A. Guinier. *X-Ray Diffraction in Crystals, Imperfect Crystals, and Amorphous Bodies* (San Francisco: W. H. Freeman, 1963). Largely theoretical and more advanced than [G.10].

***G.22** Ralph W. G. Wyckoff. *Crystal Structures*, 2nd ed. Vol. 1 (1963) to Vol. 6, Part 2 (1971) (New York: Wiley). A series devoted to the classification and description of the structures of inorganic and organic substances.

G.23 W. L. Bragg and G. F. Claringbull. *The Crystalline State. Vol. IV: Crystal Structures of Minerals* (London: George Bell, 1965). Rational and careful descriptions of mineral structures, from the simple to the complex.

G.24 J. B. Cohen. *Diffraction Methods in Materials Science* (New York: Macmillan, 1966). Includes material on the diffraction of electrons and neutrons, as well as x-rays. Studies of crystal imperfections.

G.25 Charles S. Barrett and T. B. Massalski. *Structure of Metals*, 3rd ed. (New York: McGraw-Hill, 1966). A classic book on the crystallographic aspects of physical metallurgy. Really two books in one, the first part dealing with the theory and methods of diffraction of x-rays, electrons, and neutrons; the second part with the structure of metals in the wider sense of the word. Very lucid account of the stereographic projection. Stress measurement, phase transformations, preferred orientation.

G.26 Leonid V. Azaroff. *Elements of X-Ray Crystallography* (New York: McGraw-Hill, 1968). Crystallography, diffraction theory (kinematic and dynamic), structure analysis, single-crystal and powder methods.

G.27 Leonid V. Azaroff and Raymond J. Donahue. *Laboratory Experiments in X-Ray Crystallography* (New York: McGraw-Hill, 1969). Laboratory manual detailing twenty-one experiments.

G.28 O. Johari and G. Thomas. *The Stereographic Projection and its Applications* (New York: Interscience, 1969. Vol. IIA of *Techniques of Metal Research*, R. F. Bunshah, ed.). Besides applications to x-ray and electron diffraction patterns, there are treatments of trace analysis, twinning and plastic deformation. Computer programs for plotting standard projections, and more than twenty-five standard projections of cubic and other crystals.

G.29 R. Jenkins and J. L. DeVries. *Practical X-Ray Spectrometry*, 2nd ed. (New York: Springer-Verlag, 1969). Experimental and data-handling aspects. Quantitative analysis, sample preparation, and trace analysis. Energy-dispersive methods not included.

G.30 B. E. Warren. *X-Ray Diffraction* (Reading, Mass.: Addison-Wesley, 1969). Excellent advanced treatment, in which the author takes pains to connect theoretically derived results with experimentally observable quantities. Stresses diffraction effects due to thermal vibration, order-disorder, imperfect crystals, and amorphous materials. Includes a treatment of the dynamical theory of diffraction by a perfect crystal.

G.31 L. S. Birks. *X-Ray Spectrochemical Analysis*, 2nd ed. (New York: Interscience, 1969). Clear, brief treatment of wavelength and energy dispersive methods, with emphasis on the problems of quantitative analysis. Includes electron-probe microanalysis.

G.32 H. Lipson and H. Steeple. *Interpretation of X-Ray Powder Diffraction Patterns* (London: Macmillan, 1970). An updating and expansion of that part of [G.8] that dealt with powder methods.

G.33 A. Kelly and G. W. Groves. *Crystallography and Crystal Defects* (Reading, Mass.: Addison-Wesley, 1970). Careful analysis of the departure from perfect periodicity caused by specific defects in specific, and common, structures (BCC, FCC, HCP, etc.). Describes the crystallography of dislocations, point defects, twins, martensite, and crystal interfaces.

G.34 M. M. Woolfson. *An Introduction to X-Ray Crystallography* (Cambridge: University Press, 1970). Crystallography, diffraction methods (mainly single-crystal), and structure determination.

G.35 Martin J. Buerger. *Contemporary Crystallography* (New York: McGraw-Hill, 1970). Single-crystal diffraction methods and structure determination.

G.36 A. J. C. Wilson. *Elements of X-Ray Crystallography* (Reading, Mass.: Addison-Wesley, 1970). Powder cameras and diffractometers; single-crystal cameras and diffractometers; structure determination; diffraction by imperfect crystals.

G.37 R. Jenkins and J. L. DeVries. *Worked Examples in X-Ray Analysis* (New York: Springer-Verlag, 1970). A set of forty-seven problems, graded as to difficulty, in x-ray diffractometry and spectrometry. The problems are practical ones, arising directly out of experimental work, and thus most useful.

G.38 F. C. Phillips. *An Introduction to Crystallography*, 4th ed. (New York: Wiley, 1972). A classic text on classical crystallography, in the sense that attention is directed mainly to the symmetry of outward shape of well-developed crystals. But space-group theory and the internal arrangement of atoms are also described.

G.39 Harold P. Klug and Leroy E. Alexander. *X-Ray Diffraction Procedures*, 2nd ed. (New York: Wiley, 1974). Contains a great deal of useful detail on the theory and operation of powder cameras and diffractometers. Covers the following topics in depth: chemical analysis by diffraction, parameter measurement, line-broadening analysis, texture determination, stress measurement, and studies of amorphous materials. Single-crystal methods are not included.

G.40 J. L. Amoros; M. J. Buerger; and M. L. Canut de Amoros. *The Laue Method* (New York: Academic Press, 1974). Describes the various projections used with the Laue method and the application of this method to studies of crystal symmetry and diffuse scattering.

G.41 Ron Jenkins. *An Introduction to X-Ray Spectrometry* (New York: Heyden, 1974). A more fundamental treatment than [G.29]. Describes both wavelength and energy dispersive methods. Includes a chapter on studies of chemical bonding by x-ray and electron spectrometry.

G.42 Leonid V. Azaroff, ed. *X-Ray Spectroscopy* (New York: McGraw-Hill, 1974). Advanced treatments by twelve authors of such topics as precision spectroscopy, grating spectrometers, emission and absorption spectra, photoelectron spectroscopy, and bonding effects.

G.43 Leonid V. Azaroff; Roy Kaplow; N. Kato; Richard J. Weiss; A. J. C. Wilson; and R. A. Young. *X-Ray Diffraction* (New York: McGraw-Hill, 1974). Advanced treatments of atomic scattering, kinematical and dynamical theories of diffraction, powder diffractometry, and single-crystal intensities.

G.44 Shuji Taira, ed. *X-Ray Studies on Mechanical Behavior of Materials* (Kyoto: Society of Materials Science, Japan, 1974). A collaborative account by twenty-seven Japanese investigators of x-ray studies of phenomena affecting the strength of materials. X-ray stress measurements are described, as well as texture determination, line-broadening studies, microbeam methods, pseudo-Kossel patterns, small angle scattering, and x-ray topography.

***G.45** E. Preuss; B. Krahl-Urban; and R. Butz. *Laue Atlas* (New York: Wiley, 1974). Atlas of computer-generated back-reflection Laue patterns of most elements and many important compounds.

G.46 Eugene P. Bertin. *Principles and Practice of X-Ray Spectrometric Analysis*, 2nd ed. (New York: Plenum Press, 1975). Extended treatment of apparatus and methods of quantitative analysis by wavelength and energy dispersion.

Chapter References

Chapter 1. Properties of X-Rays

1.1 K. D. Sevier. *Low Energy Electron Spectrometry* (New York: Wiley, 1971).

1.2 N. J. Taylor. *Vacuum Sci. Technol.*, **6,** 241 (1969).

1.3 R. Jenkins and D. J. Haas. *X-Ray Spectrometry*, **2,** 135 (1973) and **4,** 33 (1975).

1.4 American National Standard: *Radiation Safety for X-Ray Diffraction and Fluorescence Analysis Equipment.* National Bureau of Standards Handbook 111. Price: 30 cents. (Washington, D.C.: U.S. Government Printing Office, 1972.)

1.5 T. M. Moore; W. E. Gundaker; and J. W. Thomas, editors. *Radiation Safety in X-Ray Diffraction and Spectroscopy.* Price: $2.00. (Washington, D.C.: U.S. Government Printing Office, 1971.)

1.6 Plotted from data in *Handbook of Chemistry and Physics*, 23rd ed. (Cleveland: Chemical Rubber Publishing Co., 1939.)

1.7 C. S. Barrett. *Structure of Metals*, 2nd ed. (New York: McGraw-Hill, 1952.)

1.8 F. M. Charbonnier. *Adv. in X-Ray Analysis,* **15,** 446 (1972).

1.9 J. M. Jaklevic; R. D. Giaque; D. F. Malone; and W. L. Searles. *Adv. in X-Ray Analysis*, **15,** 266 (1972).

1.10 H. K. Herglotz. *Adv. in X-Ray Analysis,* **16,** 260 (1973).

1.11 Watkins–Johnson Co., 3333 Hillview Avenue, Palo Alto, Calif. 94304.

Chapter 2. Geometry of Crystals

2.1 C. W. Jacob and B. E. Warren. *J. Amer. Chem. Soc.*, **59,** 2588 (1937).

2.2 B. D. Cullity. *Trans. AIME,* **171,** 396 (1947).

2.3 Wulff nets and other charts useful in x-ray diffraction are available from N. P. Nies, 969 Skyline Drive, Laguna Beach, Calif. 92651.

Chapter 3. Directions of Diffracted Beams

3.1 Max von Laue. Historical Introduction, pp. 1–5 of Vol. 1 of [G.11]. See also pp. 31–56 and 293–294 of [3.2].

3.2 P. P. Ewald, ed. *Fifty Years of X-Ray Diffraction.* (Utrecht: International Union of Crystallography, 1962.) The early work of the Braggs is described on pp. 57–73 120–123, and 532–533. This interesting volume also contains the personal reminiscences of many eminent x-ray crystallographers.

Chapter 4. Intensities of Diffracted Beams

4.1 A. J. Bradley. *Proc. Phys. Soc.* (*London*), **47,** 879 (1935).

4.2 B. E. Warren. *Acta Cryst. A*, **32,** 897 (1976).

Chapter 5. Laue Photographs

5.1 Reuben Rudman, ed. *Index of Crystallographic Supplies*, 3rd ed., 1972. (International Union of Crystallography.) Available in the U.S.A. from Polycrystal Book Service, P.O. Box 11567, Pittsburgh, Pa. 15238.

5.2 Jacques M. Blum. *Science et Industries Photographiques*, **(2)29,** 211 (1958).

5.3 Lu-chia Ting; Richard Stevens; and B. D. Cullity. *Metal Progress*, March 1977.

Chapter 6. Powder Photographs

6.1 H. J. Goldschmidt. *High-Temperature X-Ray Diffraction Techniques*. Bibliography 1, 1964. (International Union of Crystallography.) Available in the U.S.A. from Polycrystal Book Service (see [5.1]).

6.2 B. Post. *Low-Temperature X-Ray Diffraction*. Bibliography 2, 1964. (International Union of Crystallography.) Available in the U.S.A. from Polycrystal Book Service (see [5.1]).

6.3 P. B. Hirsch. *X-Ray Microbeam Techniques*, pp. 278–297 of [G.13].

6.4 G. W. Brindley. *Monochromators and Focusing Cameras*, pp. 122–144 of [G.13].

6.5 P. M. de Wolff. *Acta Crystall.*, **1,** 207 (1948).

6.6 E. G. Hofmann and H. Jagodzinski. *Z. Metallk.*, **46,** 601 (1955).

6.7 J. W. Ballard; H. I. Oshry; and H. H. Schrenk. U.S. Bureau of Mines R.I. 3520.

6.8 B. D. Cullity. *Elements of X-Ray Diffraction*, 1st ed. (Reading, Mass.: Addison-Wesley, 1956.)

Chapter 7. Diffractometer and Spectrometer Measurements

7.1 R. W. Rex. *Adv. in X-Ray Analysis*, **10,** 366 (1967).

7.2 Carol J. Kelly and M. A. Short. *Adv. in X-Ray Analysis*, **15,** 102 (1972).

7.3 Armin Segmuller. *Adv. in X-Ray Analysis*, **15,** 114 (1972).

7.4 M. Slaughter and Davis Carpenter. *Adv. in X-Ray Analysis*, **15,** 135 (1972).

7.5 J. F. Croke and R. Jenkins. *Adv. in X-Ray Analysis*, **16,** 273 (1973).

7.6 R. Jenkins and R. G. Westberg. *Adv. in X-Ray Analysis*, **16,** 310 (1973).

7.7 Carol J. Kelly and E. Eichen. *Adv. in X-Ray Analysis*, **16,** 344 (1973).

7.8 W. Parrish. Pages 144–155 of Vol. 3 of [G.11].

7.9 H. Friedman. *Proc. I.R.E.*, **37,** 791 (1949).

7.10 N. Spielberg. *Adv. in X-Ray Analysis*, **10,** 534 (1967).

7.11 C. J. Borkowski and M. K. Kopp. *Rev. Sci. Instr.*, **39,** 515 (1968).

7.12 C. J. Borkowski and M. K. Kopp. *J. Appl. Cryst.*, **7,** 116 (1974). (Abstract only.)

7.13 Y. Dupont; A. Gabriel; M. Charbre; and V. Luzatti. *J. Appl. Cryst.*, **7,** 117 (1974). (Abstract only.)

7.14 R. S. Frankel and D. W. Aitken. *Appl. Spectroscopy*, **24,** 557 (1970).

7.15 G. Dearnaley and D. C. Northrop. *Semiconductor Counters for Nuclear Radiations* (New York: Wiley, 1963).

7.16 R. L. Heath. *Adv. in X-Ray Analysis*, **15,** 1 (1972).

7.17 D. A. Gedcke. *X-Ray Spectrometry*, **1,** 129 (1972).

7.18 *Energy Dispersion X-Ray Analysis: X-Ray and Electron Probe Analysis*, ASTM STP 485. (Philadelphia: American Society for Testing and Materials, 1971.)

7.19 Bill C. Giessen and Glen E. Gordon. *Science*, **159,** 973 (1968).

7.20 B. C. Giessen and G. E. Gordon. *Norelco Reporter*, p. 17, December 1970.

7.21 B. Buras; J. Chwaszczewska; S. Szarras; and Z. Szmid. Report 894/11/PS, Institute of Nuclear Research, Warsaw, March 1968.

7.22 A. P. Voskamp. *Adv. in X-Ray Analysis*, **17,** 124 (1974).

7.23 Syntex Analytical Instruments, Inc. This instrument is described by Susan B. Byram; Bui Han; G. B. Rothbart; and Roger A. Sparks. *Adv. in X-Ray Analysis*, **20,** 529 (1977).

7.24 R. E. Ogilvie. *Rev. Sci. Instr.*, **34,** 1344 (1963).

Chapter 8. Orientation and Quality of Single Crystals

8.1 A. B. Greninger. *Trans. AIME*, **117,** 61 (1935).

8.2 R. P. Goehner. *Adv. in X-Ray Analysis*, **19,** 725 (1976).

8.3 C. G. Dunn. *Trans. AIME*, **185,** 421 (1949).

8.4 Donald M. Koffman and A. N. Mariano. *Norelco Reporter*, p. 52, April–June, 1966.

8.5 Kenneth Reifsnider and Robert E. Green, Jr. *Rev. Sci. Instr.*, **39,** 1651 (1968).

8.6 Robert E. Green, Jr. *Adv. in X-Ray Analysis*, **14,** 311 (1971).

8.7 Robert E. Green, Jr. *Adv. in X-Ray Analysis*, **15,** 435 (1972).

8.8 Jonathan A. Dantzig and Robert E. Green, Jr. *Adv. in X-Ray Analysis*, **16,** 229 (1973).

8.9 C. G. Dunn and F. W. Daniels. *Trans. AIME*, **191,** 147 (1951).

8.10 A. Guinier and G. Tennevin. *Acta Cryst.*, **2,** 133 (1949).

8.11 A. Guinier and G. Tennevin. *Progress in Metal Physics*, **2,** 177 (New York: Interscience, 1950).

8.12 B. D. Cullity and Carl A. Julien. *J. Appl. Phys.*, **24,** 541 (1953).

8.13 C. A. Julien and B. D. Cullity. *Acta Met.*, **1,** 588 (1953).

8.14 C. A. Julien. Ph.D. thesis, University of Notre Dame, 1952.

8.15 J. E. White. *J. Appl. Phys.*, **21,** 855 (1950).

8.16 W. Berg. *Naturwissenschaften*, **19,** 391 (1931).

8.17 Charles S. Barrett. *Trans. AIME*, **161,** 15 (1945).

8.18 J. B. Newkirk. *Trans. Met. Soc. AIME*, **215,** 483 (1959). In Fig. 10 of this paper the indices in (c) and (d) should be interchanged.

8.19 A. R. Lang. *Acta Met.*, **5,** 358 (1957).

8.20 A. R. Lang. *Acta Cryst.*, **12,** 249 (1959).

8.21 Milena Polcarova and A. R. Lang. *Appl. Phys. Letters*, **1,** 13 (1962).

8.22 G. H. Schwuttke. *J. Appl. Phys.*, **36,** 2712 (1965).

8.23 R. L. Silver and J. C. Turner. *Adv. in X-Ray Analysis*, **15,** 123 (1972).

8.24 G. Borrmann. *Physik Z.*, **42,** 157 (1941).

8.25 H. N. Campbell. *J. Appl. Phys.*, **22,** 1139 (1951).

8.26 This photograph also appears in [8.31].

8.27 W. W. Webb. Pages 29–76 of *Direct Observation of Imperfections in Crystals*, edited by J. B. Newkirk and J. H. Wernick (New York: Interscience, 1962).

8.28 R. W. James. *Solid State Physics*, **15,** 55 (1963).

8.29 Josef Intrater and Sigmund Weissmann. *Acta Cryst.*, **7,** 729 (1954).

8.30 S. Weissmann and Z. H. Kalman. Pages 839–873 of *Techniques of Metal Research*, Vol. 2, Part 2, edited by R. F. Bunshah (New York: Interscience, 1969).

8.31 R. W. Armstrong and C. Cm. Wu. Pages 169–219 of *Microstructural Analysis, Tools and Techniques*, edited by J. L. McCall and W. M. Mueller (New York: Plenum, 1973). This paper contains a bibliography of 351 items.

8.32 Various papers in *Adv. in X-Ray Analysis*, **10,** 1–184 (1967).

8.33 A. R. Lang. Pages 1053–1063 of *Encyclopedia of X-Rays and Gamma Rays*, edited by George L. Clark (New York: Reinhold, 1963).

8.34 Stanley B. Austerman and J. B. Newkirk. *Adv. in X-Ray Analysis*, **10,** 134 (1967).

Chapter 9. The Structure of Polycrystalline Aggregates

9.1 B. E. Warren and J. Biscoe. *J. Am. Ceram. Soc.*, **21,** 49 (1938).

9.2 André Guinier. *Physics Today*, p. 25, November 1969.

9.3 B. E. Warren. *Progress in Metal Physics*, **8,** 147 (1959).

9.4 B. E. Warren and B. L. Averbach. *J. Appl. Phys.*, **20,** 885 (1949).

9.5 J. B. Cohen and J. E. Hilliard, editors. *Local Atomic Arrangements Studied by X-Ray Diffraction* (New York: Gordon and Breach, 1966).

9.6 M. S. Paterson. *J. Appl. Phys.*, **23,** 805 (1952).

9.7 I. L. Dillamore and W. T. Roberts. *Metallurgical Reviews*, **10,** 271 (1965).

9.8 A. H. Geisler. Pages 131–153 of *Modern Research Techniques in Physical Metallurgy* (Cleveland: American Society for Metals, 1953).

9.9 B. F. Decker; E. T. Asp; and D. Harker. *J. Appl. Phys.*, **19,** 388 (1948). But note that these authors define the rotation angle α as positive for counterclockwise rotation.

9.10 L. G. Schulz. *J. Appl. Phys.*, **20,** 1033 (1949).

9.11 J. B. Newkirk and L. Bruce. *J. Appl. Phys.*, **29,** 151 (1958).

9.12 K. Aoki; S. Hayami; and M. Matsuo. *Adv. in X-Ray Analysis*, **10,** 342 (1967).

9.13 L. G. Schulz. *J. Appl. Phys.*, **20,** 1030 (1949).

9.14 W. P. Chernock and P. A. Beck. *J. Appl. Phys.*, **23,** 341 (1952).

9.15 E. M. C. Huijser-Gerits and G. D. Rieck. *J. Appl. Cryst.*, **7,** 286 (1974).

9.16 Michael Field and Eugene M. Merchant. *J. Appl. Phys.*, **20,** 741 (1949).

9.17 Hsun Hu; P. R. Sperry; and Paul A. Beck. *Trans. AIME*, **194,** 76 (1952).

9.18 H.-J. Bunge and W. T. Roberts. *J. Appl. Cryst.*, **2,** 116 (1969).

9.19 Peter R. Morris and Alan J. Heckler. *Adv. in X-Ray Analysis*, **11,** 454 (1968).

9.20 ASTM Standard E81-63, "Preparing Quantitative Pole Figures of Metals," p. 137, Part 11, *1975 Annual Book of ASTM Standards.* (Philadelphia: American Society for Testing and Materials, 1975.)

9.21 Stanley L. Lopata and Eric B. Kula. *Trans. AIME*, **224,** 865 (1962).

9.22 J. J. Klappholz; S. Waxman; and C. Feng. *Adv. in X-Ray Analysis*, **15,** 365 (1972).

9.23 A. Segmüller and J. Angilello. *J. Appl. Cryst.*, **2,** 76 (1969).

9.24 Armin Segmüller. *J. Appl. Cryst.*, **2,** 259 (1969).

9.25 Hung-Chi Chao. *Adv. in X-Ray Analysis*, **12,** 391 (1969).

9.26 C. Richard Desper. *Adv. in X-Ray Analysis*, **12,** 404 (1969).

9.27 P. R. Morris and A. J. Heckler. *Trans. AIME*, **245,** 1877 (1969).

9.28 Peter R. Morris. *Adv. in X-Ray Analysis*, **18,** 514 (1975).

9.29 R. J. Roe. *J. Appl. Phys.*, **36,** 2024 (1965) and **37,** 2069 (1966).

9.30 H.-J. Bunge. *Z. Metallkunde*, **56,** 872 (1965).

9.31 R. O. Williams. *J. Appl. Phys.*, **38,** 4029 (1967) and **39,** 4329 (1968).

9.32 R. O. Williams. *Trans. AIME*, **242,** 105 (1968).

9.33 M. Matsuo; S. Hayami; and S. Nagashima. *Adv. in X-Ray Analysis*, **14,** 214 (1971).

9.34 G. J. Davies; D. J. Goodwill; and J. S. Kallend. *J. Appl. Cryst.*, **4,** 67 and 193 (1971).

9.35 August Freda and B. D. Cullity. *Trans. AIME*, **215,** 530 (1959).

9.36 L. K. Jetter; C. J. McHargue; and R. O. Williams. *J. Appl. Phys.*, **27,** 368 (1956).

9.37 W. P. Chernock; Joseph Singer; Melvin H. Mueller; and Paul A. Beck. *J. Appl. Phys.*, **27,** 1170 (1956).

9.38 B. D. Cullity and August Freda. *J. Appl. Phys.*, **29,** 25 (1958).

9.39 S. Leber. *Trans. ASM*, **53,** 697 (1961).

9.40 Vittal S. Bhandary and B. D. Cullity. *Trans. AIME*, **224,** 1194 (1962).

9.41 Leroy E. Alexander. *X-Ray Diffraction Methods in Polymer Science* (New York: Wiley, 1969).

Chapter 10. The Determination of Crystal Structure

10.1 A. W. Hull and W. P. Davey. *Phys. Rev.*, **17,** 549 (1921).

10.2 C. W. Bunn. *Chemical Crystallography* (New York: Oxford, 1961).

10.3 Polycrystal Book Service (see [5.1]) and Institute of Physics, London, England.

10.4 R. Hesse. *Acta Cryst.*, **1,** 200 (1948).

10.5 H. Lipson. *Acta Cryst.*, **2,** 43 (1949).

Chapter 11. Precise Parameter Measurements

11.1 W. Parrish and A. J. C. Wilson. *Precision Measurement of Lattice Parameters of Polycrystalline Specimens*, p. 216 of Vol. 2 of [G.11].

11.2 J. B. Nelson and D. P. Riley. *Proc. Phys. Soc. (London)*, **57,** 160 (1945).

11.3 A. Taylor and H. Sinclair. *Proc. Phys. Soc. (London)*, **57,** 126 (1945).

11.4 M. U. Cohen. *Rev. Sci. Instr.*, **6,** 68 (1935) and **7,** 155 (1936).

11.5 W. L. Bond. *Acta Cryst.*, **13,** 814 (1960).

11.6 J. Burke and M. V. Tomkeieff. *J. Appl. Cryst.*, **2,** 247 (1969).

11.7 T. W. Baker; J. D. George; B. A. Bellamy; and R. Causer. *Adv. in X-Ray Analysis*, **11,** 359 (1968).

11.8 SRM 640 may be bought in 10-gram units from the Office of Standard Reference Materials, Room B311, Chemistry Building, National Bureau of Standards, Washington, D.C. 20234.

11.9 William Parrish. *Acta Cryst.*, **13,** 838 (1960).

Chapter 12. Phase-Diagram Determination

12.1 C. S. Roberts. *Trans. AIME*, **197,** 203 (1953).

12.2 J. C. Mertz and C. H. Mathewson. *Trans. AIME*, **124,** 59 (1937).

Chapter 13. Order-Disorder Transformations

13.1 E. C. Bain. *Trans. AIME*, **68,** 625 (1923).

13.2 D. T. Keating and B. E. Warren. *J. Appl. Phys.*, **22,** 286 (1951).

13.3 D. Chipman and B. E. Warren. *J. Appl. Phys.*, **21,** 696 (1950).

13.4 Max Hansen and Kurt Anderko. *Constitution of Binary Alloys* (New York: McGraw-Hill, 1958).

13.5 F. W. Jones and C. Sykes. *Proc. Roy. Soc. (London)*, **A161,** 440 (1937).

13.6 S. C. Moss. *J. Appl. Phys.*, **35,** 3547 (1964).

13.7 B. E. Warren and B. L. Averbach. Pages 95–130 of *Modern Research Techniques in Physical Metallurgy* (Cleveland: American Society for Metals, 1953).

Chapter 14. Chemical Analysis by X-Ray Diffraction

14.1 L. Zwell and A. W. Danko. *Applied Spectroscopy Reviews*, **9(2),** 167 (1975). Contains 265 references.

14.2 J. D. Hanawalt; H. W. Rinn; and L. K. Frevel. *Ind. Eng. Chem., Anal. Ed.*, **8,** 244 (1936) and **10,** 457 (1938).

14.3 Joint Committee on Powder Diffraction Standards, 1601 Park Lane, Swarthmore, Pa. 19081.

14.4 G. G. Johnson, Jr. and V. Vand. *Adv. in X-Ray Analysis*, **11,** 376 (1968).

14.5 H. E. Swanson *et al.* *Nat. Bur. Stand. Circular 539*, Vol. 1–10 (1953–1960) and *Nat. Bur. Stand. Monograph 25*, Sections 1–12 (1962–1975). U.S. Government Printing Office, Washington, D.C.

14.6 L. E. Alexander and H. P. Klug. *Anal. Chem.*, **20,** 886 (1948).

14.7 B. L. Averbach and M. Cohen. *Trans. AIME*, **176,** 401 (1948).

14.8 J. Durnin and K. A. Ridal. *J. Iron and Steel Inst.*, **206,** 60 (1968).

14.9 G. E. Hicho; H. Yakowitz; S. D. Rasberry; and R. E. Michaelis. NBS Special Publication 260-25 (1971). Also published in *Adv. in X-Ray Analysis*, **14,** 78 (1971). The x-ray method of determining austenite, including a BASIC computer program, is given by C. J. Bechtold, NBS Technical Note 709 (1972). Both of these NBS reports are available from Superintendent of Documents, U.S. Government Printing Office, Washington, D.C., 20402.

14.10 R. R. Biederman; R. F. Bourgault; and R. W. Smith. *Adv. in X-Ray Analysis*, **17,** 139 (1974).

14.11 B. L. Averbach; M. F. Comerford; and M. B. Bever. *Trans. AIME*, **215,** 682 (1959).

14.12 Robert Lorenzelli and Pierre Delaroche. *J. Appl. Cryst.*, **5,** 267 (1972).

14.13 L. E. Copeland; S. Brunauer; D. L. Kantro; E. G. Schulz; and C. H. Weise. *Anal. Chem.*, **31,** 1521 (1959).

14.14 D. G. Feuerbacher and R. R. Clark. *Adv. in X-Ray Analysis*, **17,** 75 (1974).

14.15 A. L. Rickards. *Anal. Chem.*, **44,** 1872 (1972).

14.16 C. L. Grant and P. A. Pelton. *Adv. in X-Ray Analysis*, **17,** 44 (1974).

14.17 R. L. Miller. *Trans. ASM*, **57,** 892 (1964).

14.18 R. Gullberg and R. Lagneborg. *Trans. AIME*, **236,** 1482 (1966).

14.19 E. F. Giamei and E. J. Freise. *Trans. AIME*, **239,** 1676 (1967).

14.20 R. D. Arnell. *J. Iron and Steel Inst.*, **206,** 1035 (1968).

14.21 M. J. Dickson. *J. Appl. Cryst.*, **2,** 176 (1969).

14.22 S. L. Lopata and E. B. Kula. *Trans. AIME*, **233,** 288 (1956).

14.23 R. L. Miller. *Trans. ASM*, **61,** 592 (1968).

14.24 Charles M. Nenadic and John V. Crable. *Developments in Applied Spectroscopy*, **9,** 343 (1971).

14.25 H. E. Bumsted. *J. Amer. Industrial Hygiene Assoc.*, p. 150, April 1973.

Chapter 15. Chemical Analysis by X-Ray Spectrometry

15.1 H. Friedman and L. S. Birks. *Rev. Sci. Instr.*, **19,** 323 (1948).

15.2 Information on standard samples with specified chemical composition and physical properties, including standard samples for x-ray spectrometric analysis, is given in *Catalog of NBS Standard Reference Materials.* NBS Special Publication 260, 1975–76 ed. (Superintendent of Documents, U.S. Government Printing Office, Washington, D.C. 20402.)

15.3 A list of suppliers of standard samples can be obtained from the Society for Applied Spectroscopy, 428 E. Preston St., Baltimore, Md. 21202.

15.4 H. J. Lucas-Tooth and C. Pyne. *Adv. in X-Ray Analysis*, **7,** 523 (1964).

15.5 Benjamin S. Sanderson and James A. Yeck. *Adv. in X-Ray Analysis*, **10,** 474 (1967).

15.6 G. R. Lachance and R. J. Traill. *Can. Spectry.*, **11,** 43 (1966).

15.7 F. Claisse and M. Quintin. *Can. Spectry.*, **12,** 129 (1967).

15.8 S. D. Rasberry and K. F. J. Heinrich. *Anal. Chem.*, **46,** 81 (1974).

15.9 K. F. J. Heinrich and S. D. Rasberry. *Adv. in X-Ray Analysis*, **17,** 309 (1974).

15.10 R. Jenkins; J. F. Croke; R. L. Niemann; and R. G. Westberg. *Adv. in X-Ray Analysis*, **18,** 372 (1975).

15.11 *Adv. in X-Ray Analysis,* **19** (1976).

Chapter 16. Measurement of Residual Stress

16.1 John T. Norton. *Norelco Reporter*, p. 50, April–June, 1968.

16.2 John T. Norton. *Materials Evaluation*, p. 21A, February 1973 (Lester Honor Lecture).

16.3 M. E. Hilley; J. A. Larson; C. F. Jatczak; and R. E. Ricklefs, editors. *Residual Stress Measurement by X-Ray Diffraction*. SAE Information Report J 784a (1971). Available from the Society of Automotive Engineers, Inc., 400 Commonwealth Drive, Warrendale, Pa. 15096.

16.4 F. Gisen; R. Glocker; and E. Osswald. *Z. Tech. Physik*, **17**, 145 (1936).

16.5 A. L. Christenson and E. S. Rowland. *Trans. ASM*, **45**, 638 (1953).

16.6 R. E. Ogilvie. M.S. thesis, Mass. Inst. of Technology (1952). Work directed by John T. Norton.

16.7 D. P. Koistinen and R. E. Marburger. *Trans. ASM*, **51**, 537 (1959).

16.8 M. G. Moore and W. P. Evans. *Trans. SAE*, **66**, 340 (1958).

16.9 H. R. Letner. *Trans. ASME*, **77**, 1089 (1955).

16.10 M. R. James and J. B. Cohen. *Adv. in X-Ray Analysis*, **19**, 697 (1976).

16.11 Eric W. Weinman; Joseph E. Hunter; and Douglas D. McCormack. *Metal Progress*, p. 88, July 1969.

16.12 Donald A. Bolstad and William E. Quist. *Adv. in X-Ray Analysis*, **8**, 26 (1965).

16.13 John T. Norton. *Adv. in X-Ray Analysis*, **11**, 401 (1968).

16.14 J. T. Norton and D. Rosenthal. *Proc. Soc. Exp. Stress Analysis*, **1** (2), 77 (1943).

16.15 Leopold Frommer and E. H. Lloyd. *J. Inst. Metals*, **70**, 91 (1944).

16.16 Donald A. Bolstad; Robert A. Davis; William E. Quist; and Earl C. Roberts. *Metal Progress*, p. 88, July 1963.

16.17 R. J. Homicz. SAE Paper 670151, SAE Congress, January 1967.

16.18 M. Raefsky; N. Walter; W. Lieberman; and J. Clark. SAE Paper 670154, SAE Congress, January 1967.

16.19 Matthew J. Donachie, Jr. and John T. Norton. *Trans. ASM*, **55**, 51 (1962).

16.20 Paul S. Prevey. *Adv. in X-Ray Analysis*, **20**, 345 (1977).

16.21 Bruce M. MacDonald. *Adv. in X-Ray Analysis*, **13**, 487 (1970).

16.22 A. L. Esquivel. *Adv. in X-Ray Analysis*, **12**, 269 (1969).

16.23 B. N. Ranganathan; J. J. Wert; and W. N. Clotfelter. *J. of Testing and Evaluation*, **4**, 218 (1976).

16.24 H. R. Woehrle; F. P. Reilly, III; W. J. Barkley, III; L. A. Jackman; and W. R. Clough. *Adv. in X-Ray Analysis*, **8**, 38 (1965).

16.25 M. R. James and J. B. Cohen. *Adv. in X-Ray Analysis*, **20**, 291 (1977).

16.26 B. D. Cullity. *Adv. in X-Ray Analysis*, **20**, 259 (1977).

16.27 D. M. Vasil'ev and B. I. Smirnov. *Soviet Physics Uspekhi*, **4**, 226 (1961). (In English.)

16.28 Eckard Macherauch. *Exper. Mechanics*, **6**, 140 (1966).

16.29 David N. French. *J. Amer. Ceramic Soc.*, **52**, 271 (1969).

16.30 R. E. Marburger and D. P. Koistinen. *Trans. ASM*, **53,** 743 (1961).

16.31 W. J. McG. Tegart. *The Electrolytic and Chemical Polishing of Metals in Research and Industry*, 2nd ed. (New York: Pergamon, 1959).

16.32 M. R. James. Ph.D. thesis, directed by J. B. Cohen, Northwestern University (1977).

16.33 F. Bollenrath; V. Hauk; and W. Weidemann. *Archiv. fur das Eisenhuttenwesen*, **38,** 793 (1967).

16.34 R. H. Marion and J. B. Cohen. *Adv. in X-Ray Analysis*, **18,** 466 (1975).

Answers to Selected Problems

Chapter 1

1–1. 4.23×10^{18} sec^{-1}, 2.80×10^{-15} J; 1.95×10^{18} sec^{-1}, 1.29×10^{-15} J

1–6. 33 cm^2/gm

1–8. a) 26.97 cm^2/gm, 3.48×10^{-2} cm^{-1}
b) $I_x/I_0 = 0.50$ for 20 cm of air

1–9. 8980 volts

1–11. 1.541 Å

1–13. 8980 volts; mainly 1.54 Å (Cu $K\alpha$) and 1.39 Å (Cu $K\beta$)

1–15. 0.80, 26 to 1

1–17. 3.5 to 1

Chapter 2

2–7. A section on $(\bar{1}2\bar{1}0)$ will show this

2–11. Shear strain = 0.707

2–14. a) 20°S, 30°W; b) 27°S, 48°E; c) 39°S, 61°E

2–19. 42°N, 26°E; 19°S, 45°W; 42°S, 63°E

2–21. a) 7890 km b) 34°, 68°, 135°.

Chapter 3

3–2. 26.6°

3–5. $t = 2\lambda/(B \cos \theta)$

3–6.

t	B
1000 Å	0.11°
750	0.14
500	0.22
250	0.43

θ	B
10°	0.31°
45	0.43
80	1.76

Chapter 4

4–1. $A = A(\theta) = (1/2\mu)[1 - \exp(-2\mu t/\sin\theta)]$

4–2. 34 percent reduction for copper, 82 percent for lead

4–4. $F^2 = 0$ for mixed indices; $F^2 = 0$ for $(h + k + l)$ an odd multiple of 2; $F^2 = 64f_C^2$ for $(h + k + l)$ an even multiple of 2; $F^2 = 32f_C^2$ for $(h + k + l)$ odd.

4–6.

$h + 2k$	l	F^2
$3n$	$2p + 1$ (as 1, 3, 5, 7 . . .)	0
$3n$	$8p$ (as 8, 16, 24 . . .)	$4(f_{Zn} + f_S)^2$
$3n$	$4(2p + 1)$ (as 4, 12, 20, 28 . . .)	$4(f_{Zn} - f_S)^2$
$3n$	$2(2p + 1)$ (as 2, 6, 10, 14 . . .)	$4(f_{Zn}^2 + f_S^2)$
$3n \pm 1$	$8p \pm 1$ (as 1, 7, 9, 15, 17 . . .)	$3(f_{Zn}^2 + f_S^2 - \sqrt{2} f_{Zn} f_S)$
$3n \pm 1$	$4(2p + 1) \pm 1$ (as 3, 5, 11, 13, 19, 21 . . .)	$3(f_{Zn}^2 + f_S^2 + \sqrt{2} f_{Zn} f_S)$
$3n \pm 1$	$8p$	$(f_{Zn} + f_S)^2$
$3n \pm 1$	$4(2p + 1)$	$(f_{Zn} - f_S)^2$
$3n \pm 1$	$2(2p + 1)$	$(f_{Zn}^2 + f_S^2)$

n and p are any integers, including zero.

4–7.

Line	*hkl*	Calc. Int.
1	110	10.0
2	200	1.7
3	211	3.5
4	220	1.1

4–9. 111 and 200. The ratio is 2400 to 1.

Chapter 5

5–1. 0.67 cm for (111); 0.77 cm for (200)

5–3. a) 3rd to 18th; b) 3rd to 15th.

Chapter 6

6–1. 41 minutes

6–2. $\lambda/\Delta\lambda = 2\, R \tan \theta/\Delta S$

6–4. 81°

6–6. 4.2 cm

Chapter 7

7–1. 0.44°

7–2. 2.8 percent

7–4. a) 0.39 percent; b) 0.67 percent

7–6. a) 1.14 (Co) to 1 (Ni); b) 10.5

Chapter 8

8–1. 8°N, 23°E; 74°S, 90°E; 16°S, 64°W

8–3. 26° about beam axis, clockwise, looking from crystal to x-ray source; 3° about EW, clockwise, looking from E to W; 9° about NS, counterclockwise, looking from N to S

Chapter 9

9–2. Diffractometer

9–5. b) 0.11, 0.17, 0.28, and 0.44, listed in the order in which the incident beam traverses the layers

9–8. b) Decreased by 26 percent

Chapter 10

10–1. 111, 200, 220, 311, 222, 400, 331, 420, 422, and 511 (333); $a = 4.05$ Å

10–4. 100, 002, 101, 102, 110

10–6. 111, 220, 311, 400, 331, 422, 511 (333), 440. Diamond cubic; $a = 5.4$ Å; silicon.

10–8. 100, 002, 101, 102, 110, 103, 200, 112. Hexagonal close-packed; $a = 3.2$ Å, $c = 5.2$ Å; magnesium.

Chapter 11

11–1. ± 1.7°C

11–3. 4.997 Å

11–5. Near $\theta = 34°$

Chapter 12

12–1. ± 0.0002 Å

Chapter 13

13–2. 0.0010

Chapter 14

14–1. BaS

14–3. Mixture of Ni and NiO

14–5. 13 volume percent

Chapter 15

15–1. a) 1.2°, 1.9°, and 5.9° for 200, 220, and 420
b) 1.1°, 1.5°, and 2.4°
c) All

Chapter 16

16–1. a) $d_0 = 1.1702$ Å, $d_n = 1.1691$ Å, $d_i = 1.1716$ Å
b) -0.1 percent

16–5. -65 ksi without, and -63 ksi with, correction by the *LPA* factor

Index

Index